The Systematics Association Special Volume Series 74

Automated Taxon Identification in Systematics

Theory, Approache
and Applications

The Systematics Association Special Volume Series 74

Automated Taxon Identification in Systematics

Theory, Approaches and Applications

Edited by

Norman MacLeod

The Natural History Museum

London, UK

CRC Press is an imprint of the
Taylor & Francis Group, an **informa** business

CRC Press
Taylor & Francis Group
6000 Broken Sound Parkway NW, Suite 300
Boca Raton, FL 33487-2742

First issued in paperback 2019

ISBN-13: 978-0-8493-8205-5 (hbk)
ISBN-13: 978-0-367-38883-6 (pbk)

Library of Congress Cataloging-in-Publication Data

Automated taxon identification in systematics : theory, approaches and applications / editor, Norman MacLeod.
p. cm. -- (Systematics Association special volumes)
ISBN-13: 978-0-8493-8205-5 (alk. paper)
ISBN-10: 0-8493-8205-X (alk. paper)
1. Biology--Classification--Data processing. I. MacLeod, Norman, 1953- II. Systematics Association. III. Title. IV. Series.

QH83.A93 2007
570.1'2--dc22 2007006206

Visit the Taylor & Francis Web site at
http://www.taylorandfrancis.com

and the CRC Press Web site at
http://www.crcpress.com

Table of Contents

Chapter 1 Introduction........1

Norman MacLeod

Chapter 2 Digital Innovation and Taxonomy's Finest Hour........9

Quentin D. Wheeler

Chapter 3 Natural Object Categorization: Man versus Machine........25

Philip F. Culverhouse

Chapter 4 Neural Networks in Brief........47

Robert Lang

Chapter 5 Morphometrics and Computed Homology: An Old Theme Revisited........69

Fred L. Bookstein

Chapter 6 The Automated Identification of Taxa: Concepts and Applications........83

David Chesmore

Chapter 7 DAISY: A Practical Computer-Based Tool for Semi-Automated Species Identification........101

Mark A. O'Neill

Chapter 8 Automated Extraction and Analysis of Morphological Features for Species Identification........115

Volker Steinhage, Stefan Schröder, Karl-Heinz Lampe and Armin B. Cremers

Chapter 9 Introducing SPIDA-Web: Wavelets, Neural Networks and Internet Accessibility in an Image-Based Automated Identification System........131

Kimberly N. Russell, Martin T. Do, Jeremy C. Huff and Norman I. Platnick

Chapter 10 Automated Tools for the Identification of Taxa from Morphological Data: Face Recognition in Wasps........153

Norman MacLeod, Mark A. O'Neill and Stig A. Walsh

Chapter 11 Pattern Recognition for Ecological Science and Environmental Monitoring: An Initial Report 189

Eric N. Mortensen, Enrique L. Delgado, Hongli Deng, David Lytle, Andrew Moldenke, Robert Paasch, Linda Shapiro, Pengcheng Wu, Wei Zhang and Thomas G. Dietterich

Chapter 12 Plant Identification from Characters and Measurements Using Artificial Neural Networks 207

Jonathan Y. Clark

Chapter 13 Spot the Penguin: Can Reliable Taxonomic Identifications Be Made Using Isolated Foot Bones? 225

Stig A. Walsh, Norman MacLeod and Mark A. O'Neill

Chapter 14 A New Semi-Automatic Morphometric Protocol for Conodonts and a Preliminary Taxonomic Application 239

David Jones and Mark Purnell

Chapter 15 Decision Trees: A Machine-Learning Method for Characterizing Morphological Patterns Resulting from Ecological Adaptation 261

Manuel Mendoza

Chapter 16 Data Integration and Multifactorial Analyses: The Yeasts and the BioloMICS Software as a Case Study 277

Robert Vincent

Chapter 17 Automatic Measurement of Honeybee Wings 289

Adam Tofilski

Chapter 18 Good Performers Know Their Audience! Identification and Characterization of Pitch Contours in Infant- and Foreigner-Directed Speech 299

Monja A. Knoll, Stig A. Walsh, Norman MacLeod, Mark A. O'Neill and Maria Uther

Appendix 311

Subject Index 329

Taxon Index 337

The Systematics Association Special Volume Series

Series Editor

Alan Warren

Department of Zoology, The Natural History Museum

Cromwell Road, London SW7 5BD, UK

The Systematics Association promotes all aspects of systematic biology by organizing conferences and workshops on key themes in systematics, publishing books and awarding modest grants in support of systematics research. Membership of the Association is open to internationally based professionals and amateurs with an interest in any branch of biology including palaeobiology. Members are entitled to attend conferences at discounted rates, to apply for grants and to receive the newsletters and mailed information; they also receive a generous discount on the purchase of all volumes produced by the Association.

The first of the Systematics Association's publications *The New Systematics* (1940) was a classic work edited by its then-president Sir Julian Huxley, that set out the problems facing general biologists in deciding which kinds of data would most effectively progress systematics. Since then, more than 70 volumes have been published, often in rapidly expanding areas of science where a modern synthesis is required.

The *modus operandi* of the Association is to encourage leading researchers to organize symposia that result in a multi-authored volume. In 1997 the Association organized the first of its international Biennial Conferences. This and subsequent Biennial Conferences, which are designed to provide for systematists of all kinds, included themed symposia that resulted in further publications. The Association also publishes volumes that are not specifically linked to meetings and encourages new publications in a broad range of systematics topics.

Anyone wishing to learn more about the Systematics Association and its publications should refer to our website at http://www.systass.org

Other Systematics Association publications are listed after the index for this volume.

Foreword

The desire to classify objects is one of the most fundamental properties of the human being. The manifestation of this is reflected in everyday life in the form of the energy devoted to the classification and labelling of the population for taxation purposes, military conscription, political aspirations and marketing. A less aggressive aspect occurs in the field of hobbies – stamp collecting is perhaps the most widely spread of these, but there are many who, arising from a biological interest, make collections of various groups of insects, birds and the like. There is also a sinister side to the trait – for example, the clandestine collection of pillaged archaeological objects. This antisocial type of behaviour is much in evidence today – for example, the organized looting from the shattered museums of Iraq and, less dramatically but nonetheless equally destructive to the cultural heritage of all mankind, the seemingly uncontrollable theft of invaluable artefacts from China.

Leaving such dismal thoughts behind, let us consider the scale and scope of the topics in the present volume. Firstly, be it noted that the timing is perfect. The year 2007 is being devoted to the 300 year-jubilee of the father of classical taxonomy, the genius of Carl von Linné. Notwithstanding that the wish to stabilize the description, identification and classification was afoot in many quarters in Europe, it is to the credit of Linné that an organized, logical system for the description and recognition of plants and animals was developed and, moreover, quickly caught on. For more than a century and a half, taxonomists were largely satisfied with the Linnaean doctrine. The frequently rather boring anatomical exercise attaching to the correct identification and location of new categories seemed, for most, a small price to pay for maintaining order in the scheme of plants and animals.

An alternative (unwittingly so?) was proposed by a contemporary of von Linné, Michel Adanson, a French botanist. Adanson's ideas were more a *modus operandi*, so phrased as to incite the taxonomist to assemble as much data as reasonably possible on the organism of interest. This admirable approach was not always observed by the binomial nomenclaturists – least of all within the sphere of palaeontology. Notwithstanding von Linné's achievements with respect to the naming of species, he became, even during his lifetime, outmoded. The French biologist Count Georges Louis Leclerc de Buffon, a contemporary of von Linné, was critical of the scientific basis of what was in essence a man-made attempt at putting life forms into slots and did not reflect the results of divine creation. In the *Official Jubilee Book of the Linnean Year,* Professor Nils Uddenberg has pointed out that it was the work of Buffon that opened the way for Darwin's evolutionary theories and not the work of von Linné.

Dissatisfaction with the descriptive techniques of 'Linnaean taxonomy' led to the formation of a mechanically oriented discipline, Numerical taxonomy (NT), with Adanson as patron saint, as it were. This eventually computer-oriented approach was strongly promoted by Robert Sokal and Peter Sneath during the 1960s and 1970s. Numerical taxonomy quickly acquired a dedicated following and a jargon of its own – for example, OTU (operational taxonomic unit). This 'new taxonomy' brought to the fore the need for a way of defining species that was as unbiased as possible. It is still stimulating to consult many of the treatises that were spawned by the period of ascendancy of NT for the closely reasoned and admirably logical presentation of theses. It had, however, a negative aspect: to wit, a tendency towards degrading the skills necessary for actually studying the functional biological properties of organisms. More than a few numerical taxonomists felt this tiresome pursuit could be best left to computerized treatment. Classical taxonomy ceased to be regarded as a useful activity in many circles, with the result that many young biology doctoral candidates of today have little more than a sketchy knowledge of the panoply of plant and animal species, apart from just that particular group concerning them in their thesis work. You can test this. Just ask where brachiopods fit into the scheme of things.

Over the span of just a few years, a major upheaval has entered the field of taxonomy: notably, molecular biology and taxonomy. Well-founded species that, to all intents and purposes, were considered to be closely related have been shown not to be so. Where this will lead in the long run remains yet somewhat hazy, but rest assured, things will never be the same as they were.

Has automated taxonomy become outmoded? The results presented in the present volume gainsay this question. The subject of automated identification in systematics is presented in several respects. The old numerical taxonomy is not moribund, just modified. The collection of papers does not, however, represent a homogeneous approach to automated systematics. In fact one author explicitly states that his results are not intended to be construed as a contribution to automatized systematics in a taxonomical sense. An aspect that could well be united with this work concerns the problem of the expression of shapes in morphometrics, which, of course, in itself is not a candidate for automatically performed identification of taxa. Several of the contributions lie within the sphere of computer-based technology and will not be as readily accessible to the average taxonomist. The results presented on digital automated identification systems impress in that they seem to have been, in a few cases, well tried and tested on real, largely entomological problems. Work on the automated taxonomy of bee species has produced problems of interest to the taxonomic specialist.

The important question of biodiversity is considered in an inspiring chapter. The main factor underlying the extinction of plants and animals – overpopulation and its attendant effects – seems to have been skirted. This is, however, a subject that is difficult to address in print without the author appearing to be socially reprehensible.

Of recent years, the term 'neural networks' has been appearing in an increasingly large number of publications. For most of us, the field remains something of a mystery. It is therefore gladdening that the theme is presented in several chapters. We learn that the method is based on algorithms that are supposed to mimic the function of the (human?) brain. One of the chapters contains a very informative account of the methodology, including the 'pros and contras'. Finally, we know what neural networks are and what they can usefully do. Several contributions treat actual examples, including the identification of plants.

The chapter on geometric morphometrics is refreshing and brings to light new results and methods (noting, however, that current morphometrics does not fit in with the reigning concepts of automated taxonomy, but could, I think, well be made to do so). New, biologically more relevant aspects are introduced and possible fields for future development sketched out. Careful reading of this chapter will disclose links to the natural categorization of objects, a bond I found very interesting. The fact that French mining engineer Dr Georges Matheron's result for an optimal prediction problem, proposed some 30 years ago, is in effect the same as one of the main standard techniques of geometric morphometrics is also interesting.

Automated taxon identification is, in my opinion, a good servant, but should not be allowed to become a bad master. Statistical procedures, mainly multivariate analyses, are a feature of some of the techniques. This is as it should be, but it is necessary that these techniques are properly understood and not misused. Computer programs can do no more than they are told to do and this is where problems can, and do, arise and where manual assessments cannot be done away with. Let us briefly look at some of these where unskilled practitioners can be at fault.

In many publications and summarized in his brilliant PhD thesis of 1979, Norman A. Campbell pointed out the need for evaluating stability in vector components in canonical variate analysis and the unreliability of attempting to reify such components in mistaken geometrical analogy with principal component analysis. I know of no computer programs (except my own and Norm Campbell's) that take account of this. The factor of multivariate stability in coefficients seems to have been consistently ignored in geometric morphometrics, despite its obvious practical importance for correctly assessing the shape diagrams. (By the way, the origin of canonical variate analysis goes back to a publication by the Australian anthropologist M.M. Barnard dating from 1935. Ronald A. Fisher helped Barnard with the analysis of her Egyptian skull material by means of an ingenious time-oriented multiple discriminant analysis. True to his generous nature, he did not put his name on the paper. The 'Egyptian skulls' have been used in many textbooks on multivariate analysis, including the classical text by T.W. Anderson, 2003.)

Another problem concerns what to do about the automated recognition of inherent polymorphism. In many cases, the automated acquisition of morphological characteristics will no doubt work in a satisfactory manner, but not in all cases. Not all polymorphisms declare themselves by presenting differing ornamental details, although they could, at great expense, no doubt be recorded by elaborate bar-coding. As an example of which I have personal experience, work just concluded with my friend and colleague, Professor Kazuyoshi Endo, and his research team at Tsukuba University, Japan, indicates that the normally panmictic brachiopod species *Lingula anatina* displays in the northwest Pacific convincing evidence of molecular polymorphism. Morphological features are, nevertheless, untouched by this polymorphism.

Professor John Aitchison, Glasgow, took up the subject of estimation of statistical parameters when the data are compositional. Aitchison's work was only understood and seized upon by the geological fraternity, notwithstanding that constant-sum data are almost as common in biology, medicine, sociology, marketing, etc. A recent publication, the work of Buccianti et al. (2006), is recommended along with this volume as essential background reading for the practitioners of automated approaches to taxonomic identification.

Richard A. Reyment
Professor (emeritus), Uppsala Universitet, Uppsala, Sweden; Palaeontology Department, Naturhistoriska riksmuseet, Stockholm, Sweden

REFERENCES

Anderson, T.W. (2003) *An Introduction to Multivariate Statistical Analysis,* John Wiley & Sons, Hoboken, NJ.
Buccianti, A., Mateu-Figueras, G. and Pawlosky, G. (eds) (2006) *Compositional Data Analysis in the Geosciences: From Theory to Practice,* Geological Society of London, London.

The Editor

Norman MacLeod (BSc, MSc, PhD, FGS, FLS) is the current Keeper of Palaeontology at The Natural History Museum, London, and is responsible for all operations of The Natural History Museum's Department of Palaeontology. These include the research, curation and conservation activities of 46 staff members along with those of an approximately equal number of students, scientific associates, honorary research fellows, contract employees, volunteers, etc. MacLeod was born in Augsburg, Germany to American parents and grew up in the U.S. Midwest. He obtained a BSc degree (geology) at the University of Missouri (Columbia) in 1975. After teaching secondary school science for two years in Dallas, Texas, he returned to graduate school to pursue an MSc degree in invertebrate paleontology and paleoecology at Southern Methodist University (Dallas) under the direction of A. Lee McAlister. While at SMU, he also began publishing technical papers in vertebrate paleontology in collaboration with Bob. H. Slaughter. Upon completing this degree, MacLeod moved to a PhD programme in invertebrate palaeontology at the University of Chicago, but switched to the University of Texas at Dallas and a micropaleontology research focus in 1981. While pursuing his PhD degree, he was also employed as a consultant at Atlantic Richfield Oil and Gas Co. Research Laboratory in Plano, Texas, and owned a technical photography business.

After obtaining his PhD in 1986 (advisor: E.A. Pessagno, Jr.), MacLeod was awarded a Michigan Society Fellowship to study and teach at the University of Michigan, Ann Arbor. During this period, he began publishing on morphometrics–biostratigraphy stratigraphy of planktonic foraminifera and began a research programme into the causes of ancient mass extinctions. Dr MacLeod continued this work upon moving from Michigan to Princeton University in 1989, often collaborating on Cretaceous-Tertiary planktonic foraminiferal extinction studies with Princeton Professor Gerta Keller. In 1996 he accepted a three-year appointment at The Natural History Museum, London as a visiting scholar. In 1998 he accepted a permanent position at the NHM, from which he moved on to Associate Keeper of the Palaeontology Deptartment in 2000 and Keeper in 2001. In addition to his museum responsibilities, he holds an honorary professorship at University College, London, and is a fellow of the Geological Society of London and the Linnean Society of London.

Professor MacLeod is an internationally recognized leader in many areas of palaeontology. Prominent among these are (1) morphometrics (the quantitative analysis of form variation), (2) the causes of Phanerozoic extinction events and (3) quantitative stratigraphical data analysis. Through his work in these areas, he has made significant contributions to the punctuated-equilibrium controversy and the Cretaceous-Tertiary extinction controversy, as well as being personally responsible for the development of several new morphometric data-analysis techniques. He is the co-editor of two book-length collections of technical articles: *The Cretaceous-Tertiary Mass Extinction: Biotic and Environmental Changes* (W.W. Norton, 1996, with Gerta Keller) and *Morphometrics, Shape and Phylogenetics* (Taylor & Francis, 2002, with Peter Forey), as well as the editor of this text.

In addition to these major publications, Professor MacLeod is also the creator and executive editor of the *PaleoBase* series of electronic palaeontological data-bases (Blackwell Science 2001 to present, http://www.paleobase.com), the creator and manager of the *PaleoNet* palaeontological communications system (http://www.nhm.ac.uk/ hosted_sites/paleonet) and the cofounder and (to 2003) executive editor of the first fully electronic palaeontological journal, *Palaeontologia Electronica* (http://palaeo-electronica.org). Prof. MacLeod is a frequent contributor to the technical palaeontological literature (over 200 articles published, including many widely cited review papers), book reviewer, symposium organizer and keynote speaker at conferences. He has also appeared many times on television and radio programmes discussing paleontological and general science topics.

Contributors

Fred L. Bookstein
Department of Anthropology, University of Vienna
Vienna, Austria
Department of Statistics, University of Washington
Seattle, Washington

David Chesmore
Intelligent Systems Group, Department of Electronics
University of York
York, United Kingdom

Jonathan Y. Clark
Department of Computing
School of Electronics and Physical Sciences
University of Surrey
Guildford, United Kingdom

Armin B. Cremers
Institute of Applied Computer Science (IAI)
University of Bonn
Bonn, Germany

Philip F. Culverhouse
Centre for Interactive Intelligent Systems
School of Computing, Communications and Electronics
University of Plymouth
Plymouth, United Kingdom

Enrique L. Delgado
School of Electrical Engineering and Computer Science
Oregon State University
Corvallis, Oregon

Robert Lang
Department of Cybernetics
University of Reading
Reading, United Kingdom

Hongli Deng
School of Electrical Engineering and Computer Science
Oregon State University
Corvallis, Oregon

Thomas G. Dietterich
School of Electrical Engineering and Computer Science
Oregon State University
Corvallis, Oregon

Martin T. Do
Division of Invertebrate Zoology
American Museum of Natural History
New York, New York

Jeremy C. Huff
Division of Invertebrate Zoology
American Museum of Natural History
New York, New York

David Jones
Department of Geology
University of Leicester
Leicester, United Kingdom

Monja A. Knoll
Ecological Acoustic Research Group (EAR)
Department of Psychology
University of Portsmouth
Portsmouth, United Kingdom

Karl-Heinz Lampe
Zoologisches Forschungsmuseum Alexander Koenig (ZFMK)
University of Bonn
Bonn, Germany

David Lytle
School of Electrical Engineering and Computer Science
Oregon State University
Corvallis, Oregon

Norman MacLeod
Palaeontology Department
The Natural History Museum
London, United Kingdom

Manuel Mendoza
Department of Ecology and Evolutionary Biology
Brown University
Providence, Rhode Island

Andrew Moldenke
School of Electrical Engineering and Computer Science
Oregon State University
Corvallis, Oregon

Eric N. Mortensen
School of Electrical Engineering and Computer Science
Oregon State University
Corvallis, Oregon

Mark A. O'Neill
School of Biology, Division of Biological Sciences
University of Newcastle
Newcastle upon Tyne, United Kingdom

Robert Paasch
School of Electrical Engineering and Computer Science
Oregon State University
Corvallis, Oregon

Norman I. Platnick
Division of Invertebrate Zoology
American Museum of Natural History
New York, New York

Mark Purnell
Department of Geology
University of Leicester
Leicester, United Kingdom

Richard A. Reyment
Paleontology Department
Naturhistoriska riksmuseet
Stockholm, Sweden

Kimberly N. Russell
Division of Invertebrate Zoology
American Museum of Natural History
New York, New York

Stefan Schröder
Federal Agency for Agriculture and Food (BLE)
Bonn, Germany

Linda Shapiro
School of Electrical Engineering and Computer Science
Oregon State University
Corvallis, Oregon

Volker Steinhage
Institute of Applied Computer Science (IAI)
University of Bonn
Bonn, Germany

Adam Tofilski
Department of Pomology and Apiculture
Agricultural University of Krakow
Krakow, Poland

Maria Uther
Speech and Auditory Interaction Laboratory (SAIL)
School of Psychology
Brunel University
London, United Kingdom

Robert Vincent
Centraalbureau voor Schimmelcultures (CBS)
Royal Netherlands Academy of Arts and Sciences
Amsterdam, the Netherlands

Stig A. Walsh
Palaeontology Department
The Natural History Museum
London, United Kingdom

Quentin D. Wheeler
Entomology Department
The Natural History Museum
London, United Kingdom

Pengcheng Wu
School of Electrical Engineering and Computer Science
Oregon State University
Corvallis, Oregon

Wei Zhang
School of Electrical Engineering and Computer Science
Oregon State University
Corvallis, Oregon

1 Introduction

Norman MacLeod

CONTENTS

Acknowledgements 6
References 6

The automated identification of biological objects (individuals) and/or groups (e.g. species, guilds, characters) has been a dream among systematists for centuries. The goal of some of the first multivariate biometric methods was to address the perennial problem of group discrimination and intergroup characterization (e.g. Fisher, 1936). However, despite much optimism and preliminary work in the 1950s and 1960s, progress in designing and implementing practical systems for fully automated taxon identification proved frustratingly slow. Indeed, most practicing taxonomists still believe such systems are the stuff of science fiction. As recently as 2004, Dan Janzen updated this dream for a new audience:

> The spaceship lands. He steps out. He points it around. It says 'friendly–unfriendly—edible–poisonous—safe–dangerous—living–inanimate'. On the next sweep it says '*Quercus oleoides—Homo sapiens—Spondias mombin—Solanum nigrum—Crotalus durissus—Morpho peleides*—serpentine'. This has been in my head since reading science fiction in ninth grade half a century ago.
>
> **(p. 731)**

Regardless, the dream has never died.

Janzen's preferred solution to this classic problem involves building machines to identify species from their DNA – the so-called DNA 'barcoding' initiative. His predicted budget and proposed research team were 'US $1 million and five bright people'. However, recent developments in computer architectures, as well as innovations in software design, have placed the tools needed to realize systematists like Janzen's vision in the hands of the systematics community, not several years hence, but now; and not just for DNA barcodes, but for digital images of organisms, digital sounds, digitalized chemical data – essentially, any set of observations that can document the characteristics of organisms and be presented in digital form.

The parallels between DNA barcoding and morphological image recognition are interesting and deserve comment in order to place both in perspective. The proponents of DNA barcoding (e.g. Hebert et al., 2003; Tautz et al., 2003; Hebert et al., 2005) argue that the impressive increase in sequencing capacity and decrease in sequencing cost resulting from the Human Genome Project make both the high-throughput, automated identification of taxonomic groups and discovery of new species possible via comparatively simple analyses of one or a few standardized segments of the DNA molecule. Such segments, of course, do not embody all the genetic differences that exist between any two species, but (they argue) are sufficient to 'type' various subspecific, specific and superspecific groups on the basis of phylogenetic similarity. Moreover, adoption of a DNA-based approach for taxonomy and identification should be regarded as a community-wide priority for systematists because (they claim) such a research programme will not only support quicker and better identifications, but also capture the public imagination and so make new conceptual and financial resources available.

Over the past few years, each of these claims has been challenged in principle as well as in practice (e.g. Wheeler, 2005, and references therein; Cameron et al., 2006, and references therein; Hickerson et al., 2006, and references therein). These challenges do not deny either the power or utility of molecular methods. Rather, they argue molecular approaches should be viewed as parts of a multidisciplinary toolkit, all tools of which can and should be made available for understanding the plethora of complex structures present within systematic data. In order to be of optimal use, this toolkit needs to include traditional, qualitative morphological approaches to systematic data analysis, molecular analyses (see preceding), modern or 'geometric' morphometric approaches (see Bookstein, 1991; Zelditch et al., 2004), phylogenetic approaches (see Kitching et al., 1998) and geographic approaches (see Avis, 2000; Ebach and Tangney, 2006) – in short, all approaches currently available as well as any that can be developed to address this issue reliably in the future. The scope of the problem – that is, winning as much of the race to understand present and past biodiversity as we can before some, as yet unknown, proportion of the former turns into the latter – demands no less.

Wheeler (2005) also makes the important point that the history of systematics itself provides one of the best arguments against adopting a 'single system' approach to taxonomy and identification. Aside from the simple facts that (1) the overwhelming majority of all species known to science have been recognized on the basis of the morphological, not molecular, characteristics; and (2) fossils, for which little or no molecular data will ever be available, will inevitably play an important role in reconstructing the history of life (e.g. Donoghue, 1989), the lesson of the phenetics experiment was that systematic hypothesis tests *require* access to multiple lines of evidence (see Prendini, 2005).

Over the last several decades, systematic research programmes have embraced a multitude of different data types. The advantages and disadvantages of these data sources have been revealed by comparisons between results derived from different datasets – not decided *a priori*. No one type of data can claim priority over any other in terms of revealing 'true' relations. Given the robust state of systematics research programmes, there seems little point in repeating the mistakes of the past by arbitrarily deciding to focus on a single type of data for resolving all taxonomic problems. This is especially true when the ability of any single approach to 'deliver the goods' in terms of routine, large-scale, cost-efficient and automated taxonomic identifications (and discovery of new species), even for modern groups, remains to be demonstrated.

Whereas the claims of the superior capabilities of molecular barcoding are great and, for the most part, unsupported by empirical results, the opportunities afforded by the application of quantitative approaches to the problem of class recognition (the topic of this book) are more modest and, because of this, more likely to be realized. Class recognition is a truly generalized problem that can be used in conjunction with any types of data (e.g. morphological, auditory, chemical, molecular). The source of the information and its type are important, of course, but so is the problem of what one does with those data once they have been collected.

Traditionally, systematists have not relied on any quantitative data *per se* to identify taxa, preferring instead to rely on the visual inspection of morphology, the (mostly) qualitative assessment of characters whose patterns of variation are often complex and the comparison of these to reference specimens and/or images. While this process works, it is not quick, efficient or reliable. Anecdotes regarding disagreements between experienced specialists over the identification of even common taxa are legion within the community of systematists. These disagreements are rarely tested objectively – much less, resolved – and cause uncertainty in the application of systematic concepts within every organismal group. More seriously, there is presently no tradition of independent verification of identifications based on any objective criteria. Indeed, very few studies of consistency in the application of species concepts and identifications referenced to standardized faunas or floras have ever been published by systematists. In one recent 'blind test' of palaeontological species identification, consistencies between faunal lists generated by independent experts ranged between 0.7 and 0.2 (see MacLeod, 1998; see Culverhouse, this volume, for additional examples and discussion).

Ask most systematists why this situation exists and what can be done to improve it and you are likely to be told: (1) The problem is very complex, (2) this is the best we can do at present and/or (3) we could possibly do better if we (i.e. the systematics community) had more funding. While this is all true, such responses do not really get anyone any closer to improving the current situation. Fortunately, data-processing methods, artificial intelligence algorithms, morphometric approaches and a host of allied technologies have all made substantial progress since their inception in the 1950s and 1960s. Quietly, with surprisingly little fanfare (and even less funding), small teams of taxonomists, mathematicians, artificial intelligence experts and computer programmers have been coming together over the last decade to create the first and second generations of genuine automated identification systems.

Performance levels of these first- and second-generation systems are already impressive: typically ≥85 per cent correct identifications for small- to medium-sized species assemblages. More importantly, these systems are already more consistent and more rapid than human experts in producing identifications of some groups. As these system designs mature and continue to improve (e.g. incorporating dynamic neural net structures), they promise to deliver even better performance for a larger number of taxon groups in the future. The fact that most systematists remain unaware of these developments and their implications for the practice of systematics is the primary reason this book has been assembled.

Such advances could not have come at a better time. As many scientists – and all systematists – already know, the world is running out of specialists who can identify the very biodiversity whose preservation has become such a global concern. In commenting on this problem in palaeontology, as long ago as 1993 the palaeontologist Roger Kaesler recognized:

> We are running out of [systematists] who have anything approaching synoptic knowledge of a major group of organisms.... Paleontologists of the next century are unlikely to have the luxury of dealing at length with taxonomic problems ... [Paleontology] will have to sustain its level of excitement without the aid of systematists, who have contributed so much to its success.
>
> **(pp. 329–330)**

This expertise deficiency, which has come to be called the 'taxonomic impediment', is with us now and will only become more serious as time goes by unless some means is found to address its effects. The taxonomic impediment cuts as deeply into those commercial industries that rely on accurate identifications (e.g. agriculture, biostratigraphy) as it does into a wide range of pure and applied research programmes (e.g. conservation, biological oceanography, climatology, ecology). This expertise impediment is also not a recent development. The contemporary taxonomic literature of all organismal groups is littered with examples of inconsistent and incorrect identifications. This is due to a variety of factors, including taxonomists being insufficiently trained and skilled in making identifications (e.g. using different rules of thumb in recognizing the boundaries between similar groups), insufficiently detailed original group descriptions and/or illustrations, inadequate access to current monographs and well-curated collections and, of course, taxonomists having different opinions regarding group concepts. Peer review has little effect on non-academic studies and even then only weeds out the most obvious errors of commission or omission when an author provides adequate representations (e.g. illustrations, recordings, gene sequences) of the specimens in question. If systematics is to improve the quality as well as the quantity of the data that its practitioners provide to their colleagues, students and clients, and that they themselves use to document changes occurring in the natural world, a different approach is needed.

Many systematists appear to see research into automated identification systems as a threat to their field and their livelihood. Ironically, however, systematics has much to gain both practically and theoretically from the further development and use of such systems. It is now widely recognized that the days of systematics as a field populated by mildly eccentric – and often independently wealthy – individuals pursuing knowledge in splendid isolation from funding priorities and economic imperatives are well past. In order to attract both personnel and resources, systematics must transform itself into a 'large, coordinated, international scientific enterprise' (Wheeler, 2003, p. 4).

Many have identified use of the Internet – especially via the World Wide Web – as the medium through which this transformation can be made. While establishment, for example, of a virtual, GenBank-like system for accessing morphological data, audio clips, video files and so forth would be a significant step in the right direction, improved access to observational information and/or text-based descriptions alone will not address either the taxonomic impediment or low identification reproducibility issues successfully. Instead, the inevitable subjectivity associated with making critical decisions on the basis of qualitative criteria must be reduced or, at the very least, embedded within a more formally analytic context.

Properly designed, flexible, and robust automated identification systems organized around distributed computing architectures and referenced to authoritatively identified collections of training-set data (e.g. images, gene sequences) can, in principle, provide all systematists with access to the electronic data archives and necessary analytic tools to handle routine identifications of common taxa. Such systems not only deliver consistent identifications, but also provide accurate estimates of each identification's confidence level and can recognize when their algorithms cannot make a reliable identification (and so refer that image or specimen to a human specialist for further study). The most advanced identification systems currently available also include elements of artificial intelligence and can improve their performance the more they are used. Most tantalizingly, such systems can become true partners in systematic research, delivering high-volume, accurate and consistent identifications in literally seconds and allowing the systematist to experiment with alternative species concepts and/or models of intraspecific and interspecific variation. Automated identification systems can even be used to determine which aspects of the observed patterns of variation are being used to achieve the identification, thus opening the way for the discovery of new and (potentially) more reliable taxonomic characters.

In order to summarize the current state of the art in automated taxon-recognition systems and assess their potential to make practical contributions to systematics and taxonomy both now and into the future, the Systematics Association and The Natural History Museum (London) jointly sponsored a free, one-day symposium entitled *Algorithmic Approaches to the Identification Problem in Systematics* at The Natural History Museum, London, on 19 August 2005. The purpose of that symposium (which was part of The Systematics Association's biennial meeting) was to provide leaders of research groups, researchers and students working in or studying any area of systematics with an opportunity to (1) learn about current trends in quantitative approaches to the group-recognition problem, (2) become familiar with the capabilities of various software systems currently available for identifying systematic objects/groups and (3) evaluate various applications of this technology to present and future systematic problems. Special attention was paid to showing how different approaches to automated identification can be applied to various organismal groups and in various applied research contexts (e.g. biodiversity studies, biostratigraphy, conservation, agriculture, curation). This book represents the edited proceedings of that symposium.

The collected articles are divided into four informal sections. Background information is provided by Chapters 1 and 2. In Chapter 1, Quentin Wheeler sets the stage by reviewing the current state of systematics and placing automated taxon identification in the context of contemporary trends, needs and opportunities. Phillip Culverhouse (Chapter 2) then focuses on how humans perceive the world, with special reference to those attributes such automated systems will need to process to match human performance, which is impressive but not without its limitations.

The second section of the book addresses two technical developments at the forefront of mathematical approaches to the implementation of automated identification systems. Robert Lang (Chapter 3) reviews the concepts embodied in neural net designs to provide readers with a basis for understanding the discussions of various neural net systems and applications in the chapters that follow. In Chapter 4, Fred Bookstein reviews the issues of homology and character recognition from a morphometric point of view through presentation of a new technique (relative intrinsic warps), which can be used to locate and characterizes, within a sample-based warp system, particular regions that are undergoing localized changes, possibly reflecting character-state transitions. Bookstein suggests the intrinsic covariance structure of a set of morphologies provides a practical means for focusing automated identification systems on those aspects of shape transformation patterns that correspond to biological characters.

The third section represents five chapters that present and/or evaluate different aspects of current automated system designs. David Chesmore's chapter (Chapter 6) opens this section with a comprehensive review of neural net applications to the group- or taxon-recognition problem. This is followed by a trio of chapters devoted to describing and illustrating the use of three of the most highly developed automated identification systems currently available by key members of their system-development teams: Mark O'Neill presents and discusses the development of his PSOM-based digital image analysis system (DAISY, Chapter 7), Volker Steinhag and colleagues present their morphometrics-based automated bee identification system (ABIS, Chapter 8) and Kimberly Russell and colleagues discuss their wavelet-MLP-based species identification automated system (SPIDA, Chapter 9). My own chapter (Chapter 10, with Mark O'Neill and Stig Walsh) closes this section with a comparison of standard morphometric and PSOM-based neural net approaches that seeks to identify reasonable analytic domains for these alternative approaches and lines along which they may converge with future development.

The book's final section presents a series of eight case studies in which different practical aspects of the overall group identification problem are identified, analyzed and discussed. Eric Mortensen and colleagues (Chapter 11) review their work with automated stonefly identification as a means to assess water quality. Jonathan Clark (Chapter 12) reviews use of MLP neural nets and traditional metric characters to automate plant species identification from leaf morphology. Stig Walsh and colleagues (Chapter 13) compare and contrast both landmark and outline-based morphometric approaches with DAISY to tackle a perennial palaeontology problem: the identification of taxa (in this case, penguins) from isolated skeletal elements. David Jones and Mark Purnell (Chapter 14) use a novel outline-based morphometric approach to identify conodonts. Manuel Mendoza (Chapter 15) employs decision-trees to interpret ecological adaptations in a fossil mammal lineage based on modern ecological counterparts. Robert Vincent (Chapter 16) discusses and illustrates use of the BioloMICS system for automated identification from generalized matrices of continuous and discontinuous characters. Adam Tolfilski (Chapter 17) describes his automated honeybee wing measurement software *DrawWing*. Finally, Monja Knoll and colleagues (Chapter 18) contrast eigenshape and DAISY approaches to the analysis of human vocalization patterns.

The scope and diversity of the approaches and applications discussed herein are inspiring and fully indicative of their potential to have a real and lasting impact, not only on all areas of systematics but also on all the fields systematics serves. In the final analysis though, we are left with the realization that the technology available to collect organisms in the environment far outstrips the systematics community's ability to identify those organisms reliably irrespective of the data extracted (e.g. morphological, chemical, molecular) and it will continue to do so. This limitation on the taxonomic side of the equation represents a classic case of supply limiting demand. Throughout human history, whenever there has been a need to perform routine and well-defined tasks rapidly, accurately and consistently, the most productive solution has been to automate the procedure via application of appropriate technology. Doing this well, reliably, and for as wide a diversity of organisms as possible is, we believe, one of the fundamental challenges for systematics in the 21st century.

This is, in effect, what automated taxon identification systems do. Systematists who employ such systems and who lend a hand in their improvement are not diminishing their field or the future employment prospects of their students. Rather, they are helping ensure a future for systematics while, at the same time, helping to free systematists from the drudgery of routine identification so that their hard-won talents can be put to better use. More than this they are ensuring that systematics will be even more important to the research programmes of the future than it has been to those of the past.

Based on the evidence accumulated over the last decade and presented in this volume, the question of whether such automated taxon identification systems can be constructed using available technology has been answered. That answer is 'yes'! In reviewing the progress of over a decade spent developing such systems, the authors of these chapters ask the implicit questions, 'What will be the role of research into automated approaches to taxon identification'? 'Who will fund it'? and 'Where does improving the ability to deliver better taxonomic identifications stand in systematics' overall research agenda'? It is our hope that this volume will contribute to helping the systematics community, its client communities and those who set research-funding priorities for governments and private foundations to formulate answers to these questions, both in principle and in practice.

ACKNOWLEDGEMENTS

No book of this type can be assembled by an editor without the active collaboration and skills of numerous individuals. In particular, I would like to thank the Systematics Association and The Natural History Museum (London) for their support of the original symposium; Richard Lane (NHM director of science) for allowing me the time required to undertake the book's organization and editing; all the authors for contributing their work, persevering through the review and revision process, answering endless requests from me regarding aspects of their presentation and enduring long periods of little communication when I was working with other authors or otherwise engaged; Mark O'Neill and Stig Walsh, who contributed in ways too numerous to list; and John Sulzycki, Judith Simon and Amber Donley from Taylor & Francis for sticking with the project and shepherding the book through its production.

REFERENCES

Avise, J.C. (2000) *Phylogeography: The History and Formation of Species,* Harvard University Press, Cambridge, MA, 464 pp.

Bookstein, F.L. (1991) *Morphometric Tools for Landmark Data: Geometry and Biology,* Cambridge University Press, Cambridge, England, 435 pp.

Cameron, S., Rubinoff, D. and Kipling, W. (2006) Who will actually use DNA barcoding and what will it cost? *Systematic Biology,* 55: 844–847.

Donoghue, M.J., Doyle, J.A., Kluge, A. and Rowe, T. (1989) The importance of fossils in phylogeny reconstruction. *Annual Review of Ecology and Systematics,* 20: 431–460.

Ebach, M.C. and Tangney, R.S. (2006) *Biogeography in a Changing World,* CRC Press, Boca Ration, FL, 232 pp.

Fisher, R.A. (1936) The utilization of multiple measurements in taxonomic problems. *Annals of Eugenics,* 7: 179–188.

Hebert, P.D.N., Cywinska, A., Ball, S.L. and deWaard, J.R. (2003) Biological identifications through DNA barcodes. *Proceedings of the Royal Society of London, Series B,* 270: 313–322.

Hebert, P.D.N., Penton, E.H., Burns, J.M., Janzen, D.H. and Hallwacks, W. (2005) Ten species in one: DNA barcoding reveals cryptic species in the neotropical skipper butterfly *Astaptes fulgerator. Proceedings of the National Academy of Sciences,* 101: 14812–14817.

Hickerson, M.J., Meyer, C.P. and Moritz, C. (2006) DNA barcoding will often fail to discover new animal species over broad parameter space. *Systematic Biology,* 55: 729–739.

Janzen, D.H. (2004) Now is the time. *Philosophical Transactions of the Royal Society of London, Series B,* 359: 731–732.

Kaesler, R.L. (1993) A window of opportunity: peering into a new century of paleontology. *Journal of Paleontology,* 67: 329–333.

Kitching, I.J., Forey, P.L., Humphries, C.J. and Williams, D.M. (1998) *Cladistics: The Theory and Practice of Parsimony Analysis,* 2nd edition, Oxford University Press, Oxford, 228 pp.

MacLeod, N. (1998) Impacts and marine invertebrate extinctions. In *Meteorites: Flux with Time and Impact Effects* (eds M.M. Grady, R. Hutchinson, G.J.H. McCall and D.A. Rotherby), Geological Society of London, London, pp. 217–246.

Prendini, L. (2005) Comment on 'Identifying Spiders through DNA Barcodes'. *Canadian Journal of Zoology,* 83: 498–504.

Tautz, D., Arctander, P., Minelli, A., Thomas, R.H. and Vogler, A.P. (2003) A plea for DNA taxonomy. *Trends in Ecology and Evolution,* 18: 70–74.

Wheeler, Q. (2005) Losing the plot: DNA 'barcodes' and taxonomy. *Cladistics,* 21: 405–407.

Wheeler, Q.D. (2003) Transforming taxonomy. *The Systematist,* 22: 3–5.

Zelditch, M.L., Swiderski, D.L., Sheets, H.D. and Fink, W.L. (2004) *Geometric Morphometrics for Biologists: A Primer,* Elsevier/Academic Press, Amsterdam, 443 pp.

2 Digital Innovation and Taxonomy's Finest Hour

Quentin D. Wheeler

CONTENTS

A Gutenberg Moment for Morphology 18
Digital Tools and Taxonomic Empowerment 19
Conclusion: Taxonomy's Finest Hour? 21
Acknowledgements 22
References 23

Taxonomists face the greatest challenges and opportunities of the very long history of their science. The already daunting challenge of discovering and describing the many millions of species on Earth has a sudden urgency imposed by runaway rates of species extinction and worldwide disturbance and degradation of ecosystems (e.g. Millennium Ecosystem Assessment, 2005). Getting on with the business of describing, naming and classifying species has multiple significances. Taxonomic information is necessary for conservation biology planning and assessment; the more comprehensive our taxonomic knowledge is, the better are the phylogenetic foundation and context for comparative biology; the intellectual manifest destiny of our own curious species to understand its place in the living world rests heavily on taxonomists preserving sufficient evidence of the biological diversity – particularly that which seems now certain to be lost. Fortunately, it seems equally clear that the information revolution offers unparalleled opportunities to vastly accelerate taxonomic exploration and research (Wheeler et al., 2004). While existing information technology and digital tools are enough to convince us of their enormous potential, we must recall that we live in early days of this information revolution (e.g. Atkins et al., 2003). Peter Raven (2004) sums up the unrivalled significance of information technologies to taxonomy as follows:

> In the accumulation of knowledge about organisms and its deployment for many purposes, the efficient use of the products of the information revolution is even more important than the development of moveable type 550 years ago. The registration of all properties of organisms in efficiently constructed databases, automated identification, Web pages for different kinds of organism[s] – these and many other techniques are absolutely indispensable and need funding as amply as can be provided. There is simply no other way to achieve the aims of taxonomy broadly, as many of us have been pointing out for decades. Many more efficient ways can and are being developed to deal with the description of organisms. Fully harnessing the power of the Web clearly will and must expedite taxonomic progress in the future, building soundly on the systems of the past but expediting the accumulation and dissemination of information.

Raven (2004) also captures the human resource dimension of the challenge at a time when few taxonomists are being educated or employed who know their organisms well and are prepared to lead the arduous inventory tasks that lie ahead:

> Finally, nothing will substitute for the activities of the field naturalist. No matter how much we may speak about instant identification through DNA analysis, hand-held keys or other modern approaches, unless there are very many people who can recognize organisms, find them, go into the field and find them again, whether they be in the tropical moist forests of Congo or the chalk grasslands on the South Downs of England, nothing will work.

We neglect taxonomic and related organismal biology research and expertise at our own peril. Much of what we know or will ever learn of biological diversity, ecosystem complexity and evolutionary history came or will come as a direct result of taxonomic research. Although such taxonomic information, including Linnaean names and classifications, forms the foundation and general reference system for biology, taxonomy has received surprisingly little support in recent decades (e.g. House of Lords, 2002). This neglect, in conjunction with projected extinctions of species, has been described as a 'taxonomic crisis' (Page et al., 2005). The same (phylogenetic) theoretical advances that enable taxonomists to create biology's general reference system (Hennig, 1966) are supported today, but divorced from improvements in the formal classifications and names that are the backbone of the system. An efficient and organized approach to 'descriptive' taxonomy is, in fact, the largest missing piece of an effective response to the biodiversity crisis (Wheeler, 2004).

Taxonomy was weakened in the early twentieth century (Wheeler, 1995) because it could not yet reconstruct the relative recency of common ancestry among species in a rigorously testable way – an essential step if classifications were to reflect evolutionary-historical patterns reliably. Ironically, phylogenetic analyses are now being done more rapidly than the discovery and description of the species that would justify further reconstruction of such branching diagrams. Given the biodiversity crisis, our top priorities should be to complete an inventory of species, develop comprehensive collections and assure that taxonomic information is both reliable and openly accessible through sustainable revisionary taxonomic programs and enhanced infrastructure and tools for taxonomists. Description (circumscription) of species is a methodological prerequisite to phylogenetic analysis (Hennig, 1966); in a biodiversity crisis, it is also a priority.

The best taxonomy has long been and will remain integrative, including all relevant and available comparative evidence. Take the following, for example:

> Taxonomy is at the same time the most elementary and the most inclusive part of zoology, most elementary because animals cannot be discussed or treated in a scientific way until some taxonomy has been achieved, and most inclusive because taxonomy in its various guises and branches eventually gathers together, utilizes, summarizes, and implements everything that is known about animals, whether morphological, physiological, psychological, or ecological.
>
> **(Simpson, 1945)**

Also consider:

> Nevertheless, the data of molecular biology, when comparative at any level (either within or between populations), most decidedly enter into systematics – the distribution of the various hemoglobins is only one of innumerable strikingly pertinent examples. Systematics, in turn, does apply to whole field of molecular biology and supplies one of the several ways in which the results of that subject may be explained or meaningfully ordered.
>
> **(Simpson, 1961)**

These integrative aspects of taxonomy were also reflected by Hennig's concept of 'holomorphy' that included the totality of characters, regardless of source, but held to the same standards as morphology:

> But it is an old demand that systematic work should consider "all characters" as far as possible, and not be limited to only one sector of the fabric of characters. Consequently other characters not narrowly morphological have recently been recommended to an increasing degree. Insofar as these are intended to supplement and test the results based on morphological characters, phylogenetic systematics will profit from them. A prerequisite for this is that nonmorphological characters be used according to the same strict, theoretically incontestably based methodological principles as morphological characters are.
>
> **(Hennig, 1966, p. 101)**

While taxonomy as a science needs data from as many sources as possible (whether comparative morphology, developmental biology, palaeontology, molecular genetics, comparative ethology or other), it should not be expected that all such data be included in every study or that every taxonomist engage in the full range of possible data gathering. While phylogenetic analysis profits from a total evidence approach with its diverse data and associated benefits, studies generating new characters or critically testing existing characters need not be concerned with multiple data sources. The only criterion on which to judge the value of such studies ought to be excellence. With millions of species to discover, describe and test and many more tests of homology, synapomorphy and monophyly to complete, there is ample room for specialization. Integrating a number of excellent studies drawing upon diverse kinds of comparative data is far preferable to the superficiality sometimes seen in studies that attempt to combine treatments of, for example, molecular and morphological data. Some excellent studies are produced using these and other combinations of data, of course. In general, however, a goal of true excellence may be more directly realized through the encouragement of specialization and collaboration rather than the premium often placed on putatively interdisciplinary studies.

The current insistence that a good molecular study include (often only token) morphological data usually encourages superficial treatments, often little more than a literature review. By the same token, requiring a good morphological study to include molecular data is equally unnecessary. We should celebrate and encourage a variety of approaches and insist on uniformity only with respect to excellence, recognizing that each study, regardless of how narrow or broad, contributes ultimately to the diversity and quality of data available to taxonomy as a whole. As taxonomic information inevitably moves toward a Web-based system (Scoble, 2004) the integration of data derived from many studies and sources shall become increasingly seamless. Why not take advantage of the aptitudes, talents, motivations and interests of researchers rather than forcing compliance with a predetermined 'approved' checklist of data sources? We seem to be confusing the wants and needs of the field as a whole with those of individual researchers and students.

Recent proposals for DNA-based taxonomy (e.g. Tautz et al., 2003; Blaxter, 2004) seem to overlook the obvious benefits of the current integrative taxonomy (Will et al., 2005), as well as those of alternative data sources, including morphology. Even if a completely DNA-based taxonomy were achievable, it would not be advisable or desirable. Some obvious reasons to encourage comparative morphology include: (a) as an 'independent' test of cladistic patterns suggested by molecular (or other) data; (b) as a link between neontological and palaeontological studies; (c) as a useful, user-friendly data source for field biologists and amateur naturalists; (d) as a source of complex characters not recoverable from DNA data alone, even if whole genomes were sequenced; (e) as an entrée to complex attributes that are the objects of interest to natural selection; (f) as a necessary source of evidence to join up genetic and epigenetic effects on the phenotype; (g) as an intellectual and scientific goal in its own right, to understand the origin and history of transformations of complex characters; and (h) because it is intellectually challenging and aesthetically stimulating to study. The latter was expressed well by the nineteenth century coleopterist David Sharp:

> The aesthetic satisfaction to be derived from contemplating the mere variety of animal forms, and from tracing the order that runs through all its diversity, appeals to a very deep instinct in human nature.
>
> **(Calman, 1930, p. 10)**

At the founding of the 'new systematics', Timofeeff-Ressovsky claimed that morphology was spent.

> Since Darwin much very extensive and ingenious work has been done in the field of evolutionary studies, using palaeontological, morphological, embryological, and biogeographical data; these studies have allowed us to picture the main historical steps and events of the evolutionary process. The efficacy of these classical methods, which gave a picture of what we may call 'macro-evolution', seems now to be more or less exhausted.
>
> **(Timofeeff-Ressovsky, 1940, p. 73)**

Timofeeff-Ressovsky and fellow new systematists were interested in genetic mechanisms of mutation and inheritance rather than the historical story of evolution. They succeeded in shifting the emphasis and funding toward 'micro-evolution' instead of 'macro-evolution' (the latter including phylogenetic systematics). A scant quarter century later, the translation of Hennig's (1966) ideas into English would spur the greatest advance in macro-evolutionary theory since Darwin and renew interest in historical patterns and comparative morphology (Eldredge and Cracraft, 1980; Nelson and Platnick, 1981; Schock, 1986). Subsequent waves of technology have been accompanied by similar false predictions of the demise, or at least diminished importance, of morphology, including advocates of electrophoresis in the 1970s and of DNA sequencing today.

Morphology has the opportunity for a major revival based on a wide range of new tools and technologies. Computer-assisted tomography (CT) is making three-dimensional visualization of complex structures possible as well as opening non-destructive access to specimens and structures that were not accessible before. Applications of this technology for teaching include the potential for an instructor to 'print' a three-dimensional model of a specimen at any scale that can be handled and studied in the classroom, regardless of where in the world the actual specimen resides. Algorithms discussed by authors in this volume point to extraordinary new approaches to assess and characterize morphological characters, critically question homology assertions and further refine emerging image-based systems for automated species identifications (see also Gaston and O'Neill, 2004). A network of remotely operable high-resolution digital microscopes is being developed by entomologists at The Natural History Museum, London, the US Department of Agriculture Systematic Entomology Laboratory, Beltsville, and the Museum National d'Histoire Naturelle, Paris, in partnership with Microptics-USA, Inc. that will permit scientists and students to manipulate and photograph type and rare specimens from any broadband-connected computer in the world. Efforts are under way to merge MorphBank and MorphoBank, two recent efforts to create a repository where digitized morphological images and information may be deposited and archived in an openly accessible form. A number of groups and institutions are developing novel approaches to electronic publication of morphology-based descriptions that will eliminate traditional limitations imposed by costs of illustrations and hopefully successfully revive and transform taxonomic revisions and monographs into Web-based, continuously updated community resources. These few examples are among many fledgling steps toward the future of comparative morphology and morphology-based taxonomic applications.

Many users of taxonomic information simply want to be able to identify specimens accurately as belonging to recognizable species. While having a name associated with each species is crucial for communication and information storage and retrieval, names are ultimately only as useful and informative as the hypotheses associated with them are tested and corroborated. Through the neglect of descriptive taxonomy, such hypotheses remain infrequently tested or untested and the scientific

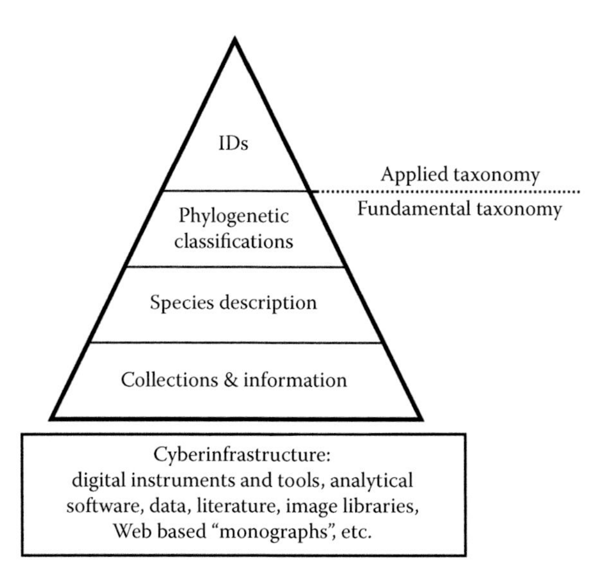

FIGURE 2.1 Cyberinfrastructure will complement and seamlessly network museum infrastructure (i.e. specimens and specimen-associated data) with digital instruments, tools, software, and people to expedite and facilitate every aspect of fundamental and applied taxonomy. This will diminish existing constraints on access (to specimens, data, literature, morphology images, analytical software, experts, etc.). It will be particularly revolutionary with respect to the documentation, analysis and communication of morphological information; the frequency and comprehensiveness of taxonomic revisions and monographs; the ways we identify species; and the ways in which we do, think about, access and use taxonomic information. (For additional insights, see Atkins at al., 2003; Wheeler, 2004; Wheeler et al., 2004; Page et al., 2005.)

names associated with them cease to reflect our current and full knowledge of patterns in nature accurately. This symptom of taxonomic neglect is contributing to a silent and growing crisis of reliability of information associated with specimens in collections (e.g. Meier and Dikow, 2004) and must be addressed alongside inventory efforts to discover new species.

Users sometimes fail to recognize the important distinction between classification and identification (Mayr, 1969); both are necessary but the latter is possible only to the extent that the former has been completed. Applied taxonomy can only exist in the presence of an accumulated body of fundamental taxonomic research (Figure 2.1). Current efforts to avoid the serious work of taxonomy can only succeed by lowering the quality of taxonomy itself (Will et al., 2005). Meanwhile, the information revolution is speeding virtually every step of taxonomic work, making such compromise of quality unnecessary.

Digital tools are revolutionizing species identification as rapidly as they are reshaping most aspects of taxonomic research. In addition to automated species identifications (e.g. Gaston and O'Neill, 2004; Mortensen et al., this volume; O'Neill, this volume; Russell, this volume, Steinhag et al., this volume) a wide range of possibilities are being opened by the information revolution. User-friendly interactive electronic keys, digitally illustrated, that require no prior morphological vocabulary are improving all the time and are downloadable to laptop or handheld devices. Experts are potentially accessed via the Web anywhere in the world to make crucial identifications, consult or teach. The US Department of Agriculture's interest in the remote microscope system mentioned before is to provide a link between taxonomic specialists and suspicious interceptions made at ports of entry among other innovative applications. Evidence provided by DNA is being effectively used to associate highly dissimilar semaphoronts (e.g. Miller et al., 2005) and to identify mere fragments of specimens, important in the enforcement of regulations governing the trade in endangered species.

Such molecular diagnostics, however, increase rather than obviate the need for taxonomic research, including continuing tests of known species and descriptions of new ones. The same unpredictable levels and patterns of variety that make comparative morphology so intellectually engaging occur at the molecular level, too. Ranges of infra- and interspecific variation can and do overlap on occasion (e.g. Meyer and Paulay, 2005), exposing the folly of assumptions made in proposals to use a 'uniform short-segment' from one gene to identify species of all or any serious percentage of animal species (Prendini, 2005; Wheeler, 2005). Armed with accurate species identifications, the taxonomy user of the future will locate a rapidly increasing wealth of biological, geographical and other information about species as Wilson's (2003) encyclopaedia becomes reality.

Considering the fragmented, uncoordinated and inadequately resourced history of taxonomy since 1758, it has been remarkably productive, documenting more than one and a half million species. This rate was acceptable, until recently. But in the context of the biodiversity crisis, this is no longer so. Existing approaches to taxonomy are, for the most part, excellent and merely need to be accelerated to keep pace with the biodiversity crisis. One of the greatest threats of the biodiversity crisis is that so much data could be lost that systematics' grand questions (Cracraft, 2002; Page et al., 2005) will never be addressed adequately. This would dramatically limit our knowledge of evolutionary history with a cascade of constraints and negative impacts on other subdisciplines of biology.

Given improved taxonomic infrastructure and research tools, taxonomists can immediately accelerate the processes of species exploration, description and classification and meet society's growing need for informed access to the world's flora and fauna. With such rapid advances in information technologies, excuses for not mapping the biological diversity of Earth are disappearing. We need now to adapt these technologies to the special needs of taxonomic research, invest in the growth of collections, and educate and support a new generation of taxonomists to realize taxonomy's research agenda (Systematics Agenda 2000, 1994; Cracraft, 2002; Page et al., 2005).

Exploring and documenting species diversity is of paramount importance. This is nowhere more urgent than in regard to hyperdiverse taxa such as insects, fungi and worms, where only a small fraction of living diversity is known. To put the magnitude of this undertaking into perspective, consider that the approximately 925,000 described insect species represent at most 25 per cent of living species (Grimaldi and Engel, 2005). Discovering new species, however, assumes that the tests and corroborations of existing species are being done. As specimens accumulate in collections and new characters are discovered, existing hypotheses of species, monophyletic groups and character distributions become outdated and the information associated with them untrustworthy in the absence of testing. Only by repeated tests can taxonomic names, concepts and classifications remain fully accurate reflections of what we know of patterns in nature.

In the past, species tests were efficiently incorporated into periodic taxonomic revisions or monographs, but such descriptive studies are less frequently completed today and few students are being educated with a firm foundation in such broadly comparative and comprehensive taxonomic knowledge. In large taxa, such revisions may appear only once or twice a century. As a result, the reliability of information associated with the estimated three billion specimens in the world's natural history collections is slowly but steadily eroding. In extreme cases, as many as 70 per cent of specimens are already wrongly identified (Meier and Dikow, 2004).

How much worse must information become before we accept the responsibility to support descriptive taxonomy? This worsening state of information credibility threatens our ability to verify identifications, store and retrieve new observations, precisely communicate findings among biologists, or confidently describe new species as they are found. Further, it increases the likelihood of economic losses resulting directly from taxonomic mistakes (e.g. Miller and Rossman, 1995). Must we wait until spectacularly costly mistakes are made before we take action to assure the quality and growth of taxonomic information?

It is a question of 'when' rather than 'if' such avoidable mistakes will happen; they are a logical consequence of the neglect of descriptive taxonomy. Museums in particular will have some explaining to do. Why were non-collections-based, university-style research programmes pursued at the expense of taxonomy when the world needed improved understanding of taxonomic diversity? We can avoid such taxonomic train wrecks – or at least the worst of them – by reinvigorating revisions and monographs in a Web-based format with immediate attention to those groups known to include species of potential economic impact.

Such electronic monographs will have many advantages. They will be continuously updatable in order to reflect the most recent and full knowledge of taxa (Scoble, 2004; Wheeler, 2004) by networks of specialists working variously alone and in collaboration. Across this vast range of possibilities one thing seems certain:

> The Internet [is] *the* fast evolving medium for providing access to information currently distributed across the published paper-based literature, in unpublished archives, in curated collections, and, increasingly, in personal or institutional databases.
>
> **(Scoble, 2004, p. 699)**

In spite of the realities of the situation, the existing body of taxonomic knowledge may appear sufficient to some. Because a disproportionate number of biologists have worked in northern latitudes, the species they most commonly encountered could be reliably identified. As biologists extended their interests into the tropics, southern hemisphere and increasingly diverse ecosystems of all descriptions, it became obvious that many species were identifiable only with great difficulty, many others not at all. As biologists now seek to better understand the full diversity and complexity of evolutionary history and ecosystems, it becomes increasingly clear that we know rather little about the tree of life and its millions of unique branches and leaves.

As the impacts of human activities spread across the Earth, new agents of human and animal diseases emerge; new agricultural pests appear and invasive species are spread greater distances with increased frequency. As biological curiosity expands, new and better model organisms are needed; in the absence of fundamental taxonomic knowledge, it is impossible to know where to begin the search for them. As we confront new and increasingly severe environmental, agricultural and medical challenges, we turn, as we always have, to nature for solutions; in the absence of good taxonomy, we increasingly find answers beyond our grasp. It is no longer possible to be content with relatively good but incomplete knowledge of the flora and fauna of Europe and North America. We need knowledge of the world's species in their full diversity; it is time that taxonomy is approached on a global scale to assure the necessary foundation of understanding for the biological sciences of the future. It is critical to recognize that descriptive taxonomy is not a one-time or provincial activity as sometimes portrayed, but rather a part of a planetary scale research programme (Cracraft, 2002).

The 'sell-by date' is rapidly expiring for our hypotheses of many species and putatively monophyletic taxa. Unless taxonomists are supported to test these species with new characters and newly collected specimens, the reliability of existing knowledge will become increasingly suspect and the detection of new species problematic. Unless taxonomists are supported to undertake such tests and to aggressively complete an inventory of Earth's species and clades, biological knowledge and environmental problem-solving options will be unnecessarily limited. The recent observation that some basal groups of dinosaurs had a reptilian-like variability in attained adult size (Gramling, 2005) refuted the common assumption that dinosaur adult size was relatively constant and illustrated the need for a broad knowledge of species and clades. Gramling quotes Thomas Holtz of the University of Maryland, College Park, who observed that 'this emphasizes the importance of tree-based thinking. We have to look at as many branches of the evolutionary tree [as possible] to get as big a picture as possible'.

The progress of taxonomy and development of collections are an investment to assure that students of all taxa have in the future as many pieces of that big picture of phylogeny as possible. This, in turn, can only be done comprehensively and efficiently in proportion to our growth in taxonomic knowledge generally. Without sound taxonomy, we have no idea how to prioritize inventory efforts or collection growth and development or any means of assessing our success in exploring biological diversity.

Taxonomy is misunderstood and maligned by some experimental biologists specifically because it is non-experimental, comparative and historical in nature. The epistemology of taxonomy – how we know what we know about species, clades, homologues and character transformations – differs dramatically from that of experimental biology. The inherent duality of descriptive and hypothesis-driven science embodied in taxonomy adds to the confusion. Scoble (2005) makes a compelling defence of the descriptive aspects of taxonomy on the grounds of its role as a necessary information infrastructure for biology. This alone is more than sufficient reason to make taxonomic research and collection improvements a high priority. Nelson and Platnick (1981) explained in detail the complex, layered hypotheses embedded in taxonomy and the appropriate and necessary distinction between taxonomy and general biology. It is ironic that taxonomy is singled out for neglect based on the descriptive aspects of its work when other descriptive sciences (e.g. many branches of physics, chemistry, astronomy and geology, as well as the Human Genome Project) are appropriately well supported.

Much of the denial of funding in the UK, for example, is justified explicitly on the basis that funding agencies like the National Environmental Research Council are concerned with support for hypothesis-driven science only (see House of Lords, 2002). Like it or not, as E.O. Wilson (2003) has noted:

> Biology is primarily a descriptive science...for it is defined uniquely by the particularity of its elements. Each species is a small universe in itself, from its genetic code to its anatomy, behaviour, life cycle and environmental role, a self-perpetuating system created during an almost unimaginably complicated evolutionary history.

Our failure to seek out and describe Earth's species incalculably diminishes our understanding of life, for every species yields a substantial and unique insight.

> In a purely technical sense, each species of higher organism – beetle, moss, and so forth – is richer in information than a Caravaggio painting, a Mozart symphony, or any other great work of art. Consider the house mouse, *Mus musculus*. Each of its cells contains four strings of DNA, each of which comprises about a billion nucleotide pairs organized into 100,000 structural genes. If stretched out fully, the DNA would be roughly 1 meter long. But this molecular is invisible to the naked eye because it is only 20 angstrom units in diameter. If we magnified it until its width equalled that of wrapping string, the fully extended molecular would be 600 miles long. As we travelled along its length, we would encounter some 20 nucleotide pairs of 'letters' of genetic code per inch. The full information contained therein, if translated into ordinary-sized letters of printed text, would just about fill all fifteen editions of the *Encyclopaedia Britannica* published since 1768.
>
> **(Wilson, 1985, p. 22)**

Having recognized that we must address our overwhelming ignorance about most of the unique components of evolutionary and ecosystem complexity, we sense that we are in danger of accomplishing too little, too late. Rather than learning obvious lessons from 250 years of successful taxonomy and seeking tools to further accelerate and build upon good practice, we now seriously entertain expedient substitutes for good taxonomy (e.g. Tautz et al., 2003). Second-rate alternatives are justified, by non-taxonomists, on the basis of being fast, easy, new or just obviating the need to learn about the species we seek to discover and their morphology. Where else in science would we entertain the proposition that the avoidance of learning is a virtue?

We are asked to believe that our current ignorance of Earth's species is due to inherent shortcomings or weaknesses in taxonomy rather than the continuing wholesale neglect of taxonomy and our inaction in adapting new technologies to support taxonomic work. We are asked to sacrifice willingly the remarkable theoretical advances in taxonomy (from Hennig, 1966, to the present) and related fields of evolutionary biology (since the 'new synthesis' in the 1930s) in the name of expediency; to abandon comparative morphology, the most intellectually rich and rewarding aspect of taxonomy; to indulge arrogant molecular workers for whom there is no distinction between science and mere technology and for whom the costly lessons of phenetics have been forgotten (Lipscomb et al., 2003; Prendini, 2005; Wheeler, 2005).

For what other scientific challenge do we shy away from hard work, choosing to sacrifice excellence and depth of understanding for mere ease? We are asked to avoid the heavy lifting of good taxonomy and the hard work of educating a new generation of taxon experts and to settle for an inferior approach that replaces testable species concepts with non-scientific conveniences simply because it is cheaper and faster to do so. There are times when excellence is worth paying for. With one chance to get the exploration of Earth's species right, this is clearly such a time. The reality is that with a little ingenuity and courage, we can have both speed and excellence, both molecular and morphological, both good taxonomy and experimental biology. It is possible to retain taxonomic theory and practice at the highest levels of excellence *and* to accelerate the speed of doing taxonomy by orders of magnitude. Those who would deny this simple reality seemingly do so from some motive other than advancing knowledge.

I emphasize descriptive taxonomy because of its comparative neglect and the fact that its theoretical strengths are often overlooked. Completion of an inventory of species is of the utmost importance, but this cannot be done at the expense of phylogenetics. Bertalanffy (1932) emphasized that taxonomy stands at the beginning of biological understanding because 'its goal is to produce as complete and accurate a catalog as possible of plant and animal species'. To this Hennig (1966) responded, 'Such a systematics would not be a science' (p. 7). Taxonomy aims to produce a complete catalogue, but also much more. A catalogue plus a phylogenetic context (in the form of a classification and names) for species goes far beyond a mere inventory, much as the periodic table in chemistry serves as more than a mere list of the elements.

While all comparative data should be considered, all data are not equal. Some data sources provide unique or additional information. For example, fossils provide us with minimum dates of origin for particular characters, species and clades; developmental biology helps us to understand gene interactions, epigenetic factors and the pathways between genotype and phenotype; and morphology provides us with the polygenic complex characters that are the objects of natural selection and ultimately a primary source of our fascination with biological diversity and motivation to study evolution.

Face it: If all species looked more or less the same, we would soon understandably tire of naming them and trying to figure out their phylogeny. Phylogenies are fascinating precisely because they help us interpret and understand complex character transformations and geographical distributions. That character analysis has been neglected for two decades, while disproportionate emphasis placed on tree construction and choices among equal-length trees has skewed our view. In this narrow context, characters are faceless data points in matrices and the historical information content in complex characters is underestimated. Framed in a phylogenetic context, however, complex characters beg to be explained by evolutionary processes; this was, after all, the sequence of events that led to evolutionary theory in the first place (Nelson and Platnick, 1981).

The current dominance of molecular techniques over comparative morphology, palaeontology and ontogeny results more from a preference for new technology and trends in funding than any defensible theoretical or practical preference. To paraphrase a comment made by E.O. Wilson (AAAS Meeting, Washington, 2005), molecular studies do not receive more funding because the data they produce are better; rather, the data are perceived by some to be better because they receive greater funding. We sell taxonomy short when we neglect potential data sources. No responsible systematic biologist would advocate the neglect of comparative studies from any credible source of data that could contribute to integrated taxonomy (Will et al., 2005).

There is no historical information contained in an amino acid. Instances of guanine are identical anywhere and any time in the universe. An amino acid, therefore, contributes to phylogenetic information only in respect to its position relative to other amino acids in a sequence. Some suppose molecular data to be more objective because amino acids may simply be read. In other words, we can supposedly simply observe the sequence of amino acids directly and thereby avoid what is perceived to be an imprecise and subjective business of hypothesizing homologies among complex characters such as morphology. This argument is enticing but false. Such claims of unfettered empiricism are also not new.

> If our knowledge is limited to those things we can (supposedly) observe directly, the task of studying the history of life is indeed fraught with insurmountable difficulties, since we can hardly observe directly the past history of present-day organisms. But such a philosophical view characterizes (if anything) technology, not science; it denies to science precisely those processes that are most characteristic of it: the proposal and testing of hypotheses.
>
> **(Nelson and Platnick, 1981, p. 7)**

The interpretation of molecular or any other comparative data as indicative of phylogeny – that is, of history – is necessarily hypothetical and only as reliable as it is testable and tested and corroborated. There is every scientific reason to want additional sources of data as (relatively independent) tests of phylogenetic patterns. There is special value in interpreting complex characters that carry with them the potential for greater historical information and which can be teased apart and critically re-examined at successive levels of complexity. In the case of morphology, when reciprocal illumination questions the validity of a homology statement following upon a cladistic analysis, we have many recourses: finer level anatomical observations, assessment of developmental (ontogenetic) patterns, study of roles of genes coding for expression of the character and, ultimately, DNA sequencing of those genes. Why abandon such rich indicia of history for an 'objectivity' that is largely sterile?

A GUTENBERG MOMENT FOR MORPHOLOGY

For centuries, morphologists have attempted to describe and communicate complex three-dimensional structural details through cumbersome jargon and costly and sometimes ineffective drawings and photographs. Digital tools offer a widening array of innovative tools that can revolutionize how morphological data are gathered, analyzed and communicated. These digital capabilities will change, too, the relationship between taxonomists and users of taxonomic information, who will be increasingly empowered to make identifications and openly access information about detailed attributes shared by species and higher taxa (Wilson, 2003; Raven, 2004). The digital medium has begun a revolution in comparative morphology, storing and communicating complex visual knowledge more effectively and widely than moveable type spread the written word.

Several recent authors draw attention to the need to reinvest effort in the analysis of individual morphological characters. For example, Williams, Wägele and Rieppel (all in Williams and Forey, 2004) each made such a recommendation, from quite different points of view. The failure to pay attention to individual characters diminishes our understanding of evolutionary history and adds to the perception that a technological quick-fix may be a viable substitute for morphological knowledge.

As impressive as recent leaps in digital instrumentation, bioinformatics, digital communications and cyberinfrastructure have been, the most impressive advances are yet to come (Atkins et al., 2003). The taxonomic community has begun to position itself to take advantage of the next generation of cyberinfrastructure through efforts such as LINNE (Page et al., 2005) and EDIT (Scoble, 2004) and envision ways forward (e.g. Wheeler and Valdecasas, 2005). Improved cladistic analysis algorithms and computers, in addition to new sources of data such as DNA sequences, have enabled cladistic analyses to progress much faster than just a few years ago. Digital innovations now make it possible to envision a similar acceleration of morphology-based taxonomic work.

Most of the technical requirements for this futuristic world of descriptive taxonomy exist already but have not been made available or specially modified to meet the needs of taxonomy.

This digital divide between the present and the future of taxonomic research must be closed as rapidly as possible. In so doing, we will also address the other, better known digital divide. By making specimens, data, information, tools and knowledge openly accessible in digital form, no one benefits more immediately or profoundly than those historically denied access to the great museums and libraries of the world. Children in isolated rural or inner-city schools will have newfound access to what we know; scientists working in biodiversity-rich developing nations will have the specimens and knowledge extracted from their nations over several centuries at their fingertips; access for all researchers will open from data-bases to digital libraries and even to specimens themselves.

Large comprehensive collections will always be the most important and useful research tools for taxonomy. They facilitate broadly comparative studies and enable verification of species identifications. That said, we are fortunate that we can now begin to envision 'virtual' collections brought together in cyberspace rather than physical space. Thus, our efforts to expand and enrich physical collections should continue unabated. At the same time, we must take full advantage of digital tools that make it possible to synthesize collections from dozens of locations around the world and to virtually repatriate species and information.

The Global Biodiversity Information Facility (GBIF) in Copenhagen has made strides in building a portal through which the information contained in collections can be made easily accessible. Studies are emerging that positively demonstrate how it is possible to collate information from a large number of distributed collections to complete analyses that would have been practically impossible before (e.g. Soberon and Peterson, 2004).

Digital tools are key to reversing the decline of taxonomy (e.g. House of Lords, 2002; Page et al., 2005). As these tools reform taxonomy into an efficient and productive modern enterprise and as taxonomic knowledge empowers scientists and society to understand the natural world, new sources of support will be found. Taxonomy has declined so long that even its strongest supporters are dispirited; the tipping point seems to be impossibly distant to many. But with the digital revolution, this is not so; it is closer than most think. Taxonomy can literally transform overnight. After teaching taxonomy for 24 years at Cornell University, I know for a fact that every generation produces a significant number of bright young scientists who fall in love with groups of animals just as countless generations have before. A capable, willing, enthusiastic workforce can be assembled rapidly if the fundamental needs of taxonomy are met.

Most people shy from the notion of completing an inventory of Earth's species, thinking it an impossible task, but this is not so. There are two phases to such a planetary taxonomic agenda. First is a concerted push to fill the greatest voids in our ignorance of species; this is the initial mapping of the biosphere and the composition of the first edition of the encyclopaedia of life advocated by Wilson (2003). The returns on such an investment would be of incalculable immediate and enduring value to human and environmental welfare. The second phase is maintaining an ongoing steady state of taxonomic research proportionate to the need to test and improve our species and classifications. Discovering and describing species is not a one-time activity as some ecologists have naively suggested. Approaching taxonomic revisions as teams of researchers and institutions can speed the work as demonstrated by the US National Science Foundation's Planetary Biodiversity Inventory (PBI) projects. Equipping such teams with state-of-the-art digital tools can further increase productivity by orders of magnitude.

DIGITAL TOOLS AND TAXONOMIC EMPOWERMENT

Taxonomy's digital revolution will directly benefit far more people and communities than taxonomists themselves. While the core of this revolution should address the infrastructure needs of a transformed descriptive taxonomy, it is equally important that the fruits of a highly productive taxonomic community are openly available to all who can benefit from them. This includes especially applied taxonomy; making accurate identifications is the necessary first step for anyone seeking to study biological diversity. For hyperdiverse taxa such as fungi and insects, such identifications have often required access to synoptic collections or the time of a specialist. If we are able to survive the absurd current proposal for DNA barcoding (Wheeler, 2005; Will et al., 2005) and accept that DNA evidence is merely an (admittedly very useful) additional tool for good taxonomy but not a surrogate for it, progress will be rapid.

On the digital front, several classes of new identification tools can empower anyone to make identifications. Several well-illustrated interactive online keys have appeared already for a range of taxa and the software for such utilities is being rapidly improved. It is easy to forecast handheld devices linked via satellite to powerful libraries of taxonomic information and images. A single such device would enable the nature enthusiast to identify whatever plant or animal she encountered in the woods, a farmer to determine whether an unfamiliar insect is a pest and an ecologist to determine what species he has unexpectedly encountered. Image-recognition systems are also being perfected that will permit a digital photo of an unknown species to be submitted to a data-base of images and intelligent software to return an identification along with a probability of its accuracy; see chapters in this volume by MacLeod et al. and Russell et al., for example. As these and the other chapters in this volume attest, transformative digital tools are no longer science fiction but reality.

It should be obvious that the more such tools we have for applied taxonomy, the better. Different users, for a variety of reasons, may need or prefer one or another tool for making identifications. Empowering anyone interested in exploring the world around them to make accurate identifications will have profound impacts on taxonomy. Charting the species of the biosphere is an enormous challenge that is well suited to the contributions of able and energetic amateurs as well as those of professionals. Serious amateurs deserve the resources they need to educate themselves and to sustain their interest by not becoming frustrated over an inability to recognize what they have collected.

Enabling people to use taxonomic information and knowledge will result in a wider appreciation and support for taxonomy and in benefits to science and society that we cannot yet imagine. Releasing taxonomists from providing all but the most demanding identification services will liberate precious expertise to inventory, describe and classify Earth's species; to test known species; and to improve and expand digital resources. This is the optimal division of labour: taxonomists generating knowledge and validating information, organizations and software like GBIF providing ready access and one-stop shopping, and clever users having access to just the information and identification tools they need to pursue their interests. Realizing the potential of this digital revolution will change the reputation of taxonomy overnight from an impediment to an empowerment.

Armed with a new research infrastructure, it is not too ambitious to aim to discover and describe most, say 80 per cent, of Earth's species within an intense period of planetary exploration, say 50 years. Such a 50-year inventory would not be an end, but a beginning – an effort to triage evidence of as many species, clades and components of ecosystems as possible for the benefit of scientists, including taxonomists, of tomorrow. Even if all species could be documented, an unrealistic but laudable goal, the need to sample and collect; reanalyze; apply new theories, methods and tools; and test and refine taxonomic concepts and Linnaean names and classifications would continue indefinitely.

Like any science, taxonomy is more a process than a product. Every answer raises new questions. Every new specimen or character presents a new opportunity to learn. Continuing the intellectual journey set upon by Linnaeus will have countless scientific, aesthetic and practical benefits. Beyond what improved taxonomic knowledge will do to prepare us to face the challenges of rapidly changing ecosystems, the specimens and information preserved will be an endless source of amazement, inspiration and instruction to future generations. Considering the endless debates over dinosaurs and human ancestors, just imagine what science will do with documented evidence for millions of species! Further, we have only begun to explore the interfaces between taxonomy and ecosystem science, genomics, developmental biology and comparative ethology; an inventory of Earth's species only lays the foundations for such interdisciplinary studies.

I imagine that the current arguments to redirect funding into taxonomy would not be necessary had the Mars 'rover' found complex life forms during its recent romp over the surface of that planet, along with evidence that 25 per cent of them would soon become extinct. It is fair to guess that we as a society would have enthusiastically appropriated billions to do a rapid and thorough inventory. And that would have been the right decision. The fact that millions of life forms new to our solar system happen to live on the third planet instead of the fourth should not dissuade us from pressing ahead with our exploration and documentation of them. What they can teach us about evolution and ecosystems is no less interesting; the solutions they might provide to countless environmental and practical problems are no less valuable.

CONCLUSION: TAXONOMY'S FINEST HOUR?

Reasserting taxonomy in time to bequeath a legacy of specimens and information about Earth's species, clades and ecosystems to posterity is a tall order. Although no generation would choose to face such a challenge, we have the means, opportunity and methods to succeed. We have opportunity, as the last generation with access to many of Earth's species and clades (Wheeler, 2004); the means through emerging digital tools and cyberinfrastructure (Page et al., 2005); and the motive – in the face of the biodiversity crisis – to reconceive taxonomy as a digitally enabled 'big' science and to assure that its theoretical advances and excellent practices are not sacrificed on the altars of popularity and expediency. We do not need to compromise our standards or abandon the rigorous taxonomic practices we have worked so hard to gain in the past 250 years in order to realize an acceleration of taxonomy. To the contrary, we can utilize existing strengths of taxonomy and its collections to make remarkably fast progress.

If taxonomists come together and reinvent their science to take full advantage of the information revolution, digital tools and instruments and cyberinfrastructure, it is possible for taxonomy to enjoy its finest hour and to create a legacy of knowledge and of specimens that will forever expand the horizons of science and of the human experience. Imagine how intellectually impoverished our lives would be had there been no fossilization of life forms from earlier geological times. The comparatively few species preserved in the fossil record have educated and inspired and continue to educate and inspire us in too many ways to list. Specimens that we collect, describe, classify, and preserve today will be nothing short of perfectly preserved fossils for future generations to study. Given appropriate associated data, these specimens will permit future generations to continue to probe the complexities of ecosystems, the changes of biodiversity through time and space, the transformation of characters that is the wonder of evolution, and the details of the unparalleled three-billion year history of life on Earth.

What would it be worth to discover a well-preserved collection of a large proportion of the species of the Jurassic including dinosaurs and early ancestors of birds and mammals? We have the capacity to bequeath such comprehensive collections of Neocene animal and plant groups to future taxonomists, evolutionary biologists and environmental scientists at a remarkably small expenditure. Imagine how such collections will be valued and studied. It will cost us little to do this great and selfless thing that will never be repeated in the future at any price. If we succeed, future generations will look back at the twenty-first century as not merely a time of great environmental upheaval and species decimation but also as the time of taxonomy's finest hour. In our age of cynical self-interest, it will not be easy to bypass easily funded pop science in order to pursue the hard work of exploring a changing planet's species. The clarity of purpose and courage of conviction needed to succeed are rarely seen in today's scientific institutions.

We should take courage from the impressive progress that has been and continues to be made. As a community, taxonomists and institutions engaged in taxonomy should resolve over the period of a decade to transform the infrastructure of taxonomy, including access to existing resources (literature, data in collections, etc.) – and especially development of cyberinfrastructure (hardware and software) for taxonomic research and in support of the full range of taxonomic activities from

TABLE 2.1
A Comparison of 'Grand Challenge' Questions of Taxonomy

Cracraft (2002)	Page et al. (2005)
1. What is a species?	1. What are Earth's species, and how do they vary?
2. How many species are there?	2. How are species distributed in geographical and ecological space?
3. What is the tree of life?	3. What is the history of life on Earth, and how are species interrelated?
4. What has been the history of character transformation?	4. How has biological diversity changed through space and time?
5. Where are Earth's species distributed?	5. What is the history of character transformations?
6. How have species distributions changed over time?	6. What factors lead to speciation, dispersal and extinction?
7. How is phylogenetic history predictive?	

Sources: Cracraft, J., 2002, *Annals of the Missouri Botanical Garden,* 89: 127–144; Page, L.M., 2005, LINNE: Legacy infrastructure network for natural environments. Champaign, Illinois, Natural History Survey.

field work to e-publication – and including basic descriptive taxonomy, phylogenetics and applied taxonomy. Users will appreciate taxonomy more when they are empowered to make species identifications and effortlessly access information about all species on demand. Taxonomists will, in turn, generate more new knowledge, test more existing hypotheses, and finally get on with the grand challenges of taxonomy (Table 2.1) that have been sidelined by fashionable alternatives.

The chapters in this book reveal innovative, promising directions in the imaginative development and use of digital tools and algorithms to aid in identification work and in the documentation and analysis of variation. This is an essential ingredient in a vibrant, successful future for taxonomy and should give heart to those who recognize the great test facing us. It is an important part of an increasingly repeated call for a new infrastructure for descriptive taxonomy (e.g. Wheeler et al., 2004). Will taxonomy rise to meet its greatest challenge? Can the appropriate priorities for science funding be adopted before it is too late to document biological diversity at its various levels of complexity? Can taxonomists manage to set their differences aside, establish priorities and speak with one effective political voice in time? Do we have the vision and courage to harness the incredible power in emerging digital tools to revitalize taxonomy? Are there museums, botanical gardens and universities with the confidence in taxonomy to be leaders in this digital infrastructure revolution? I hope so.

ACKNOWLEDGEMENTS

I thank my colleagues Dr Norman MacLeod (The Natural History Museum, London), for his leadership in organizing this important meeting, envisioning exciting and novel approaches to the analysis and visualization of morphological data, and for his kind invitation to contribute to the symposium and book; Dr Michael Schauff (USDA, Beltsville), Dr Thierry Bourgoin (Museum National d'Histoire Naturelle, Paris), and Mr Roy Larimer (Microptics-USA, Inc.) for their dedication to making the dream of a network of remotely operable digital microscopes for entomology a soon-to-be reality; and Dr Antonio Valdecasas (Museum Nacional de Ciencias Naturales, Madrid) for stimulating discussion and encouragement.

REFERENCES

Atkins, D.E., Droegemeier, K.K., Feldman, S.I., Garcia-Molina, H., Klein, M.L., Messerschmitt, D.G., Messina, P., Ostriker, J.P. and Wright, M.H. (2003) Revolutionizing science and engineering through cyberinfrastructure. Report of the National Science Foundation Blue-Ribbon Advisory Panel on Cyberinfrastructure. US National Science Foundation, Arlington, VA.

Blaxter, M.L. (2004) The promise of a DNA taxonomy. *Philosophical Transactions of the Royal Society of London, Series B,* 359: 669–679.

Calman, W.T. (1930) *The Taxonomic Outlook in Zoology,* British Association for the Advancement of Science (Section D: Zoology), Bristol, England, pp. 1–10.

Cracraft, J. (2002) The seven great questions of systematic biology: An essential foundation for conservation and the sustainable use of biodiversity. *Annals of the Missouri Botanical Garden,* 89: 127–144.

Edwards, J. (2003) Research and societal benefits of the Global Biodiversity Information Facility. *BioScience,* 54: 485–486.

Gaston, K.J. and O'Neill, M.A. (2004) Automated species identification: Why not? *Philosophical Transactions of the Royal Society of London, Series B,* 359: 655–667.

Gramling, C. (2005) How fast does your dinosaur grow? *Science,* 310: 1751.

Hennig, W. (1966) *Phylogenetic Systematics.* Urbana, University of Illinois Press.

House of Lords. (2002) *What on Earth? The Threat to the Science Underpinning Conservation,* Select Committee on Science and Technology, Third Report. HL Paper 118(i). HMSO, London.

Millennium Ecosystem Assessment. (2005) *Ecosystems and Human Well-being: Biodiversity Synthesis,* World Resources Institute, Washington, DC.

Miller, D.R. and Rossman, A.Y. (1995) Systematics, biodiversity, and agriculture. *BioScience,* 45: 680–686.

Nelson, G. and Platnick. N. (1981) *Systematics and Biogeography: Cladistics and Vicariance,* New York, Columbia University Press.

Page, L.M., Bart, H.L., Jr., Beaman, R., Bohs, L., Deck, L.T., Funk, V.A., Lipscomb, D., Mares, M.A., Prather, L.A., Stevenson, J., Wheeler, Q.D., Woolley, J.B. and Stevenson, D.W. (2005) LINNE: Legacy infrastructure network for natural environments. Champaign, Illinois, Natural History Survey.

Prendini, L. (2005) Comment on 'Identifying spiders through DNA barcodes'. *Canadian Journal of Zoology,* 83: 498–504.

Raven, P.H. (2004) Taxonomy: Where are we now? *Philosophical Transactions of the Royal Society of London, Series B,* 359: 729–730.

Scoble, M. (2004) Unitary or unified taxonomy? *Philosophical Transactions of the Royal Society of London, Series B,* 359: 699–710.

Simpson, G.G. (1945) The principles of classification and a classification of mammals. *Bulletin of the American Museum of Natural History,* 85: 1–350.

Simpson, G.G. (1961) *Principles of Animal Taxonomy,* New York, Columbia University Press.

Soberon, J. and Peterson, A.T. (2004) Biodiversity informatics: Managing and applying primary biodiversity data. *Philosophical Transactions of the Royal Society of London, Series B,* 359: 689–698.

Tautz, D., Arctander, P., Minelli, A., Thomas, R.H. and Vogler, A.P. (2003) A plea for DNA taxonomy. *Trends in Ecology and Evolution,* 18: 70–74.

Timofeeff-Ressovsky, N.W. (1940) Mutations and geographical variation. In (ed J. Huxley), *The New Systematics,* Oxford University Press, London, pp. 73–136.

Wheeler, Q.D. (2004) Taxonomic triage and the poverty of phylogeny. *Philosophical Transactions of the Royal Society of London, Series B,* 359: 571–583.

Wheeler, Q.D. (2005) Losing the plot: 'DNA barcoding' and taxonomy. *Cladistics,* 21: 405–407.

Wheeler, Q.D., Raven, P.H. and Wilson, E.O. (2004) Taxonomy: impediment or expedient? *Science,* 303: 285.

Wheeler, Q.D. and Valdecasas, A.G. (2005) Ten challenges to transform taxonomy. *Graellsia,* 61: 151–160.

Will, K.P., Mishler, B.D. and Wheeler, Q.D. (2005) The perils of DNA bar-coding and the need for integrative taxonomy. *Systematic Biology.*

Wilson, E.O. (1985) The biological diversity crisis: A challenge to science. *Issues in Science and Technology,* Fall, 1985: 29–29.

Wilson, E.O. (1992) *The Diversity of Life,* New York, Norton.

Wilson, E.O. (2003) The encyclopedia of life. *Trends in Ecology and Evolution,* 18: 77–80.

3 Natural Object Categorization: Man versus Machine

Philip F. Culverhouse

CONTENTS

Introduction 25
Visual Recognition of Objects 26
Psychophysical Issues 27
How Good Are People at Sorting? 28
The Process of Recognition: Context, Categorization, Biases 31
Context 32
Categorization 32
Biases 33
Human Performance 34
Machine Performance 36
An Example: HAB Buoy 37
Machine-Vision Issues 38
Expert-Opinion Issues 38
A Consensus Experiment 39
Speed of Recognition 41
Conclusions 41
The Need 42
Acknowledgements 43
References 43

INTRODUCTION

This chapter is concerned with natural object categorization. Whether done by man or machine, the outcome is a labelled object. The object will have been presented to the categorizer as a drawing or as a natural specimen or an image of a specimen. The categorizer will have been asked to apply prior knowledge to the analysis of visible features that are in some way characteristic of the object class. Systematics *per se* is outside the scope of this discussion, but much scientific work in many fields is concerned with labelling specimens. Much remains to be done using manual methods, but recently, advanced pattern-recognition methods have been successful in automating some categorization tasks.

In this chapter two questions are posed: 'How good are experts at categorization?' and 'How do experts compare with current machine methods?' To answer these questions, factors that affect human performance must be understood. These may then be compared to machine performance to give a reasonably balanced view of the potential of people and machines at natural object categorization. Surprisingly, people can be less than perfect at giving class labels to objects if there are lots of objects to label.

Visual Recognition of Objects

Visual scene analysis and object recognition are abilities shared by many animal vision systems. It is clearly an important survival characteristic to be able to differentiate between food and foe, obstacle and path, etc. What neural processes are involved remains unclear, although progress has been made through investigations of insect, fish, amphibian, bird and mammalian (including human) vision systems (e.g. Cerella, 1979, 1980; Ewert, 1984; Proffitt, 1993; Srinivasan, 1998; Hubel and Wiesel, 2004). Motion provides strong cues for scene segmentation and object recognition (Bertenthal, 1992). Object colour and shape are also important (Gibson, 1969; Jacobs, 1981).

There are two closely related, but different, psychophysical models of object recognition currently being debated: recognition by components (RBC) (Biederman, 1987) and a chorus of prototypes (Edelman, 1999). Both models have supporting and conflicting evidence. However, there are strong limitations on the scope of objects that RBC can represent and recognize because mechanisms for dividing objects into parts and representing parts are not general purpose and the RBC model has difficulty in representing closely related objects that we find no difficulty in discriminating – for example, different makes of car or two different golden retrievers (Tarr and Bültoff, 1995).

Current computational models used by the machine vision community offer a wide variety of possible solutions to the object recognition problem, depending on the number of constraints applied to limit the possible scene interpretations. For example, highly constrained, production-line inspection, where a known set of objects is imaged and analysed for features, allows prior two- and three-dimensional models to be compared with features derived from live-image cameras. Under these constraints, recognition is reduced to matching a sufficient set of features to reveal object defects. Fine detail can then be compared to confirm the object category: a good part, malformed, and so on (Lowe, 1987). As the recognition problem becomes more open ended, it also becomes increasingly ill posed (Bertero et al., 1998). At some point, any number of scene interpretations is possible, leading to a combinatorial explosion of possibilities.

An emerging view of perception as statistical decision making, originated by Gibson (1957) as the ecological theory of perception, has led to much interest in the statistics of natural scenes (Hancock et al., 1992; Ruderman, 1994) as well as the supposition that a Bayesian approach can constrain the search for scene interpretations (Duda and Hart, 1973). In most cases, contextual knowledge is crucial for successful visual object recognition.

The outcome is that, for most people, scene segmentation into objects of interest and background clutter (also known as figure–ground separation) happens quickly and mostly within 1 or 2 seconds of scene presentation. Sometimes the segmentation and object recognition task is easy even in static images where colour and high-contrast objects simplify figure–ground separation and are aided by contextual knowledge. Figure 3.1 highlights this with an image of marine fish and coral, where a sense of up and down suggests sky and ground (correlating with water and coral ground) and a prior knowledge of fish will provide most readers with a sense of understanding of this scene. However, even contextual knowledge can be insufficient, especially if the image is information limited. This can make the figure–ground separation and subsequent recognition difficult. Figure 3.2 illustrates this with a low-contrast, high-clutter tree scene. The photograph was taken with the sun low in the sky. Yet even this scene yields information to the observer, with a suggestion that the central tree has ivy growing up its trunk. But, how many trees and species are in the scene?

Humans normally interpret their visual world as if the sun is in the sky and the ground is beneath their feet. Unusual occlusion or unusual view angles delay scene interpretation and understanding (Johnson and Olshausen, 2005; see Figure 3.3). Monochromatic images are harder to interpret than colour images, presumably as there is less information in them to assist segmentation. Figure 3.4 shows this for the same image as in Figure 3.3.

FIGURE 3.1 (Color Figure 3.1 follows page 110.) Sometimes, scene segmentation and object recognition are easy.

FIGURE 3.2 (Color Figure 3.2 follows page 110.) Sometimes, low contrast and high clutter make object recognition hard.

PSYCHOPHYSICAL ISSUES

There is a wealth of experimental evidence gained from psychophysical studies on humans (e.g. Duchowski, 2002) describing the way in which we scan a visual scene. Our eyes are controlled by series of short, fast movements known as saccades. These can be reflex actions derived from a stimulus in the field of view (bottom up) or guided by a conscious (or unconscious) desire to search a particular part of the field of view (top down). Part of the reason for saccadic eye motion is that mammalian eyes do not have even visual acuity across the whole field of view. In fact, the central part of the light-sensitive retina (the fovea) has the highest resolution. So, an entire microscope field of view, for example, can be seen by a series of search-based saccades that direct the eyes towards attractors in the scene. Reflex eye movements tend toward colour and high-contrast areas of the visual field.

Human vision also has a property called selective attention (Broadbent, 1958). This allows us to tend to objects in particular parts of the visual field to the exclusion of others. It has been likened to a spotlight moving in a darkened room, partially illuminating different areas as it is moved, and is known as the 'spotlight of attention' (Treisman, 1982; Crick, 1984). Shifts of attention can be overt or covert (Pashler, 1996).

FIGURE 3.3 (Color Figure 3.3 follows page 110.) Unusual scenes delay recognition.

Related to attention is pop-out. Pop-out occurs against a background of distracters (Treisman, 1985, 1986). The item that pops out, and to which subsequent attention is directed, is variant in some manner from distracters. For example, in a field of horizontal lines a vertical line will pop out. Figure 3.5 demonstrates pop-out for a more complex array of distracters and attractor: oriented seagulls. Pop-out can be context driven (i.e. directed by a prior conscious feature selection) (Reimann et al., 2003), with colour appearing to be the strongest cue.

Unusual occurrence can be ignored if it constitutes a distracter of attention or can be the focus of attention if other features in the scene are judged as background. This is probably the case in Figure 3.6, where the eye tends towards the coffee cup as it is the only recognizable object in the scene. Alternate interpretation is possible, where the background is the focus of attention and only after a few seconds does the viewer sense the coffee cup. Yet, inverting the scene in Figures 3.4, 3.5 and 3.6 reveals, in Figure 3.7, a range of objects and cues such as shadows, allowing depth perception and correctly assigning ground at the bottom of the photograph (Verera et al., 2002).

Prior knowledge is essential for specialist search and identification tasks. Figure 3.8 highlights this. For those who have not seen phytoplankton before, the image can plausibly hold one object, whose contortions appear strange. Those with more experience will see three specimens overlapping.

HOW GOOD ARE PEOPLE AT SORTING?

Experience plays an important part in sorting objects. Try a test yourself. Look at the central image in Figure 3.9 and measure the time it takes you to count the objects listed on the right-hand side. The more familiar you are with a type of scene, the quicker sorting can be. However, trying to be too fast can increase errors as the eye skips whole sections of the field of view. This is probably due to the properties of saccadic eye movement and the spotlight of attention mentioned earlier.

FIGURE 3.4 Monochrome images present reduced information to the viewer.

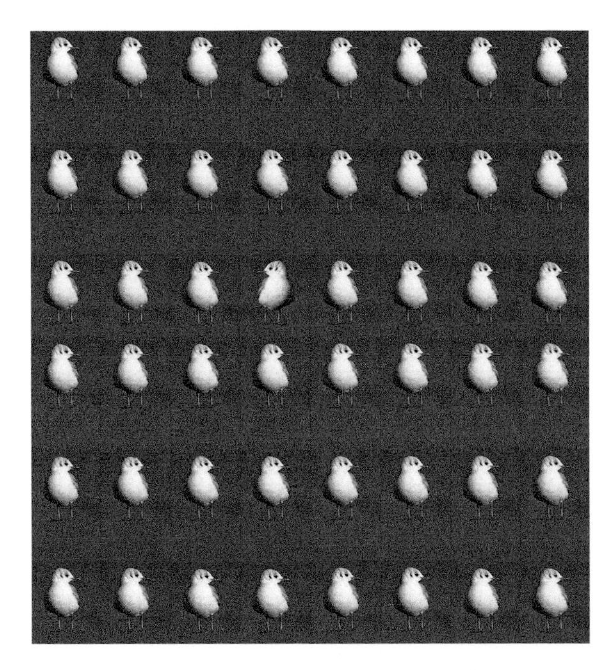

FIGURE 3.5 Pop-out can occur, highlighting a unique or a repetitive feature according to circumstances.

FIGURE 3.6 **(Color Figure 3.6 follows page 110.)** An unusual item can be ignored: the cup is the focus of attention and the background is ignored.

FIGURE 3.7 **(Color Figure 3.7 follows page 110.)** Correct orientation facilitates scene analysis and recognition.

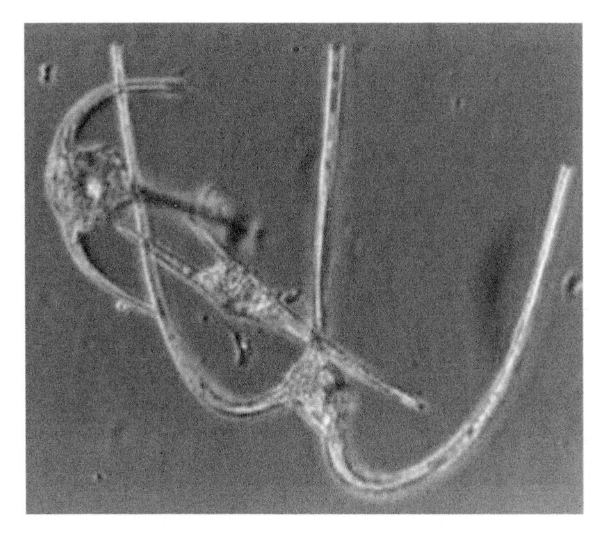

FIGURE 3.8 Is this one object or many?

FIGURE 3.9 Sorting a scene can take a long time.

Pop-out cannot be used reliably in these situations, since it cannot be guaranteed that the specimen features causing it are consistently 'popped out'. A careful serial search of the scene is required, which is both slow and tiring. If there is any pop-out or attentional effect in Figure 3.9, it is the larger objects, which correspond to zooplankton: copepods and decapod larvae.

The Process of Recognition: Context, Categorization, Biases

In general, object recognition can be viewed as the application of constraints to a visual scene. The constraints are context and category. Context can prime search categories and define expectations. Category relies on salient features characteristic of an object class; a sufficient set of features should provide recognition. Biases affect the quality of the judgements made during recognition.

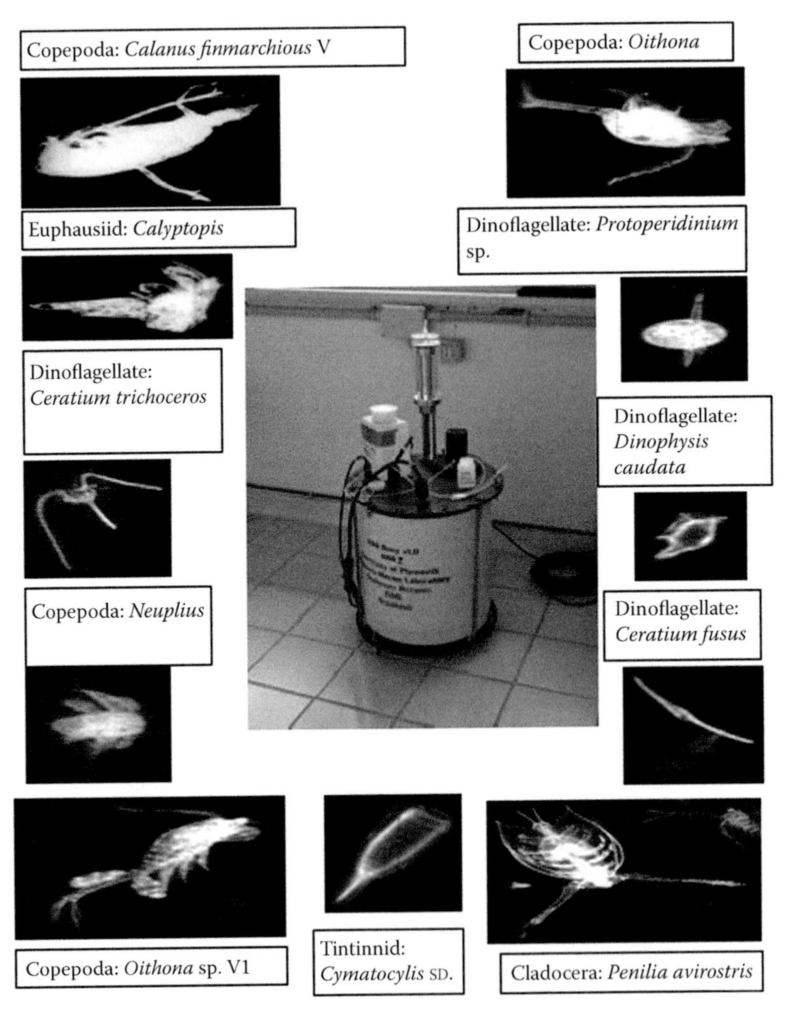

FIGURE 3.10 HAB Buoy. An example image set of natural objects that currently requires expert recognition for labelling. Note: All images were acquired *in flow*.

Context

Context and other prior cues to category speed recognition significantly (Oliva et al., 2003). For example, because a marine scientist will know from prior experience of manually sorting plankton that samples taken from specific areas are likely to include particular species, Atlantic Ocean samples will be expected to have Atlantic Ocean species present. Figure 3.10 indicates the diversity of species that can be found in one sample. Other species may occur in low abundance, but this contextual knowledge allows the expert to predict what specimens are likely to be found. This knowledge primes their visual search of a scene, perhaps by expected species frequency, and hence primes the use of discriminant features known to be associated with those species. This, unfortunately, also constitutes bias (see following).

Categorization

Over the years since Linnaeus developed his key classification methodology, the systematics community has arrived at a stable consensus relating species labels to characteristic features that have been described in the scientific literature for recognized flora and fauna. Recent advances in

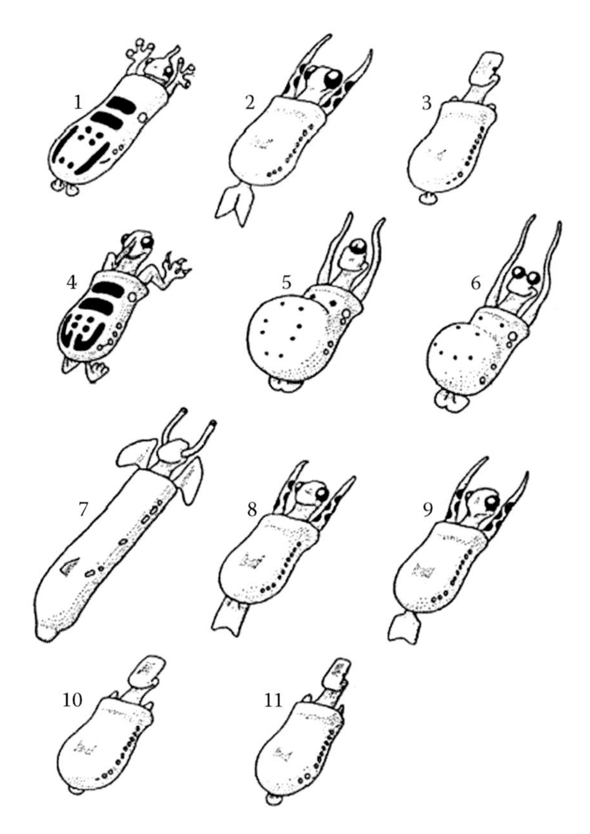

FIGURE 3.11 Example of *Caminacules*. (Reprinted with permission.)

DNA analysis have also led to the use of DNA to support or reclassify species (e.g. Bucklin and Kann, 1991; Bucklin et al., 1996; Hill et al., 2001). New taxa are identified and organized according to this established framework. Expert knowledge is learnt through rules and examples.

Over time taxonomic experts will add to their personal knowledge base through evidence-based reasoning. This allows an expert to economize on the effort required to identify a particular specimen when engaged in routine identification tasks. It is generally acknowledged that taxonomists are either 'groupers' or 'splitters' and will preferentially either add an unknown specimen to an existing group found in the sample or create a new group in which to place that specimen. In a series of particularly revealing studies, Sokal (see 1966, 1974 and 1983) explored how experts gather specimens into groups. He used synthetic creatures he called *Caminalcules* (after Joseph H. Camin, the taxonomist) that possessed both simple and complex features.

Figure 3.11 depicts some examples of these *Caminalcules*. Sokal discovered that experts make up their own rules for grouping. They even do not agree totally on the feature conjunctions that are best employed in the categorization. However, even after we accept that experts differ in the way they group things, they still return similar groupings. This is a remarkable result, yet not error free, since the various groupings did not agree totally. This idiosyncratic rule creation of human expert judgement suggests a potential for long-term bias errors, when groups of experts who use idiosyncratic rules to identify objects do not form a consensus.

Biases

Human performance in identifying and sorting organisms is affected by several psychological factors: (a) the human short-term memory limit of five to nine items, (b) fatigue and boredom, (c) recency effects where a new classification is biased toward those in the set of most recently used labels and (d) positivity bias, where specimen identification is biased by one's expectations of the species likely to be present in the sample (Evans, 1987).

Short-term memories are very volatile. Studies have shown that most people have difficulty remembering even three items after 18 seconds (Peterson and Peterson, 1959). Even more surprising, Marsh et al. (1997) suggest that short-term memory can decay within 2 seconds. The nature of working memory also appears biased to favour gain in certain risk-averse decision-making situations (Toth and Lewis, 2002). It is possible that this bias might affect routine sampling in time-pressured situations. When ambiguous objects are labelled, the bias may reveal itself by favouring a predicted outcome rather than the actual outcome of the sample analysis.

Colquhoun (1972) suggests that fatigue sets in quickly on highly repetitive tasks. In an earlier study (Colquhoun, 1959), he established that ambient noise, high ambient temperature, difficulty of signal detection and loss of sleep decrease performance. In some circumstances, very high levels of mistakes can arise in short periods of time (e.g. less than 30 minutes). He also noted that rest periods can help. Fox (1971) observed that within 30 minutes radar operators could show a 70 per cent drop in efficiency. It seems this is an issue of boredom, rather than fatigue. Visual inspection is particularly problematic, with Megaw (1979) showing that many factors affect accuracy and Welford (1968) describing fatigue as resulting from overloading the operator. Welford also noted that under-loading an operative can be just as damaging through monotony.

Monotony can occur when tasks require little cognitive processing or attention, are highly repetitious, are not complex or have been over-learned. Under these conditions, Welford states that performance may be generally poor, as with overloading, but that the nature of performance degradation is fundamentally different. When the operator is overloaded, actions are confused and judgements tend to be unclear. In contrast, when the operator is under-loaded, attention can drift, signals are missed and performance becomes poor. This is why histopathology laboratories limit an expert's serial inspection of cytology slide material to a maximum of 4 hours per day, with frequent breaks to minimize the effects of fatigue and monotony.

Finally, it is generally acknowledged that expertise takes time to acquire and is domain specific. However, studies into expert performance suggest that the mere number of years of experience is only weakly related to performance, but that deliberate practice (i.e. reinforcement learning) is crucial (Ericsson and Lehmann, 1996). The suggestion that experts may become less effective as they get older is, therefore, theoretically possible. Consider the case of an expert's deliberate practice that includes the idiosyncratic creation of rules (c.f. Sokal's experiments discussed earlier) having a neutral or even a negative effect on discrimination. This will not improve the expert's skill. Experts working alone are probably more susceptible to this condition, where new rules are developed and tested without external reference. This is often the norm in biological science. So, how do experts continue to improve when they often are the only local expert in a particular domain? A solution may be regular intercalibration within a peer group, aiming to achieve group consensus.

HUMAN PERFORMANCE

It is a tacit assumption that a trained, expert taxonomist is an error-free identifier when engaged in identifications of taxa that he or she is expert in. However, the biases and beliefs described previously, together with operational problems (e.g. speed of labelling, length of time performing the task), contribute to errors. A study that compared experts for self- and mutual consistency (Simpson et al., 1991) revealed inconsistencies.

The task was discriminating between field-collected specimens of *Ceratium longipes* and *Ceratium arcticum*, two closely related species of dinoflagellate phytoplankton (see Table 3.1 and Figure 3.12). Taxonomic experts demonstrated 94–99 per cent self-consistency, whereas experienced marine ecologists given 'book' descriptions of the two species and a small training set to hone their skills (this group are referred as 'book experts' later) demonstrated wider variance with 67–83 per cent self-consistency. The experiment required experts to identify, but not tally, the

TABLE 3.1
Expert and Mutual Consensus Performance

Categorization task	Self-consistency within panel	Panel consistency across individuals	Machine performance
Ceratium longipies and *C. arcticum*	N/a	43–95%	99%
Tintinidae (5 spp.)	N/a	91% (six experts)	87%
Fish larvae (3 spp.)	N/a	N/a	70%

specimens in the sample. Combining the labelling results of these two sets of skilled scientists, a panel consistency measure of 43–95 per cent was obtained for all the specimens in the trial. This suggests that consensus can be difficult to obtain in practice. Yet, consensus is the basis of scientific progress. A computer-based analysis of the same data attained a consistency of 99 per cent.

The same team carried out further consistency experiments (Culverhouse et al., 1992, 1994, 1996; Simpson et al., 1992). Results are summarized in Table 3.1. It can be seen that expert consistency within a panel is never 100 per cent, but varies between 83 and 95 per cent, depending on the task. Difficulty is essentially defined by the number of categories the panel was required to consider and keep in mind; it is possible that the short-term memory limit of five to nine items impacted directly

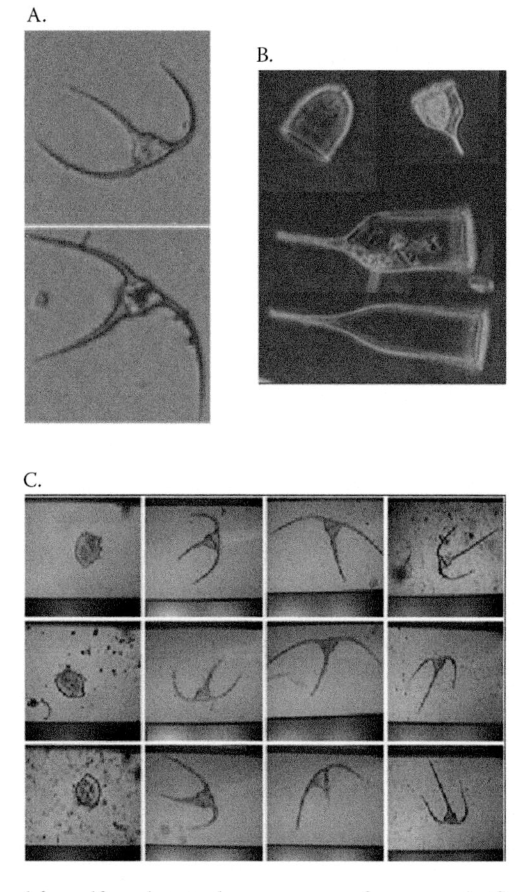

FIGURE 3.12 Plankton used for self- and mutual consensus performance. A. *Ceratium longipes* (upper) and *C. arcticum* (lower). (From R. Williams, PML, UK) B. Tintinidae. (From J. Turner, U. Mass, USA) C. Fish larvae. (From A. Lindley, PML, UK)

on their performance when asked to label 23 phytoplankton species. Machine performance attains similar performances to humans, but uses non-human metrics to carry out the identifications.

In quality-control studies, Kelly (2001) used benthic diatom specimens to compare performance of a trained 'counter' and an 'auditor'. The Trophic Diatom Index counts for 58 UK river samples varied by as much as 30 per cent between counters, with 13 of the 58 counts having more than 10 per cent discrepancy. It seems in these circumstances that the errors are non-linear, with the most error-prone counts obtained from those samples with low degrees of species Bray–Curtis similarity (defined in Gauch, 1982; i.e. most variety within the sample).

MACHINE PERFORMANCE

All current machine-vision categorization methods employ a form of template, or feature, matching during the recognition process. Computer-based three-dimensional models can generate two-dimensional templates at different object poses to match through correlation or convolution with an object in a two-dimensional visual scene (Goad, 1986; Lowe, 1987). However, these techniques suffer from reduced performance when the object is occluded or noisy. Also, models and profiles have to be created for each object or object pose, and this is a significant drawback for natural object recognition where objects have natural morphological variation arising from genetic or environmental factors, or from structural damage due to predation or accident.

More recently, it has become common to use a plethora of machine-extracted features to correlate with object class. A training set of objects is used to establish the pool of features and their prior distributions. Statistical and other pattern-classification methods are then used to cluster the feature occurrences in test specimens and hence derive an identification. Thus, in the Automatic Diatom Identification and Classification (ADIAC) system (Du Buf and Bayer, 2002), a large set of morphological measurements (e.g. specimen length, width, aspect ratio) is made of each specimen placed under the microscope. Some of these measurements are similar to those made by taxonomic experts and similar to ZooSCAN (Grosjean et al., 2004), used for zooplankton recognition and counting where a forest of classifiers is used. DiCANN (Dinoflagellate Identification by Artificial Neural Network) (Culverhouse et al., 1996; Toth and Culverhouse, 1998; Culverhouse et al., 2004), a tool for dinoflagellate phytoplankton species recognition, analyses low-resolution shape, texture and size characteristics, but uses the machine to discover how these features correlate with object classes through support vector machine (SVM) clustering.

Shadow Image Particle Profiling Evaluation Recorder (SIPPER) (Samson et al., 2001; Remsen et al., 2004) has employed SVM categorizers fed from shape moments (Hu, 1962), granulometric and domain-specific features to recognize five classes of plankton. The Video Plankton Recorder (VPR), developed at Woods Hole Oceanographic Institute, has been used as a test bed for a number of analysis protocols (Tang et al., 1998; Hu and David, 2005). The most recent VPR system demonstrated recognition through texture analysis and categorization. Table 3.2 summarizes recent performance results of these systems.

A distinct advantage of automation is the speed of operation. Published data show that Zooscan, SIPPER and VPR can process many thousands of objects per hour (and, being automata, without fatigue too). This is clearly an order of magnitude improvement in identification rate compared to taxonomic experts, though these machines can only perform generic-level discriminations at present.

All these machines have been tested with field-collected data, either *in situ* or in the laboratory. However, they all suffer from the basic problem that they do not work like expert taxonomists or ecologists. This means that they can only operate with data that fit their training set. Extrapolations from the training set are undefined, although having a training bin labelled 'reject' can help reduce false positive identifications of detritus and so on. New categories cannot always be added automatically. It is important to recognize that a correspondence between machine and manual results has to be established to ensure that the machines operate with the desired clustering characteristics. This may be because the machines are not using the normal taxonomic features described in the

TABLE 3.2
Performance Summary for Marine Plankton Recognition Systems

Systems	Group size	Correct identifications
ADIAC	37 taxa	75–90%
Zooscan	29 groups	75–85%
SIPPER	5 groups	75–90%
DICANN	3–23 species	70–87%
VPR	7 groups	72%

TABLE 3.3
Summary of Arbitrary Categorization of Marine Plankton Data

	Gliders	Blobs	Triangles	Cups
Gliders	25	1	3	8
Blobs	3	21	5	3
Triangles	0	0	1	0
Cups	0	0	0	1
Totals	28 (89%)	22 (95%)	9 (11%)	12 (51%)

taxonomic literature. Often the exact combination of features is obscure or hidden within data arrays in the machine. Future machines will need to operate with features more consistent with established taxonomic knowledge.

The arbitrary nature of category is highlighted by the example shown in Table 3.3 and Figure 3.13 using the Harmful Algal Bloom (HAB) Buoy imaging system together with the DiCANN clustering and labelling software. The categories chosen are not derived from biological taxonomy, yet the machine can operate with them. The patterns derived from the DiCANN feature extraction stage can be linked to any categorical collections, which in this case are 'gliders', 'blobs', 'triangles' and 'cups'. It has been shown previously that performance is related to training set size and feature variance with the classes and limited by information extracted from features (Simpson et al., 1991). The 'confusion matrix' is the normal way of summarizing performance. It can be seen from Figure 3.13 that the class 'cups' had 12 examples. In a 'leave-one-out' training protocol (used for small data-sets), eight of these were miscategorized as 'gliders', three as 'blobs' and only one identified correctly. The task was, however, difficult, since the data-set was small, with less than 40–100 items in each category. View angle was unconstrained, giving rise to large variances in feature view angle (or pose) in the data-set.

An Example: HAB Buoy

As an example of current machines for object recognition, HAB Buoy represents a new class of automatic *in situ* instruments that possess embedded artificial intelligence to carry out the identification of specimen images. Its camera resolution is 1 μm, with a dynamic range of 20–4000 μm. The buoy is designed for *in situ* static sampling and analysis of micro- and mesoplankton. Identification is accomplished through automatic feature extraction and multichannel SVM. As shown earlier (Table 3.3), group clustering and identification are flexible, allowing plankton specimens to be grouped by ecological function or taxonomy.

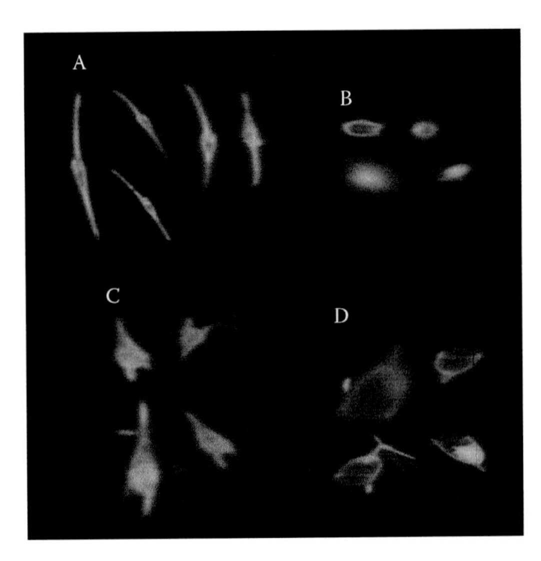

FIGURE 3.13 Morphologies used for categorization of marine plankton test. A: gliders; B: blobs; C: triangles; D: cups.

Accuracy is proportional to training data variance and HAB Buoy identifies at a speed of approximately 250–500 msec per object. HAB Buoy can image more than 10,000 specimens in 40 minutes, taking another few hours to process them. Once fully commissioned, this type of instrument will be used to analyse the common easily distinguished species in water samples, freeing experts up for more complex analyses and reducing their burden of routine, perhaps tedious sample analysis, and thus removing human bias from routine sample analysis.

Machine-Vision Issues

It is clear that automatic, natural-object categorization is feasible and useful, yet still in its early years. However, already a number of issues have become apparent. These must be considered in the future design and wider use of such machines. The most important of these is the quality of the training-set data. If, as is absolutely necessary at present, experts select the training specimens, then the quality of their decision making must be questioned. It is normal practice to establish a 'gold standard' data-set that constitutes an authoritatively identified reference specimen set. As we have seen before, however, human errors can arise and be high (as much as 27% disagreement). Unfortunately, it is common practice to train machine-vision systems with 'trial data-sets' constructed by an individual that have not been subject to any quality assurance process. This is not good practice.

Quantity of training data is also an issue, especially if you need to ask an expert to label 1000 or more specimens accurately. The nature of statistical and other pattern classifiers is that they require enough examples of a class to demonstrate the natural variations expected in future test data. Experiments have shown that 40–100 examples are normally sufficient. Yet the norm in most areas of taxonomy is to select and describe a small number of type specimens for reference. This is not enough for current machine-recognition methods.

Expert-Opinion Issues

Humans are biased, fatigue easily and are thus are imperfect tools for creating machine-recognition training data. To minimize errors in machine recognition, it is clear that expert consensus is required for training data. The accepted methodology is the Delphi protocol (Linstone and Turoff, 1975), where experts agree on a label through a mediated consensus. Obviously, this might be impractical

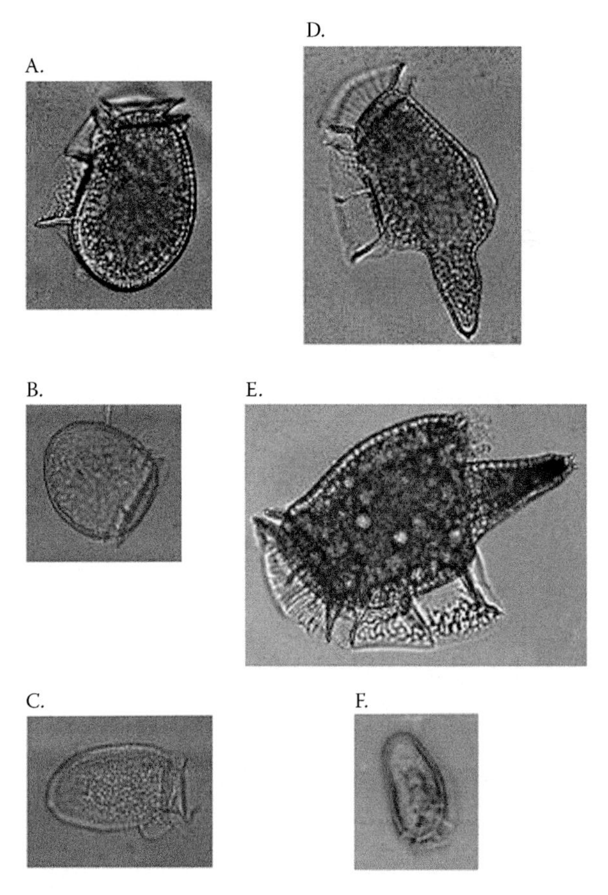

FIGURE 3.14 Example images of six species used in the six-species consensus study. A. *Dinophysis fortii*; B. *D. rotundata*; C. *D. acuminata*; D. *D. caudata*; E. *D. tripos*; F. *D. sacculus*. A total of 310 images were given to 16 marine taxonomists and ecologists. (Sources: B. Reguera, IEO, ES: A, D, E; S. Fonda, LBM, IT: B, C, F.)

for many species where the lack of taxonomic expertise has created an impediment to scientific research (see Darwin Declaration, 1998). The question of consensus is difficult and can be confrontational for experts, many of whom are used to working alone.

A Consensus Experiment

A six-phytoplankton species experiment was constructed to test expert opinion with a difficult taxonomic group from the microplankton: the *Dinophysis* dinoflagellate species group. The data, 310 images of closely related dinoflagellates, were shown to 16 marine ecologists and phytoplankton taxonomists across Europe via the World Wide Web. Each expert identified each image and assigned a probability of confidence to each identification. Figure 3.14 shows an example of each species.

Specimens were presented as high-quality photomicrograph images displayed on computer screens, shown to each expert through active Web pages via the Internet. The task was rated as difficult, since several of the species exhibit polymorphism and morphological variance. Example images are shown in Figure 3.15. Complexity was added to the task because experts normally view plankton through a microscope where they can focus up and down through the image to gain more information.

Figure 3.16 shows the difficulty for machine and human classifiers, where four species display adjacent clustering in a linear discriminant analysis plot of morphometric data for specimens. This highlights potential confusions between certain species. Results are presented in Figure 3.17.

A.

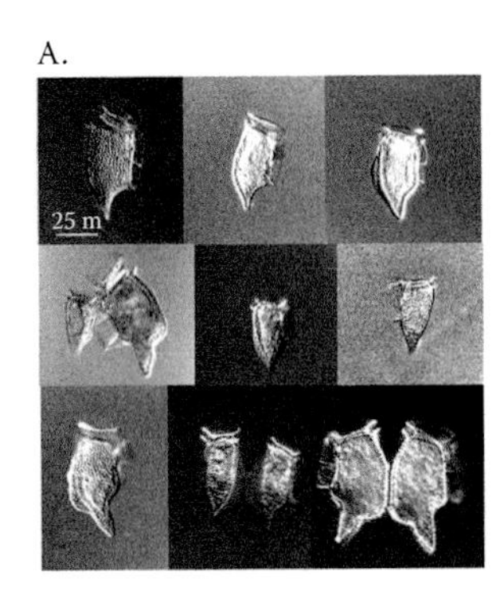

B.

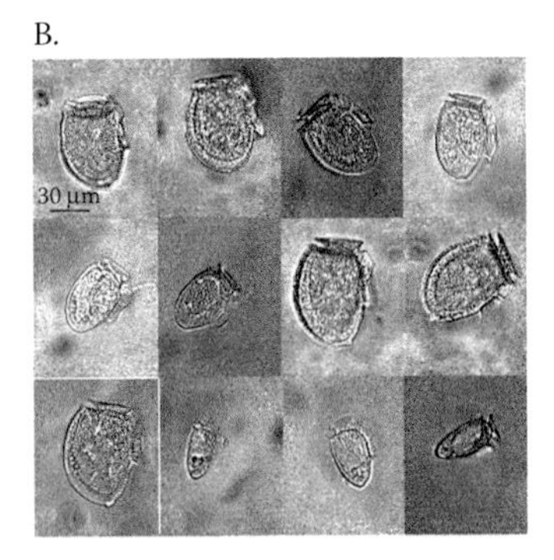

FIGURE 3.15 A. polymorphism in *Dinophysis caudata*. B. Morphological variation in *D. acuminata*. (Source: B. Reguera, IEO, ES.)

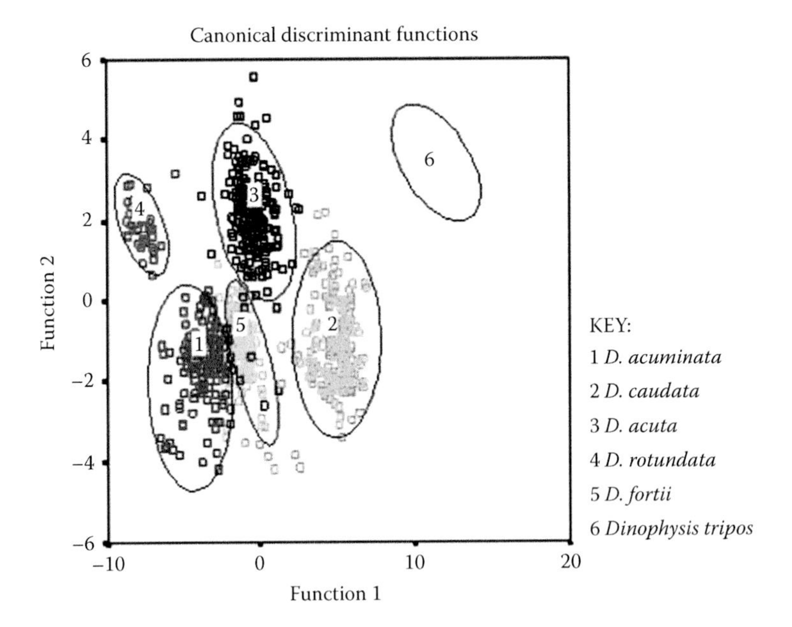

FIGURE 3.16 Linear discriminant analysis plot of morphometric features. (From Culverhouse, 2003. With permission.)

Agreement between the HAB taxonomists was 88 per cent (range of 83–94%) compared to 72 per cent between marine ecologists. As before, there was less agreement among ecologists. Also, only 245 of the 310 images (79%) attained better than 90 per cent consensus across all experts, with only 177 of 310 image (57%) achieving the important 'gold standard' criteria of 100 per cent agreement between experts. Some 11 per cent of the specimens had only 66 per cent agreement across the entire expert panel.

It would be folly to attempt a ranking of expert judgements and hence assign weights to individual expert performance since expert errors can be both systematic and random. The construction of a noise model may therefore be difficult and the desire to pool opinion unfeasible (Pennock and Wellman, 1999). Nevertheless, these results highlight a serious issue, since if only gold standard data are to be used for training machines, only 57 per cent of the preceding data-set could be used. But, to use the remaining data introduces uncertainty to the training data-set. It is unclear at present how this impacts or degrades machine-vision performance.

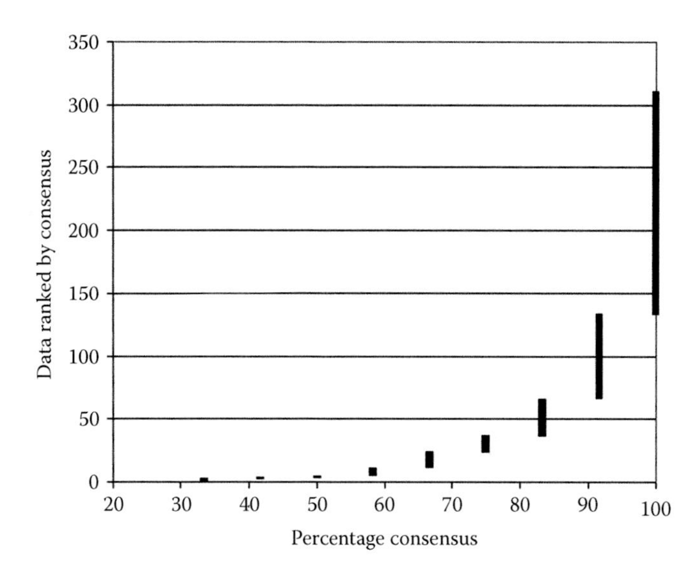

FIGURE 3.17 Agreement between HAB experts.

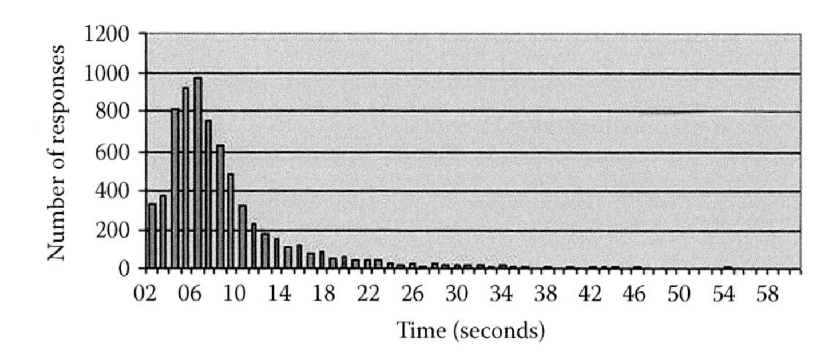

FIGURE 3.18 Speed of expert labelling (per specimen).

Speed of Recognition

It is rather satisfying to know that experts are quite fast at recognizing plankton, even in difficult circumstances. Figure 3.18 shows that range of recognition times across the image set and all subjects at the mode is 7 seconds, with the range varying between 3 and nearly 60 seconds. However, this performance is below that of existing machine vision systems; ADIAC, HAB Buoy, SIPPER, VPR and Zooscan can all identify specimens at a rate of from 1 per second to 500 per second. It is certain that future machine identification systems will be very much faster than humans.

CONCLUSIONS

There are a number of conclusions that can be drawn from the information and results presented in this chapter; perhaps most surprisingly is the fact that human factors are important. Experts are as error prone as anyone. They suffer from the same psychological biases of positivity and recency effects. They fatigue and find some tasks monotonous. Over time, they hone their skills by making up their own rules to improve their perceived effectiveness, perhaps to make the task less arduous. Unless individual competence is benchmarked, there is no easy way of knowing if an individual's improvement is real or illusory. Although evidence is scant (one experiment), it is plausible to state that taxonomic experts are more self-consistent than book experts when categorizing objects. In addition, book experts are not consistent with other experts. So, what is the value of using individuals who are not trained in a specialist categorization task? How accurate are graduate students? How accurate are people who might find the task very tedious?

Specialists who have honed their taxonomic decision-making skills through repeated practice can be up to 95 per cent self-consistent and are also able to form a high-level consensus of opinion with other experts (83% in the consensus study when they regularly inter-calibrated), even on repeated tasks. But both types of expert still suffer from cognitive biases that can degrade their performance. The task of identifying biological specimens is error prone and this should be acknowledged – especially as this type of activity is the basis of much science, yet often based on reference to small numbers of exemplar 'type' specimens.

The experiments described here have used difficult species to explore expert performance to highlight human problems with taxonomic identification tasks. Nevertheless, these human failings are prevalent in all sample-analysis situations. Of particular importance is the common 'one expert is the identifier' scenario, which can introduce significant systematic errors to sample analysis through individual differences in performance. Indeed, anecdotal evidence from the *Cymatocylis* categorization experiment was uncovered in which one specialist taxonomist admitted to mislabelling, for many years, several of the species used in the study. It was only when the subject discussed his labels with the experimenters that the problem came to light. Additional performance degradations can occur through tiredness or a lack of vigilance – sometimes within 30 minutes of starting a task, which can cause very high levels of degradation.

Thankfully, there are solutions to these problems. Where experts collude (i.e. discuss their findings and reach agreement), the error rate drops significantly. In the Plymouth experiment discussed previously, one major difference between the taxonomy and ecology groups was that the taxonomists (who worked for health monitoring laboratories) routinely 'intercalibrated' both within and between laboratories. That is, they discussed the difficult specimens and also cross-checked an individual's performance in a supportive environment.

For field-based work, interlaboratory calibration is probably the only reasonable approach to achieving high accuracies among human teams. This is probably rather rare at present, but perhaps the ease of sharing data over the Internet might encourage such cooperation in the future.

The Need

Irrespective of current human practice, in the future it is impossible not to envisage high-throughput machines analysing routine samples of biological specimens. Widespread use of automation will not be immune from the introduction of human biases into the results. Errors will be due in part to the inadequacies of the machines and software themselves, which, after all, are both human constructs, but a proportion will be due to incorrect expert-labelled training data. To improve the entire process, these errors must be minimized and also quantified. It is possible that error rates as high as 25 per cent will be the norm for large-scale analysis. These error rates may well be similar to current and historical human errors in field sample analysis. Some of these errors will be due to human factors, short-term memory, fatigue, monotony and biases. Some errors will be due to a lack of consensus on the label given to difficult taxonomic specimens and species.

Accordingly, there is now, and will be in the future, a need to place error bars on all taxonomic results, human as well as machine generated. It is possible that within a sample, the errors will vary by species or taxon. This is quite acceptable, since the recording of error will allow explicit discussion of the sources of variance, and allow a focus of effort to reduce those errors. Uncertainty is pervasive in the natural world. Having a record of uncertainty in species recognition and categorization analyses is just good scientific practice. There will also be a need to develop machine recognition methods that adopt the taxonomic approaches employed by systematics experts.

ACKNOWLEDGEMENTS

I thank Bob Williams (Plymouth Marine Laboratory) for his suggestions, advice and guidance. I also thank Beatriz Reguera, Jeff Turner, Marina Cabrini and Serena Fonda for photomicrographs.

REFERENCES

Bertenthal, B.I. (1992) Origins and early development of perception, action, and representation. *Annual Review of Psychology,* 47: 431–459.

Bertero M., Poggio, T. and Torre, V. (1998) Ill-posed problems in early vision. *Proceedings of the Institute of Electrical and Electronics Engineers,* 76: 869–889.

Biederman, I. (1987) Recognition by components: A theory of human image understanding. *Psychological Review,* 94: 115–147.

Broadbent, D.E. (1958) Perception and communication. London, Pergamon.

Bucklin, A. and Kahn, L. (1991) Mitochondrial-DNA variation of copepods – markers of species identity and population differentiation in *Calanus. Biological Bulletin of Woods Hole* 181(2): 357–357.

Bucklin, A., Sundt, R.C. and Dahle, G. (1986) The population genetics of *Calanus* finmarchicus in the North Atlantic. *Ophelia* 44: 29–45.

Cerella, J. (1979) Visual classes and natural categories in the pigeon. *Journal of Experimental Psychology: Human Perception and Performance,* 5: 68–77.

Cerella, J. (1980) The pigeon's analysis of pictures. *Pattern Recognition,* 12: 1–6.

Colquhoun, W.P. (1959) The effect of a short rest pause on inspection efficiency. *Ergonomics,* 2: 367–372.

Colquhoun, W.P. (1982) Biological rhythms and performance. In *Biological Rhythms, Sleep, and Performance* (ed. W.B. Webb), Wiley, Chichester, pp. 59–86.

Crick, F. (1984) Function of the thalamic reticular complex: the search light hypothesis. *Proceedings of the National Academy of Science U.S.A.,* 81(14): 4586–4590.

Culverhouse, P.F., Ellis, R.E., Simpson, R., Williams, R., Pierce, R.W. and Turner, J.T. (1994) Categorization of 5 species of *Cymatocylis* (Tintinidae) by artificial neural network. *Marine Ecology Progress Series,* 107: 273–280.

Culverhouse, P.F., Simpson, R.G., Ellis, R., Lindley, J.A., Williams, R., Parasini, T., Requera, B., Bravo, I., Zoppoli, R., Earnshaw, G., McCall, H. and Smith, G. (1996) Automatic categorization of 23 species of dinoflagellate by artificial neural network. *Marine Ecology Progress Series,* 139: 281–287.

Culverhouse, P.F., Williams, R., Reguera, B., Herry, V. and González-Gil, S. (2003) Do experts make mistakes? *Marine Ecology Progress Series,* 247: 17–25.

The Darwin Declaration, Australian Biological Resources Study, Environment Australia, Canberra, Australian Biological Resources Study, 1998 (http://www.biodiv.org/programmes/cross-cutting/taxonomy/darwin-declaration.asp).

Du Buf, H. and Bayer, M.M. (eds). (2002) *Automatic Diatom Identification,* World Scientific Series in Machine Perception and Artificial Intelligence, World Scientific Publishing Company, Hackensack, NJ, vol. 51, 328 pp.

Duchowski, A.T. (2002) *Eye Tracking Methodology: Theory and Practice,* Springer–Verlag, UK, 295 pp.

Duda, R. and Hart, P. (1973) *Pattern Classification and Scene Analysis,* John Wiley & Sons, New York, 482 pp.

Edelman, S. (1999) *Representation and Recognition in Vision,* MIT Press, Cambridge, MA, 402 pp.

Ericsson, K.A. and Lehmann, A.C. (1996) Expert and exceptional performance: Evidence of maximal adaptation to task conditions. *Annual Review of Psychology,* 47: 273–305.

Evans. J. St. B.T. (1987) *Bias in Human Reasoning: Causes and Consequences,* Laurence Erlbuam Associates, Hove, 160 pp.

Ewert, J.P. (1984) Tectal mechanisms that underlie prey-catching and avoidance behaviors in toads. In *Comparative Neurology of the Optic Tectum* (ed H. Vaneges), Plenum Press, New York, pp. 247–416.

Fox, J.G. (1971) Background music and industrial efficiency – A review. *Applied Ergonomics,* June 1971: 70–73.

Gauch, H.G. (1982) *Multivariate Analysis in Community Ecology,* Cambridge University Press, Cambridge, 314 pp.

Gibson, E.J. (1969) *Principles of Perceptual Learning and Development*, Appleton–Century–Crofts, New York, 537 pp.

Gibson, J.J. (1957) Survival in a world of probable objects. *Contemporary Psychology,* 2: 33–35.

Goad, C. (1986) Fast 3D model-based vision. In *From Pixels to Predicates* (ed A.P. Pentland), Ablex Norwood, Norwood, NJ, pp. 371–391.

Grosjean, Ph., Picheral, M., Warembourg, C. and Gorsky, G. (2004) Enumeration, measurement, and identification of net zooplankton samples using the ZOOSCAN digital imaging system. *ICES Journal of Marine Science,* 61: 518–525.

Hancock, P.J.B., Baddeley, R.J. and Smith, L.S. (1992) The principal components of natural images. *Network Computation in Neural Systems,* 3: 61–70.

Hill, R.S., Allen, L.D. and Bucklin, A. (2001) Multiplexed species-specific PCR protocol to discriminate four N. Atlantic *Calanus* species, with an mitCOI gene tree for ten *Calanus* species. *Marine Biology,* 139: 279–287.

Hu, M.K. (1962) Visual pattern recognition by moment invariants. *IEEE Transactions on Information Theory,* 8: 179–187.

Hu, Q. and David, C. (2005) Automatic plankton image recognition with co-occurrence matrices and support vector machine. *Marine Ecology Progress Series,* 295: 21–31.

Hubel, D.H. and Wiesel, T.N. (2004) *Brain and Visual Perception: The Story of a 25-Year Collaboration.* Oxford University Press, Oxford, 744 pp.

Jacobs, G. (1981) *Comparative Color Vision,* Academic Press, New York, 209 pp.

Johnson, J.S. and Olshausen, B.A. (2005) The earliest EEG signatures of object recognition in a cued-target task are post sensory. *Journal of Vision,* 5: 299–312.

Kelly, M.G. (2001) Use of similarity measures for quality control of benthic diatom samples. *Water Research,* 35: 2784–2788.

Linstone, H.A. and Turoff, M. (1975) *The Delphi Method: Techniques and Applications.* Addison–Wesley, Reading, MA, 620 pp.

Lowe, D. (1987) Three-dimensional object recognition from single two-dimensional images. *Artificial Intelligence,* 31: 355–395.

Marsh, R.L., Sebrechts, M.M., Hicks, J.L. and Landau, J.D. (1997) Processing strategies and secondary memory in very rapid forgetting. *Memory and Cognition,* 25: 173–181.

Megaw, E.D. (1979) Factors affecting visual inspection accuracy. *Applied Ergonomics,* March 1979: 27–32.

Oliva, A., Torralba, A.B., Castelhano, M.S. and Henderson, J.M. (2003) Top-down control of visual attention in object detection. *IEEE International Conference on Image Processing,* pp. 1253–1256.

Pashler, H. (1996) *The Psychology of Attention,* MIT Press, Cambridge, MA, 480 pp.

Pennock, D.M. and Wellman, M.P. (1999) Graphical representations of consensus belief. In *Proceedings of the 15th Conference on Uncertainty in Artificial Intelligence* (eds K. Laskey and H. Prade), Morgan Kaufmann, San Francisco, pp. 531–540.

Peterson, L.R. and Peterson, M.J. (1959) Short-term retention of individual verbal items. *Journal of Experimental Psychology,* 58: 193–198.

Proffitt, D. (1993) A hierarchical approach to perception. In *Foundations of Perceptual Theory* (ed S.C. Masin), North Holland, New York, pp. 75–111.

Reimann, B., Krummenacher, J. and Müller, H.J. (2003) Visual search for singleton feature targets across dimensions: Stimulus and expectancy-driven effects in dimensional weighting. *Journal of Experimental Psychology/Human Perception and Performance,* 29: 1021–1035.

Remsen, A., Hopkins, T.L. and Samson, S. (2004) What you see is not what you catch: A comparison of concurrently collected net, optical plankton counter, and shadowed image particle profiling evaluation recorder data from the northeast Gulf of Mexico. *Deep-Sea Research,* 51: 129–151.

Ruderman, D.L. (1994) The statistics of natural images. *Network Computation in Neural Systems,* 5: 517–548.

Samson, S., Hopkins, T., Remsen, A., Langebrake, L., Sutton, T. and Patten, J. (2001) A system for high-resolution zooplankton imaging. *IEEE Journal of Ocean Engineering,* 26: 671–676.

Simpson, R., Culverhouse, P.F., Ellis, R. and Williams, R. (1991) Classification of *Euceratium gran.* In *Neural Networks.* IEEE International Conference on Neural Networks in Ocean Engineering, Washington, DC, August, pp. 223–230.

Simpson, R., Williams, R., Ellis, R. and Culverhouse, P.F. (1992) Biological pattern recognition by neural networks. *Marine Ecology Progress Series,* 79: 303–308.

Simpson, R., Williams, R., Ellis, R. and Culverhouse, P.F. (1993) Classification of Dinophyceae by artificial neural networks. In *Toxic Phytoplankton Blooms in the Sea* (eds T.J. Smayda and Y. Shimizu), Elsevier Science, New York, pp. 183–190.

Sokal, R.R. (1966) Numerical taxonomy. *Scientific American*, 215(6): 106–116.

Sokal, R.R. (1974) Classification: Purposes, principles, progress, prospects. *Science,* 185: 1115–1123.

Sokal, R.R. (1983) A phylogenetic analysis of the Caminucales. IV. congruence and character stability. *Sytematic Zoology,* 32(3): 259–275.

Srinivasan, M.V. (1998) Insects as gibsonian animals. *Ecological Psychology,* 10: 251–270.

Tang, X., Stewart, W.K., Vincent, L., Huang, H.E., Marra, M., Gallager, S.M. and Davis, C.S. (1998) Automatic plankton image recognition. *Artificial Intelligence Review,* 12: 177–199.

Tarr, M.J. and Bülthoff, H.H. (1995) Is human object recognition better described by geo-structural descriptions or by multiple views? *Journal of Experimental Psychology: Human Perception and Performance,* 21: 1494–1505.

Toth, J. and Lewis, M. (2002) The role of working memory and external representation in individual decision making in diagrammatic representation and reasoning. In *Diagrammatic Representation and Reasoning* (eds M. Anderson, B. Meyer and P. Olivier), Springer–Verlag, Berlin, Heidelberg, pp. 207–222.

Toth, L. and Culverhouse, P.F. (1999) Three-dimensional object categorization from static 2D views using multiple coarse channels. *Image and Vision and Computing,* 17: 845–858.

Treisman, A. (1982) Perceptual grouping and attention in visual search for features and for objects. *Journal of Experimental Psychology Human Perception and Performance,* 8: 194–214.

Treisman, A. (1985) Preattentive processing in vision. *Graphics- and Image-Processing,* 31: 156–177.

Treisman, A. (1986) Features and objects in visual processing. *Scientific American,* 255: 114B–125.

Vecera, S.P., Vogel, E.K. and Woodman, G.F. (2002) Lower region: a new cue for figure–ground assignment. *Journal of Experimental Psychology,* 131: 194–205.

Welford, A.T. (1968) *Fundamentals of Skill*, Methuen, London, 426 pp.

4 Neural Networks in Brief

Robert Lang

CONTENTS

Introduction 47
The Example 48
Types of Neural Networks 48
Benefits of Neural Networks 48
Drawbacks of Neural Networks 49
A Typical Artificial Intelligence System 49
Data Capture 50
Encoding 50
The Neural Network and Lookup Table 50
Traditional Neural Networks 51
The Multilayered Perceptron (MLP) 51
Self-Organizing Map (SOM) 57
Dynamic Neural Networks 61
The Plastic Self-Organizing Map (PSOM) 61
Discussion and Summary 67
References 67

This chapter serves as a short introduction to the field of artificial neural networks, or simply neural networks. Its aim is to provide readers with a basic understanding of these systems. Two main types of neural networks are dealt with: traditional neural networks and dynamic neural networks.

INTRODUCTION

An artificial neural network is a computer algorithm that mimics the manner in which the brain processes and stores information. Artificial neural networks are said to have properties similar to those of the brain, such as the ability to learn, adapt and automatically group similar 'things' together. Furthermore, neural networks can perform these functions on massive and complex data-sets. Haykin (1999) is a comprehensive source, while the work of Beale and Jackson (1990) is a lighter read.

As daunting as it may seem, the reality of applying neural networks is no Herculean task. When dissected, the different algorithms follow similar ideas and can be applied using simple principles without a deep knowledge of the underlying mathematics. So much of the power of neural networks is stored within the complexity of the network as a whole that, for non-trivial tasks, it is practically impossible to divine *why* the network has produced the output. Despite some individuals' innate desire to know how the answer to a particular problem was determined, for most tasks there is no need to understand this in order to make use of the neural network's power.

Neural networks are best described using examples. The following trivial problem is designed to highlight the properties of different types of neural networks. Other chapters in this volume

FIGURE 4.1 Example dinosaur images *Diplodocus* (left), *Tyrannosaurus rex* (centre) and *Stegosaurus* (right).

contain more realistic systematic problems. Our simple example will also be used to guide readers through some of the terminology that can be found in standard neural network literature.

The Example

This simple problem concerns processing images of three different dinosaurs: *Tyrannosaurus rex, Diplodocus* and *Stegosaurus.* For each of these dinosaurs, we have a number of pictures (Figure 4.1) and a large amount of (mostly fabricated) data. Each dinosaur is called a *class* and each picture of the dinosaur is called an *input* or *pattern.* The quantitative data include statistics common to all three, such as weight, height and number of legs. Initially, we want the neural network to do two things: (1) recognize the dinosaur and (2) compare the different dinosaurs. Recognizing a particular dinosaur from data is called *classification,* whereas comparing different dinosaurs (and thus grouping similar ones) is called *clustering.* Classification is the ability to recognize a previously unseen input (new dinosaur pictures), having been trained to 'see' that input. Clustering is the ability to draw comparisons between different classes (dinosaurs) automatically. The broad field of neural networks encompasses many other abilities, but classification and clustering are the most common.

Types of Neural Networks

Neural networks may be divided into static (or traditional) and dynamic types. Although both types solve the same sorts of problems, they do so in different ways. Static neural networks do not change their structure once they have been created and operate on a fixed number of classes (e.g. types of dinosaurs). Dynamic neural networks can change their structure and can operate in an environment where the number of classes is not fixed. Which type to use will depend on the problem that is to be solved.

Benefits of Neural Networks

1. *Nonlinearity.* Most classification and clustering problems are nonlinear; there is rarely a straight line that can be drawn among inputs to split up the classes. Different classes might share some inputs (e.g. both the *Diplodocus* and *Stegosaurus* have four legs), but this poses no problem for non-linear data analysis methods like neural networks.
2. *Robust data acceptance.* Input data are rarely perfect, but with enough examples of a class, a neural network will automatically focus on the parts of the data it needs to perform the classification and clustering operations.
3. *Input–output mapping.* This is the core function of classification. A neural network can be used to map an input (dinosaur picture) to an output (the name of the dinosaur).
4. *Adaptivity.* A neural network can automatically adapt to new inputs, so you can use the same network architecture (and computer program) on different problems.
5. *Generalization.* This is the ability of the neural network to classify a previously unseen example of a class. For example, if you are given a new picture of a *T. rex* (this time, baring teeth), a properly trained neural network then will recognize it as a *T. rex.*

6. *Confidence.* A neural network can provide a confidence estimate of its classification. For example, the neural network can say it is 20 per cent sure the input is a *T. rex.*
7. *Fault tolerance.* If the architecture of the network is damaged, the network will retain some functionality. This is more important if the neural network is implemented in custom hardware rather than on a desktop computer.

Drawbacks of Neural Networks

1. *Large amounts of data required.* For non-trivial systems, neural networks only function well when they are trained on many examples of each class. The more examples of a class you have, the better the network will learn. Unfortunately, automatic systems are rarely required to recognize the common inputs, but instead tend to be asked to recognize the rare patterns for which examples are scarce. There are ways to mitigate these problems (detailed later). The more dimensions your data have (the more different aspects about your dinosaur you want the network to learn), the more data you need. This is sometimes referred to as the 'curse of dimensionality'.
2. *Few analytical aids.* Deciding on network architecture – what types of data to train the network on and which network to use – can be difficult. There are no set rules that match particular problems with network settings. As shown later, the use of a typical neural network is something of a trial-and-error process. Furthermore, it can be difficult to decide before the network has trained whether a neural network can make sense of the problem at all. This is a particularly important downside if it takes a long time to train the network.
3. *Over-fitting.* This occurs when a neural network learns the precise patterns it is provided with too well and thus loses the ability to generalize. Over-fitting is a side effect of too much learning and can be mitigated by stopping the learning process early. However, when to stop training is a difficult question to answer beforehand.
4. *Non-trivial tasks can be computationally expensive.* On non-trivial problems, a well coded neural network can take many hours to learn the patterns. Typically, large systems have a lot of classes. Having lots of classes requires many patterns. Further, the more you want the network to learn, the bigger the structure is likely to be and the longer it will take to process each pattern. It is normally wise to test a network on a subset of a large number of classes before scaling it up to its full size.

A Typical Artificial Intelligence System

Neural networks cannot be used on their own in a system. Data are transformed before being submitted into the network and the network output is normally transformed into a user-friendly format. A simple system is shown in Figure 4.2. In this system, photographic data are collected using a camera and digitization process. This input is then transformed into something that the neural network will understand (e.g. brightness detection values). Only then are the data fed into the network. After processing by the network, the output is back-transformed using a lookup table.

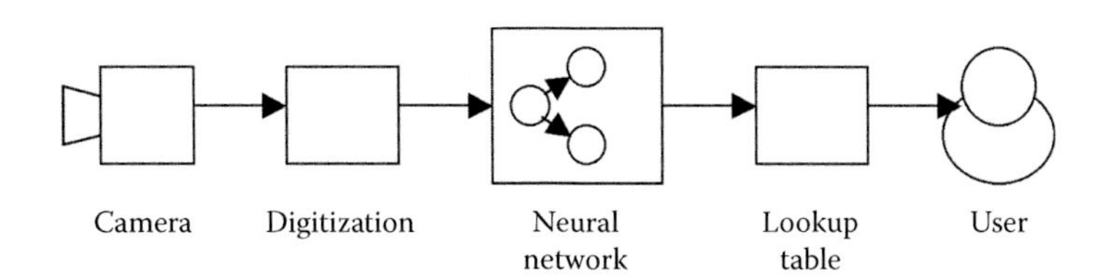

FIGURE 4.2 A typical neural network system.

Data Capture

Neural networks often require a prohibitive amount of data for non-trivial tasks. This is because many different patterns are required for each class: lots of *T. rex* pictures, for example. If your problem includes some rare classes (represented by only a few patterns), it is not uncommon to take these patterns and duplicate them many times while adding noise to each duplicate. This will make each input slightly different. As long as each input still represents your class (still looks like a *T. rex*), the neural network will learn the pattern. The noise is important because it will help the network generalize.

Encoding

Data often need to be transformed into a format that will help the neural network learn. This transformation is called *encoding* and is a very important step. If the neural network is given data in a poorly encoded state, it may not be able to learn the features of each class. For example, you could take each dinosaur picture and encode it as a series of black-and-white pictures. You could then submit the individual black-and-white pixel values into the neural network and let it decide what features and shapes to choose from. For some problems, this is fine. However, better results would be obtained if the network were given all the lines and shapes of the dinosaur from the picture and then trained on these shapes. This shape-based encoding might yield better results than a pixel-based encoding.

All data need to be converted into a numeric scale before they are submitted to the network. The manner of this scale will affect the way the network learns and might give preference to some patterns over others. For example, the dinosaur data might include qualitative information such as 'skin type' with values such as soft tissue, scales, feathers, etc. These should be encoded into numbers: soft tissue = 10, scales = 20, feathers = 30.

The Neural Network and Lookup Table

This is where the real work is done. Once the neural network is set up (see later discussion), a pattern (dinosaur picture) is fed into the network and a number produced. This number is then translated by the lookup table into user-friendly equivalent output.

Setting Up a Neural Network

Typically, neural networks are used in the following manner:

1. *Problem identified.* The problem the network is to solve must first be identified.
2. *Network selection.* Based on the problem, appropriate neural network size and design are chosen.
3. *Data collection.* Neural networks perform best with a large amount of data. The more data, the better the network's performance is likely to be.
4. *Data segregation.* The data must be split into two sets. First, a 'training set' must be assembled. This is the set of data that the network will use to build up its base of knowledge. The second set is the validation set. This is used to make sure that the network has learnt correctly.
5. *Decide on network topology.* The topology of the network is the number and arrangement of neurons and weights (see later details). With traditional neural networks, this is decided by the data analyst.
6. *Perform training.* This is where the learning algorithm is applied to the network using the training data-set. The learning algorithm changes a set of numbers called the 'weights' of the network so that future presentations of the same input yield the same result.

7. *Perform validation.* The validation set is now used to measure how well the network has learnt. If the network performs within the desired tolerances it can be used from this point on and the development process can terminate here.
8. *Iteration.* If the output error of the network has not met the desired tolerance, then a number of options are available:
   a. *Change training set.* It is possible that the training set did not contain adequate exemplar patterns for all the classes. Return to step 4 and hand-pick a number of good examples of each class for the training set.
   b. *Choose new topology.* This is the usual way of solving network problems. Increasing the network size might allow more classes to be learnt but the training set might need to be increased. Return to step 5 and continue.
   c. *Tune network parameters.* Each type of neural network has a number of parameters that can be changed to modify its overall performance. In this case, only training and validation are needed.

Traditional Neural Networks

Traditional neural networks can be divided into two distinct groups by how they learn. *Supervised learning* networks primarily perform the task of classification. As the name suggests, the neural network is supervised and told what each input is. Unsupervised learning networks tend to perform clustering. They are not told what each input is; the neural network must decide how to process the data. A classic example of a supervised learning network is the multilayered perceptron (MLP). A classic example of an unsupervised learning network is the self-organizing map (SOM). In both cases, the knowledge of the network is stored in a series of numbers (either in a vector or just single values) and they are usually called the *weights.* Learning is implemented as a process for modifying these weights.

The Multilayered Perceptron (MLP)

The MLP (Rosenblatt, 1958) is a network structure consisting of a number of input and output network nodes called *neurons* connected together with weights (Figure 4.3). Each neuron contains a simple processing unit that performs a simple mathematical function and each weight is a simple number. The knowledge of the network is stored within these weights and organized according to the arrangement of neurons.

The MLP is divided into layers of neurons, input, hidden and output (left to right in Figure 4.3). The input layer does not perform any processing as such; these neurons simply hold the data

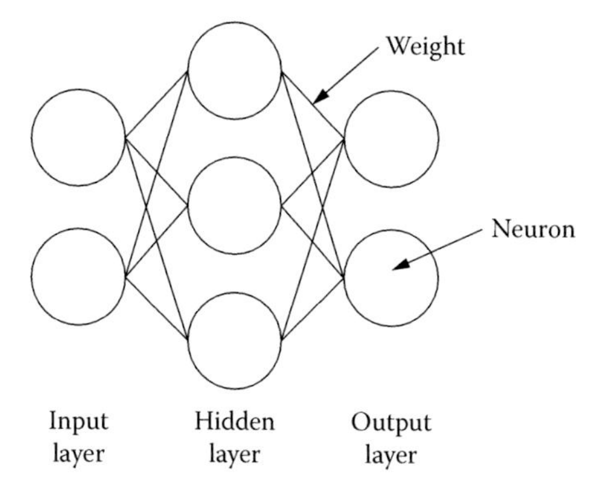

FIGURE 4.3 A typical multilayered perceptron (MLP) design.

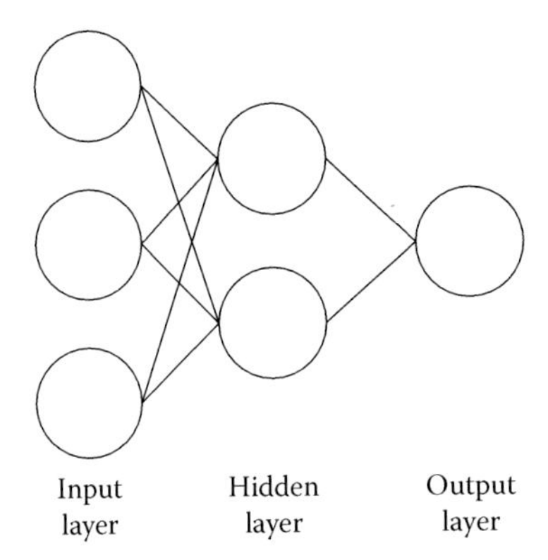

FIGURE 4.4 The example MLP network setup.

that will be submitted to the network. The hidden layer provides the network's processing power. The output layer contains the result. Typically, any neuron is connected to all the neurons in the next layer by a weight. This design is called a *fully connected* network. The MLP has two algorithms, one for practical use and one for training. The training algorithm is an extension of the practical use algorithm.

Setting Up the MLP Network

The best method of demonstrating the way the MLP functions is by example. The steps taken here mimic the numbered steps described before in 'Setting Up a Neural Network' (numbers included in parentheses).

For this example, we will assume the problem (1) is to identify a dinosaur from a picture. This is a classification task, so the MLP is best suited (2). Let us say data collection (3) included only the *Diplodocus* and the *T. rex* images. We will leave *Stegosaurus* aside for simplicity. Our data-set would contain many different pictures of these two dinosaurs – for example, 100 of each. These pictures are put through an arbitrary transform that turns each one into three numbers. Each of these numbers is a *dimension* of the input. We will also number the classes to help the network learn. Class 0 will be the *T. rex* and class 1 the *Diplodocus*.

Once these data have been collected they need to be segregated (4) into a training set and a validation set. The training set is what the network will learn from and the validation set will be used to test what the network has learned. A good rule of thumb is to select two thirds of the whole data-set as training, leaving one third for the validation. Therefore, in this example the training set will be constructed of 66 *T. rex* pictures and 66 *Diplodocus* pictures. Once it has been selected, you should check that the training set does indeed include good examples of the class you are asking it to learn. Finally, randomize the order of the training patterns.

Deciding on network topology (5) is a difficult step for the MLP. For this example, the network will be set up arbitrarily small to facilitate explanation. It will consist of three input nodes (one for each dimension of our dinosaur), two nodes in the hidden layer, and one output (Figure 4.4). The weights of the network are initially set to random numbers between 0 and 1.

Training Algorithm

Now the network is ready to be trained (6). Take the first input dimensions from the training set and submit them to the network (Figure 4.5). In our hypothetical case, the randomly selected dinosaur is a *T. rex*. The *activation* of each neuron is then calculated. Figure 4.6 shows a simple activation of a neuron.

On the left of the diagram are two input neurons (A and B) with two of the values from the input pattern (the dinosaur statistics), 2 and 1. These two neurons are connected to neuron C by two weights (0.3 and 0.6). Neuron C is set to use a threshold of 1 as its activation function. This

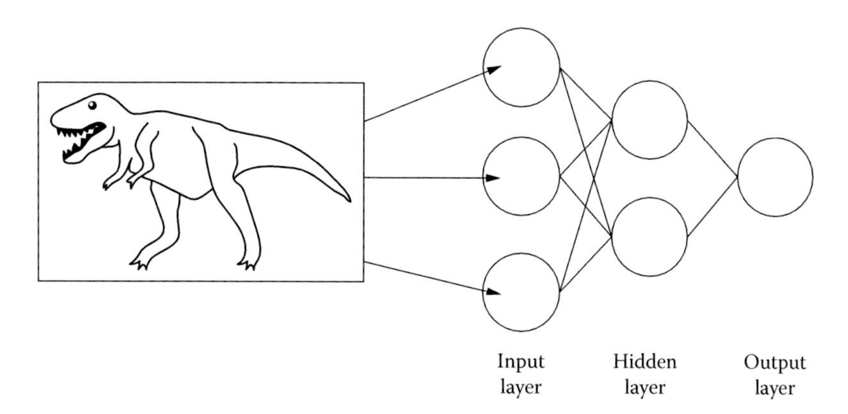

FIGURE 4.5 The first training pattern (a *T. rex*) is shown to the network.

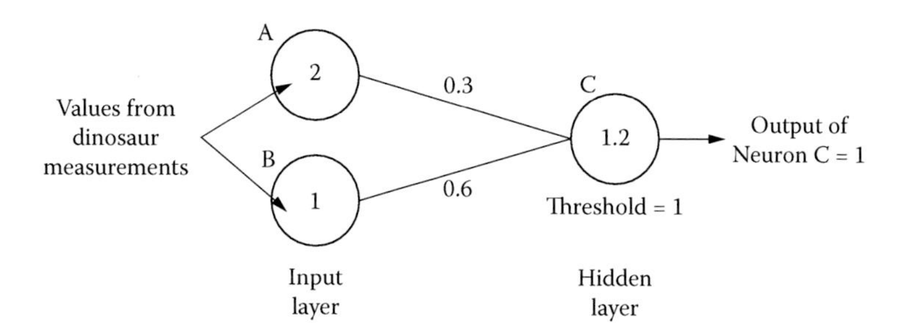

FIGURE 4.6 An example of a neuron firing (see text for details).

means that, provided the sum product of its inputs is larger than 1, it will activate, giving a 1 as its output.

The input to the neuron is calculated by summing the product of input and weight feeding into it. In this case, there are two inputs, so the sum product is simply calculated: $(2 * 0.3) + (1 * 0.6) = 1.2$. As 1.2 is larger than the global threshold of 1, the neuron is said to *fire*. The output of this neuron then becomes the input to another neuron located deeper in the MLP structure.

There are lots of different activation functions. In the preceding example, the activation function was a linear function, but the most popular function for this purpose is the sigmoid function. A sigmoid function is a smooth S-shaped curve around a value of 0.5. It is not necessary to change this value as the changing weights of network provide a tuning process. Above this value a positive output is given and below it a negative value is given. Between the two extremes of ±1 is a smooth change. All the neurons in the network fire this way, each neuron taking its value from either the input pattern (the dinosaur) or from the firing of other nodes. Using the sigmoid improves the network's ability to distinguish between dinosaurs. Intuitively, this can be explained as the difference between asking 'Does this look like a *T. rex,* yes or no?' and 'How much do you think this looks like a *T. rex,* as a per cent?' Clearly, the second question gives a more detailed answer, which is better for understanding the nuances in a complex data-set.

Once all the neurons have fired, a final value will be located on the output neuron. At the start of training, this value is set to a random number. Therefore, with our example, at the start of training, the network will look like Figure 4.7. The activation values of the different neurons in the network are hidden for simplicity.

In Figure 4.7, the network produced an output of 0.4. However, to classify properly we need an output of 0 as the number 0 is tied to the class '*T. rex*'. So that future presentations of *T. rex* images will yield a 0, the weights need to be changed. This task is performed by taking the error between the result obtained by the first pass through the network (0.4) and the result we wanted

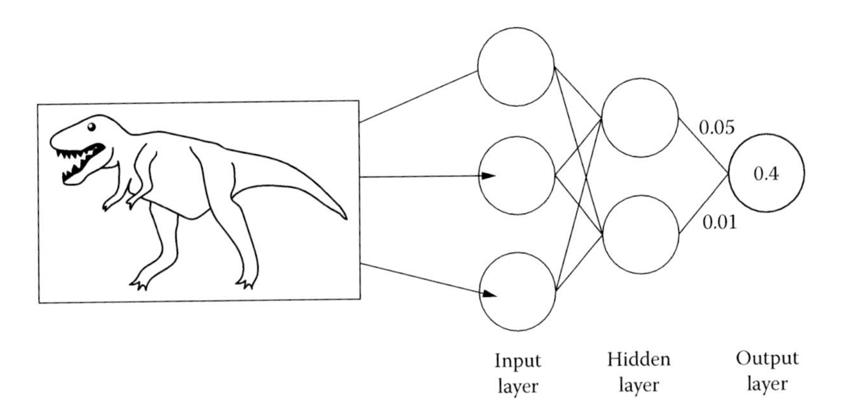

FIGURE 4.7 The weights for the output neuron during training (before update).

(0) by propagating calculations back through the network, changing the weights starting at the output and moving from right to left to the input. This aspect of the training procedure is called back-propagation.

How do we change the weights and by how much? This depends on where we are located within the network. If we are on the output layer (right-most set of weights) or earlier on in the network (left-hand set of weights), we will need to calculate the error first and then change the weights depending on this error. The amount we change the weights is called the *learning rate*. The larger the learning rate is, the more effect the error will have.

Output layer calculations. We start by calculating the error for the output layer:

$$e = z * (1 - z) * (y - z) \tag{4.1}$$

where e is the error, z is the value on the node and y is the value we wanted.

In our example:

$$e = 0.4 * (1 - 0.4) * (0 - 0.4)$$

$$e = 0.4 * (0.6) * (-0.4)$$

$$e = -0.096$$

This is the raw error that we are going to apply to each of the weights. Now, we need to reduce this a little by using a learning rate. Learning rates tend to be around 0.1 for a starting point. A higher learning rate means the network learns more quickly and uses fewer example pictures to learn a dinosaur pattern. However, the network is more sensitive to patterns corrupted with noise (bad pictures of dinosaurs) and if your classes (types of dinosaurs) are subtly different, the network will not distinguish between them. The opposite is also true. It is best to choose more example patterns and a lower learning rate than fewer patterns and a high learning rate.

$$\theta = \lambda * e \tag{4.2}$$

where λ is the learning rate (we have decided 0.1) and θ is the change we are going to make in each weight.

In our example:

$$\theta = 0.1 * -0.096$$

$$\theta = -0.0096$$

Now we can update the weights in each of the nodes of the output layer. This is done by multiplying the weight by this change. Figure 4.7 shows our example with some values on the weights (for the purpose of this example). Each of those weights will need to be updated using the following equation:

$$w(new) = w + (w * \theta) \tag{4.3}$$

where w is one of the weights on the output layer.

Using just the top weight in Figure 4.7 (0.05), the example would be as follows.

$$w(new) = 0.05 + (0.05 * -0.0096)$$

$$w(new) = 0.05 + (-0.00048)$$

$$w(new) = 0.04952$$

Now we would do the same for each of the weights coming into the output node. In this case, only one other weight (0.01) needs to be adjusted. Using the previous equation, this weight would be reset to 0.0096 (0.01 + (0.01 * –0.0096)). Once all the weights connecting to the output layer are complete, we step back to the hidden layer.

Hidden layer calculations. To update the left-hand layer of weights (between input and hidden layers), we need the information we have just calculated. Figure 4.8 shows the results of the last calculation on the network. For the sake of simplicity, the dashed nodes and weights in the network will be ignored. In practice, you would perform the following process on those neurons as well. The hidden node shown gave an output of 0.5 on the first pass through the network.

To update the left-hand layer of weights, we first need the error coming out of the node. We only have one weight coming out of this node (labelled 0.04952). The total error is calculated as follows:

$$g = \Sigma\, m * e \tag{4.4}$$

where g is the total error output of the neuron, m represents each of the output weights of the neuron and e is the error on the output of the next layer.

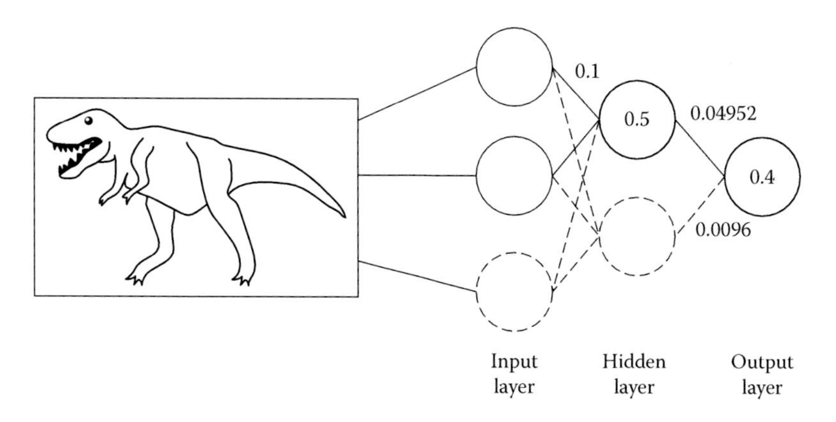

FIGURE 4.8 Updating the hidden layer: results from the update of the output layer (see text for discussion).

In our example:

$$g = 0.04952 * -0.096$$

$$g = -0.00475$$

Now we can work out the error at this hidden neuron:

$$ehidden = z * (1 - z) * g \tag{4.5}$$

where *ehidden* is the error on the hidden neuron we are looking at now and z is the value on this neuron.

For our example, the calculations are:

$$ehidden = 0.5 * (1 - 0.5) * -0.00475$$

$$ehidden = 0.5 * (-0.5) * -0.00475$$

$$ehidden = -0.0011875$$

Now that we have the error, we are on familiar ground for updating all the weights coming into the neuron. We work out the change on each input weight using Equation 4.2. For this hidden neuron, our calculations would be as follows:

$$\theta = 0.1 * -0.0011875$$

$$\theta = -0.00011875$$

The nearer we get to the input, the smaller are the changes to the weights. Now that we have the change to each input neuron, we calculate the new value to weight with Equation 4.3. Using just the top input weight in Figure 4.8 (0.1), the example calculations would be as follows:

$$w(new) = 0.1 + (0.1 * -0.00011875)$$

$$w(new) = 0.1 + (-0.000011875)$$

$$w(new) = 0.0999$$

We then update all the other input weights to this layer and the training cycle is complete. This procedure is then repeated for the next training pattern until the entire training set has been submitted.

Stopping Training

You have to stop training at some point or the network will not be able to generalize for inputs it has not previously seen. The best way to judge when to end is to track the square of the error on the output node. When this value becomes sufficiently small, it is best to stop showing the network new patterns and begin validation. How small these stopping criteria should be depends on your problem, so an iterative approach is the best method of determining it; there is no typical stopping value. One way to determine the best stopping value is to train the network over a small number of patterns for each class, validate the network (see the next step) and record this error. Then continue the training process for a few more steps and validate once more. Initially, this validation error will drop, but after time, it will begin to increase once more. The point at which you achieve the smallest error in the validation set is your stopping point.

Validation

This is checking of the operation of the network (7). Before training began, part of the data-set was set aside to form the validation data-set. Once training has finished, this validation data-set is submitted to the network. These patterns have not been seen by the network during training, so they can be used to assess the network's performance.

Validating an MLP is relatively simple. Submit the previously unseen pattern to the network through the input node, propagate the values through the network and produce a value on the output node (this is the forward pass in the learning algorithm). You know what the value should be, so the difference is the error. If the error is great, then the network has not learned properly. By adding the errors for all of the validation patterns, you can get a measure of how well the network has learnt. Once the network has been validated (i.e. acceptable validation error statistics have been obtained), the network may be used to process unknown inputs by presenting a pattern and then converting the number on the output node to something more useful, such as a dinosaur name.

Self-Organizing Map (SOM)

The self-organizing map (SOM) (Kohonen, 1982) is a clustering network that works differently from the MLP. The SOM is made up of a grid of neurons. Each neuron contains a *reference vector,* which is a set of numbers the same size as the input.

In our example, there are three dimensions to the input (e.g. height, weight, number of legs), so the reference vector will contain three numbers. The numbers in reference vectors are often called the *weights.* As before, the knowledge of the network is stored in these reference vector weights.

Some confusion may be caused between use of the term 'weights' in an MLP and in a SOM. In an MLP, the weights hold information by connecting neurons together. Training is performed by changing these weights. In a SOM, the weights still hold information and are still changed during training, but the difference is that a reference vector of weights represents a pattern in the input space – a dinosaur in our example.

The SOM grid is two dimensional (see Figure 4.9) and can contain any number of neurons. The more classes you have, the bigger the network will need to be. The neurons are not directly connected together. Rather, neurons are associated given their place in the network: nearby neurons will be affected by any activity and far away neurons will not.

Setting Up the Network

Once again, the best way of demonstrating the manner in which the SOM functions is by example. The steps taken here mimic the numbered steps given earlier in the Introduction (numbers included in brackets).

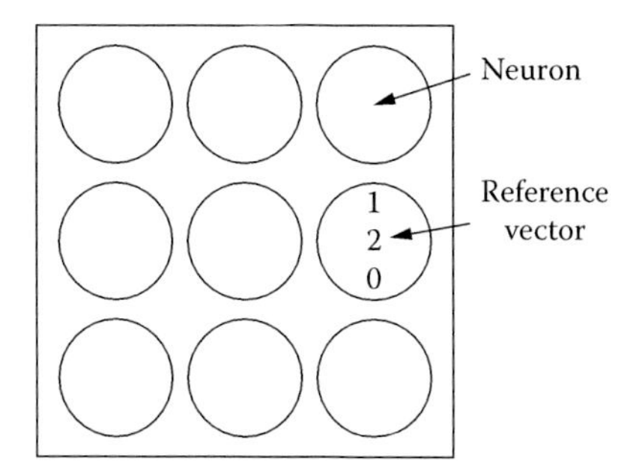

FIGURE 4.9 A self-organizing map (SOM) structure with a single neuron and a reference vector highlighted.

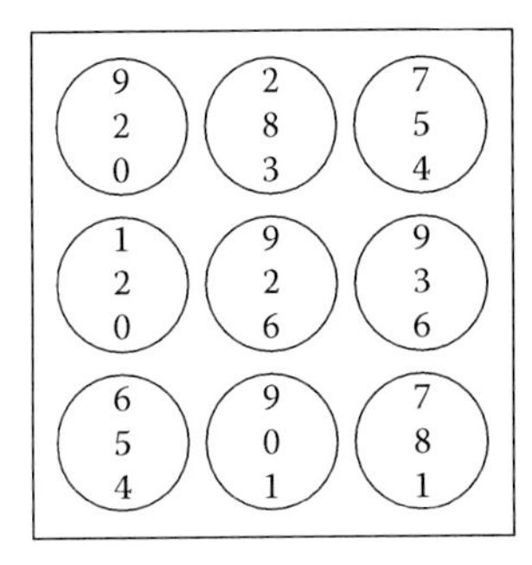

FIGURE 4.10 The SOM example with nine randomly initialized reference vectors.

For this example, we will assume that the problem (1) is to cluster the dinosaurs, given a picture. This is a clustering task, so a SOM is well suited to solving the problem (2). Data collection (3) includes all three dinosaurs: *Diplodocus*, *T. rex* and *Stegosaurus*. Our data-set would again contain many different pictures of these three dinosaurs – for example, 100 of each. Much like the MLP, the pictures are put through an arbitrary transform that turns each into three numbers, the dimensions of the input. Class 0 will be the *T. rex,* class 1 the *Diplodocus* class and class 2 the *Stegosaurus* class.

Deciding on network topography (5) is difficult for the SOM. It is normally best to start large and reduce the size of the network until its ability to cluster is reduced. This iterative process can be time consuming if you are trying to solve for a large number of classes. The weights in the reference vectors are initially set to random values – normally between 0 and 1. For simplicity, this example will use a grid of nine neurons, arranged in a 3×3 grid (Figure 4.9). The weights of the network are initialized to random values (Figure 4.10). In this example, we have used integers between 0 and 9. There are also two additional parameters that need to be established at the outset: the *learning rate* and the *neighbourhood size*.

Like the MLP, the SOM learning rate represents a measure of the how much the weights will change for each input. For our example, we will use the same learning rate of 0.1.The neighbourhood size is a measure of how many neurons will be affected by each pattern. Initially, the neighbourhood size is set to more than half the longest dimension of the network. In this case, our square network has dimensions of 3×3, so the neighbourhood size will begin at more than half this value, 2.

The Training Algorithm

Now the network is ready to be trained (6). Take the first input from the training set and apply it to the network. In our case, the randomly selected dinosaur image is a *T. rex*.

Finding the focus. The network is scanned to find the neuron that holds a reference vector closest to that of the *T. rex*. This 'winning neuron' is called the *focus*. Figure 4.11 shows the focus for the first presentation of a *T. rex* input pattern. In this example, a quick inspection of the nine nodes can provide the reader with the focus. For more complex examples, the focus is found by calculating the *Euclidean distance* between the input vector, u, and each neuron, z. The neuron with the smallest Euclidean distance is the focus.

Calculating Euclidean distance. This is a standard 'straight-line' distance between two sets of values (i.e. weights) calculated as follows:

$$e = \sqrt{(z-u)^2} \tag{4.6}$$

where u is the input pattern reference vector (1, 2, 1), z is any given neuron, l is the size of the input vector (3 in our example), and e is the Euclidean distance, an error between input and neuron.

For example, the Euclidean distance between the focus and the *T. rex* input pattern is as follows:

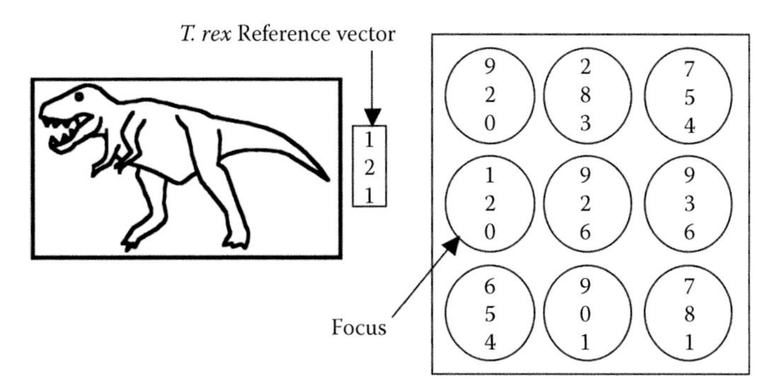

FIGURE 4.11 The focus is found for the *T. rex* input pattern. The focus is the node that has a reference vector most similar to the input pattern.

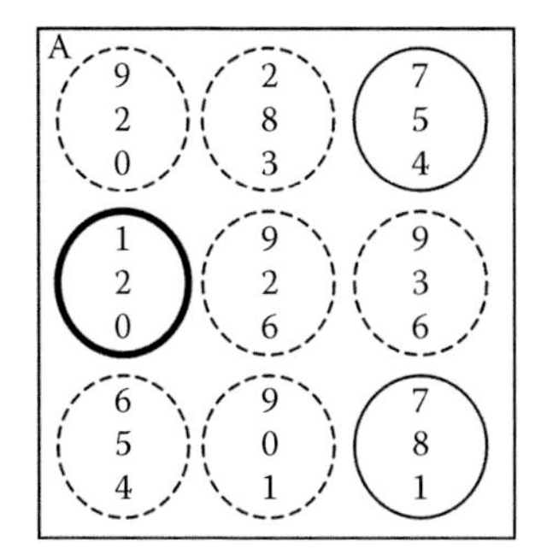

FIGURE 4.12 Updating the neighbourhood around the focus. The initial neighbourhood is the neurons with dashed outlines. Node A is used for the numerical example.

$$e = (1 - 1)2 + (2 - 2)2 + (0 - 1)2$$

$$e = (0)2 + (0)2 + (-1)2$$

$$e = 1$$

To find the focus, you would need to make this calculation for all the neurons in the network and find the smallest distance e.

Updating the neighbourhood. The next step is to update all the neurons in the neighbourhood so that they are more similar to the input pattern (and therefore the focus). The initial neighbourhood size that was decided at the start was 2 (more than half the width). Figure 4.12 shows the neurons that will be affected by the initial neighbourhood size. Those nearer the focus are updated more than those far away. There are lots of ways of updating these neurons. For simplicity, we will use a linear neighbourhood function as follows:

$$z = z + \lambda * (h/r) * (f - z) \tag{4.7}$$

where h is the neighbourhood size, f is the focus reference vector, z is the reference vector of the neuron that is going to be updated, λ is the learning rate, and r is the distance between the focus and neuron to update.

To demonstrate this update, we shall update the neuron A (in Figure 4.12). Starting with the top number in the vectors,

$$z_{\text{updated}} = 9 + 0.1 * (2/1) * (1 - 9)$$

$$= 9 + (0.1 * 2 * -8)$$

$$= 9 + (-1.6)$$

$$= 7.4$$

Running through this for the rest of the vector will give a new neuron A of (7.4, 2, 0). The other values of the reference vector did not need to change. All of the neurons in the neighbourhood are updated this way. Once the neighbourhood is updated, we move on to the next input pattern.

Reducing the neighbourhood function. After showing the network many patterns, the neighbourhood size should be reduced. This allows the network to cluster similar classes tightly together and push apart dissimilar patterns. The speed at which the neighbourhood function is reduced depends entirely on the domain and will need to be tuned to get the network's best performance.

Stopping Training

The training process should be stopped when the error between the focus and the input patterns becomes small. This is a delicate value to set because, if the training continues too long, the network will over-fit to the data and will lose its ability to generalize. If the network has over-fit to the training data, then it might not correctly classify the validation patterns.

Validation

Validation is performed by applying the validation patterns one by one to the network and then recording the location of the focus for that pattern. For each dinosaur picture, you should expect that the focus locations will be in the same area on the grid of neurons.

Result of the SOM. The result of the training of the SOM can be seen in Figure 4.13. Three neurons are labelled B, C and D. Each of these neurons represents the centre of the cluster for each animal: *T. rex, Diplodocus* and *Stegosaurus,* respectively. Among these three neurons is a gentle gradient. The network does not indicate which neurons are associated with which dinosaur. To find out which cluster is associated with which dinosaur, example patterns from each class can be shown to the network to see which neuron becomes the focus. That node can then be labelled.

In this example, each dinosaur was different enough for none to be classified as similar to each other. However, in more complex problems with more dinosaurs added to the training set, similar dinosaurs would be grouped together in the network automatically. This is a powerful clustering technique that can be used to find groupings within your data-set that may not be immediately obvious. The SOM does require some inspection after training to extract this information.

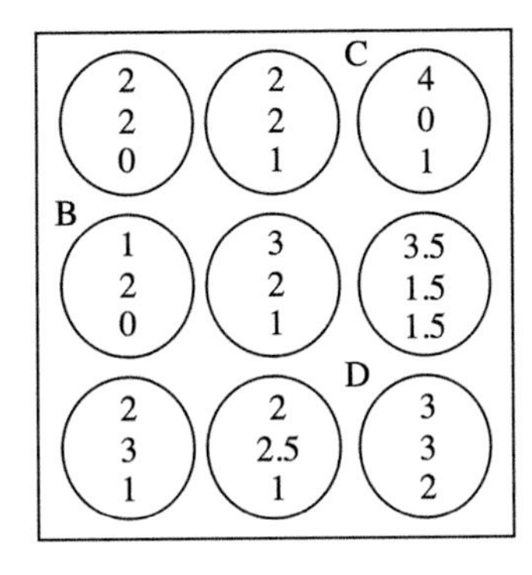

FIGURE 4.13 The result from training. Neurons B, C and D represent the *T. rex, Diplodocus* and *Stegosaurus,* respectively.

Dynamic Neural Networks

A dynamic neural network is a neural network that can alter its structure so that it can always learn new classes and forget old classes. With traditional neural networks you must decide the structure before training. This can be very difficult as the network structure is often needed to determine the statistical properties needed to set the structure in the first place! A dynamic neural network's structure will grow or shrink in size and complexity to meet the needs of the problem.

The second benefit of dynamic neural networks is that the training set is not fixed. In our example, if you wanted to add a pterosaur to the classes held by the network, then you would have to reset the topology and begin training again. For small problems (such as our example), this is a valid process. But for large problems with many classes and patterns in each class, retraining may take months. Dynamic neural networks can modify their own structure to accept the new class without the need of retraining from the start.

Good examples of dynamic neural networks are the cascade correlation (Fahlman and Lebiere, 1990), which extends the MLP to add structure to its network; the growing neural gas with utility (GNG-U) (Fritzke, 1997), which uses a SOM-like structure to learn the shape of the data; and the plastic self-organizing map (PSOM) (Lang and Warwick, 2002), which is very similar to the SOM except that the neurons are not fixed in a grid and new neurons may be added and removed. The PSOM will be described here.

The Plastic Self-Organizing Map (PSOM)

The plastic self-organizing map (PSOM) is a SOM-like structure consisting of neurons and *links*. Like the SOM, a neuron contains a reference vector, but unlike the SOM each neuron is not bound in a grid. Instead, each neuron is connected to other neurons using a dynamic link (Figure 4.14).

Like the SOM, a neuron represents one pattern in the input space (a particular picture of a dinosaur). The link represents the similarity between the two neurons it connects. A smaller value link means that the two neurons represent patterns that are similar (e.g. two different pictures of a *T. rex*) and large link values mean that the neurons represent dissimilar patterns (e.g. one picture of a *T. rex* and one of a *Stegosaurus*). The lengths of the links can be altered by the network and, if any links grow very long, they may be broken. This has the result of breaking dissimilar groups of neurons apart, giving very distinct clusters.

When a new class (dinosaur) is presented to the network for the first time, a new group of neurons is added to the network to represent this new class. The network does not know that the new class is called a *T. rex,* of course, so it labels the cluster numerically. Later, an observer (either human or another computer algorithm) can inspect the neurons by showing a pattern of a class they know and then label the neuron from 1 to *T. rex.*

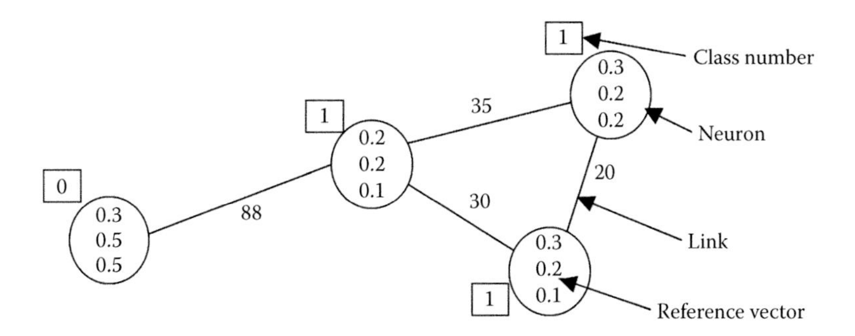

FIGURE 4.14 The structure of the plastic self-organizing map (PSOM) showing four neurons, four links and their reference vectors. Class numbers are shown in boxes next to the neuron.

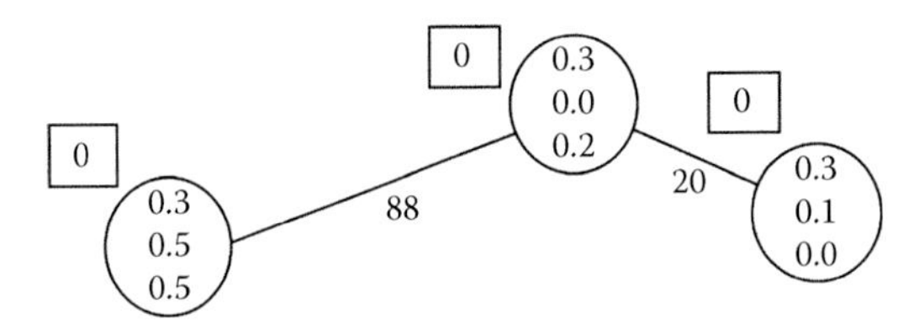

FIGURE 4.15 The initial structure of the PSOM, the reference vectors inside the neurons and the link lengths are all set to random values. The class numbers are all set to 0, signifying neurons that existed at the start.

Setting Up the Network

The steps given here mimic the numbered steps given in the Introduction (numbers included in brackets). For this example, we will assume that the problem (1) is to cluster the dinosaurs, given a picture, and to work out their own classes. This is both a clustering and classification task (2). Data collection (3) includes two dinosaurs initially, *T. rex* and *Diplodocus*. The data-set would include many different examples of these dinosaurs. The classes do not need to be numbered as the PSOM will give each class its own number. Also, the data-set needs to be scaled between 0 and 1, as the data are scaled between 0 and 10 (i.e. divide each value by 10).

The PSOM topology (5) is easy to set up as it is the same for any problem domain. Figure 4.15 shows the initial PSOM setup of three neurons and two links. For the typical PSOM, each link is randomly set between 20 and 90 and, for our example, the reference vectors are randomized between 0 and 9. The class label on these neurons starts off as 0.

PSOM networks need five parameters to operate. The *node-building parameter, a_n,* controls the ease at which neurons are added to the network. There are some heuristics for setting this parameter, but a discussion of these is beyond the scope of this chapter (see Lang, 2007). However, a good way of finding a node-building parameter is to (1) choose a value in the middle of the range 0–1, (2) train the network on a number of representative classes and (3) determine whether the network forms itself into a sufficient number of distinct groups (one for each class). If the required number of groups is not formed, reduce the parameter value by 10 per cent and train again. For our example, the node-building parameter will be 0.3.

The next parameter to set is the *cluster threshold, a_{cl}*. This controls what specifies the size of a class. Initially, this value can be set to 90 per cent of the node-building parameter. In our example, this is 0.27 (90% of 0.3). The cluster threshold affects the *maximum link length, a_r* which is the length at which links will be cut. This parameter can be set to any value except 90, but it is not very important because link tuning can be performed using other parameters. Finally, the *link-ageing parameter, b_a,* controls the rate at which the links grow with each training iteration. If in doubt, set this value to be less than 1 and then increase it through experimentation. For the sake of this example, the link-ageing parameter for our example will be set to 3.

Training Algorithm

Now the network is ready to begin training (6). Take the first input from the training set and apply it to the network. In this case, the randomly selected dinosaur is the *T. rex*. The training algorithm is as follows:

1. Find the focus.
2. Check for node building and either:
   a. Update the focus and its neighbourhood
   b. Add more neurons
3. Age all the links.
4. Remove long links.
5. Remove unlinked neurons.

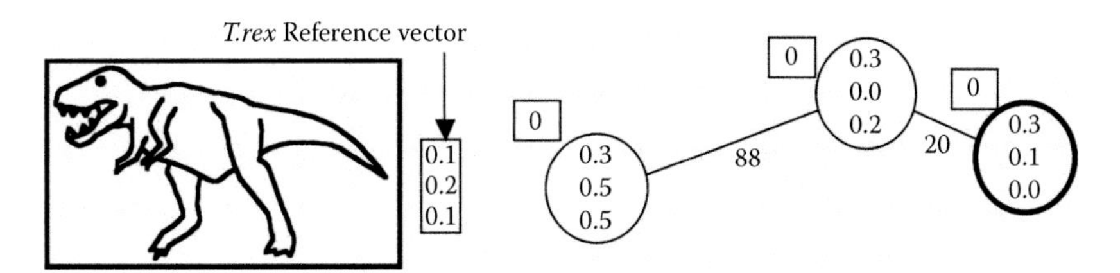

FIGURE 4.16 The first *T. rex* pattern is shown in the network. The focus is found as the neuron most similar to the *T. rex* and is outlined in black.

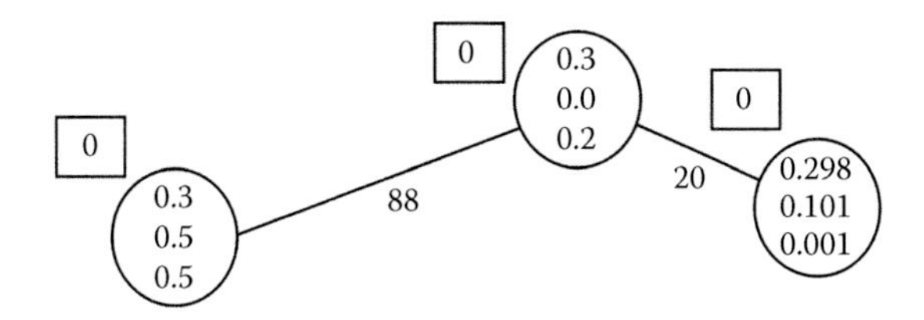

FIGURE 4.17 The network before the neighbourhood update, showing an updated focus. The focus only has one neuron in its neighbourhood (the central one), so it is the only one to be updated.

Finding the focus. This step is exactly the same as with the SOM. The Euclidean distance, e, between the input and each neuron is calculated and the neuron with the smallest distance is called the *focus* (shown in Figure 4.16). The Euclidean distance between the focus and the first *T. rex* input vector is approximately 0.25.

Checking for node building. The Euclidean distance between the input and the focus is now compared to the node-building parameter, a_n. If the Euclidean distance is larger than a_n, a new group of neurons is created. If the Euclidean distance is smaller than a_n, no neuron is added to the neighbourhood. For clarity, adding new neurons is dealt with later. In our example, the Euclidean distance is 0.25 and the node-building parameter is 0.3. Therefore, no neurons are added at this time.

Updating the focus. The focus is now updated to be made more similar to the input (Figure 4.17). For this, a *learning rate* is used. The learning rate operates in a similar way to the PSOM's static counterparts. If in doubt, use a small learning rate.

$$\Delta z = \lambda(u - z) \tag{4.8}$$

where Δz is the change in the focus, u is the input, z is the focus and λ is the learning rate.

An example calculation follows:

$$\Delta z = 0.01 * \left(\begin{bmatrix} 0.1 \\ 0.2 \\ 0.1 \end{bmatrix} - \begin{bmatrix} 0.3 \\ 0.1 \\ 0.0 \end{bmatrix} \right)$$

$$\Delta z = 0.01 * \left(\begin{bmatrix} -0.2 \\ 0.1 \\ 0.1 \end{bmatrix} \right)$$

$$\Delta z = \begin{bmatrix} -0.002 \\ 0.001 \\ 0.001 \end{bmatrix}$$

This is the change to the focus; now we can update the focus as follows:

$$z_{new} = z_{current} + \Delta z \tag{4.9}$$

The example calculation continues as follows:

$$z_{new} = \begin{bmatrix} 0.3 \\ 0.1 \\ 0.0 \end{bmatrix} + \begin{bmatrix} -0.002 \\ 0.001 \\ 0.001 \end{bmatrix}$$

$$z_{new} = \begin{bmatrix} 0.298 \\ 0.101 \\ 0.001 \end{bmatrix}$$

Updating the neighbourhood. As the neurons are not held in a grid, the neighbourhood for the PSOM comprises those neurons connected to it via links. In this example (Figure 4.18), the focus has only one neighbour, shown by the dotted outline. The neighbourhood update is performed in two steps. First, all the links between the focus and its neighbours are updated. Then, the reference vectors in each of the neighbourhood neurons are updated. The new link length is a proportion of the maximum link length and Euclidean distance between the focus and the neighbourhood neuron.

$$c = a_r * e_{zx} \tag{4.10}$$

where c is the new link length, a_r is the maximum link length and e_{zx} is the Euclidean distance between the focus and the neighbourhood neuron.

The example calculation continues as follows:

$$c = 90 * 0.22$$

$$c = 20$$

If there were any more neighbours of the focus, the links would be updated in the same way.

The second step is to update the reference vector of the neuron. This is where the cluster threshold is used. If the Euclidean distance between the focus and the neighbourhood neuron is

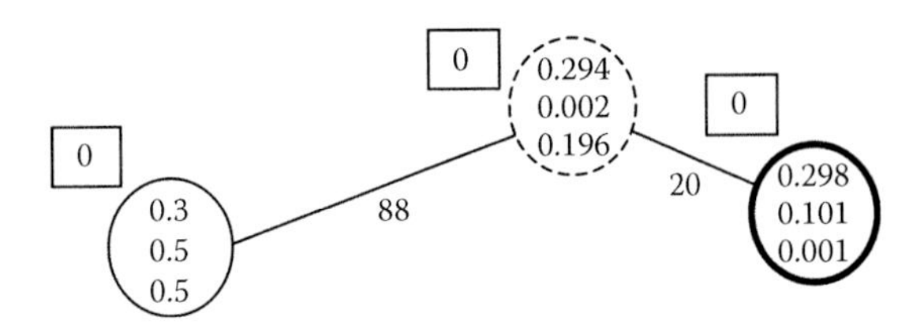

FIGURE 4.18 The network after the neighbourhood update.

above the cluster threshold, then we push the neuron away; otherwise, we move the neighbourhood neuron towards the input.

For our example, the cluster threshold is 0.27 and the Euclidean distance between the focus and its neighbour is 0.22. Therefore, the neighbourhood neuron will be pulled together. First, let us look at this 'pull-together' equation:

$$\Delta x = \frac{c}{100} * \left(z - u\right) \tag{4.11}$$

where x is the neighbourhood neuron to be updated, Δx is the change in the neighbourhood neuron to be updated, z is the focus, and c is the value of the link between the focus and x.

Continuing the example calculation,

$$\Delta x = 0.2 * \left(\begin{bmatrix} 0.298 \\ 0.101 \\ 0.001 \end{bmatrix} - \begin{bmatrix} 0.3 \\ 0.0 \\ 0.2 \end{bmatrix} \right)$$

$$\Delta x = 0.2 * \left(\begin{bmatrix} -0.02 \\ 0.101 \\ -0.199 \end{bmatrix} \right)$$

$$\Delta x = \begin{bmatrix} -0.006 \\ 0.002 \\ -0.04 \end{bmatrix}$$

This is the change in the neighbourhood neuron; the new value of the neighbourhood neuron is given by an equation analogous to Equation 4.9:

$$x_{new} = x_{current} + \Delta x \tag{4.12}$$

Accordingly, the example calculation is given as

$$x_{new} = \begin{bmatrix} 0.3 \\ 0.0 \\ 0.2 \end{bmatrix} + \begin{bmatrix} -0.006 \\ 0.002 \\ -0.04 \end{bmatrix}$$

$$x_{new} = \begin{bmatrix} 0.294 \\ 0.002 \\ 0.196 \end{bmatrix}$$

This process is repeated for all of the neurons connected to the focus. The pulled neighbourhood neurons also take on the class number of the focus. In this case, all the class numbers are 0, so there is no change. When the Euclidean distance between the focus and the neighbourhood neuron

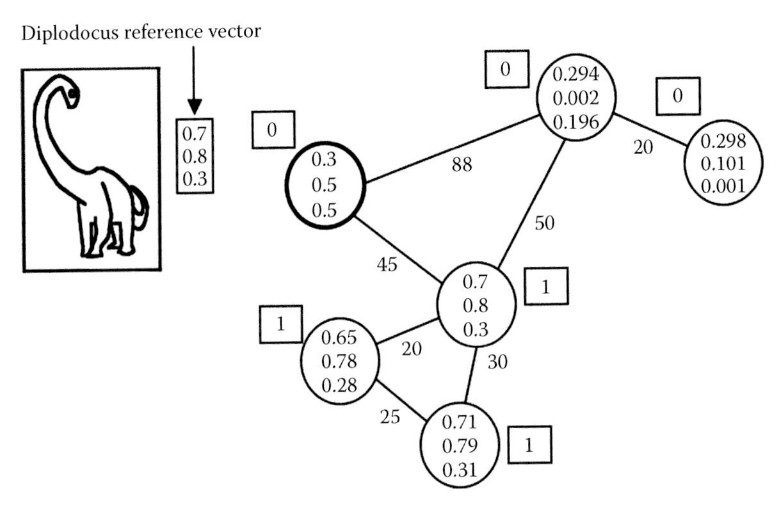

FIGURE 4.19 A new node group is added to the network, connected to the focus and a neighbour of the focus. The new group is given the class '1'.

is larger than the cluster threshold, the neuron to be updated is pushed away by changing the node update, using the inverse of Equation 4.12:

$$x_{new} = x_{current} - \Delta x \tag{4.13}$$

Adding neurons. After finding the focus, if the Euclidean distance between the focus and the input is larger than the node-building parameter, then a neuron group is added to the network. If the *Diplodocus* was the next class shown to the network, a new neuron group would be added, as shown in Figure 4.19. A group of three neurons and five links has been added to the network. The new neurons are created with reference vectors similar to that of the input. The new links are created and their lengths set as if they were being updated in the neighbourhood update. Each new neuron is labelled automatically with the next class number. In this case, it is the class number 1 as it is the first new class to be added to the network.

Link ageing. Link ageing is the next step of the training process. All the links in the network are increased by b_a. Then, all of the links longer than the maximum link length, a_r (90 in our example), are removed. Any neurons that are then unconnected are removed.

Removing neurons. Neurons are removed when there are no links attached to them. This serves to remove those neurons that are associated with noise.

Validation

Validation is performed by manual inspection of the network after the training has been completed. A selection of known patterns is shown to the network to help the user understand how the network is formed.

Using the PSOM for Classification Alone

For data-sets that do not have much noise (i.e. you know that every pattern you show to the network is valid), it might be desirable to set the link ageing parameter to 0. This will reduce the network's capability to reject noise but will allow the network to retain classes that only have a few example patterns in the input space.

Result of the PSOM

The PSOM does not have a finished state like a static network does. Instead, training of the network is paused to analyse the structure. If further patterns were to be shown to the network, it would

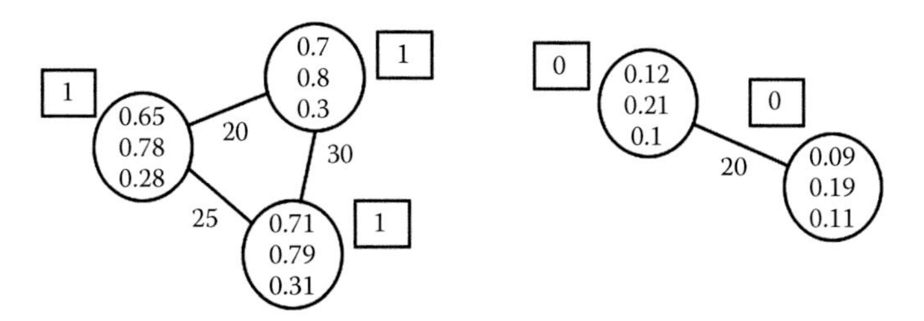

FIGURE 4.20 The result of training the PSOM. Two groups of neurons representing the two classes shown to the network. The two neurons of class 0 represent the *T. rex* and the neurons of class 1 represent the *Diplodocus*.

continue learning the new patterns. Figure 4.20 shows the result of our example after 150 pattern presentations. The two classes are represented as distinct groups in the network as the links joining these patterns have been broken through neurons being pushed apart and link ageing. If, in the future, we wish the network to learn the *Stegosaurus* class as well, we would not need to retrain with the *T. rex* and the *Diplodocus* classes. In this case, we would set the link ageing parameter to 0 (otherwise, we might lose the existing knowledge in the data-set) and show the *Stegosaurus* patterns to the network.

In more complex problems, where the classes are more similar, the groups might not be distinct and similar patterns would be joined together. The links in this case would be of medium length. This is a desirable result because it is often interesting to find classes that are similar within the data-set. If dissimilar classes are expected, adjustment of the node-building parameter and the cluster threshold would be the first priority.

For data-sets that do not have much noise (you know that every pattern you show to the network is valid), it might be desirable to set the link ageing parameter to 0. This will reduce the network's capability to reject noise but will allow the network to retain classes that only have a few example patterns in the input space.

DISCUSSION AND SUMMARY

In this chapter, we have examined three examples of neural networks both analytically and numerically. Before data can be used by these networks, they will need to be transformed to facilitate the learning process. Neural networks can be used to classify and cluster large data-sets. A feed-forward network (such as an MLP) can be used for classifying input classes so that an unknown pattern shown to the network can be classified as one of the learned classes. A clustering neural network, such as the SOM, groups similar classes together. Static neural networks are useful for problems where the data-set is finite and fixed before training. However, a dynamic neural network, such as the PSOM, does not stop learning and can be used for open data-sets where not all the data are found at time of training.

To be useful for most real-world problems, the neural network must be embedded as part of a larger system and, quite often, the algorithm is modified to better fit the problem. Not every problem can be solved with neural networks, but they have been shown to provide interesting results where other analytical techniques have failed.

REFERENCES

Beale, R. and Jackson T. (1990) *Neural Computing – An Introduction,* Taylor & Francis, London, 256 pp.

Fahlman, S.E. and Lebiere, C. (1990) The cascade correlation learning architecture. In *Advances in Neural Information Processing Systems,* Vol. 2 (ed D.S. Touretzky), Morgan Kauffman Publishers Inc, San Francisco, pp. 524–532.

Fritzke, B. (1997) A self-organizing network that can follow non-stationary distributions. In *International Conference on Neural Networks* (eds W. Gerstnre et al.), Springer–Verlag, Lausanne, pp. 613–618.
Haykin, S. (1999) *Neural Networks, a Comprehensive Foundation,* 2nd edition, Prentice Hall, New York, 852 pp.
Kohonen, T. (1982) Self-organized formation of topologically correct feature maps. *Biological Cybernetics,* 43: 59–69.
Lang, R.I.W. (2007) *Dynamic Neural Networks.* PhD thesis, University of Reading.
Lang, R.I.W. and Warwick, K. (2002) The plastic self-organizing map. In *WCCI Proceedings of the International Joint Conference on Neural Networks* (ed D. Fogel), IEEE, Honolulu, pp. 727–732.
Rosenblatt, F. (1958) The perceptron: a probabilistic model for information storage and organization in the brain. *Psychological Review,* 65: 386–408.

5 Morphometrics and Computed Homology: An Old Theme Revisited

Fred L. Bookstein

CONTENTS

Introduction 69
How We Got Here: A Whig History of Morphometrics 70
But We Have No 'Null Model' 71
Equation-Free Sketch of a Miraculous Mathematical Rescue 72
A Biometric Speculation: Intrinsic Warps 75
Implications for the Notion of a Computed Homology 78
An Example 78
Concluding Optimistic Postscript 79
Acknowledgements 80
Notes 80
References 80

INTRODUCTION

Arising in numerical taxonomy in the 1960s, the modern toolkit of geometric morphometrics migrated (by random walk, it seems, in retrospect) into crainiofacial biology, then statistical science. It is now centred mostly in computational anatomy and physical anthropology. During these *Wanderjahre,* a solid consensus developed about the handling of simplicial data such as landmark points, curves and surfaces for a variety of classic biometric purposes. Yet, taxonomy has not noticeably benefited. It is time to bring the diaspora back home to systematics, by reformulating morphometrics' statistical foundations to align with the demands of computer-assisted intelligence for taxonomy. This chapter, which reports work in progress, is intended as a sharp initial provocation to that end.

Specifically, we now have quite good graphics for shape variation and shape change for two- and three-dimensional data, but our models of what is uninformative – of noise – lag badly. The Procrustes distance metric, in particular, embodies profound geometric symmetries that engender well-known pathologies and paradoxes when applied to the typical problems of biometric systematics. Evidently, their symmetry is not a particularly biological one. This chapter will introduce a strikingly different sort of symmetry, a *self-similarity of noise* – symmetry of scale rather than of digitizing error that seems quite a bit more suited to systematics' domain. Its statistical implementation, based on a profound original insight by John Kent and Kanti Mardia (University of Leeds), aligns very closely with the thin-plate splines on which morphometricians have become accustomed to rely for some of their core interpretive graphics.

The term used in the stochastic literature for these models is *intrinsic random fields* (IRFs); they appear under that name in textbooks of geostatistics (e.g. Cressie, 1991; Chilés and Delfiner, 1999). This chapter also introduces the *intrinsic warps* (IWs) intended to replace my principal and partial warps of 1989 for data analysis of landmark-like configurations under symmetries of scale. These IWs could well implement the movable 'potentially homologous characters' that may someday underlie computer-assisted algorithms for stable ordinations or ecophenotypy surveillance much better than existing techniques could likely do.

HOW WE GOT HERE: A WHIG HISTORY OF MORPHOMETRICS

Presented by the stratagem of Whig history (the present as culmination of the past), the technical path to the morphometrics of 2005 appears to have begun about 40 years ago, in a fusion of R.E. Blackith's 1965 book chapter, 'Morphometrics', with P.H.A. Sneath's (1967) suggestion 2 years later of a formal trend-surface analysis for transformation grids. The 1971 book by Blackith with Richard Reyment crystallized the early adaptive radiation of this approach using the existing multivariate toolkit.

This was too early, but not by much. Independent developments over the 1970s and 1980s in stochastic processes and in interpolation theory – David Kendall's 1977 Riemannian shape space (which most of us learned about from Kendall, 1984) along with the underlying process of shape diffusion on which it was based, and Duchon and Meiriguet's thin-plate spline of 1975 for globally optimal point-driven interpolation in higher dimensions, which most of the field learned about from me (Bookstein, 1989) – combined with biometrics in the *morphometric synthesis* emerging in the mid-1990s that unified a huge range of applications seemingly in every field *except* systematics (e.g. craniofacial biology, computational neuroanatomy, physical anthropology and paleoanthropology; for reviews, see Bookstein,1998, or the chapters of Marcus et al., 1996, and, for anthropology in particular, Slice, 2004).

Here in 2007, we can now handle data in the form of discrete named landmark points, curves and surfaces from a wide variety of geometric data sources (photographs, solid medical images, surface scans) in two or three dimensions. We can accommodate the special cases, quite common in practice, for which landmark points arise from curves, or landmark curves from surfaces. Indeed, this approach, the *simplicial* decomposition of a single form, has become the most common way of building a coherent geometric representation (see Bookstein, 2004) that compromises between the enormous over-representation by surface meshes and the equally severe under-representation by conventional discrete point schemes.

This issue was central to the main German-speaking tradition of anthropometrics (c.f. Martin, 1928), but then lay fallow until the last few years, when computing caught up with the anthropologist's nuanced eye. The key was using the bending energy of the thin-plate spline to generate a computed homology on curves between landmarks or surfaces between curves (Bookstein, 1997a; Gunz et al., 2004). By this maneuver, the Blackith–Reyment multivariate toolkit that translated from length and angle measures to shape coordinates continues to apply to the much richer domain of arbitrarily complete simplicial decompositions without any substantial modification.

Techniques of innovative graphics and biologically informed creative play arise very conveniently by exploring this simplicial synthesis with algebraic or geometrical ingenuity. One can exploit a new thin-plate spline that treats growth gradients as part of the trend rather than the warping term, or animate shape change by grids in three-dimensional navigating along homologous trajectories. I review many of my current favorites from this newly adaptive toolkit in Bookstein (2004): for instance, a powerful new approach to description of asymmetry, or the spectacular animation sequences that, freeze-framed, serve very nicely as cover images for peer-reviewed journals (Bookstein et al., 2002) and as workbenches for 'virtual dissection' and other physically impossible approaches to gross and fine biological structure more generally.

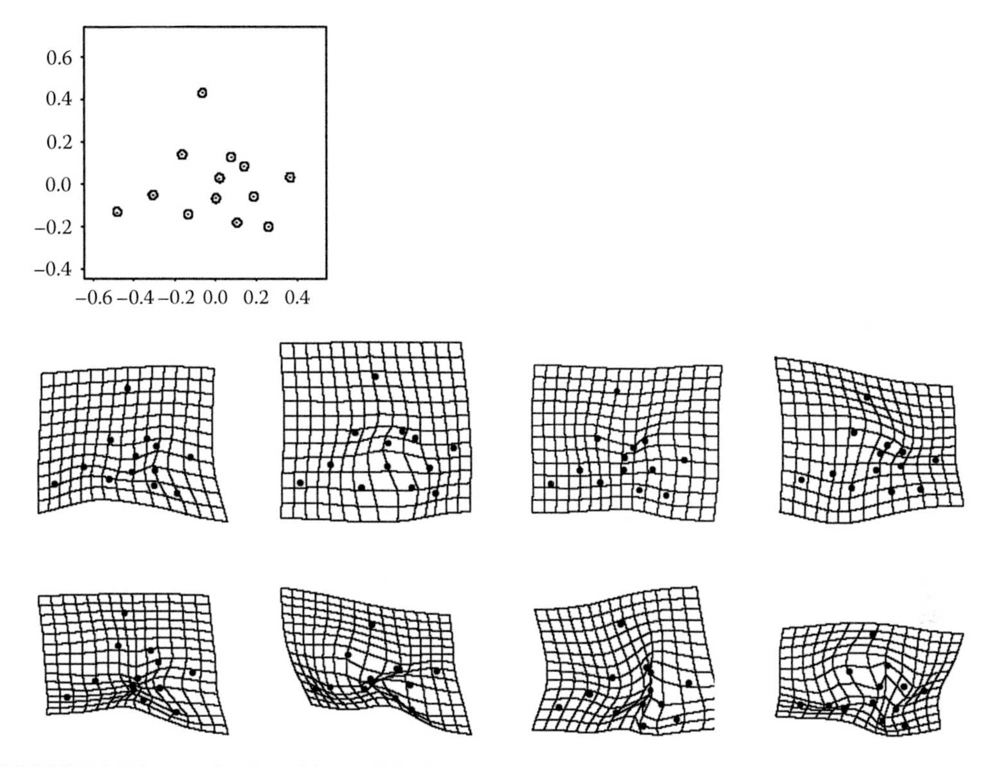

FIGURE 5.1 Diagram for the critique of the isotropic Mardia–Dryden distribution as a basis for biomathematical interpretation. Upper left: the distribution: independent isotropic (circular) Gaussians at each landmark separately. The radius of the circles, standard deviation of the variance of each shape coordinate, is taken as 0.02 (on a scale for which the centroid size of the mean form is 1.0). Second row: deformations between four randomly selected pairs of forms from the distribution above, drawing too much attention to the region in the middle of the form. Lower row: the same for twice the standard deviation. The behavior of the grids is now clearly unacceptable for biological interpretations.

BUT WE HAVE NO 'NULL MODEL'

Yet, there is a paradox underlying all these successes: the metric we use for assessing signal strength of the *Procrustes distance* whose least-squares decomposition is at the core of all our linear models and singular-value decompositions does not meaningfully represent the complementary space, the space of shape noise. Figure 5.1 is a sketch of the underlying dilemma. The Procrustes metric per se corresponds to a model for variation (absence of signal) in the form of an isotropic (circular) noise distribution at each landmark point independently. Here that model is drawn out in the immersive Cartesian space (the 'original digitizing plane') for a textbook example, the 13 midsagittal brain landmarks of the deQuardo–Bookstein schizophrenia study. These points are 13 landmark locations from magnetic resonance images of a thick slice up the middle of the adult human brain (see DeQuardo et al., 1996, and other references). At the upper left, a graphic lays out the independent circular Gaussian variation postulated at each of the 13 mean landmark locations, independently.

This is a distribution of shapes. If we draw the corresponding distribution of wholly unpatterned (uninformative) shape differences, we get a somewhat disturbing pattern of 'focal' deformations, as in the four independent realizations of the middle row. Our systematist's eye is drawn too often to the region near the centre, where the original landmarks were densest. This is an aspect of the cognitive psychology of the scientist that should not be permitted to bias the equivalent sort of descriptions whether pursued by man or by machine (see other chapters of this book). Under less

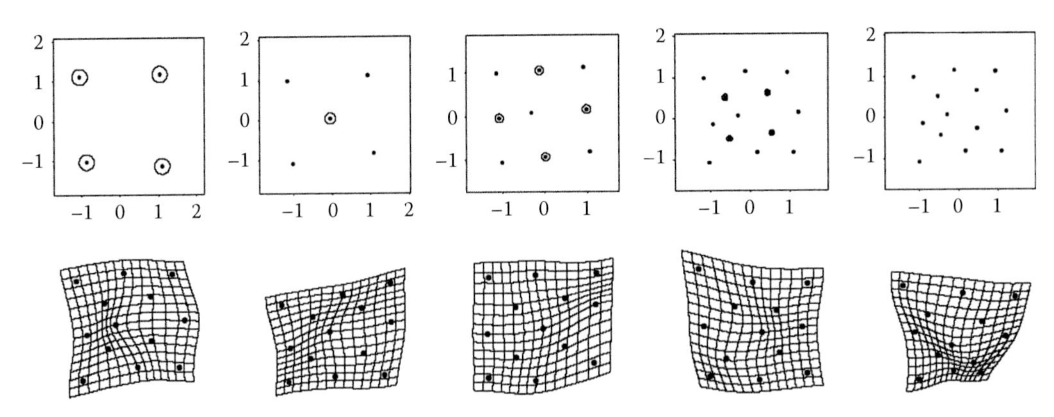

FIGURE 5.2 A possible remedy: decline of the scale of variations in keeping with the spacing of landmarks. The diagram shows this in a hierarchical fashion, even though the resulting shape distribution can be written out quite simply as a non-isotropic Mardia–Dryden. Upper row: sketch of four steps for a systematically refined square grid. Lower row: five deformations from the template to five forms sampled from this distribution. The variability 'feels' much more biological, showing features of comparable import at a variety of scales.

favorable noise models, the situation is even worse: in the lower row are the same simulations for twice the digitizing error sketched at left. Now there is a substantial chance that the induced splines are not even diffeomorphisms. The resulting writhing is unlikely to be of any particular systematic value.

Evidently there is a problem here. Whereas the organism, in managing its developmental, evolutionary or ecophenotypic shape changes, necessarily pays considerable attention to issues of landmark spacing, our null model pays them no attention at all. To represent truly uninformative variability in a biologically informed way, we need to constrain variances at landmarks to the spacing of other landmarks in their vicinity. Such a step-down model, because it would induce correlations among neighboring landmarks arising strictly from their adjacency in the mean form, could be transcribed into the class of general *Mardia–Dryden* models by breaking the symmetries of the Procrustes metric. It is still a content-free interpretation, a null model but a more promising one. Figure 5.2 shows, in the top row, a sequence of stages in the construction of such a simulation, based on successive centred squares (quincunxes). A sample of the corresponding simulations, following, is much more biologically realistic, with differences at many different scales, and without as much of that horrid crushing or folding.

Pursuing this notion naïvely, one might imagine a sort of hierarchical scheme, analogous to wavelets, such that digitizing noise was simulated in descending order of spatial neighborhood size. These figures hint at the explicit possibility of signals emerging at large and small scales at the same time. There thus begins to emerge the kernel of a feasible approach to the modeling of noise in a non-Kendallian, but biologically more realistic way. In my view, this is precisely what is needed for the computer-aided systematics applications: a mathematical formalism that corresponds to the actual cognitive psychology of the working systematist, who naturally looks for characters at whatever scale they can be seen to emerge – from tiny details of function or ornament all the way up to gross aspects of shape. We had not created such a noise model before mainly because nobody working in this domain had asked for one.

EQUATION-FREE SKETCH OF A MIRACULOUS MATHEMATICAL RESCUE

When we got to this mathematical-yearning stage in the construction of transformation grids, fresh insight came from an unexpected mathematical source. One day in 1985, after a colloquium presentation of mine, a kindly Seattle mathematician, David Ragozin, noted helpfully that the interpolant we were using at that time (a piecewise finite-element scheme) really should be replaced by a principled new approach from interpolation theory, the thin-plate spline, that had far greater mathematical elegances and symmetries. This was itself new mathematics, not previously applied in systematics (or indeed anywhere in biology).

For geometric morphometrics, and indeed this chapter, to become possible, an analogous transfer had to be made from a different subfield. In an obscure proceedings contribution of 1994, Professors John Kent and Kanti Mardia of the University of Leeds noted that the thin-plate spline, which was already proving so mysteriously helpful for our morphometric visualizations, was actually the solution of an optimal *prediction* problem, and was furthermore the unique solution to that problem that was *self-similar,* with the same noise spectrum at every spatial scale. The mathematics in this case originated (again in France) with Georges Matheron, around 1970; again, it had not previously been applied anywhere in biomathematics, let alone in systematics.

About 5 years ago Mardia repeated this statement to me – this time a little more pointedly – noting, in particular, that the algebra we had been using under the spline for a full 20 years actually includes the covariance structure of this noise model in the explicit kernel of the spline bending energy. With the usual choice $U(r) = r^2 \log r$, the matrix

$$K = \begin{pmatrix} 0 & U_{12} & \cdots & U_{1k} \\ U_{21} & 0 & \cdots & U_{2k} \\ \vdots & \vdots & \ddots & \vdots \\ U_{k1} & U_{k2} & \cdots & 0 \end{pmatrix} \tag{5.1}$$

right in the centre of the spline is (in one of those classic mathematical puns that lies at the root of all major inventions – the Gaussian distribution, for example) at *the same time* the covariance kernel of an optimal kriging predictor for an *intrinsic random field* (IRF) in-between the landmarks, a prediction field independent of the overall affine drift term (in morphometrics, the uniform term). This prediction is self-similar, with the same covariance structure at every spatial scale.

The word 'intrinsic' in this context, means 'independent of large-scale drift'. Just as a random walk is examined without keeping track of where zero actually was, these deformations should be examined without any attention being paid to what the linear drift (the uniform term) actually was. They are *purely* local and hence 'intrinsic'. In other words, there is a price to be paid for adopting this noise model. The information about the uniform term must be sequestered to be dealt with separately. This price seems fair as we are well used to thinking in this way from many earlier applications (e.g. Rohlf and Bookstein, 2003).[1]

I will skip all the algebra, which is written out in full in a manuscript (Mardia et al., 2006) currently under review. What I *can* show is a collection of useful and suggestive graphics: simulations of pure noise, and then an analogue for the spline's 'partial warps', which I would like to call 'intrinsic warps', that may help serve as a bridge from conceptualizing entities behaving like theorems to entities behaving like parts of organisms – in short, to possible characters underlying a mostly automated taxonomy and systematics.

One can draw out the intuitive consequences of self-similarity on simple and regular grids. Figure 5.3 is a sample of twelve deformations of a starting square 10×10 grid. What 'self-similar' means here is, roughly, that the distribution of the non-uniform part of shape of *every* square of the original square grid, whether on side 1, $\sqrt{2}$, 2, 4 or whatever, is the same. You may wish to examine this figure carefully. It is the equivalent of the 'bell curve' for a pure shape noise process that begins with a starting form that is exactly square. To permit the drawing, one must specify the uniform part of the shape deformation. Here that was done by fixing the isosceles right triangle at lower left, upper right and lower right corners of the square. The intuition that these stand for

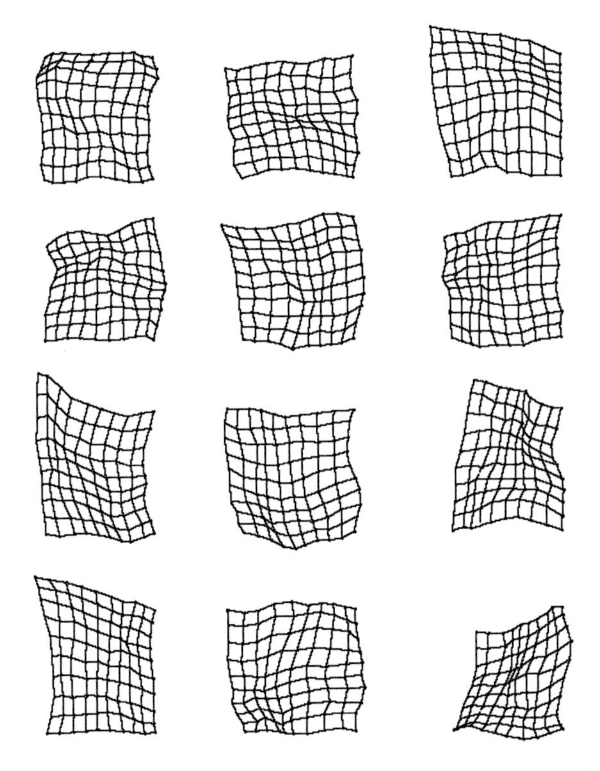

FIGURE 5.3 Twelve realizations of the same intrinsic random field on a 10×10 grid. Registration is to the lower left, upper left and upper right corners. At this level of complexity, every grid appears to have reportable features, but all are meaningless by explicit construction.

taxonomic variability is almost overwhelming; yet, I assure you, these simulations are pure noise – the same shape variation in every small square regardless of scale. But how reminiscent they are of real image data-sets (see other chapters of this volume)! Features are *suggested* at so many different geometric scales that if we did not know *a priori* that this variation is all the result of induced noise, we would likely be ready to begin preparing taxonomic keys.

Even more compelling, because its size encompasses possibilities of greater complexity, is the 20×20 grid example in Figure 5.4. Here your eye cannot help but see the overall shape change from square to elongated trapezoid, the pinching at upper centre and a couple of foci of almost isotropic expansion at upper right and at lower centre. *Nevertheless, none of these are real; none of the parameters of the visible local features are encoded in the programming in any way.* By explicit construction of the transformation (which includes no local information whatever), all are pure noise. It is this model that I commend to the systematist reader as the proper equivalent of the Gaussian noise model for image-driven deformation data.

With this many grid points, we can explicitly verify the self-similarity by actual computation. In Figure 5.5 are the shape distributions of the non-uniform part of the little squares of size 1, 2 and 4 from the grid preceding. The distribution is of four corners at the same time, with each form being a deviation from the evident average parallelogram shown. Do not worry about their orientation; the theorems underlying this process are completely rotatable, so the distribution applies to diamonds, to squares of knight's-moves, and to anything else that can be put down in the Minkowski geometry. Clearly, these distributions are the same, regardless of scale. In other words, the 'overall' trapezoid was as likely to apply at any lower spatial scale instead, and likewise the growth centre, etc. that our systematist's eyes could not help seeing.

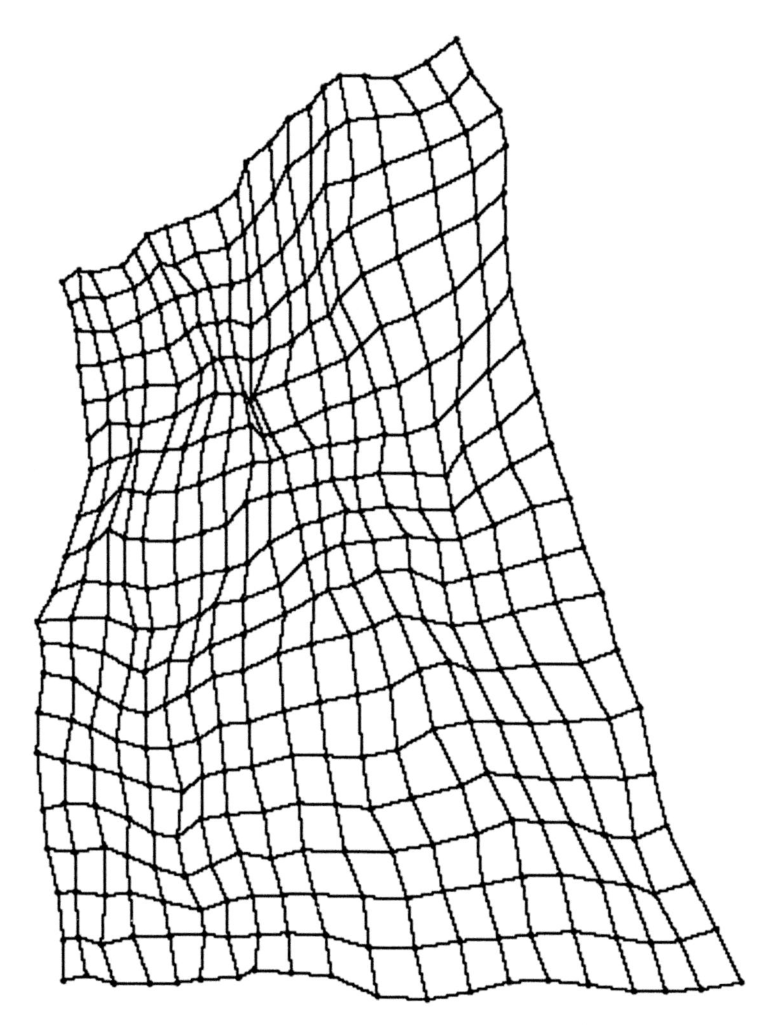

FIGURE 5.4 An example with even more realistic complexity: one 20×20 realization of the same process. It is very difficult to convince oneself that a structure this strongly patterned can nevertheless represent pure noise.

A BIOMETRIC SPECULATION: INTRINSIC WARPS

So we have a noise model. What kinds of scientific descriptions can we pursue with its aid? By analogy with the bending energy of the thin-plate spline, which led to useful formalisms such as time partial warps, the gauge metric of these self-similar processes likewise has an eigenstructure, which I have tentatively named the *intrinsic warps* (IWs) of the mean point configuration. Keep in mind that these exist only in the non-uniform subspace of the overall shape space, though they apply at all scales.

The morphometric synthesis of the 1990s exploits the principal warps – eigenfunctions (which are not scale free) of the bending energy matrix – as a basis in which to verify sphericity of the non-uniform part of a landmark-borne deformation and also to describe deviations from that sphericity that connote systematic factors of change at different spatial scales. That analysis was not multiscale. It was, in fact, conditioned on the actual landmark mean locations, but we often forgot to say that in our applications papers. The more appropriate statistic, which I hope will be found suited for routine applications to systematic data-sets, compares the observed variation instead to the noise model described earlier. If my guess is correct, both the search for characters and the

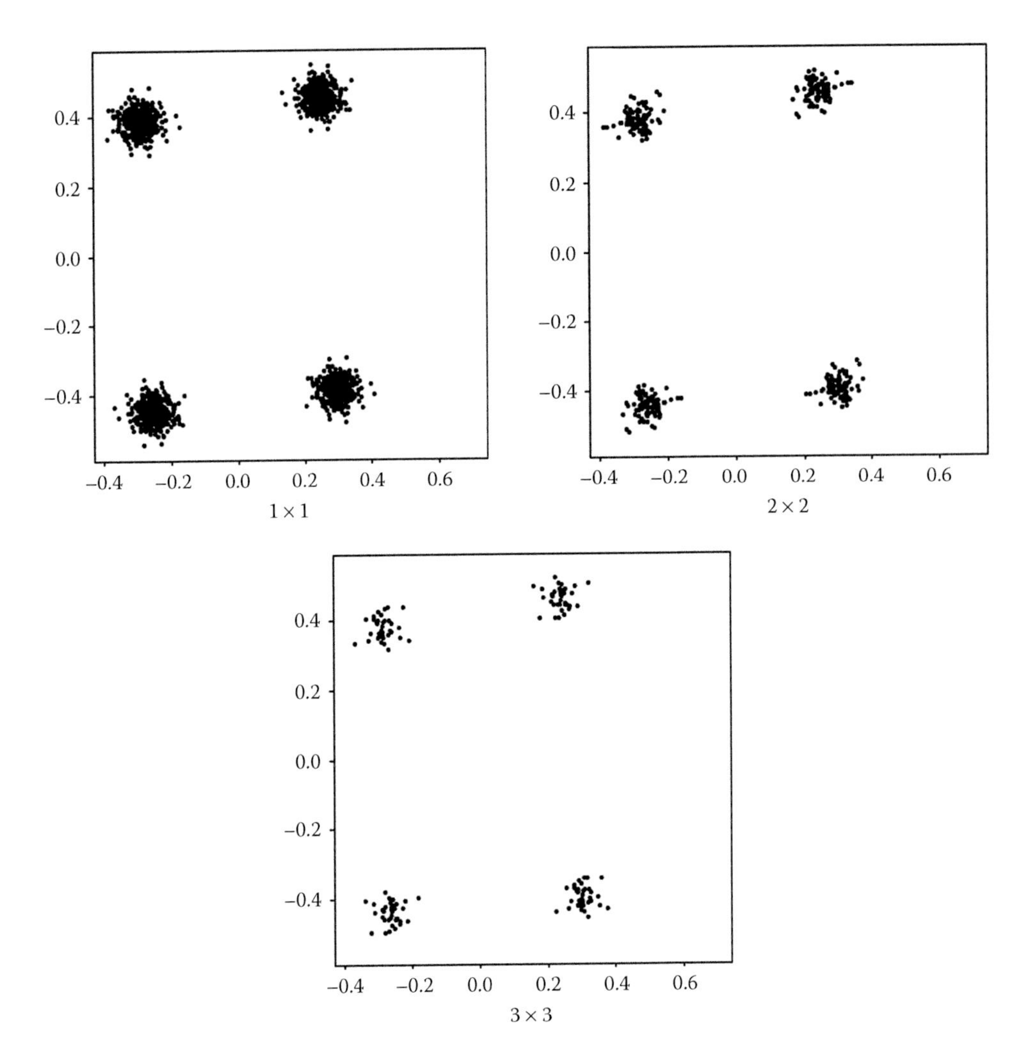

FIGURE 5.5 Demonstration of the self-similarity of the model underlying the preceding figure: identity of the shape distributions of squares of side 1, 2, amid 4 from the grid.

modeling of prior knowledge in bench applications of automated taxon recognition systems ought to be conditioned on neither the bending energy of the spline nor the Procrustes geometry of the same mean locations, but instead on *this* covariance structure.

Figure 5.6 shows all 10 of these eigenvectors for the configuration of mean landmark locations in Figure 5.1. As the 'drift' (affine) term needs to be registered (c.f. Figure 5.3), it is registered to a large triangle with fixed vertices at top, left margin and right margin of the mean form. In other words, the shape and orientation of this specific triangle are unchanging over the panels of the figure. Each deformation has been drawn as if applying to the x-coordinate of the landmark configuration, but as there was room in the figure matrix for two additional panels, I have added the representation of the first pair as applied to the y-coordinate as well. Panel labels set out the eigenvalue for the underlying IRF model, dimension by dimension, as scaled to a value of unity for the first. These warps are vaguely similar to the more familiar *partial warps* for the same configuration (see Bookstein, 1997b), but they are not in fact the same.

The interpretation of the IWs likewise differs from that of the partial warps. Whereas the largest scale partial warp, for instance, represents a pattern of singularly low bending energy for displacements in a gradient across the width of this configuration (Bookstein, 1991), the IW of largest

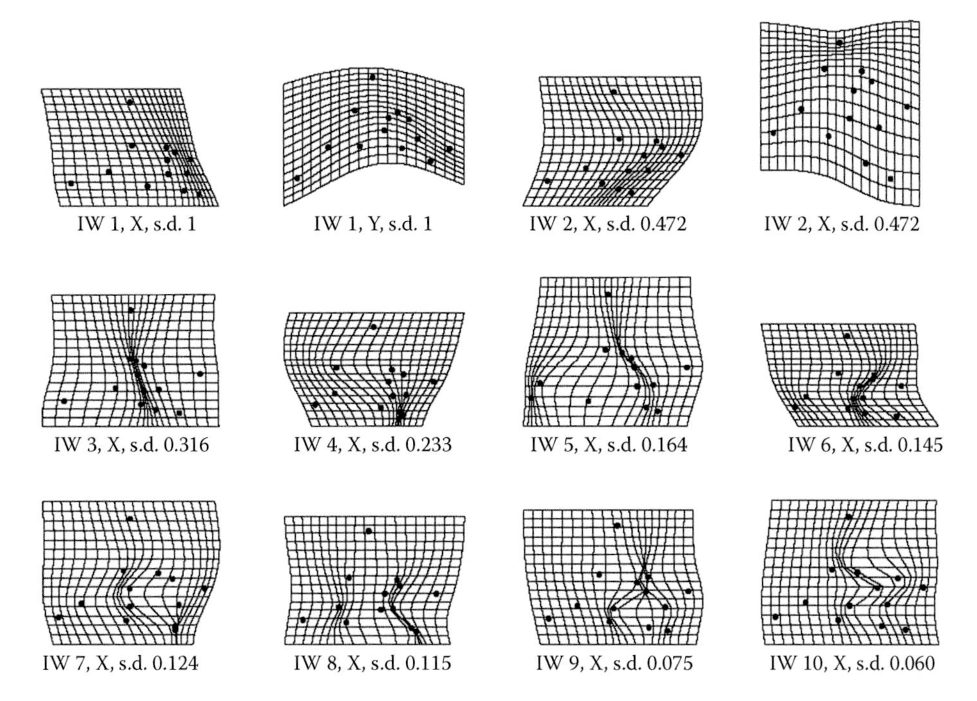

FIGURE 5.6 All 10 intrinsic warps (IWs) for the mean form driving the simulations in Figure 5.1. Each is drawn in the horizontal direction except for the first and second, which are drawn both horizontally and vertically.

expected variance (Figure 5.6, first row, first two panels) represents the pattern of greatest variance on the self-scaling IRF model. This is, after registration, the relative displacement of the middle of the mean configuration with respect to the ends, but the feature is a pattern of covariance, not a bending. Indeed it represents the long-sought linear statistic for the Pinocchio effect!

Similar language applies to any other panel in Figure 5.6. For instance, the third IW (second row, left panel) conveys a pattern of relative enlargement at left and right, together with relative compression centrally, because this is something that self-scaling patterns will do at the amplitude indicated by the label under the panel: namely, about one third of the amplitude of the end-to-end gradient. The fifth IW offers, at roughly half the amplitude (0.164 as against 0.314), the pattern at smaller scale (hence the smaller amplitude) involving expansion of the left end against the middle, and so forth, down to the patterns of smallest variance of all, which, like the last partial warps, deals with the highly damped rearrangements of the landmarks at closest mean spacing, just as in Figure 5.2.

Gratifyingly enough, these warps may have a clear systematic meaning. *They may actually look like morphometric characters*. Even though multiscale, they are 'character-istic'. As a rough generalization, where the conventional partial warps are analogues of growth gradients trying to integrate changes smoothly over larger and larger regions of the form, these intrinsic warps are rather the opposite, trying to concentrate covariation spatially in just a few domains of the form's pattern space. The intrinsic warps of Figure 5.6, in fact, emerge in an order of increasing complexity corresponding to the number of different landmark-like features (points, edge elements [Bookstein and Green, 1993; Bookstein, 1999] and curves) that could emerge as features of the corresponding deformations. Each is a description of a *potential form of systematic organization* of the system of identified points as a whole, but that is, more or less, exactly what systematists have always hoped a morphometric character could be. The neural-net procedures reviewed elsewhere in this volume would have such features emerge out of the self-organization of a net; here they emerge, at least as a description space, from the corresponding self-organization of self-similarity over multiple spatial scales.

IMPLICATIONS FOR THE NOTION OF A COMPUTED HOMOLOGY

The title of this chapter refers to a topic that has dogged morphometrics ever since the early days of computational image analysis. A deformation grid is a mapping between extended regions of a Cartesian space, but the data available to the biologist do not often take the form of such a map. There are some exceptions, as reviewed in Bookstein (1978); for instance, sheets of dividing cells such as plant meristems can sometimes be followed in full detail in an effectively continuous geometry. Regardless, the biotheoretical notion of homology is fundamentally centred on the discretization of an organism into finite parts. Re-use of this word to apply to a mapping function is a semantic subterfuge, a tacit extension of language, which my Michigan Morphometrics Group in 1985, following a suggestion of Sneath's (1967), called a *computed homology*.

There is then a very tight connection between the IRF models of the present discussion and the idea of a limit to the precision that can be claimed for any computed homology between the landmarks, at points where there is no actual landmarked information. Inside the smallest cells of a landmark configuration, the IRF models imply irreducible uncertainty of how points correspond between specimens. Maps cannot be interpolated to any claimed precision better than what the residual stochastic models supply. In effect, everywhere except precisely at the landmark locations, the grid hues of a thin-plate spline are 'thick' hues – not mathematical curves, but fuzzy ribbons. This corresponds to the current epistemology of applied geometry in many of its other applications, where the corresponding lower bound is set by the aperture of the imaging device (e.g. Koenderink, 1990). The literature of morphometrics has long needed a vocabulary for attending to the empirical absurdity that every point landmark is actually claiming to represent.

It is one implication of this approach that more care needs to be taken with the selection of landmark points than is presently the case. The allocation of a landmark halfway between two others, or in the middle of a surface path bounded by some curves, adds information to a data-set *even if the landmark is located exactly where a spline driven* by *the remaining information would place it.* It adds, precisely, the information that *that* location was observed with only digitizing error, so points nearby have a prediction error that is lower simply by virtue of the additional data. Where landmarks are widely spaced, the deformations predicted by, say, a thin-plate spline are far less reliable than where landmarks are precise *regardless of the statistics of the landmark locations themselves.*

The IRF models here might be considered the first feasible quantitative representations for the uncertainty of a deformation grid, a computed homology, as a whole. Models of the uncertainty of imputed parameters (uniform terms, growth gradients, etc.) are not equivalent, as they fail to accord with the self-scaling property that gives this particular model its power and coverage. For decades, the attention of mathematical biologists has been distracted from fundamental issues of descriptive anatomy, such as the nature of claimed correspondences across multiple instances of the 'same' developmental program or its evolutionary changes. One purpose of this chapter is to encourage other theoretical biologists to attend likewise to this foundational issue, with an eye toward patching it up for the evo–devo advances of the twenty-first century, through all the corresponding mathematics.

AN EXAMPLE

Attend again to the increment space in Figure 5.6. These were components of variance on the null (IRF) model, but there is an actual data-set here, combining two groups of human subjects (14 schizophrenic patients and 14 normal employees of the University of Michigan Hospitals at the time). An appropriate empirical eigenanalysis would summarize the shape dimensions that show variance most in excess of what is predicted for their scale. The equations (omitted here) match those of Bookstein (1991) in combining both x- and y-components of the shape coordinates. Computation is by a classic relative eigenanalysis, empirical covariance structure

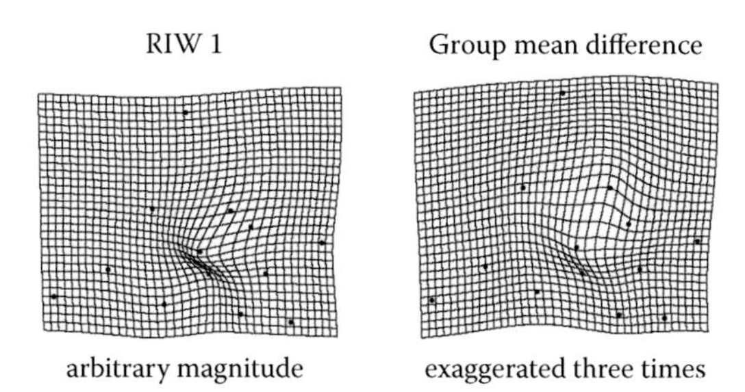

FIGURE 5.7 The first relative intrinsic warp (RIW) for the real data of DeQuardo et al. (1996) for two groups of brain images compared to the actual group mean difference between schizophrenics and their psychiatric ward staff as found there. The RIW emphasizes the last few IWs from Figure 5.6.

against modeled covariance structure, in the resulting 20-dimensional space of non-affine shape for these 13 landmarks. (For a note on relative eigenanalysis in morphometrics, see also Bookstein et al., 2002.)

The first three relative eigenvalues (for observed covariance structure of the increments as a ratio to IRF1-modeled) are in the proportions 65:33:28, indicating that we should pay most attention to the very first. Figure 5.7 shows, at left, the plot of this particular relative eigenvector, the first relative intrinsic warp (RIW), and, at right, the mean difference of average shape between the two groups. The similarity of these two panels is indeed remarkable. The first RIW precisely centres on the region that is responsible for the group difference, and gets almost the right increment (the extra variance in the shape of that particular little triangle). In other words, if we choose to abuse the term 'taxonomic character' by applying it to an ordination like this one, the IRF method has found the correct character at high signal amplitude (ratio of 2:1 of those first two relative eigenvalues) and with great spatial accuracy. One can ask nothing more of an empirical ordination (see Blackith and Reyment, 1971).

No earlier multivariate analysis of this data-set has ever located this crucially important region by such a simple linear method. See DeQuardo et al. (1996), Bookstein (1997b, 1997c) and even Dryden and Mardia (1998), in all of which this signal is produced not by algebra, but by visual inspection of spline diagrams. Only the method of creases (Bookstein, 2000) has teased it out before, but crease analysis is a complex computation involving the location of a hierarchy of extremal derivatives all over the coordinate mesh in which the landmarks are embedded. The technique also resembles the approach to relative warps with $a = -1$, the exponential parameter for weighting by bending energy introduced in Bookstein (1996). When α is negative, there is no problem with the uniform term, as it is multiplied by $0^{-\alpha}$ and hence vanishes.

CONCLUDING OPTIMISTIC POSTSCRIPT

So here is my working hypothesis: if there is landmark-like covariance structure in your warped image data-set, these new relative intrinsic warps (relative now with respect to the IWs instead of the old PWs), may have a fair chance of finding it. In other words, *ordinary biometric principal components with respect to the new 'featureless' covariance structure – relative intrinsic warps* might be the nearest thing yet that morphometrics can offer as a formalism based in image deformation that nevertheless refers to the sorts of features systematists have typically found useful in thinking about their taxonomic task.

If biometricians other than I can likewise unearth examples where this highly modified relative warp analysis leads to valuable ordinations, the further areas of extension would become obvious: to (a)symmetry, to growth and development, to ecophenotypy, to evolutionary selection gradients, ... to all the kinds of explanatory factors we nowadays apply to shape space covariance structures 'by hand'. This characterization of landmark covariations, then, may represent a core of the next morphometrics toolkit, the fusion of computerized image processing with morphometrics in the new computer-assisted systematics.

Of course, Darwin 'is in the details'. Still, if my guess is right, this is a potentially major new technical research programme. Volunteers are sought, and readers should feel free to speculate on any or all of this as they peruse the rest of this volume's text and *its* applications to pressing taxonomic problems. By insisting on the priority of systematics thought styles – in particular, on the relative irrelevance of scale to characters – we may end up reshaping morphometrics as radically as systematics did once before, in the 1960s, with the initial thrusts of Blackith, Reyment and Sneath.

I commend the self-similar model to your attention as a potentially central new tool in the toolkit corresponding to your purpose as collated in the meeting of which these are the proceedings. Not without irony, I also put it forward as yet another area in which Europe must take the lead over its ex-colonies toward the salvaging of our systematic and biometric heritage.

ACKNOWLEDGEMENTS

This work could not have been carried out without the extraordinarily wise and generous support of my good friend Kanti Mardia. It was partially sponsored, however inadvertently, by grant P01 CA 87634 from the U.S. National Institutes of Health to Charles R. Meyer, University of Michigan, for studies of cancer therapy. I am grateful to Horst Seidler, longtime chair of the Department of Anthropology, University of Vienna, for generously supporting this and many other initiatives in new morphometric techniques under the rubric of grant P200.093/1-VI/2004 from the Austrian Council for Science and Technology to the Institute for Anthropology, University of Vienna, Austria, and to the departments of statistics and psychiatry, University of Washington, for jointly supplying an American home complementary to the European home in Vienna.

NOTES

1. In passing, we may have also solved a puzzle famously called to our attention by Cohn Goodall, at the Princeton Shape Theory Workshop of 1989, when he noted that the thin-plate spline is obviously quite useful for a variety of biometrical applications, but that nobody had the slightest idea *why*. This might be the underlying reason why: it is capable of tracking meaningful changes at any or all spatial scales at the same time.

REFERENCES

Blackith, R.E. (1965) Morphometrics. In *Theoretical and Mathematical Biology* (eds T.H. Waterman and H. Morowitz), Pergamon, Oxford, pp. 225–249.

Blackith, R.E. and Reyment, R.A. (1971) *Multivariate Morphometrics,* Academic Press, London, 412 pp.

Bookstein, F.L. (1978) The measurement of biological shape and shape change. *Lecture Notes in Biomathematics,* 24: 1–191.

Bookstein, F.L. (1989) Principal warps: thin-plate splines and the decomposition of deformations. *IEEE Transactions on Pattern Analysis and Machine Intelligence,* 11: 567–585.

Bookstein, F.L. (1991) *Morphometric Tools for Landmark Data: Geometry and Biology,* Cambridge University Press, Cambridge, 435 pp.

Bookstein, F.L. (1996) Combining the tools of geometric morphometrics. In *Advances in Morphometrics* (eds L.F. Marcus, M. Corti, A. Loy, G.J.P. Naylor and D.E. Slice), NATO ASI Series A: Life Sciences, volume 284. Plenum, New York, pp. 131–151.

Bookstein, F.L. (1997a) Landmark methods for forms without landmarks: localizing group differences in outline shape. *Medical Image Analysis,* 1: 225–243.

Bookstein, F.L. (1997b) Biometrics and brain maps: the promise of the morphometric synthesis. In *Neuroinforniatics: An Overview of the Human Brain Project. Progress in Neuroinforniatics* (eds S. Koslow and M. Huerta), volume 1, Lawrence Erlbaumn, Mahweh, NJ, pp. 203–254.

Bookstein, F.L. (1997c) Shape and the information in medical images: A decade of the morphometric synthesis. *Computer Vision and Image Understanding,* 66: 97–118.

Bookstein, F.L. (1998) A hundred years of morphometrics. *Acta Zoologica,* 44: 7–59.

Bookstein, F.L. (1999) Linear methods for nonlinear maps: procrustes fits, thin-plate splines, and the biometric analysis of shape variability. In *Brain Warping* (ed A. Toga), Academic Press, London, pp. 157–181.

Bookstein, F.L. (2000) Creases as local features of deformation grids. *Medical Image Analysis,* 4: 93–110.

Bookstein, F.L. (2002) Creases as morphometric characters. In *Morphology. Shape, and Phylogeny* (eds N. MacLeod and P. Forey), Systematic Association Special Volume Series 64, Taylor & Francis, London, pp. 139–174.

Bookstein, F.L. (2004) After landmarks. In *Modern Morphometrics in Physical Anthropology* (ed D.E. Slice), Kluwer, Amsterdam, pp. 49–71.

Bookstein, F.L., Chernoff, B., Elder, R., Humphries, J., Smith, G. and Strauss, R. (1985) *Morphometrics in Evolutionary Biology. The Geometry of Size and Shape Change, with Examples from Fishes.* Academy of Natural Sciences of Philadelphia, 277 pp.

Bookstein, F.L. and Green, W.D.K. (1993) A feature space for edgels in images with landmarks. *Journal of Mathematical Imaging and Vision,* 3: 231–261.

Bookstein, F.L., Sampson, P.D., Connor, P.D. and Streissguth, A.P. (2002) The midline *corpus callosum* is a neuroanatomical focus of fetal alcohol damage. *The Anatomical Record the New Anatomist,* 269: 162–174.

Chilés, J.-P. and Delfiner P. (1999) *Geostatistics,* Wiley, New York, 720 pp.

Cressie, N. (1991) *Statistics for Spatial Data,* Wiley, New York, 928 pp.

DeQuardo, J.R., Bookstein, F.L., Green, W.D.K., Brurnberg, J. and Tanidon, R. (1996) Spatial relationships of neuroanatomic landmarks in schizophrenia. *Psychiatry Research: Neuroimaging,* 67: 81–95.

Dryden, I.L. and Mardia, K.V. (1998) *Statistical Shape Analysis,* Wiley, New York, 376 pp.

Gunz, P., Mitteroecker, P. and Bookstein F.L. (2004) Semilandmarks in three dimensions. In *Modern Morphometrics in Physical Anthropology* (ed D.E. Slice), Kluwer, Amsterdam, pp. 73–98.

Kendall, D.G. (1984) Shape-manifolds, procrustean metrics, and complex projective spaces. *Bulletin of the London Mathematical Society,* 16: 81–121.

Kent, J.T. and Mardia, K.V. (1994) The link between kriging and thin-plate splines. In *Probability, Statistics, and Optimization: A Tribute to Peter Whittle* (ed F. Kelly), Wiley, New York.

Koemderink, J.J. (1990) *Solid Shape,* MIT Press, Cambridge, MA, 715 pp.

Marcus, L.A., Corti, M., Loy, A., Naylor, C.J.P. and Slice, D.E. (1996) *Advances in Morphometrics,* NATO ASI Series A: Life Sciences, volume 284, Plenum, New York, 578 pp.

Mardia, K.V., Bookstein, F.L. and Kent, J.T. (2006) Intrinsic random fields and image deformations. *Journal of Mathematical Imaging and Vision,* 26: 59–71.

Martin, R. (1928) *Lehrbuch der Anthropologie in Systematischer Darstellung,* C. Fischer Verlag.

Rohlf, F.J. and Bookstein, F.L. (2003) Computing the uniform component of shape variation. *Systematic Biology,* 52: 66–69.

Slice, D.E. (2004) *Modern Morphometrics in Physical Anthropology,* Springer, Berlin, 384 pp.

Sneath, P.H.A. (1967) Trend surface analysis of transformation grids. *Journal of Zoology, London,* 151: 65–133.

6 The Automated Identification of Taxa: Concepts and Applications

David Chesmore

CONTENTS

Introduction 83
Structure of an Automated Taxon Identification System 84
Sensor 85
Acoustic Sensors 85
Visual Sensors 86
Active Sensors 87
Preprocessor 87
Feature Extractor 87
Classifier 88
Practical ATI Systems 90
Applications of ATI Systems 91
Acoustic Applications 91
Image/Visual Applications 91
Applications Involving Other Sensors 92
The Future of ATI 92
Summary 93
References 94

INTRODUCTION

The automated identification of objects is not a new concept, only a concept that is new to many systematists. The overwhelming majority of research in automated object identification is being carried out in engineering, computing and robotics (e.g. Pasquariello et al., 1998; Yakcun and Bozma, 1998; Koha et al., 2002; Nilubol et al., 2002). There is a large body of research in these fields spanning the past four decades (Moganti et al., 1996; Desouza and Kak, 2002; Egmont-Peterson et al., 2002), but only relatively recently have truly automated approaches to identification been applied to non-human biological problems.

Early work on automated identification concentrated on image processing applications such as recognition of Hymenoptera (Daly et al., 1982) and pollen using scanning electron microscope (SEM) images (Langford et al., 1986, 1990). More recently, there has been an increased research effort in this area, partially due to advances in technology and also because of the 'taxonomic impediment'. It is becoming increasingly evident that the shortfall in professional taxonomists is not being addressed with sufficient vigour (Gaston and O'Neill, 2004). Automated identification systems have the potential to aid taxonomists and parataxonomists in sorting tasks, and for other routine applications such as rapid biodiversity assessment, long-term conservation/ecological monitoring and enhanced ecological studies (Riede, 1993). Acceptance of automated taxon identification (ATI) within the systematics community has been slow and generally thought too difficult to achieve or a threat to employment (Gaston and O'Neill, 2004).

The development of ATI is an inherently multidisciplinary task, requiring knowledge of application areas, taxonomic groups, computing and electronics technology. ATI is part of 'computer-aided taxonomy', a definition for which was proposed at the inaugural meeting of the BioNET–INTERNATIONAL Group for Computer-aided Taxonomy as 'the application of any computer or computer technique for taxonomic purposes' (Chesmore, Yorke, Bridge and Gallagher, 1997, p. 3) and is divided into three branches as follows.

1. Automated taxon identification (ATI) (originally designated automated species identification).
2. Computer-aided identification keys, including Internet-based keys.
3. Numerical techniques for systematics such as cladistics, genome searching and numerical taxonomy.

This chapter deals exclusively with ATI, which has many potential applications that may be divided, in broad terms, into the following categories:

1. Rapid sorting of samples from traps or vacuum samplers (ATI can provide identification to order, family or genus to aid parataxonomists in sorting).
2. Rapid biodiversity assessment.
3. Long-term unattended monitoring (low-power field deployable ATI systems can provide new data at a much higher sampling frequency than is possible with human surveys; systems capable of operation continuously for months are feasible with identifications occurring in near real time).

STRUCTURE OF AN AUTOMATED TAXON IDENTIFICATION SYSTEM

An ATI system can take two forms: fully automated and semi-automated (Chesmore, 2000). Fully automated ATI does not require any user interaction and is suitable for long-term unattended operation – for example, in forest biodiversity studies. Semi-automated ATI is simpler to achieve and produces a partial identification or an identification that is verified by the user. The latter is likely to be more acceptable in the short term as it keeps the user 'in the loop'. Semi-automated systems also allow for a degree of user interaction – for example, to select objects within an image field of view for identification. In terms of schematic design, an ATI system is basically a standard pattern recognition system, as shown in Figure 6.1, and is divided into four distinct functional blocks: sensor, preprocessor, feature extractor and classifier. Each stage is described next in more detail.

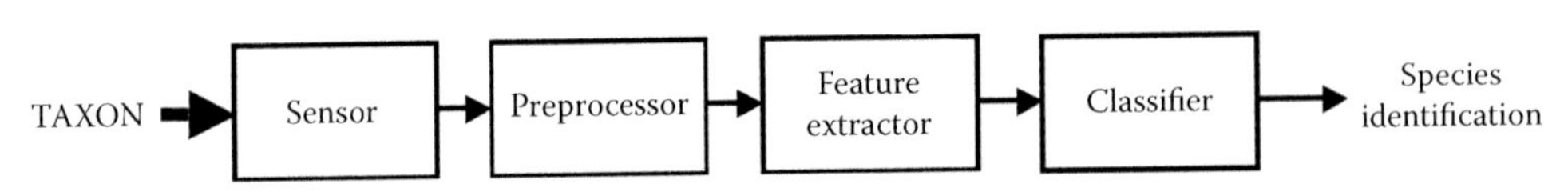

FIGURE 6.1 Block diagram of and automatic taxon identification system based on classical pattern-recognition system design.

Sensor

A sensor uses real-world signals as input and converts these into electrical energy. The range of sensors available for measurement of real-word signals is very wide and includes temperature, light, acoustic, vibration, chemical, humidity, acceleration, radar, sonar, cameras, flow, etc. However, when considering ATI, the range of sensors is more restricted. For example, a humidity measurement is not likely to provide any useful signal capable of distinguishing between taxa. The following sections describe the most common sensors currently employed for ATI. This list is not exhaustive and new developments – for example, in chemical sensors in the form of electronic noses (Hines et al., 1999) – may lead to new applications such as distinctive odour and pheromone detection.

Acoustic Sensors

Many animals produce acoustic signals as a means of communications or as a by-product of activity such as flying, eating and locomotion. The majority of acoustic signals occur within the human auditory frequency range (20 Hz–20 kHz), but some occur at lower frequencies known as infrasound (e.g. African elephants, *Loxondonta africana,* and various species of whale) (McComb et al., 2003) and higher frequencies termed ultrasound (20–200 kHz or higher; e.g. bats, bush-crickets, some cetaceans). Many sounds are substrate borne (e.g. Hemiptera; see Cokl and Virant-Doberlet, 2003) and wood-boring beetle larvae (Farr and Chesmore, 2005); these require a vibration sensor, or accelerometer, to detect signals.

Selection of an appropriate acoustic sensor is dependent on the application, the taxonomic group to be identified and the type of signal generated, whether infrasound, ultrasound, human hearing range or vibration. There is a wide variety of commercially available sensors for airborne signals in the human hearing range (microphones); sensors for other frequency ranges are more specialized.

The most important factors to be considered when choosing a microphone are its frequency response and directionality. Selection of appropriate sensors will always be application dependent. Highly directional microphones are useful for maximizing signal level, but need to be oriented toward the sound source, which usually requires human operation. Parabolic dishes are also used for increasing directional sensitivity, but affect the frequency response, which may be disadvantageous in some applications. Ultrasonic sensors, particularly piezoelectric transducers, tend to be highly resonant, but have poor sensitivity at frequencies away from the resonant frequency. These may be detuned to remove resonance, as is often employed in low-cost bat detectors. But sensitivity is greatly reduced and the signal often must be amplified to a level where noise is significant. Wideband ultrasonic microphones are available, but tend to be very expensive.

Substrate-based signals can be considered as acoustic signals since these are also pressure waves travelling through the medium. Detection of substrate-based signals depends on the signal intensity and characteristics of the medium through which the signal travels; the most common media are wood, plant stems (either woody or soft) and the ground. Sensors can be simple piezoelectric devices that generate a voltage when flexed or accelerometers such as micro-electrical mechanical system (MEMS) integrated circuits, which have micromachined strain gauges to measure acceleration or velocity. While very sensitive, accelerometers are expensive. A third type of vibration sensor is the laser vibrometer, which detects changes in reflection of a laser beam. It is the most sensitive technique and is capable of detecting vibrational signals in leaf miners up to 26 kHz, well beyond the capabilities of the other sensors (Bacher et al., 1996).

A major consideration in acoustic sensing is the rate at which the signal is to be digitized, which should be greater than two times the maximum frequency in the signal; this is known as the Nyquist sampling criterion (Cohen, 1995). Digitization at rates below the Nyquist frequency will result in aliasing (folding of frequencies above the Nyquist frequency onto lower frequencies) and the signal cannot be reproduced. In practical systems, an anti-aliasing filter is included to remove signal energy above half the Nyquist rate. However, practical filters are not perfect and there is always some signal leakage above the cutoff frequency. It is therefore usual practice to sample a signal at a rate higher than the Nyquist rate in order to reduce the likelihood of aliasing.

It is also necessary to consider the accuracy to which the signal's amplitude is to be reproduced (quantization). Quantization is carried out by an analogue to digital (A–D) converter, which produces a digital output as a number of binary bits, n. The number of levels is given by 2^n, so a 16-bit A–D converter will divide its input range into 65,536 amplitude levels. The higher the number of bits is, the greater is the accuracy; however, the overall memory requirements for storage (or, equivalently, transmission) is a function of both sample rate (R) and number of bits as $R \times n$.

A 16-bit signal sampled at 44.1 kHz (standard audio sample rate for CD-quality sound) will require 352,800 bytes of memory per second (twice this for stereo). A bat sound with maximum frequency of 140 kHz will require a sample rate typically of 350 kHz, which is sufficiently greater than the Nyquist rate to overcome leakage due to non-ideal filtering, requiring 2.8 Mbytes per second of memory for 16-bit samples. It is possible to compress acoustic signals using Windows Media Audio (WMA), MPEG3 or minidisc (MD) algorithms, but it is important to note that compression algorithms rely on the mechanics of human hearing to remove parts of the signal and thus may not be suitable if the waveform has to be preserved accurately for the purposes of characterization and/or analysis. It has, however, been shown that use of such compression algorithms does not necessarily have a deleterious effect on identification accuracy for some systems (Chesmore and Ohya, 2004).

Acoustic sensing can be affected significantly by interference, mainly from other animal sounds. In many parts of the world, anthropogenic sound is also becoming more prevalent and poses a significant problem. Some acoustic ATI systems overcome this problem to an extent by recognizing anthropogenic sounds as well as those of the target taxa (Chesmore and Ohya, 2004). It is also possible to filter the signal in situations where the interfering sound has a much higher intensity if its spectrum differs significantly from the target taxa, but significant frequency overlap cannot be overcome easily. Separation of multiple simultaneous sources can be achieved to a certain extent, using more than one sensor, if they are spatially separated and independent component analysis or blind source separation (Choy et al., 2005).

Visual Sensors

Visual sensing requires a monochrome or colour camera to obtain images of individuals. Of importance here is the image resolution in pixels since a higher resolution for a given magnification will enable smaller features to be observed. The same sampling and quantization issues exist as those for acoustic signals, but now in two dimensions. Other factors to consider are the sharpness of focus, depth of field and contrast. Colour is often preferable to monochrome images, especially when taxa are highly or inconsistently coloured and exhibit a high contrast with the background. Colour images are generally considered as three independent images: red, green and blue (RGB), with processing being carried out separately. With special equipment, it is possible to extend the colour range beyond that of human vision (400–700 nm) and record images in near infra-red (IR, >750 nm) and ultraviolet (UV, 200–400 nm) parts of the spectrum. The insensitivity of normal silicon photosensors to UV means they cannot be used to image UV, so UV-enhanced sensors must be employed. Silicon sensors are, however, very sensitive to near-IR (around 1000 nm). Some photosensors are multispectral in that they are able to sense a wide range of spectral bands; these are useful, for example, in the detection of vegetation and vegetation type from the ratio of red to near-IR reflectance characteristic of green plants. Such sensors are used most widely in remote sensing satellites, but are increasingly used in aircraft to obtain higher resolution.

There are many visual sensors to select from, ranging from low-resolution monochrome sensors to high-resolution colour and multispectral sensors. Webcams, digital cameras and scanners are all suitable. The first consideration is the number of spectral bands required – for example, whether it sufficient to use greyscale or colour required. A colour image needs three times the storage requirements of a greyscale image. The second consideration is resolution in pixels (i.e. the number of pixels in the image for a given optical configuration). High-resolution (e.g. 6 or 12 Mpixel) colour images are very memory intensive and so are expensive computationally. Compression algorithms such as JPEG can reduce the size of an image, but care must be taken since information will usually be lost during compression.

Active Sensors

Acoustic and visual sensors are known as passive sensors in that they do not interact with the object being sensed. An active sensor transmits electromagnetic or acoustic energy and observes reflections from the object. Radar uses electromagnetic energy at radio frequencies and has been employed for identification of birds and insects, although it has proven difficult to correlate reflected energy with species (Weber et al., 2005). This is due to the inter-relationship among returned echo size, distance, effective radar cross-section (size of object as seen by the radar), and operating wavelength. It is, however, possible to obtain wingbeat frequency and mass using a nutating beam (Chapman et al., 2003).

It is not practical to use electromagnetic energy in water since penetration is only significant below a few hundred hertz and at visible wavelengths (400–700 nm). However, sound propagates well in water and sonar has been successfully used to discriminate between fish species. Another active sensor employed in flow cytometry uses lasers of different wavelengths and scattered energy from cells is picked up. The amount of energy scattered is a function of the wavelength of incident radiation and the physical structure of the cell (i.e. a cell with many projections will have different scattering characteristics from those of a smooth, spherical cell). Other sensors include scanning electron microscope (SEM) and computer tomography (CT) scans, though the outputs of these devices are usually processed as standard digital images. Specific applications are discussed later in this chapter.

Preprocessor

Once the signal has been converted into electrical form, a number of preprocessing operations are often required. These may include amplification to improve the signal-to-noise ratio (SNR), filtering to remove noise and unwanted signals, filtering to enhance signals (e.g. edge detection) and other, more complex functions such as comb filtering (multiple bandpass filters). The type of preprocessing necessary is entirely dependent on the type of sensor employed. For example, an acoustic sensor may require amplification and often filtering to remove unwanted frequencies, whereas an image might need contrast enhancement or histogram equalization to improve signal levels. Many preprocessing functions are carried out while the signal is still in its analogue form, but there is an increasing emphasis on performing these operations after digitization for two main reasons: it is possible to create very complex filters, which would require many analogue components to realize practically (Mulgrew et al., 1999), and the cost of high-speed computing is continually being reduced. One disadvantage of digital filtering is that the processing functions are subject to mathematical rounding errors that may be large enough to distort signals; they may also be computationally expensive.

Feature Extractor

Feature extraction is the most important aspect of any ATI system. In simple terms, a good feature set is one in which each taxon has a 1:1 correspondence with a set of features, which is another way of saying 'no feature overlap'. If feature overlap occurs among taxa, then an erroneous identification may occur.

The concept of recognizable taxonomic unit (RTU) is frequently used to place taxa into morphologically similar groups (Riede, 1993). In some cases – particularly in visual sensing – such groups will map well to the morphological characteristics of the taxonomic group. However, in generalized ATI, a feature set is not likely to bear a direct resemblance to any morphological

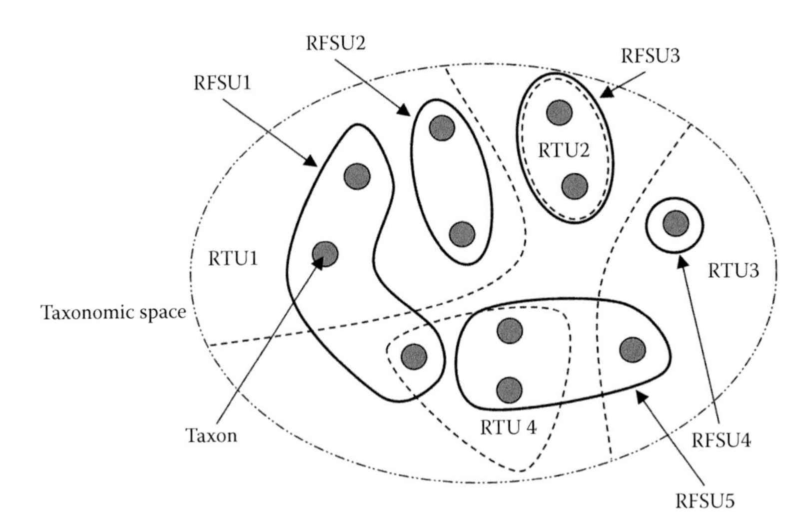

FIGURE 6.2 Representation of recognizable feature space unit. Each taxon occupies a location in taxonomic space with morphologically similar species grouped in RTUs. Here there are four taxa in RTU1, two in RTU2, two in RTU3 and three in RTU4. RFSUs form different groupings, depending on similarity in the feature space. Here, RFSU1 contains three taxa, two of which belong to RTU1 and one to RTU4; similarly for RFSU2 to RFSU5. An ideal ATR system will have a 1:1 mapping between taxon and RFSU.

characters. For example, flow cytometry has features based on scattering parameters of different wavelengths of light (Balfoort et al., 1992; McCall et al., 1996). Here, the amount of light scattered is a function of wavelength and the structure of the organism (number, size of spines, surface texture, etc.); these features are different from features obtained by visual observation.

Figure 6.2 is a diagrammatic representation of a taxonomic space where taxa (grey circles) are grouped according to morphological similarity (RTU) as indicated by dashed lines. In this scenario, there are four taxa in RTU1, two in RTU2, two in RTU3 and three in RTU4. However, for a given feature set, taxa may be grouped differently according to the similarity between taxa in terms of the feature space. Such groupings might be termed 'recognizable feature space units' (RFSU). In this example, RFSU1 contains three taxa, two of which belong to RTU1 and one to RTU4 (similarly for RFSU2 to RFSU5). Note that RFSU3 is the same as RTU2. Selection of a more appropriate feature set (if possible) should reduce the number of taxa in each RFSU, with a perfect ATI system having one taxon per RFSU.

Feature selection is the most important aspect of an ATI system. Choice of features is also dependent on the sensor employed, as indicated in Table 6.1, which is a list of some of the most commonly used features. In acoustics, frequency components are most widely used and can be generated by fast Fourier transform (FFT) (see Walker, 1996) or wavelet transform (Chan, 1995; Addison, 2002). These are computationally intensive and the FFT exhibits a trade-off between the time and frequency domains (Duhamel and Vetterli, 1990); high accuracy in the frequency domain (large FFT) will result in a poor time resolution and vice versa. There are many features available for images (see Parker, 1997; Gonzalez and Woods, 2002; Javidi, 2002). A change of sensor and/or feature set may dramatically improve or degrade the reliability and accuracy of identifications (see O'Neill, this volume), so it is important to consider all options before deciding how best to approach the design of an ATI system fitted for any particular identification problem.

Classifier

The term 'classification' as used in computing and engineering is defined as the assignment of a signal or pattern to one of a number of prespecified classes based on features extracted from the

TABLE 6.1
Examples of Features for Different Sensors

Sensor type	Features	Comments/advantages/drawbacks
Passive acoustic	Frequency components	Calculated by fast Fourier transform (FFT). Computationally intensive. A trade-off between temporal and spectral resolution (fine time resolution means poor frequency resolution and vice versa) (Duhamel and Vetterli, 1990).
	S- and A-matrices	Time domain signal coding (TDSC) based on zero-crossings and wave shape. S-matrix is a one-dimensional matrix of the frequency of occurrence of waveshapes. A-matrix is a two-dimensional matrix of the frequency of occurrence of pairs of waveshapes (Chesmore, 2001; Chesmore and Ohya, 2004). Computationally simple to implement and can be implemented in real-time at well over 1 MHz.
	Autocorrelation, cross-correlation	Gives information on self-similarity (periodic structure) and similarity between two waveforms, respectively. Computationally intensive if long samples are used. Also memory intensive (Kondoz, 2004).
	Linear predictive coefficients	All pole model of sound production mechanism (common method for speech analysis and synthesis). Good if signal has strong resonant structure (e.g. humans, birds); can also be used for spectral smoothing (Kondoz, 2004).
	Wavelets	Has variable spectral resolution which is an advantage for many biological signals. Computationally intensive (similar to FFT) (Chan, 1995; Addison, 2002).
	Amplitude probability density functions	Only provides statistical information on amplitude. Useful for determining optimum detection threshold.
Active acoustic (sonar) and active electromagnetic (radar)	Echo size	Provides information about object size in conjunction with time of flight of transmitted pulse.
	Echo shape	Shape of echo is a function of the object's size and texture in relation to the wavelength of the acoustic signal (scattering parameters).
	Time delay of pulse	Measures distance to object.
Visual	Edges (monochrome or RGB)	There are a large number of different edge detectors. Obtain information about sudden changes in colour or brightness values across the image (Gonzalez and Woods, 2002).
	First- and higher order moments	Standard statistical measures of mean, variance, etc. Higher order moments give information relating to orientation (Gonzalez and Woods, 2002).
	Texture	There are a large number of different texture measures that give information on image structure (e.g. Langford et al., 1986).
	Image histogram mean, SD, median	Information relating to image composition (Gonzalez and Woods, 2002).
	Shape indices	Eccentricity, Hough transform, snakes. Give information on general shape or whether a particular shape is present (Gonzalez and Woods, 2002).
Flow cytometry	Scattering parameters	Objects (e.g. cells) have wavelength dependent scattering functions that are related to the surface texture (e.g. Balfoort et al., 1992; McCall et al., 1996).
	Fluorescence	Cells containing chlorophyll fluoresce red with short wavelength light, thus providing an indication of the presence of chlorophyll.
Chemical (e.g. electronic nose)	Response of multiple chemical sensors	Electronic noses use arrays of sensors to detect different chemicals (Hines et al., 1999). Has little application in ATI at present.

signal; recognition is defined as the ability to classify (Schalkoff, 1992). Schalkoff divided pattern recognition systems into three types: statistical (or decision theoretic) pattern recognition, syntactic pattern recognition, and neural pattern recognition. The first and last are widely used in ATI, whereas syntactic pattern recognition has not been a popular choice, though it has been successful in human speech recognition (Allerhand, 1987) and may be suitable for complex bioacoustical signals such as bird song. Syntactic pattern recognition can also be applied to non-temporal signals and is being used to generate wing-venation descriptions automatically for identification of dipteran and hymenopteran species (Dai and Chesmore, 2005).

Statistical pattern recognition is based on the statistical nature of signals and extracted features are represented as probability density functions (Schalkoff, 1992). It therefore requires knowledge of *a priori* probabilities of measurements and features. Statistical approaches include linear discriminant functions, Bayesian functions and cluster analysis and may be unsupervised or supervised. Supervised classifiers require a set of exemplars for each class to be recognized; they are used to train the system. Unsupervised learning, on the other hand, does not require an exemplar set.

Artificial neural networks (ANNs) are good at classifying non-linearly separable data. There are at least 30 different types of ANNs, including multilayer perceptron, radial basis functions, self-organizing maps, adaptive resonance theory networks and time-delay neural networks. Indeed, the majority of ATI applications discussed later employ ANNs – most commonly, MLP (multilayer perceptron), RBF (radial basis function) or SOM (self-organizing map). A detailed treatise of neural networks for ATI is beyond the scope of this chapter and the reader is referred to the excellent introduction to ANNs in Haykin (1994) and neural networks applied to pattern recognition in Looney (1997) and Bishop (2000). Classifiers for practical ATI systems are also described in other chapters of this volume.

PRACTICAL ATI SYSTEMS

An ideal ATI system should be capable of recognizing all taxa with 100 per cent accuracy, be immune to all interfering signals, have a resolution down to the lowest taxonomic level required by the problem under consideration (even to the individual in terms of small populations), and be scalable so that new taxa can be added without reduction in identification accuracy. Accuracy and resolution are dependent on the taxa being recognized and the system's characteristics. For example, some taxa may be easily separable and therefore will have a high degree of identification. An ideal ATI system should be able to provide data continuously over a spatial scale defined by the user or the application and on a real-time basis.

In practice, spatial resolution will be determined primarily by cost (i.e. the number of devices that can realistically be deployed). The rate at which identification takes place may be a critical factor for systems designed to provide continuous data, where rapid changes in local situations must be tracked. The limiting factor is a function of the total processing time between sensing and identification; this is given by the sum of preprocessing time, feature extraction time and classifier time. In many applications, the heaviest processing requirement is in feature extraction, which may place an upper limit on the rate of identifications. Taylor et al. (1996) developed a system for real-time identification of frogs in Australia. The system could sample for only 75 per cent of the time; the remaining 25 per cent was missed due to memory restrictions and the time required for calculation of features based on a FFT.

All sensors have inherent limitations in dynamic range, resolution (if digital) and bandwidth. A sensor with a wide dynamic range can cope with large variations in the SNR (i.e. loud/quiet sounds or bright/faint images). In some applications – particularly acoustic – it is possible to improve the SNR using multiple sensors. Multiple sensors can also overcome interference to a degree by subtracting signals, thereby eliminating common components. Interference is likely to be a significant problem in all ATI systems. Problems with acoustic systems have already been mentioned; interference in image-based systems can take a number of forms, including multiple, possibly overlapping organisms in the same visual field; a background of similar characteristics (e.g. similar colours or textures) to the object being recognized; inconsistencies in lighting, pose; etc. It is important to be able to quantify many of the preceding parameters in order to achieve a robust system with repeatable data in which the user can have confidence.

APPLICATIONS OF ATI SYSTEMS

ACOUSTIC APPLICATIONS

Acoustic methods have been used for many years to identify insects, birds and mammals, and are still employed for manual species surveys (Baillie, 1995; Fischer et al., 1997; Gardiner et al., 2005). Simple detection of the presence of insects for phytosanitary and quarantine purposes has been the focus of most research, including beetle larvae in rice grains (Shuman et al., 1993, 1997), *Rhizopertha dominica* in wheat kernels (Hagstrum et al., 1990), subterranean insect pests and stem-borers (Mankin and Weaver, 2000; Mankin et al., 2000), termites (Matsuoka et al., 1996), larvae feeding inside cotton bolls (Hickling et al., 2000) and detection of the Asian longhorn beetle (*Anoplophora glabripennis*) in live trees and solid wood packing materials (Haack et al., 1997; MacLeod et al., 2002).

Recent research into automated identification has concentrated on quarantine wood-boring Coleoptera such as *A. glabripennis, A. chinensis* and *Hylotrupes bajulus* (Farr and Chesmore, 2005) and the development of non-invasive tools for locating stag beetle (*Lucanus cervus*) larvae *in situ* (Farr et al., 2005). These systems make use of low-cost vibration sensors to detect bites and stridulation from larvae, with identification achieved using time domain signal processing for feature extraction. Other acoustic applications for insect identification include Orthoptera (Chesmore, Swarbrick and Femminella, 1997; Chesmore and Nellenbach, 2001; Dietrich et al., 2003; Ohya and Chesmore, 2003; Schwenker et al., 2003; Chesmore, 2004; Chesmore and Ohya, 2004; Dietrich et al., 2004), cicadas (Ohya, 2004) and mosquitoes (Campbell et al., 1996).

Relatively little work has been carried out on amphibian acoustic identification. One notable project developed a field-deployable system to identify the presence of cane toads (*Bufo marinus*) in Australia (Taylor et al., 1996). Birds are an obvious choice for acoustic ATI, but there has been relatively little progress on comprehensive systems due to the complex nature of bird song. Systems exist for identification of migrating birds (Mills, 1995) and birds in general (McIlraith and Card, 1995; Chesmore, 2001), and locating individual elements of birdsong (Anderson et al., 1996; Kogan and Margoliash, 1998; Terry and McGregor, 2002). One interesting application employed a hand-held computer capable of automatically recognizing a small number of bird species to teach Japanese children about soundscapes (Oba, 2004).

Identification of mammals leads to the possibility of intraspecific identification in addition to interspecific identification. Research in this area includes individual fallow (*Dama dama*) and roe deer (*Capreolus capreolus*) (Reby et al., 1997, 1998a, 1998b), domesticated cows (*Bos taurus*) (Jahns et al., 1997), Gunnerson's prairie dogs (*Cynomys gunnisoni*) (Placer and Slobodchikoff, 2000) and false killer whale (*Pseudorca crassidens*) (Murray et al., 1998a, 1998b). Bats are another target for ATI; ultrasonic echolocation calls are generally species specific. Several semimanual bat identification systems have been tested with good success (Vaughan et al., 1996; Parsons and Jones, 2000; Parsons, 2001). At the other end of the acoustic spectrum, infrasound has recently been used for identification of elephant calls (Clemins and Johnson, 2002).

IMAGE/VISUAL APPLICATIONS

There are potentially many more applications of visual ATI than acoustic ATI, ranging from bacteria to insects and birds. In the plant kingdom, applications have tended to focus on problem species such as weeds (Shulin and Runtz, 1996; Critten, 1996; Hemming and Rath, 2001; Manh et al.,

2001), pests on cotton (Zhigang et al., 2003), quarantine fungal pathogens (Lane et al., 1998) and pollen (Langford et al., 1986, 1990; France et al., 2000). Other applications include mushrooms (van de Vooren et al., 1992), fungal pathogens (Chesmore et al., 2003), plant variety (Felföldi et al., 2001), trees based on leaf identification (White and Prentice, 1988; Clark and Warwick, 1998; Clark 2000, 2003), and blue-green algae (Thiel, 1994).

Identification of marine zooplankton using image processing was one of the earliest ATI applications (Katsinis et al., 1984). Bacteria have also been the object of ATI research (Blackburn et al., 1998; Dörge et al., 2000; Walker and Kumagai, 2000; Foreroa et al., 2004).

Lepidoptera are an obvious choice for image-based identification; however, relatively little work in this area has been carried out (Chesmore and Monkman, 1994; Watson et al., 2003; White and Winokur, 2003; Kipling and Chesmore, 2005). More work has been carried out on Hymenoptera such as braconid wasps (Weeks et al., 1997a, 1997b; Gauld et al., 2000), honeybees (Daly et al., 1982; Schröder et al., 1995; Steinhage et al., 1997; see also Steinhage et al., this volume), solitary bees (Roth et al., 1999), ichnumonid wasps (Yu et al., 1992), parasitic wasps (Angel, 1999) and leafhoppers (Dietrich and Pooley, 1994). Other invertebrate research involves the location and description of wing venation – for example, in Diptera (Dai and Chesmore, 2005) – and identification of spiders via genitalia images (Do et al., 1999; see also Russell et al., this volume).

Applications Involving Other Sensors

Extant, active-sensor ATI systems include those based on sonar, radar, flow cytometry and optical detection. Sonar has been shown to be effective in identifying single species fish shoals via backscatter parameters (Haralabous and Georgakarakos, 1996; Scalabrin et al., 1996; Simmonds et al., 1996) and radar can identify certain large flying insects (Chapman et al., 2003). Acoustic backscatter has been used for classification of zooplankton (Martin et al., 1996). Flow cytometry has been used extensively for identification of microbial populations (Jonker et al., 2000), algae (Balfoort et al., 1992; Wilkins et al., 2003), phytoplankton (Boddy et al., 1994; McCall et al., 1996; Pech-Pacheco and Alvarez-Borrego, 1998; Wilkins et al., 1999; Morris, Autret, et al., 2001; Morris, Boddy, et al., 2001; Wilkins et al., 2001; Luo et al., 2002; Wilkins et al., 2003), zooplankton (Jeffries et al., 1984), diatoms (Du Buf and Bayer, 2002) and dinoflagellates (Culverhouse, Williams, Reguera, Herry, et al., 2003; Culverhouse, Williams, Reguera and Herry, 2003). An interesting active optical system has been developed that detects reflections of a light source from flying insects and has successfully been tested on aphids (Moore and Miller, 2002).

THE FUTURE OF ATI

ATI is becoming more practical as processing power increases and cost decreases, and more acceptable to taxonomy as the taxonomic impediment worsens. Major improvements are contingent on development of accurate, scalable classifiers capable of operating with many taxa. Recent work on plastic self-organizing map-based neural networks (Lang and Warwick, 2002; Watson et al., 2003) may lead to systems that are not only efficient but also scalable due to their ability to accept new taxa without requiring complete retraining.

Additional information relating to a taxon's autecology, geographical distribution, host plants and other parametric data may be used to further improve reliability of the classification step. For example, a classifier may give two possible identifications, only one of which is likely at a given geographical location. There are a number of approaches to providing such metadata. Perhaps the most effective is through an associated expert system where the metadata are stored in the form of rules (Jackson, 1990; Luger and Stubblefield, 1993). Figure 6.3 shows how an expert system might be combined with an ATI system. An expert system has an additional advantage in that it is capable of providing an explanation of the reasoning behind the answer, whereas an ANN, for

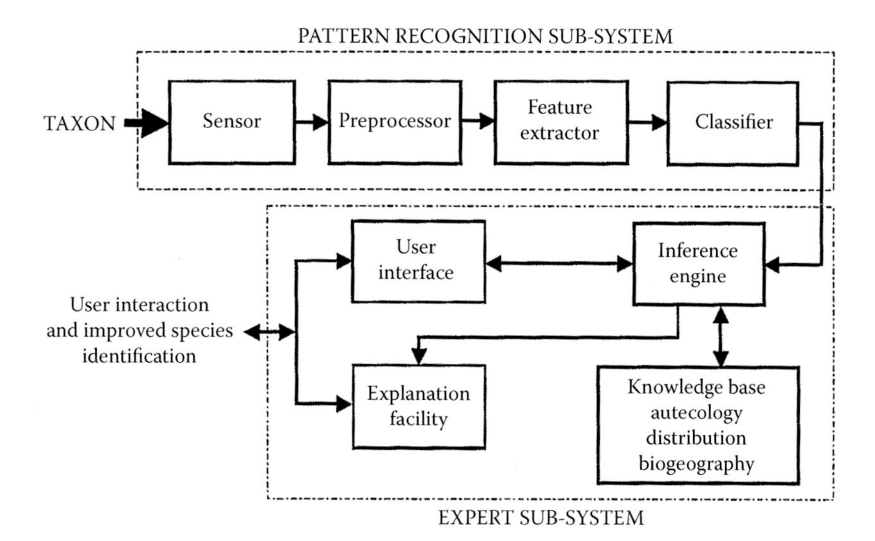

FIGURE 6.3 Design for improving an ATI system by adding an expert system.

example, does not have an explanation facility and must be treated as a black box. This approach has been used by Walker and Kumagai (2000) for image-based identification of cyanobacteria. It must be stressed, however, that this is only one of a range of possible alternatives.

There are many other approaches to knowledge-based pattern recognition that should be explored. An alternative expert system architecture that has potential for ATI is the blackboard system, which is suited to complex data analysis (Hallam, 1990; Craig, 1995) and has been used for speech recognition in the well-known Hearsay II system (Erman et al., 1980). A simplified blackboard system has been tested on British Orthoptera using time-domain signal coding and demonstrated the potential of this approach for acoustic signals (Chesmore, Swarbrick and Femminella, 1997).

It is now possible to implement ATI on handheld computers and, more significantly, mobile phones, which now have high-quality cameras, large memory and significant processing power. This has many significant advantages for in-field use, including automatic location fixing using GPS, provision of additional data such as images and ecological information, and the ability to transmit images (or sounds) via radio. There is also much potential for integrating ATI systems with the Internet. A combination of both approaches means that it will be possible for a field researcher to identify a taxon immediately or, if a specimen cannot be recognized, an image transmitted via mobile radio or satellite link to a server where more powerful identification engines are available. An answer could then be relayed back to the user. All the technology required for implementation of such a system is already available with the radio infrastructure present in many countries and expanding in others.

SUMMARY

Automated taxon identification is a rapidly emerging science requiring a multidisciplinary approach that is becoming more achievable due to increases in processing power and significant reductions in memory and processing costs. An ATI system, like any pattern-recognition system, comprises four functional units: sensor, preprocessor, feature extractor and classifier. While there are a wide range of sensors available for signal measurement, only a relatively small subset is useful for ATI; these can be grouped into three primary categories: acoustic, visual and active sensors.

To date, the majority of ATI systems use acoustic or visual sensors, although it is anticipated that others, such as chemical sensors, may play an increasing role in the future. The sensor's output is then modified by the preprocessor to be suitable for the feature extractor which is the most important part of any pattern recognition system. Selection of features is a function of the sensor and taxonomic group; good feature set is one in which each taxon has a 1:1 correspondence with a set of features (i.e. there is no feature overlap). If an overlap does take place, then an erroneous identification may occur. The classifier is the final stage and is designed to place the input signal into one of a number of categories, namely, taxa. There is a large range of classifiers available; the majority employed to date are statistical or neural and the latter have the advantage of non-linear reparability.

An ATI system should ideally have an identification accuracy of 100 per cent and be scaleable (i.e. the same accuracy for small or large taxonomic groups). In reality, this is difficult to achieve, especially when dealing with 'real' signals, which may have poor SNR or high levels of interference. It is important that these potential limitations are appreciated at the design stage so that a robust system with repeatable data in which the user can have confidence can be achieved.

The number and range of applications is increasing, particularly for image-based approaches. It is expected that the number of practical field-deployable systems, rather than research systems, will increase rapidly in the near future for three reasons: (1) availability of new, powerful mobile technology such as mobile camera phones, (2) the worsening taxonomic impediment and (3) an increasing acceptance of ATI as a viable tool.

REFERENCES

Addison, P.S. (2002) *Illustrated Wavelet Transform Handbook,* IOP Publishing Ltd., 400 pp.

Allerhand, M. (1987) *Knowledge-Based Speech Pattern Recognition,* Kogan Page, London, 240 pp.

Anderson, S.E., Dave, A.S. and Margoliash, D. (1996) Template-based automatic recognition of birdsong syllables from continuous recordings. *Journal of the Acoustical Society of America,* 100: 1209–1217.

Angel, P.N. (1999) *Multiscale Image Analysis for the Automated Localization of Taxonomic Landmark Points and the Identification of Species of Parasitic Wasp.* PhD thesis, University of Glamorgan.

Bacher, S., Casas, J. and Dorn, S. (1996) Parasitoid vibrations as potential releasing stimulus of evasive behaviour in a leafminer. *Physiological Entomology,* 21: 33–43.

Baillie, S.R. (1995) Monitoring terrestrial bird populations. In *Monitoring for Conservation and Ecology* (ed F.B. Goldsmith), Chapman and Hall, London, pp. 112–132.

Balfoort, H.W., Snoek, J., Smitts, R.M., Breedveld, L.W., Hofstraat, J.W. and Ringelberg, J. (1992) Automatic identification of algae: neural network analysis of flow cytometric data. *Journal of Plankton Research,* 14: 575–589.

Bishop, C.M. (2000) *Neural Networks for Pattern Recognition,* Oxford University Press, Oxford, 504 pp.

Blackburn, N., Hagström, Å., Wikner, J., Cuadros-Hansson, R. and Bjørnsen, P.K. (1998) Rapid determination of bacteria abundance, biovolume, morphology and growth by neural network-based image analysis. *Applied and Environmental Microbiology,* 64: 3246–3255.

Boddy, L., Morris, C.W., Wilkins, M.F., Tarran, G.A. and Burkhill, P.H. (1994) Neural network analysis for flow cytometric data for 40 marine phytoplankton species. *Cytometry,* 15: 283–293.

Campbell, R.H., Martin, S.K., Schneider, I. and Michelson, W.R. (1996) Analysis of mosquito wing beat sound, *132nd Meeting of the Acoustical Society of America,* Honolulu.

Chan, Y.T. (1995) *Wavelet Basics*, Kluwer Academic Publishers, New York, 148 pp.

Chapman, J.W., Reynolds, D.R. and Smith, A.D. (2003) Vertical-looking radar: a new tool for monitoring high-altitude insect migration. *Bioscience,* 53: 503–511.

Chesmore, E.D. (2000) Methodologies for automating the identification of species. In *Proceedings of Inaugural Meeting of the BioNET–INTERNATIONAL Group for Computer-Aided Taxonomy (BIGCAT) 1997* (eds E.D. Chesmore, L. Yorke, P. Bridge and S. Gallagher), pp. 3–12.

Chesmore, E.D. (2001) Application of time domain signal coding and artificial neural networks to passive acoustical identification of animals. *Applied Acoustics,* 62: 1359–1374.

Chesmore, E.D. (2004) Automated bioacoustic identification of species. *Anais da Academia Brasileira de Ciencias,* 76: 435–440.

Chesmore, E.D., Bernard, T., Inman, A.J. and Bowyer, R.J. (2003) Image analysis for the identification of the quarantine pathogen. *Tilletia indica. EPPO Bulletin,* 33: 495–499.

Chesmore, E.D. and Monkman, G. (1994) Automated analysis of variation in Lepidoptera. *The Entomologist,* 113: 171–182.

Chesmore, E.D. and Nellenbach, C. (2001) Acoustic methods for the automated detection and identification of insects. *Acta Horticulturae,* 562: 223–231.

Chesmore, E.D. and Ohya, E. (2004) Automated identification of field-recorded songs of four British grasshoppers using bioacoustic signal recognition. *Bulletin of Entomological Research,* 94: 319–330.

Chesmore, E. D., Swarbrick, M.D. and Femminella, O.P. (1997) Automated analysis of insect sounds using TESPAR and expert systems – A new method for species identification. In *Information Technology, Plant Pathology and Biodiversity* (eds P. Bridge et al.), CAB International, Wallingford, UK, pp. 273–287.

Chesmore, E.D., Yorke, L., Bridge, P. and Gallagher, S. (1997) *Proceedings of the Inaugural Meeting of the BioNET–INTERNATIONAL Group for Computer-Aided Taxonomy (BIGCAT)*, BioNET–INTERNATIONAL Technical Secretariat.

Choy, S., Cichocki, A., Park, H-M. and Lee, S-Y. (2005) Blind source separation and independent component analysis: a review. *Neural Information Processing – Letters and Reviews,* 6: 1–57.

Clark, J.Y. (2000) *Botanical Identification and Classification Using Artificial Neural Networks.* PhD thesis, University of Reading, UK.

Clark, J.Y. (2003) Artificial neural networks for species identification by taxonomists. *Biosystems,* 72: 131–47.

Clark, J.Y. and Warwick, K. (1998) Artificial keys for botanical identification using a multilayer perceptron neural network (MLP). *Artificial Intelligence Review – Special Issue on Applications in Biology and Agriculture,* 12: 105–115.

Clemins, P.J. and Johnson, M.T. (2002) Automatic type classification and speaker verification of African elephant vocalizations. *Animal Behaviour 2002,* Vrije Universiteit, Amsterdam.

Cohen, L. (1995) *Time-Frequency Analysis*, Prentice Hall Signal Processing Series, Englewood Cliffs, NJ, 320 pp.

Cokl, A. and Virant-Doberlet, M. (2003) Communication with substrate-borne signals in small plant-dwelling insects. *Annual Review of Entomology,* 48: 29–50.

Craig, I. (1995) *Blackboard Systems,* Intellect Books, Bristol, 256 pp.

Critten D.L. (1996) Fourier-based techniques for the identification of plants and weeds. *Journal of Agricultural Engineering Research,* 64: 149–154.

Culverhouse, P.F., Williams, R., Reguera, B. and Herry, V. (2003) Expert and machine discrimination of marine flora: a comparison of recognition accuracy of field-collected phytoplankton. *Institution of Electrical Engineers International. Conference on Vision Information Engineering*, May 23–25 Guildford. UK, pp. 177–183.

Culverhouse, P.F., Williams, R., Reguera, B., Herry, V. and González-Gil, S. (2003) Do experts make mistakes? A comparison of human and machine identification of dinoflagellates. *Marine Ecology Progress Series,* 247: 17–25.

Dai, J. and Chesmore, D. (2005) Identification of Diptera species by image analysis of wing venation. *Royal Entomological Society National Meeting: Entomology 2005*, 12–14 September 2005, University of Sussex Royal Entomological Society.

Daly, H.V., Hoelmer, K., Norman, P. and Allen, T. (1982) Computer-assisted measurement and identification of honeybees (Hymenoptera. Apidae). *Annals of the Entomological Society of America,* 75: 591–594.

Desouza, G.N. and Kak, A.C. (2002) Vision for mobile robot navigation: a survey. *IEEE Transactions on Pattern Analysis and Machine Intelligence,* 24: 237–267.

Dietrich, C.D. and Pooley, C.D. (1994) Automated identification of leafhoppers (Homoptera: Cicadellidae: *Draeculacephala* Ball). *Annals of the Entomological Society of America,* 87: 412–423.

Dietrich, C., Palm, G., Riede, K. and Schwenker, F. (2004) Classification of bioacoustic time series based on the combination of global and local decision. *International Journal of Pattern Recognition,* 37: 2293–2305.

Dietrich, C., Palm, G. and Schwenker, F. (2003) Decision templates for the classification of bioacoustic time series. *Information Fusion,* 4: 101–109.

Do, M.T., Harp, J.M. and Norris, K.C. (1999) A test of a pattern recognition system for identification of spiders. *Bulletin of Entomological Research,* 89: 217–224.

Dörge, T., Carstensen, J.M. and Frisvad, J.C. (2000) Direct identification of pure *Penicillium* species using image analysis. *Journal of Microbiological Methods,* 41: 121–133.

Du Buf, H. and Bayer, M.M. (eds). (2002) *Automatic Diatom Identification,* volume 51, World Scientific Series in Machine Perception and Artificial Intelligence, World Scientific Publishing Company, Hackensack, NJ, 328 pp.

Duhamel, P. and Vetterli, M. (1990) Fast Fourier transforms: a tutorial review and a state of the art. *Signal Processing,* 19: 259–299.

Egmont-Petersen, M., de Ridder, D. and Handels, H. (2002) Image processing with neural networks – A review. *Pattern Recognition,* 35: 2279–2301.

Erman, L.D., Hayes-Roth, F., Lesser, V.R. and Reddy, D.R. (1980) The Hearsay-II speech-understanding system: integrating knowledge to resolve uncertainty. *Computing Surveys,* 12: 213–253.

Farr, I., Chesmore, D., Harvey, D., Hawes, C. and Gange, A. (2005) Bioacoustic detection and recognition of stag beetle (*Lucanus cervus*) larvae underground using vibration sensors. *Royal Entomological Society National Meeting: Entomology 2005,* 12–14 September 2005, University of Sussex, Royal Entomological Society.

Farr, I.J. and Chesmore, E.D. (2005) Acoustic detection and recognition of wood-boring insects. *Royal Entomological Society National Meeting: Entomology 2005,* 12–14 September 2005, University of Sussex, Royal Entomological Society.

Felföldi, J., Fekete, A., Györi, E. and Szepes, A. (2001) Variety identification by computer vision. *Acta Horticulturae,* 562: 341–345.

Fischer, F.P., Schulz, U., Schubert, H., Knapp, P. and Schmoger, M. (1997) Quantitative assessment of grassland quality: acoustic determination of population sizes of orthopteran indicator species. *Ecological Applications,* 7: 909–920.

Foreroa, M.G., Sroubek, F. and Cristóbal, G. (2004) Identification of tuberculosis bacteria based on shape and color. *Real Time Imaging,* 10: 251–262.

France, I., Duller, A.W.G., Duller, G.A.T. and Lamb, H.F. (2000) A new approach to automated pollen analysis. *Quaternary Science Review,* 19: 537–546.

Gardiner, T., Hill, J. and Chesmore, D. (2005) Review of the methods frequently used to estimate the abundance of Orthoptera in grassland ecosystems. *Journal of Insect Conservation,* 9: 151–174.

Gaston K. and O'Neill, M.A. (2004) Automated species identification – Why not? *Philosophical Transactions of the Royal Society, Series B,* 359: 655–667.

Gauld, I.D., O'Neill, M.A. and Gaston, K.J. (2000) Driving Miss Daisy: the performance of an automated insect identification system. In *Hymenoptera: Evolution, Biodiversity and Biological Control* (eds A.D. Austin and M. Dowton), CSIRO, Collingwood, Victoria, pp. 303–312.

Gonzalez, R.C. and Woods, R.E. (2002) *Digital Image Processing,* Prentice Hall, Englewood Cliffs, NJ, 793 pp.

Haack, R.A., Law, K.R., Mastro, V.C., Ossenbruggen, H.S. and Raimo, B.J. (1997). New York's battle with the Asian long-horned beetle. *Journal of Forestry,* 95: 11–15.

Hagstrum, D.W., Vick, K.W. and Webb, J.C. (1990) Acoustic monitoring of *Rhizopertha dominica* (Coleoptera: Bostrichidae) populations in stored wheat. *Journal of Economic Entomology,* 83: 625–628.

Hallam, J. (1990) Blackboard architectures and systems. In *Artificial Intelligence, Concepts and Applications in Engineering* (ed A.R. Mirzai), Chapman and Hall, London, pp. 35–64.

Haralabous, J. and Georgakarakos, S. (1996) Artificial neural networks as a tool for species identification of fish schools. *ICES Journal of Marine Science,* 53: 173–180.

Haykin, S. (1994) *Neural Networks, a Comprehensive Foundation,* Macmillan, New York, 842 pp.

Hemming, J.J. and Rath, T. (2001) Computer-vision-based weed identification under field conditions using controlled lighting. *Journal of Agricultural Engineering Research,* 78: 233–243.

Hickling, R., Lee, P., Velea, D., Denneby, T.J. and Patin, A.I. (2000) Acoustic system for rapidly detecting and monitoring pink bollworm in cotton bolls. In *Proceedings of the Beltwide Cotton Conference,* San Antonio, National Cotton Council and the Cotton Foundation, pp. 1–10.

Hines, E.L., Llobet, E. and Gardner, J.W. (1999) Electronic noses, a review of signal processing techniques. *IEEE Proceedings on Circuits, Devices and Systems,* 146: 297–310.

Jackson, P. (1990) *Introduction to Expert Systems,* 2nd edition, Addison–Wesley Publishers Ltd, London, 560 p.

Jahns, G., Kowalczyk, W. and Walter, K. (1997) An application of sound processing techniques for determining condition of cows. In *4th International Workshop on Systems, Signal and Image Processing,* May 28–30, Poznan, Poland, Institution of Electrical and Electronic Engineers, pp. 1–4.

Javidi, B. (2002) *Image Recognition and Classification, Algorithms, Systems and Applications,* Marcel Dekker, New York.

Jeffries, H.P., Berman, M.S., Poularikas, A.D., Katsinis, C., Melas, I., Sherman, K. and Bibins, L. (1984) Automatic sizing, counting and identification of zooplankton by pattern recognition. *Marine Biology,* 78: 329–334.

Jonker, R., Groben, R., Tarran, G., Medlin, L., Wilkins, M., Garcia, L., Zabala, L. and Boddy, L. (2000) Automated identification and characterization of microbial populations using flow cytometry in the AIMS project. *Scientia Marine,* 64: 225–234.

Katsinis, C., Poularikas, A.D. and Jeffries, H.P. (1984) Image processing and pattern recognition with applications to marine biological images. In *SPIE 28th Annual International Technical Symposium on Optics and Electro-optics,* San Diego, Society of Photo-optical Instrumentation Engineers, pp. 150–155.

Kipling, M.L. and Chesmore, D. (2005) Automated recognition of moth species using image processing and artificial neural networks. In *Royal Entomological Society National Meeting, Entomology 2005*, 12–14 September University of Sussex, Royal Entomological Society.

Kogan, J.A. and Margoliash, D. (1998) Automated recognition of bird song elements from continuous recordings using dynamic time warping and hidden Markov models, a comparative study. *Journal of the Acoustical Society of America,* 103: 2185–2196.

Koha, L.H., Ranganath, S. and Venkatesh, Y.V. (2002) An integrated automatic face detection and recognition system. *Pattern Recognition,* 35: 1259–1273.

Kondoz, A.M. (2004) *Digital Speech: Coding for Low Bit Rate Communication Systems,* Wiley, London, 458 pp.

Lane, C.R., Jeuland, H., Hall, A.G. and Chesmore, E.D. (1998) Image analysis as an aid to discrimination between *Colletotrichum acutatum* and *C. gloesporioides*. In *Seventh International Conference of Plant Pathology (ICPP98),* Edinburgh, British Society for Plant Pathology, Paper No. 4.3.2.

Lang, R. and Warwick, K. (2002) The plastic self organizing map. In *Proceedings of the International Joint Conference on Neural Networks* (ed C. Society). Institute of Electrical & Electronics Engineers, Honolulu, HI, pp. 727–732.

Langford, M., Taylor, G.E. and Flenley, J.R. (1986) The application of texture analysis for automated pollen identification. In *Proceedings of the Conference on Identification and Pattern Recognition,* 2, University Paul Sabatier, Toulouse, France, pp. 729–739.

Langford, M., Taylor, G.E. and Flenley, J.R. (1990) Computerized identification of pollen grains by texture. *Review of Palaeobotany and Palynology,* 64: 97–203.

Looney, C.G. (1997) *Pattern Recognition Using Neural Networks,* Oxford University Press, Oxford, 415 pp.

Luger, G.F. and Stubblefield, W.A. (1993) *Artificial Intelligence, Structures and Strategies for Complex Problem Solving,* Benjamin/Cummings Publishing Company, Harlow, 1000 pp.

Luo, T., Kramer, K., Goldgof, D., Hall, L.O., Scott Samson, S., Remsen, A. and Hopkins, T. (2002) Learning to recognize plankton. In *IEEE International Conference on Man, Systems and Cybernetics*, 5–8 October 2003, Washington, DC, USA, Institution of Electrical and Electronic Engineers, pp. 888–893.

MacLeod, A., Evans, H.F. and Baker, R.H.A. (2002) An analysis of pest risk from an Asian longhorn beetle (*Anoplophora glabripennis*) to hardwood trees in the European community *Crop Protection,* 21: 635–645.

Manh, A-G., Rabatel, G., Assémat, L. and Aldon, M-J. (2001) In-field classification of weed leaves by machine vision using deformable templates. In *Proceedings of the Third European Conference on Precision Agriculture* (eds G. Grenier and S. Blackmore), June 2001, Montpellier, France, pp. 599–604.

Mankin, R.W., Brandhurst-Hubbard, J., Flanders, K.L., Zhang, M., Crocker, R.L., Lapointe, S.L., McCoy, C.W., Fisher, J.R. and Weaver, D.K. (2000) Eavesdropping on insects hidden in soil and interior structures of plants. *Journal of Economic Entomology,* 93: 1173–1182.

Mankin, R.W and Weaver, D.K. (2000) Acoustic detection and monitoring of *Cephus cinctus* sawfly larvae in wheat stem. *International Congress of Entomology,* Brazil, 2000.

Martin, L.V., Stanton, T.K., Wiebe, P.H. and Lynch, J.F. (1996) Acoustic classification of zooplankton. *ICES Journal of Marine Science,* 53: 217–224.

Matsuoka, H., Fujii, Y., Okumura, S., Imamura, Y. and Yoshimura, T. (1996) Relationship between the type of feeding behaviour of termites and the acoustic emission (AE) generation. *Wood Research,* 83: 1–7.

McCall, H., Bravo, L., Lindley, J.A. and Reguera, B. (1996) Phytoplankton recognition using parametric discriminants. *Journal of Plankton Research,* 18: 393–410.

McComb, A., Reby, D., Baker, L., Moss, C. and Sayialel, S. (2003) Long-distance communication of acoustic cues to social identity in African elephants. *Animal Behaviour,* 65: 317–329.

McIlraith, A.L. and Card, H.C. (1995) Birdsong recognition with DSP and neural networks. In *IEEE WESCANEX'95 Proceedings*, Institution of Electrical and Electronic Engineers, pp. 409–414.

Mills, H. (1995) Automatic detection and classification of nocturnal migrant bird calls. *Journal of the Acoustical Society of America,* 97: 3370–3371.

Moganti, M., Ercal, F., Dagli, C.H. and Tsunekawa, S. (1996) Automatic PCB inspection algorithms, a survey. *Computer Vision and Image Understanding,* 63: 287–313.

Moore, A. and Miller, R. (2002) Automated identification of optically sensed aphid (Homoptera, Aphidae) wingbeat waveforms. *Annals of the Entomological Society of America,* 95: 1–8.

Morris, C.W., Autret, A. and Boddy, L. (2001) Support vector machines for identifying organisms – A comparison with strongly partitioned radial basis function networks. *Ecological Modelling,* 146: 57–67.

Morris, C.W., Boddy, L. and Wilkins, M.F. (2001) Effects of missing data on RBF neural network identification of biological taxa, discrimination of microalgae from flow cytometry data. *International Journal of Smart Engineering System Design,* 3: 195–202.

Mulgrew, B., Grant, J. and Thompson, J. (1999) *Digital Signal Processing: Concepts & Applications,* Macmillan Press, London, 408 pp.

Murray, S.O., Mercado, E. and Roitblat, H.L. (1998a) Characterizing the graded structure of false killer whale (*Pseudorca crassidens*) vocalizations. *Journal of the Acoustical Society of America,* 104: 1679–1688.

Murray, S.O., Mercado, E. and Roitblat, H.L. (1998b) The neural network classification of false killer whale (*Pseudorca crassidens*) vocalizations. *Journal of the Acoustical Society of America,* 104: 3626–3633.

Nilubol, C., Mersereau, R.M. and Smith, M.J.T. (2002) A SAR target classifier using radon transforms and hidden Markov models. *Digital Signal Processing,* 12: 274–283.

Oba, T. (2004) Application of automated bioacoustic identification in environmental education and assessment. *Anais da Academia Brasileira de Ciencias,* 76: 445–451.

Ohya, E. (2004) Identification of *Tibicen* cicada species by a principal components analysis of their songs. *Anais da Academia Brasileira de Ciencias,* 76: 441–444.

Ohya, E. and Chesmore, E.D. (2003) Automated identification of grasshoppers by their songs. *Annual Meeting of the Japanese Society of Applied Entomology and Zoology,* Iwate University, Morioka, Japan, 2003, Japanese Society of Applied Entomology and Zoology.

Parker, J.R. (1997) *Algorithms for Image Processing and Computer Vision,* John Wiley, New York, 432 pp.

Parsons, S. (2001) Identification of New Zealand bats (*Chalinolobus tuberculatus* and *Mystacina tuberculata*) in flight from analysis of echolocation calls by artificial neural networks. *Journal of Zoology (London),* 253: 447–456.

Parsons, S. and Jones, G. (2000) Acoustic identification of 12 species of echolocating bat by discriminant function analysis and artificial neural networks. *Journal of Experimental Biology,* 203: 2641–2656.

Pasquariello, G., Satalino, G., la Forgia, V. and Spilotros, F. (1998) Automatic target recognition for naval traffic control using neural networks. *Image and Vision Processing,* 16: 67–73.

Pech-Pacheco, J.L. and Alvarez-Borrego, J. (1998) Optical-digital system applied to the identification of five phytoplankton species. *Marine Biology,* 132: 357–365.

Placer, J. and Slobodchikoff, C.N. (2000) A fuzzy-neural system for identification of species-specific alarm calls for Gunnison's prairie dogs. *Behavioural Processes,* 52: 1–9.

Reby, D., Hewson, A.J.M., Cargnelutti, B., Angibault, J-M. and Vincent, J-P. (1998a) Use of vocalizations to estimate population size of roe deer. *Journal of Wildlife Management,* 62: 1342–1348.

Reby, D., Joachim, J., Lauga, J., Lek, S. and Aulagnier, S. (1998b) Individuality in the groans of fallow deer (*Dama dama*) bucks. *Journal of the Zoological Society of London,* 245: 79–84.

Reby, D., Lek, S., Dimopoulos, I., Joachim, J., Lauga, J. and Aulagnier, S. (1997) Artificial neural networks as a classification method in the behavioural sciences. *Behavioural Processes,* 40: 35–43.

Riede, K. (1993) Monitoring biodiversity, analysis of Amazonian rainforest sounds. *Ambio,* 22: 546–548.

Roth, V., Pogoda, A., Steinhage, V. and Schröder, S. (1999) *Integrating Feature-Based and Pixel-Based Classification for the Automated Identification of Solitary Bees.* Informatik aktuell, pp. 120–129.

Scalabrin, C., Diner, N., Weill, A., Hillion, A. and Mouchet, M. (1996) Narrowband acoustic identification of monospecific fish shoals. *ICES Journal of Marine Science,* 53: 181–188.

Schalkoff, R. (1992) *Pattern Recognition, Statistical, Structural and Neural Approaches.* Wiley, New York, 384 pp.

Schröder, S., Drescher, W., Steinhage, V. and Kastenholz, B. (1995) An automated method for the identification of bee species (Hymenoptera, Apoidea). In *Proceedings of the International Symposium on Conserving Europe's Bees, International Bee Research Association and Linnean Society, London*, International Bee Research Association, pp. 6–7.

Schwenker, F., Dietrich, C., Kestler, H.A., Riede, K. and Palm, G. (2003) Radial basis function neural networks and temporal fusion for the classification of bioacoustic time series. *Neurocomputing,* 51: 265–275.

Shulin, D. and Runtz, K. (1996) Image processing methods for identifying species of plants. In *IEEE WESCANEX'95 Proceedings,* Institution of Electrical and Electronic Engineers, pp. 403–408.

Shuman, D., Coffelt, J.A., Vick, K.W. and Mankin, R.W. (1993) Quantitative acoustical detection of larvae feeding inside kernels of grain. *Journal of Economic Entomology,* 86: 933–938.

Shuman, D., Weaver, D.K. and Mankin, R.W. (1997) Quantifying larval infestation with an acoustical sensor array and cluster analysis of cross-correlation outputs. *Applied Acoustics,* 50: 279–296.

Simmonds, E.J., Armstrong, F. and Copland, P.J. (1996) Species identification using wideband backscatter with neural network and discriminant analysis. *ICES Journal of Marine Science,* 53: 189–195.

Steinhage, V., Kastenholz, B., Schröder, S. and Drescher, W. (1997) *A Hierarchical Approach to Classify Solitary Bees Based on Image Analysis.* Informatik aktuell Springer, pp. 419–426.

Taylor, A., Grigg, G., Watson, G. and McCallum, H. (1996) Monitoring frog communities, an application of machine learning. In *Eighth Innovative Applications of Artificial Intelligence Conference* (eds H. Shrobe and E. Senator), AAAI Press, Menlo Park, CA, pp. 1564–1569.

Terry, A.M.R. and McGregor, P.K. (2002) Census and monitoring based on individually identifiable vocalizations, the role of neural networks. *Animal Conservation,* 5: 103–111.

Thiel, S.U. (1994) *The Use of Image Processing Techniques for the Automated Detection of Blue-Green Algae.* PhD thesis, University of Glamorgan.

Vaughan, N., Jones, G. and Harris, S. (1996) Identification of British bat species by multivariate analysis of echolocation call parameters. *Bioacoustics,* 7: 189–207.

van de Vooren, J.G., Polder, G. and van de Heijden, G.W.A.M. (1992) Identification of mushroom cultivars using image analysis. *Transactions of ASAE,* 35: 347–350.

Walker, J.S. (1996) *Fast Fourier Transforms,* CRC Press, Boca Raton, FL, 464 pp.

Walker, R. and Kumagai, M. (2000) Image analysis as a tool for quantitative phycology, a computational approach to cyanobacterial taxa identification. *Limnology,* 1: 107–115.

Watson A.T., O'Neill, M.A. and Kitching I.J. (2003) A qualitative study investigating automated identification of living macrolepidoptera using the digital automated identification system (DAISY). *Systematics and Biodiversity,* 1: 287–300.

Weber, P., Nohara, T.J. and Gauthreaux, S., Jr. (2005) Affordable, real-time, 3-D avian radar networks for centralized North American bird advisory systems. In *Bird Strike Conference,* 14–18 August 2005, Vancouver, Canada, Bird Strike Committee USA, pp. 1–8.

Weeks, P.J.D., O'Neill, M.A., Gaston, K.J. and Gauld, I.D. (1997a) Automating the identification of insects, a new solution to an old problem. *Bulletin of Entomological Research,* 87: 203–211.

Weeks, P.J.D., O'Neill, M.A., Gaston, K.J. and Gauld, I.D. (1997b) Development of an automated insect identification system. *Journal of Applied Entomology,* 123: 1–8.

White, R.J. and Prentice, H.C. (1988) Comparison of shape description methods for biological outlines. In *Classification and Related Methods of Data Analysis* (ed H.H. Bock), Elsevier (North–Holland), Amsterdam, pp. 395–402.

White, R.J. and Winokur, L. (2003) Quantitative description and discrimination of butterfly wing patterns using moment invariant analysis. *Bulletin of Entomological Research,* 93(4): 361–374.

Wilkins, M.F., Boddy, L. and Dubelaar, G.B.J. (2003) Identification of marine microalgae by neural network analysis of simple descriptors of flow cytometric pulse shapes. In *Ecological Informatics* (ed F. Recknagel), Springer, Berlin, pp. 355–367.

Wilkins, M.F., Boddy, L., Morris, C.W. and Jonker, R.R. (1999) Identification of phytoplankton from flow cytometry data by using radial basis function neural networks. *Applied and Environmental Microbiology,* 65: 4044–4410.

Wilkins, M.F., Hardy, S.A., Boddy, L. and Morris, C.W. (2001) Comparison of five clustering algorithms to classify phytoplankton from flow cytometry data. *Cytometry,* 44: 210–217.

Yakcun, H. and Bozma, H.I. (1998) An automated inspection system with biologically inspired vision. In *IEEE Proceedings of the 1998 IEEE/RSJ International Conference on Intelligent Robots and Systems,* Victoria, B.C., Canada, October, 1998, Institution of Electrical and Electronic Engineers, pp. 1808–1813.

Yu, D.S., Kokko, E.G., Barron, J.R., Schaalje, G.B. and Gowen, B.E. (1992) Identification of ichneumonid wasps using image analysis of wings. *Systematic Entomology,* 17: 389–395.

Zhigang, L., Zetian, F., Yan, S. and Tiehua, X. (2003) Prototype system of automatic identification cotton insect pests and intelligent decision based on machine vision. In *ASAE Annual Meeting*, 2003. American Society of Agricultural and Biological Engineers Las Vegas, Paper No. 031007.

7 DAISY: A Practical Computer-Based Tool for Semi-Automated Species Identification

Mark A. O'Neill

CONTENTS

Introduction 101
Implementations of the DAISY System 102
DAISY Using a PCA-Based Approach 102
DAISY Using a Nearest-Neighbour/Kohonen/PSOM Approach 103
Application of NNC/NVD DAISY (DAISY II) to Pattern-Recognition Tasks 105
Limitations of NNC/NVD DAISY as a Pattern-Recognition Engine 106
Effects of Pose Scale and Background Noise 106
Effect of Cluster Geometry in Pattern Space on Identification Accuracy 106
Use of DAISY to Infer Class Membership Statistically 110
The Future: Can We Improve the Function of the DAISY System? 111
Acknowledgements 112
Notes 113
References 113

INTRODUCTION

DAISY (digital automated identification system) is based on an idea for a versatile, general-purpose identification system for biological species first mooted by the ecologist Kevin Gaston and the taxonomist Ian Gauld and me in 1993. While stranded at Juan Santa Maria Airport in San Jose, Costa Rica, they started thinking about how a technologically advanced nation (e.g. Japan) would approach the problem of biological species identification – and therefore the taxonomic impediment – from first principles. Gaston and Gauld realized that computer power was approaching the level where it might be possible to build a fully automated tool that could initially augment and perhaps eventually replace human experts for routine identifications.

Such tools would have a profound impact on both taxonomy and those areas of the biological sciences (e.g. ecology, biogeography) that are dependent on taxonomic services. Firstly, they would effectively take the knowledge of the taxonomic expert out of the museums and the universities located in the first world and place it where it is most sorely needed: in the hands of those third-world institutions struggling to catalogue tropical biodiversity before it disappears. Secondly, they would permit government officials and others to rapidly identify those species that threaten agriculture (e.g. *Anastrepha* fruitflies in the tropics) or health (e.g. *Aedes aegypti,* the vector of yellow fever) in an accurate and time-effective manner. Overworked human experts can take days to weeks to identify an unknown specimen, by which time it may be difficult to limit damage.

Although Gaston and Gauld were among the first to think seriously about utilizing computer technology to overcome the taxonomic impediment, the idea has caught on over the past decade. Today many eminent biologists – for example, the tropical ecologist Daniel Janzen (see Janzen, 2004) – have added their voices to the growing chorus for computer-aided taxonomy (CAT), as have a number of organizations at both national and international levels tasked with cataloguing biodiversity (e.g. *BioNET* and *INBIO*). Now, after more than a decade of development, the biological community is on the threshold of being able to use tools like DAISY, SPIDA (Do et al., 1999, Russell et al., 2005), ABIS (Schröder et al., 1995) and the many other systems discussed in this volume as *production identification systems* for species identification in the field, at ports of entry and in schools. Free access to such systems will indeed contribute to 'taxonomy's finest hour' (Wheeler, this volume). The ability to identify species, and gain consequent access to the information that goes with such identifications will be available to all – to the schoolboy in Newcastle who has found an interesting beetle, to the hotel keeper in Zikanthos whose ornamental *Nerium oleander* bushes are being devoured by a strange caterpillar with startling eyespots, to the customs official who has found a maggot in a shipment of oranges at Felixstowe – species identification will be made available to those who need it, where they need it and when they need it.

IMPLEMENTATIONS OF THE DAISY SYSTEM

DAISY Using a PCA-Based Approach

In its original form, DAISY was implemented in 1995 under the aegis of UK government's Agriculture and Food Research Council funding. In order to reduce the problem to one of tractable size, given the limitation in available computer power, a decision was made to concentrate on identifying insects using optical imagery of their wings. There were several reasons for this decision:

- Wings are two dimensional. This means that perspective does not have to be taken into account; thus, preprocessing is simplified.
- Wings are easily detached from both museum training material and specimens caught in the field and may be robustly mounted on slides for analysis.
- Much of the computer-based identification work reported at that time had used wings (for the preceding reasons). If the DAISY project used wings, it would allow like to be compared with like when assessing the system performance.
- Arthropods, especially insects, are good indicators of both changes and the health of the ecosystems they inhabit. From the scientific perspective, insects are, therefore, a good exemplar group on which to base prototype organismal identification systems.

A survey of techniques then being used in allied fields showed that, for fully automated species identification (where the user simply supplies the system with an image of the unknown, presses a button and gets an answer), approaches being developed for the machine identification of human faces provided an ideal initial starting point.

The initial DAISY implementation was particularly influenced by the work of Matthew Turk and Alex (Sandy) Pentland at MIT (Turk and Pentland, 1991). With training sets of 8–10 (wing) specimens per species, a principal component analysis (PCA)-based DAISY exemplar implementing Turk and Pentland's algorithm was able to recognize five species of parasitic wasp (Weeks et al., 1997) and some 30+ species of biting midge (Weeks et al., 1999a, 1999b) with a high level of accuracy (>95% of the material presented to the system was correctly identified to species) using slide-mounted wings. While the level of (identification) performance delivered by this system was similar to that of (then) extant systems using manual measurement of character sets derived from wing venation geometry

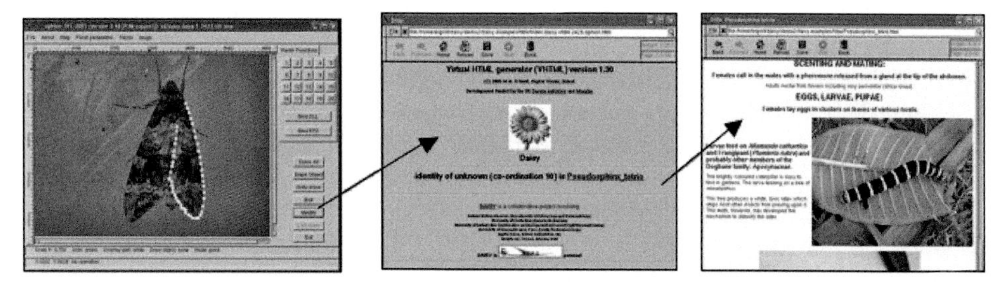

FIGURE 7.1 Identification of *Psuedosphinx tetrio* using DAISY.

(Lane, 1981; Yu et al., 1992), it possessed the potential for greater ease of use, as the holistic PCA-based approach obviated the time-consuming step of manually measuring wing characters.

The 'push-button' nature of this first-generation system – albeit tested on very few taxa – suggested that it might be possible to achieve the dream of Gaston, Gauld, Janzen and many others to implement a fully automated biological species recognition system. As a consequence, further funding was sought and obtained from the UK government's Darwin Initiative to build a fully functional DAISY exemplar. This work, done in collaboration with the University of Costa Rica and INBIO, was designed to test the feasibility of a computer-based system that would allow non-specialists to identify a difficult, but ecologically important, group of Costa Rican insects: parasitic wasps in the genus *Enicospilus*. This genus was chosen as an exemplar because of the following considerations:

- It is moderately specious with a Costa Rican fauna of some 50 species.
- *Enicospilus* species are all very similar in appearance. Therefore, it is difficult for non-specialists to differentiate between them. *Enicospilus* thus represents a *difficult* test case for an automated species identification system.
- *Enicospilus* wing venation is relatively complex. Thus, there are many potential wing-based characters that may have the potential to diagnose species.
- The ability to identify members of this genus rapidly and accurately would be of significant utility to ecologists.

Although the PCA-based DAISY system was capable of differentiating a substantial proportion of the members of this difficult group, the exercise of extending the exemplar to a practically useful number of species highlighted a number of serious shortcomings in the PCA approach. One such problem was linearity. For simple PCA approaches, the morphospace containing the pattern classes must be linear. Although a non-linear morphospace is problematic for the PCA approach, it is soluble: it is possible to extend PCA to non-linear morph spaces by partitioning space into a set of linear regions, each with its own principal components.[1]

A much more serious problem inherent in the PCA approach is the need to recompute the transform from the high-dimensional pattern space to a low-dimensionality component space each time material is added or removed from the system. This is clearly a very expensive option if the system is non-closed and in situations where training data are liable to be added and/or removed at regular intervals. Worse still, the speed at which the system computes principal components (e.g. trains) is $O(n^2)$ where n is the number of taxa in the system.[2] Thus, the time taken to train the system increases at a rate proportionate to the square of the number of species it contains. This is clearly not ideal if we want to build a general-purpose system capable of scaling to tens of thousands of taxa.

DAISY Using a Nearest-Neighbour/Kohonen/PSOM Approach

In order to tackle the issues of non-linearity and scaling, a second version of DAISY was recoded from scratch and implemented (Figure 7.1). This new version (NNC/NVD DAISY) was based on nearest-neighbour classification (NNC), a simple, yet very powerful classification scheme first advocated by Igor Aleksander and Graham Stonham (1979). Nearest-neighbour classification is easily understood and trivial to implement, but classification algorithms derived from it deliver at least 95 per cent of the accuracy of *much* more complex pattern-matching algorithms (see Lucas, 1997).

In its simplest form, NNC reduces to comparing an unknown (U) with a set of pattern vectors {P}. If U is correlated with each member P_i of {P}, U is assumed to belong to the same class as pattern vector P_m for which it has the highest correlation affinity $\mathbf{a}_{u,m}$. This is equivalent to a so-called 'first-past-the-post' (FPTP) classification. Despite its simplicity, NNC has *massive* advantages over more computationally intensive approaches such as PCA, for example:

- When new material is added to training sets there is no need to recompute a high-dimensionality to low-dimensionality transform. In fact, like Kohonen self-organizing maps (Kohonen, 2001) and the closely related plastic self-organizing map (PSOM) (Lang, this volume), NNC effectively trains *in real time*.
- The methodology is readily extended to incorporate dynamic learning; for example, the techniques used by Lang in the PSOM algorithm can be adapted (Lang, 2005).
- NNC is relatively easy to implement in hardware should the need arise. This was, in fact, done by Stonham in the Wisard facial recognition machine. The hardware implementation of NNC and similar algorithms is provided by Austin (1998).
- The simple and robust nature of NNC and the linear nature of its pattern-correlation algorithms mean that it is well suited to geometric parallelization.[3] This is exploited within NNC/NVD DAISY to permit the system to scale seamlessly to tens of thousands of taxa, making optimal use of available computing resources. An appropriate micro-cluster hardware architecture for efficient geometric parallelization of DAISY and similar systems is discussed by O'Neill et al. (2003).
- NNC copes well with non-linear morphospaces. In addition, the training data for NNC need not be ordered into putative species classes. In the case of non-linear PCA, for example, training data need to be ordered as a consequence of dividing the morphospace into a number of approximately linear regions. PCA and back propagation ANN-based individual species classifiers (e.g. SPIDA; see Do et al., 1999) also require training data to be ordered. Servicing this requirement adds significantly to algorithm complexity.

In order to optimize the performance of the NNC algorithm, DAISY II does not use standard cross-correlation to compute the pattern–pattern correlation affinity. Experiments revealed this measure to be far too blunt a comparator when dealing with classes whose interclass pattern differences are very small. Although the Kendall-*t* non-parametric correlation statistic (see Press et al., 1992) used initially to compute this gave significantly better results than cross-correlation, the approach is also constrained by the $O(n^2)$ scaling relation (in the number of elements in the pattern vectors compared) and therefore too slow to be of practical use in a production system. Consequently, a novel form of pattern–pattern correlation was developed: the normalized vector difference (NVD) metric, which proved to be a statistically optimal method of comparing patterns. For patterns that are very similar to each other, NVD gives a far superior result to traditional cross-correlation, yet it is faster to compute. Furthermore, the accuracy of NVD results either equals or exceeds those obtained using the Kendall-*t* method.

The NVD algorithm has the following form:

$$a_{x,y} = \sum \left| x_i^2 - y_i^2 \right| \tag{7.1}$$

In this equation, $a_{x,y}$ is the NVD correlation coefficient for patterns X and Y, x_i is the *i*th component of pattern X, and y_i is the *i*th component of pattern Y. A schema of the NNC implementation for DAISY II is shown in Figure 7.2.

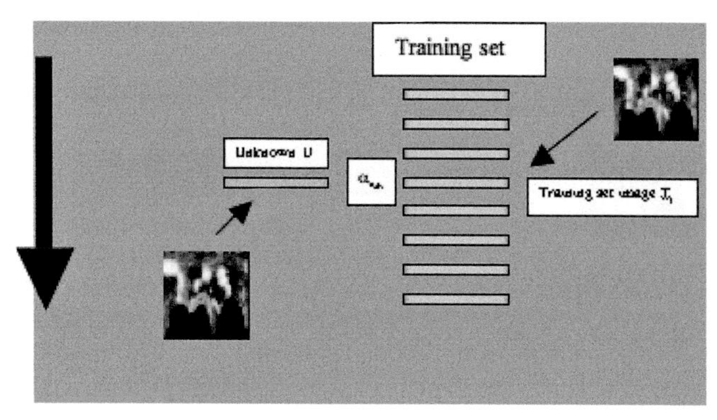

FIGURE 7.2 Operation of the NNC pattern correlation algorithm. The unknown U is compared with each training set image in turn. The identity of the unknown is the same as the training set image T_j for which the affinity $a_{u,j}$ is maximized.

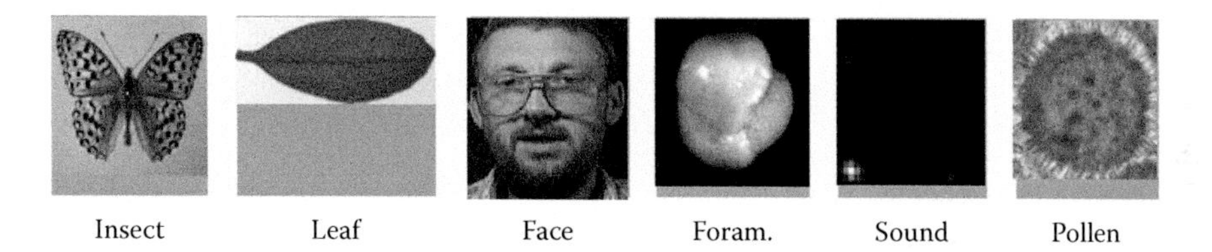

FIGURE 7.3 Examples of the various biological of patterns that have been identified successfully using the DAISY system.

APPLICATION OF NNC/NVD DAISY (DAISY II) TO PATTERN-RECOGNITION TASKS

NNC/NVD DAISY or DAISY II has been applied to a wide range of pattern-recognition tasks with significant success. These include the following (see also Figure 7.3):

- insect identification using wing pattern and shape (butterflies and moths, bees, parasitic wasps; see Gauld et al., 2000; Watson et al., 2003) and parasitic wasp faces (MacLeod et al., this volume);
- plant identification via leaf pattern and shape (and also from pollen grains);
- planktonic microfossil identification (MacLeod et al., 2003);
- vertebrate bone systematic (Walsh et al., 2004, this volume) and functional morphological (MacLeod et al., 2004) categorization;
- human faces (e.g. The Cambridge University/Olivetti Research facial data-base); and
- speech patterns (Knoll et al., this volume).

In all cases, a high level of identification accuracy has been achieved (see also Gaston and O'Neill, 2004); typically, greater than 85 per cent of material presented to the system is correctly identified to class (>95% for easy data-sets such as butterflies and moths) and, in most cases, greater than 50 per cent of specimens are identified to a high level of confidence (e.g. when prior statistics derived from the training set are used to assess confidence of identification). In order to achieve these high-confidence identifications, DAISY II extends the basic NNC approach by incorporating a highly conservative, empirically tested certainty measure based on the demography of local neighbourhoods in pattern space. If the coordination C is ≥3 (i.e. ≥3 nearest neighbours in pattern space of some unknown pattern U are in class Y, see Figure 7.4), then it is highly likely that U is also a member of class Y.[4]

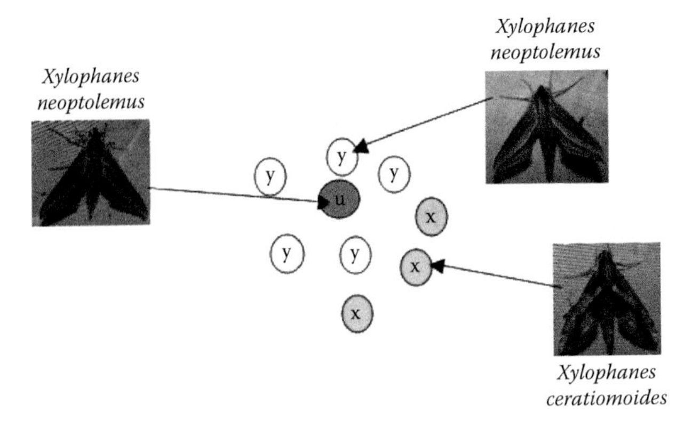

FIGURE 7.4 Operation of the DAISY coordination certainty measure.

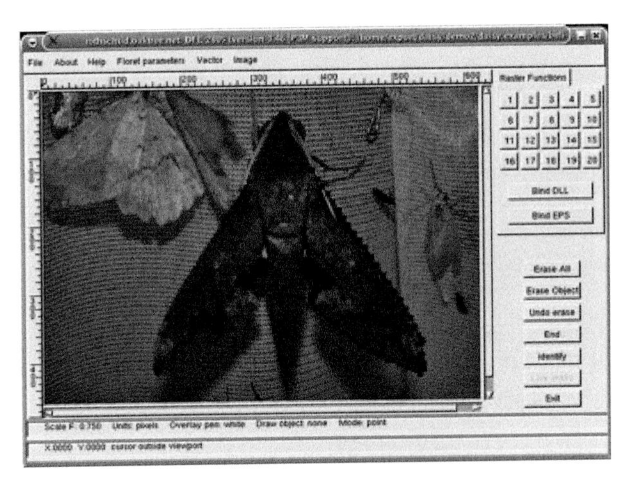

FIGURE 7.5 Use of PolyROI landmarking to outline the wing of a Belizian specimen of *Xylophanes chiron*.

LIMITATIONS OF NNC/NVD DAISY AS A PATTERN-RECOGNITION ENGINE

Effects of Pose Scale and Background Noise

Based on these results, DAISY II seems to be relatively successful as a generalized pattern-classification engine. Does DAISY II have any generic limitations? Of course. If images are supplied to the system in arbitrary poses and at an arbitrary scale, 'landmarking' is required (Figure 7.5). For example, insect wings need to be outlined in a specific manner in order to achieve identifications that are both accurate and repeatable. This is especially true when dealing with very similar taxa and/or gross differences in pose and/or scale. The purpose of landmarking is twofold: (1) to permit object rotation and scale to be automatically compensated for and (2) to remove the object from any background that may otherwise confuse the system. Examples of the sort of background clutter seen in typical images processed by the DAISY II system are shown in Figure 7.6. This background clutter must be digitally removed in order for DAISY II to make accurate identifications.

Effect of Cluster Geometry in Pattern Space on Identification Accuracy

There is also a limit to the amount of overlap that can be tolerated between two or more classes if substantial amounts of material in those classes are to be identified with high level of confidence.

Xylophanes titania

Adhemarius gannascus

FIGURE 7.6 Images of live Belizian sphingids in a collecting net (left) and in the wild (right) showing typical background clutter.

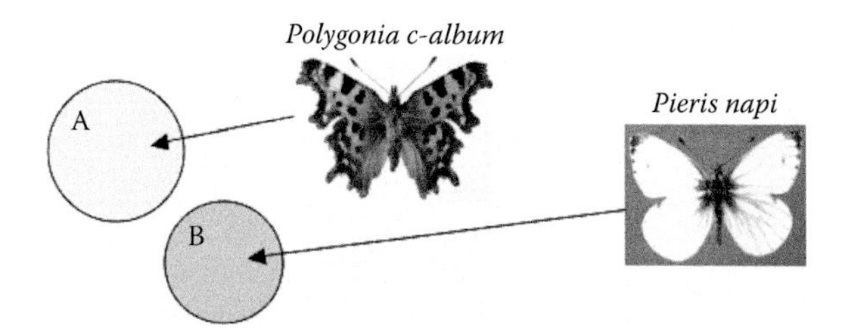

FIGURE 7.7 (Color Figure 7.7 follows page 110.) Hypothetical diagram of a populated morphological space with distinct convex hulls.

Classes that form distinct convex hulls in the morphospace and have no overlap (Figure 7.7) are, of course, easy to characterize. The DAISY II approach can even deal with cases where the classes overlap to a modest extent (Figure 7.8); fortunately, many species identification problems conform to this model. In this case, an unknown U will be identified unambiguously as A or B with a high level of confidence if it lies in one of the *non-intersecting regions* of morphospace for classes **A** or **B**, respectively. But quite a few biological pattern-identification problems are more complex and exhibit almost complete overlap between the two classes in question (Figure 7.9).[5] Unfortunately, if this situation arises, it is intractable. The best any system can do in this case is to return an answer that the unknown, U, is A|B; that is, the unknown belongs to either class A or class B.

Disambiguation of A|B to classes A or B in this situation requires human intervention. The best that the system can accomplish here is to provide a screening function. A screening set of possible but non-resolved species is produced that then have to be keyed out to precise identities using a (limited) key. Note that there is another possibility here. If the set of unresolved possibilities {R} for unknowns of a given species {U} is small and approximately constant, a back propagation ANN (e.g. a multilayer perceptron) could be used as a post-processing stage. This will effectively resolve U_i down to some member R_j by using character information. For example, the character formed by the distinct white spots on the upper forewing of *Aricia artaxerxes* can be used to resolve it from *Aricia agestis*.

Use of multilayer perceptrons has been championed by Chesmore (2004) to distinguish closed sets of closely related British orthoptera using audio patterns. His methodology could be extended to encompass the closed intractable species complexes generated by DAISY II on the fly. Alternatively, it may be possible disambiguate intractable species sets by acquiring the corresponding patterns using a different sensor technology or, in the case of images, by acquiring different views using the same sensor. In all of these cases, the idea is to obtain pattern sets for

TABLE 7.1
Effect of Sensor on FPTP Accuracy for Images of Planktonic Foraminifera Acquired Using Conventional Optical Microscopy and Using Confocal Microscopy

FPTP accuracy (optical microscopy)				FPTP accuracy (depth maps)			
Species	**Fptp 24**	**Fptp 32**	**Fptp 48**	**Species**	**Fptp 24**	**Fptp 32**	**Fptp 48**
all:	75	75	75	all:	95	95	95
Gl. conglobatus	80	80	60	*Gl. conglobatus*	100	100	100
Gl. quadrilobatus	100	100	100	*Gl. quadrilobatus*	100	100	100
Gl. ruber	100	100	100	*Gl. ruber*	100	100	100
Gn. aequilateris	60	40	40	*Gr. aequilateris*	100	100	100
Gr. truncatulinoides	60	60	60	*Gr. truncatulinoides*	60	60	60
Gr. tumida	40	60	80	*Gr. tumida*	100	100	100
O. universa	100	100	100	*O. universa*	100	100	100
S. dehiscens	60	60	60	*S. dehiscens*	100	100	100

Note: In each case, results are given for polar thumbnails[8] of 24 × 24, 32 × 32 and 48 × 48 pixels, respectively.

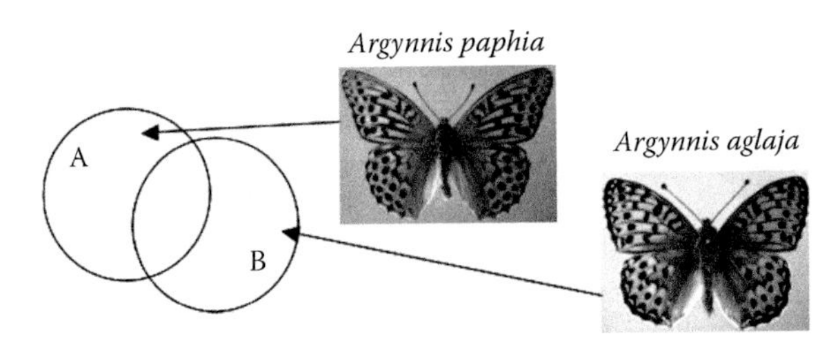

FIGURE 7.8 (Color Figure 7.8 follows page 110.) Hypothetical diagram of a populated morphological space with overlapping convex hulls.

the non-resolved classes that (ideally) possess the convex clustering characteristics illustrated in Figure 7.7. The effect of sensor type on classification accuracy is given in Table 7.1, which compares the identification accuracy attained using images of planktonic foraminifera acquired using a conventional optical microscope with that attained using depth images acquired using a confocal microscope system.

In this particular instance, it is clear that the level of accuracy obtained can be much higher in terms of both FPTP correctly classified percentage and for the corresponding certainty in the case of the depth images where artefacts due to lighting, patterning from staining and defocus have been removed (Figure 7.10; see also MacLeod et al., this volume). An equally dramatic example is afforded by the *Aricia artaxerxes*/*Aricia agestis* pair (Figure 7.11). The undersides of the wings of these two species are sufficiently different to reduce the interclass morphospace overlap from nearly total to partial: this means that at least some specimens of *Aricia artaxerxes* and *Aricia agestis* will be identified with a high level of accuracy and certainty if the ventral as opposed to the dorsal surfaces of the wings are used to train the system. For many species identification tasks, this is all that is required.

In practice, of course, there can be many more than two classes within the pattern cloud. A striking example of this is afforded by the British bumblebee fauna as illustrated in Figure 7.12.[6] With appropriate training set composition, DAISY is able to deal effectively with situations where the pattern clusters are non-convex or even situations where a single class maps to multiple clusters in morphospace (Figure 7.13).

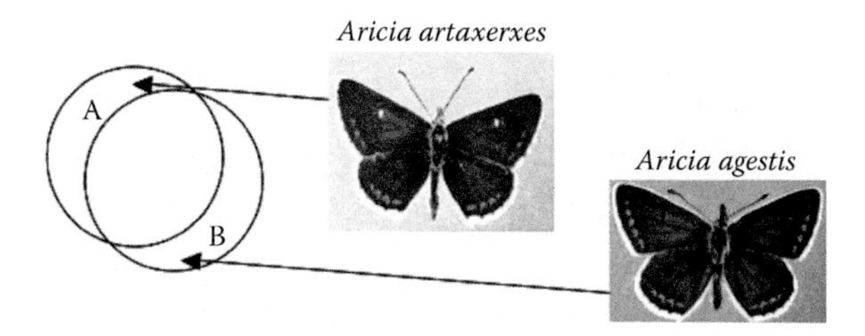

FIGURE 7.9 **(Color Figure 7.9 follows page 110.)** Hypothetical diagram of a populated morphological space with almost totally overlapping convex hulls.

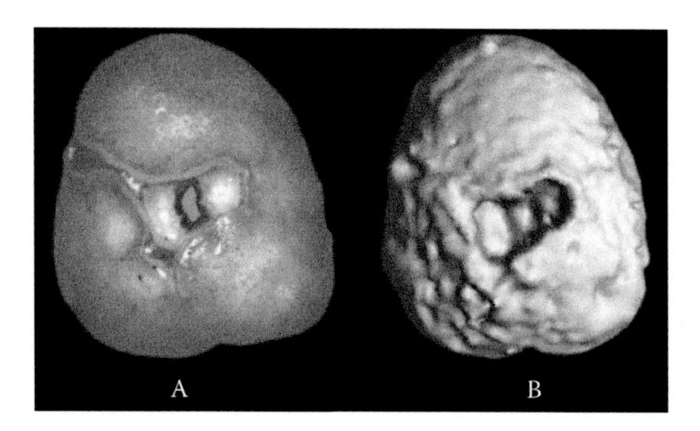

FIGURE 7.10 The effect of sensor type. Here, the same specimen of planktonic foraminifera (*Sphaeroidinella dehiscens*) is shown imaged (A) via conventional optical microscopy and (B) as a depth map using confocal microscopy. In (B), diagenetic staining is removed, and lighting is standardized between all images in the training set (see also MacLeod, 2006).

FIGURE 7.11 The undersides of the wings of *Aricia artaxerxes* and *Aricia agestis* (see also Figure 7.9). These species are readily distinguished if the ventral, as opposed to dorsal, wing surfaces are used to train DAISY.

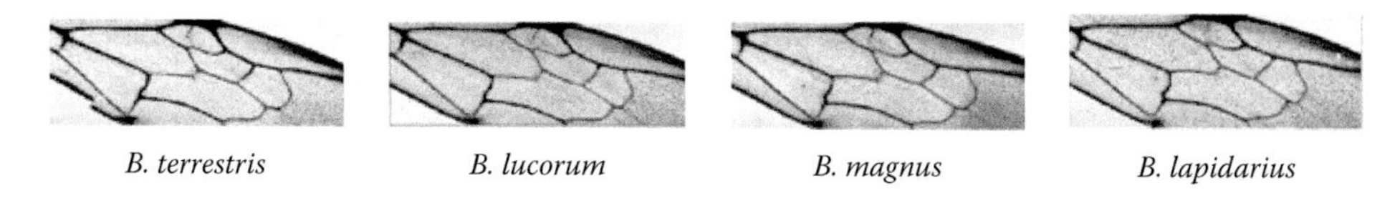

FIGURE 7.12 An example of a species cloud: (wing venation of) British *Bombus* sp.

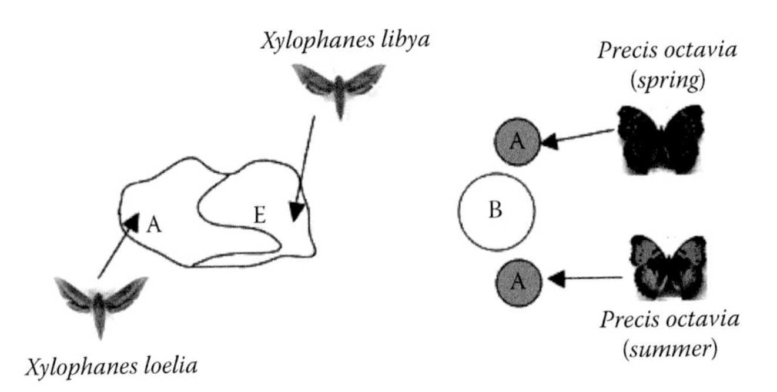

FIGURE 7.13 An illustration of difficult morphological space configurations that DAISY is able to resolve.

USE OF DAISY TO INFER CLASS MEMBERSHIP STATISTICALLY

Given the potential barriers to successful species identification, some of which have been raised in the previous section, it is clear work remains to be done if DAISY and similar systems are to realize their full potential. For example, work is currently in progress to determine whether clustering statistics derived from jacknife testing of training material can be used to (better) identify the classes to which this training material belongs. This will be especially useful in those situations where high-confidence identification of individual object patterns to class is impossible or where we wish to test the hypothesis that a group of patterns are in fact clustered into a set of distinct classes.

This case is well illustrated by the problems encountered when identifying pollen grains in ill-constrained pose imaged using conventional optical transmission microscopy. Although FPTP classification of individual pollen grains is relatively poor, clustering the identification statistics of the pollen grains show inter- and intraspecies regularities of a level of significance far greater than may be expected by chance alone. Unfortunately, unless the variation in the population from which the pollen grains are drawn matches that of the training sets, the method is of limited utility.[7] But if the variation of the training sets does approximate to that of the real populations, it may be possible to perform impressive 'generalized matching' analyses (e.g. taking a pollinator such as a bee and unambiguously identifying the plant species represented within its pollen load). This would be of significant utility, as a DAISY-based system could be used to identify both pollinators and the pollen loads they carry, facilitating the investigation of pollination networks. These may be subsequently analysed to identify key providers of pollination services within a given target ecosystem (see Watson, 2005).

A preliminary example of the end point of this type of DAISY-based network is shown in Figure 7.14. Here DAISY was used to establish the identities of the sphingid moths using wing shape and pattern and the host plants on which they feed using leaf shape and pattern, thus establishing the connectivity of the network.

Another hypothetical example of the power of the statistical approach is illustrated by the example of putative speciation in the sphingid *Xylophanes libya*. Analysis of both molecular data and field studies by tropical ecologist Dan Janzen supports the idea that *Xylophanes libya* is, in fact, two species: a lowland form (*X. libya*) and a montane form (*X. cryptolibya*). In this situation, DAISY can be used to examine training sets containing the putative species, $\{T_{Libya}\}$ and $\{T_{Cryptolibya}\}$. If Janzen's hypothesis is correct, jacknife testing these putative species training sets should show intratraining set affinities $a_{Libya,Libya}$; $a_{Cryptolibya,Cryptolibya}$ are higher than the intertraining set affinities $a_{Libya,Cryptolbya}$ by a magnitude significantly greater than the null hypothesis of random grouping.

FIGURE 3.1 Sometimes, scene segmentation and object recognition are easy.

FIGURE 3.2 Sometimes, low contrast and high clutter make object recognition hard.

FIGURE 3.3 Unusual scenes delay recognition.

FIGURE 3.6 An unusual item can be ignored: the cup is the focus of attention and the background is ignored.

FIGURE 3.7 Correct orientation facilitates scene analysis and recognition.

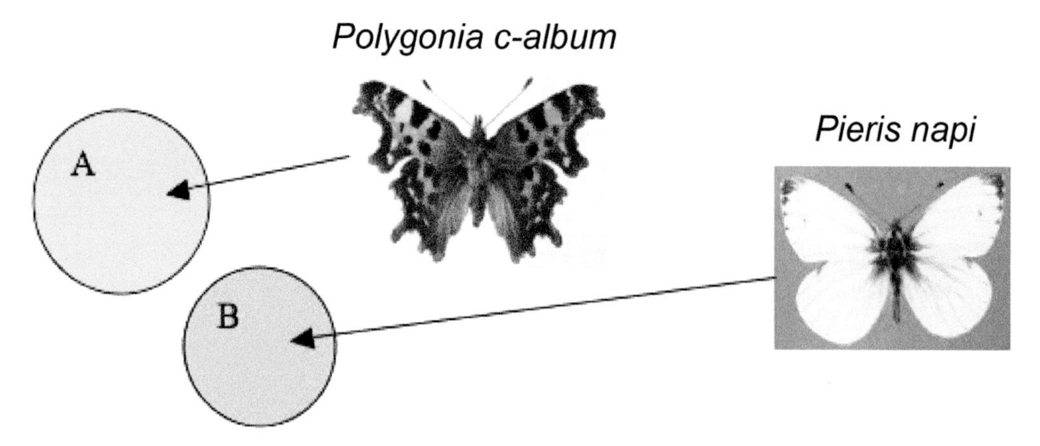

FIGURE 7.7 Hypothetical diagram of a populated morphological space with distinct convex hulls.

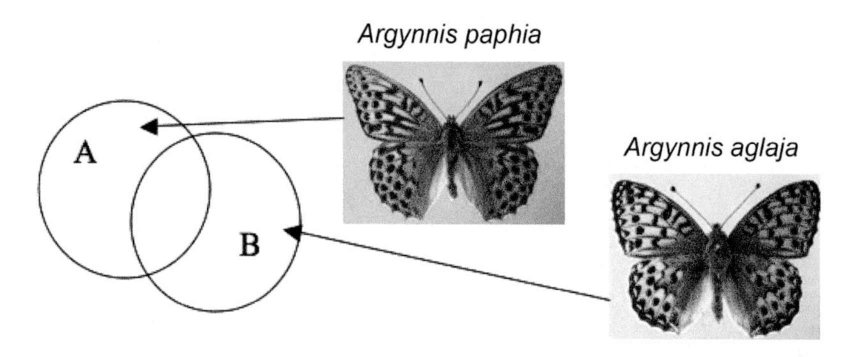

FIGURE 7.8 Hypothetical diagram of a populated morphological space with overlapping convex hulls.

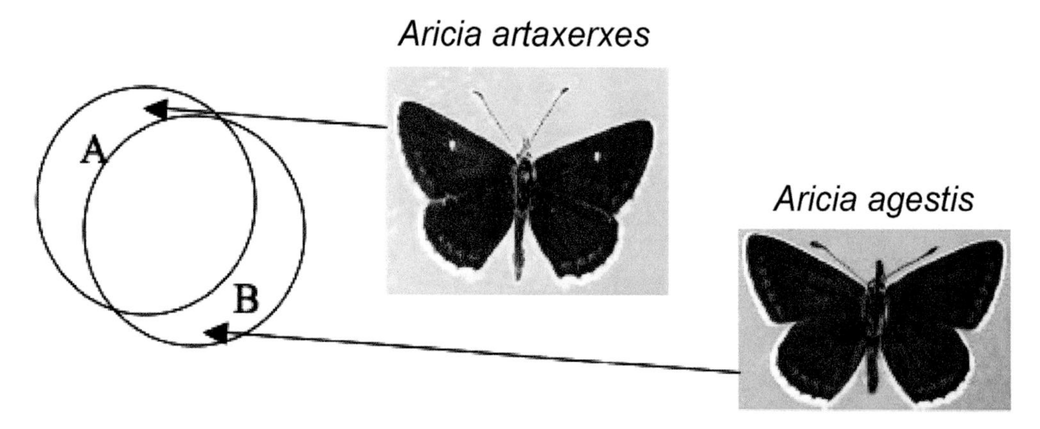

FIGURE 7.9 Hypothetical diagram of a populated morphological space with almost totally overlapping convex hulls.

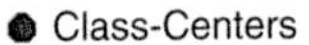

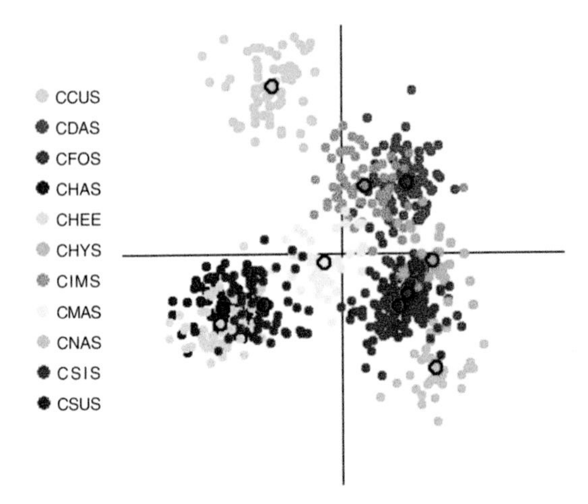

FIGURE 8.11 Visualization of species complexes as clusters in the feature space. Note: This is a two-dimensional projection of a high-dimensional feature space.

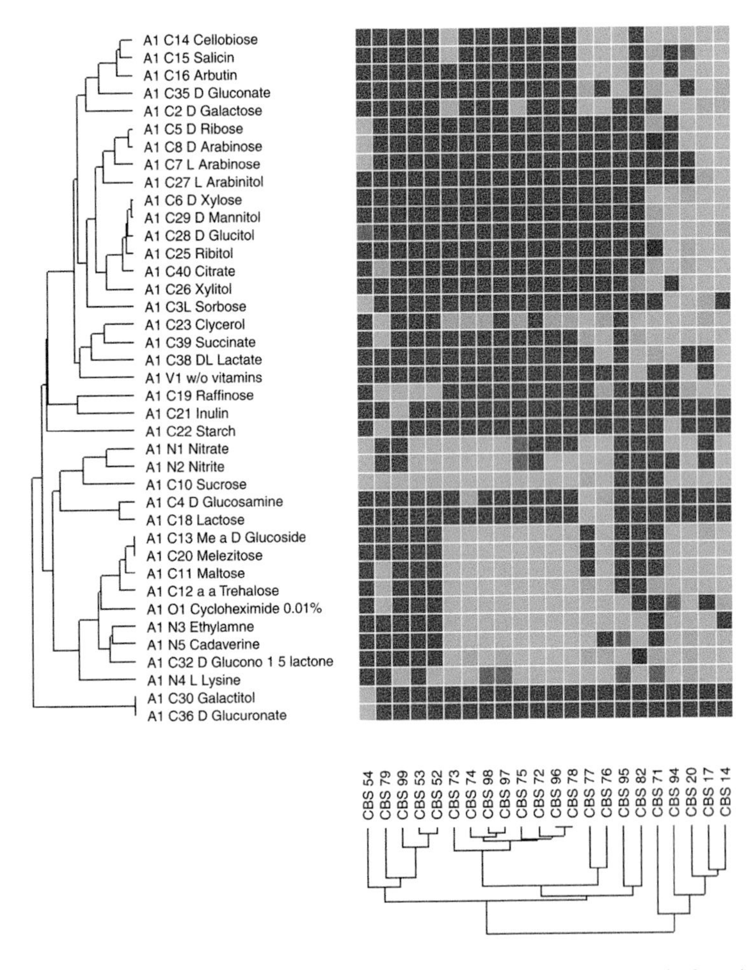

FIGURE 16.5 Functional analysis of a series of strains of yeast (UPGMA left-vertical tree) based on a set of physiological features (UPGMA top-horizontal tree). Normalized and reduced states of the above characters are displayed in red (negative result or absence of activity), in green (positive result or presence of activity), or in an intermediate shade for intermediate results.

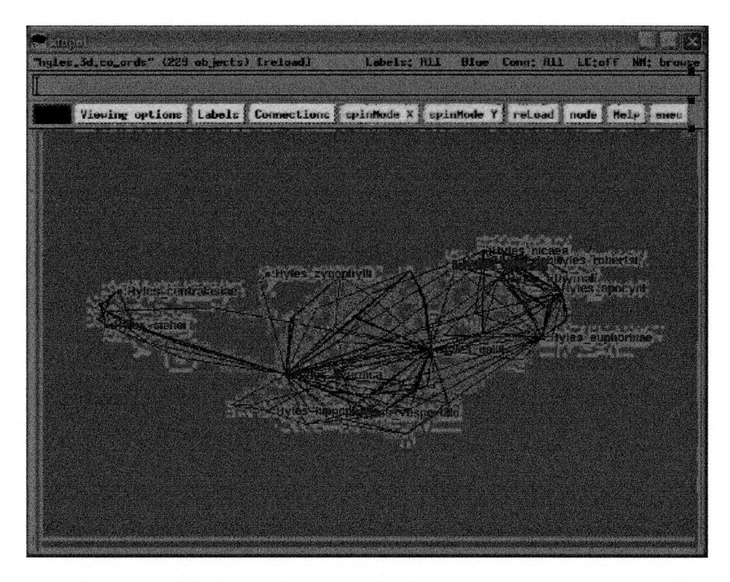

FIGURE 7.14 *Hyles* host plant interaction network constructed using material identified in part using DAISY. The moth nodes are labelled in this network for clarity only.

THE FUTURE: CAN WE IMPROVE THE FUNCTION OF THE DAISY SYSTEM?

Despite the many successes achieved by DAISY and similar systems, there remains a significant wish list that must be realized if a robust, scaleable identification system is to be realized. The major items outstanding include the following:

- Speeding up the basic DAISY algorithms. In the case of DAISY, this can be achieved in a number of ways – for example, by caching training set data or by extending and improving the distribution facilities provided by the P3 middleware under which DAISY operates (O'Neill et al., 2001; O'Neill et al., 2003).
- Investigation of methods to optimally extract characters to disambiguate classes that are members of pattern clouds. One possible approach would be to incorporate back-propagation algorithms into the DAISY architecture. For example, a multilayer perceptron could be added as a post-processing stage to DAISY.
- Looking at optimal conditions (and sensor technologies) for pattern acquisition. In the case of optical imagery, for example, the effects of lighting, pose scale, etc. also merit thorough investigation.
- Achieving platform independence. To be acceptable to a wide range of potential users, CAT systems must be able to run on industry-standard hardware and software systems such as Apple Macintosh systems running Mac OS X or PCs running Microsoft Windows XP. A start has already been made here. DAISY can now run seamlessly on the XP desktop within a CoLinux parasite (Aloni, 2004) which runs cooperatively as a user process within the XP host and communicates with it via an embedded X-server. DAISY is also being ported to PDAs (e.g. the Pocket PC). In this case, DAISY runs in a pocket Tux-Linux parasite (see http://www.dsitri.de/wiki.php?page=PocketTux), communicating with the host PocketWindows OS via an embedded X server and/or a VNC client (Richardson et al., 1998). Screen shots of these parasitic implementations are shown in Figures 7.15 and 7.16.

FIGURE 7.15 DAISY running on the PocketWindows desktop under pocket Tux-Linux parasite on an Xda II i Pocket PC.

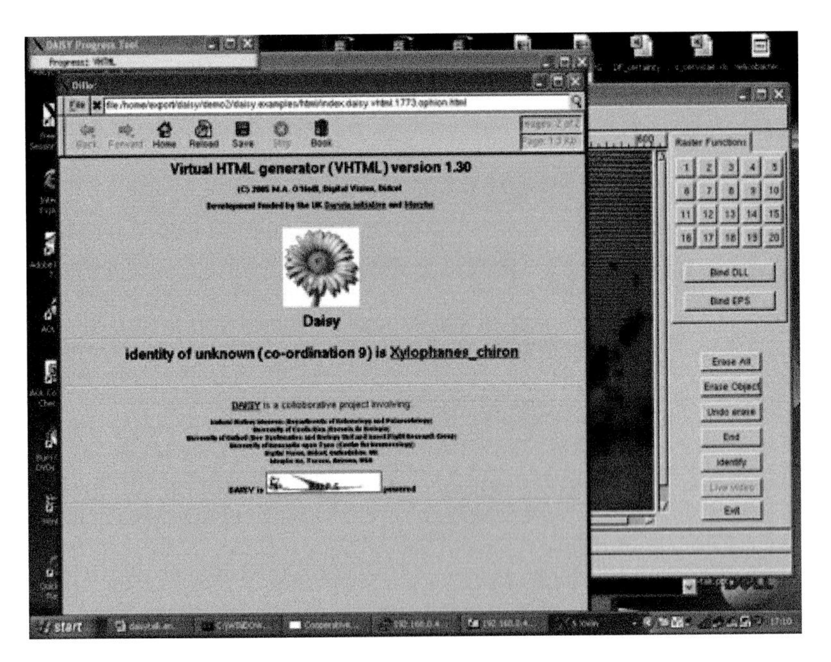

FIGURE 7.16 DAISY running on the Windows XP desktop from within a CoLinux parasite.

The building of generic systems that can identify a wide variety of biological objects remains *very* difficult. It will be taxing, but not impossible, to produce a fast, scaleable, fully automated generic solution to the biological species recognition problem in the next decade. Several very promising starts have already been made.

ACKNOWLEDGEMENTS

Without the help of people too numerous to mention individually, the DAISY would simply not have happened. Many people have contributed to the project over the last decade. I am particularly indebted to Professor Kevin Gaston (University of Sheffield) and Dr Ian Gauld (Entomology, NHM) without whose help DAISY simply would not exist. I also wish to thank Chris O'Toole (Oxford Bee Company), Mark Boddington (University of Oxford), Anna Watson (University of St Andrews), Ian Kitching (Entomology, NHM), Stig Walsh and Norman MacLeod (Palaeontology, NHM) for their considerable help in developing and testing the DAISY system over the last 5 years. Mia

Denos gave unflinching help acquiring test data from INBIO and UCR in Costa Rica, and also provided detailed feedback that vastly improved the efficiency of the DAISY GUI. Professor Malcolm Young (University of Newcastle-upon-Tyne), Dr Claus Hilgetag (I.U. Bremen), Dr Gully Burns (University of Southern California) and Shyam Kapadia (University of Southern California) also gave their time generously, helping to implement and test the P3 middleware on which DAISY depends. Thanks are also due to Professor Steve Simpson and Dr George McGavin, for tolerating my presence in the Hope entomological collections for the better part of a decade while DAISY was slowly gestating.

NOTES

1. In fact, this was done for the PCA version of DAISY. Each species had its own PCA-based classifier; the unknown was deemed to belong to that species to which it had the highest affinity. A full description of this version of DAISY is given in Weeks et al. (1997).
2. This inherent non-scalability is also a feature of back-propagation artificial neural networks (ANNs) – for example, multilayer perceptrons.
3. This is not a contradiction. NNC is a linear (i.e. first-order) algorithm capable of handling pattern spaces that are themselves non-linear.
4. Typically, $C \geq 3$ implies a certainty of correct identification ≥ 95 per cent for data-sets processed to date.
5. There is some evidence that adding more training material may resolve the situation by creating small regions of convexity within overlapped regions (e.g. MacLeod et al., this volume).
6. Actually, DAISY was able to classify ~60 per cent of British bumblebees presented to a high level of certainty (for local neighbourhood coordination ≥ 3, the system was at least 95 per cent certain that its classification of unknowns was correct).
7. This situation also pertains in many morphometric approaches.
8. Both training set and unknown images are subsampled to a standard size in a rectangular polar coordinate system. The resulting image is known as a polar thumbnail.

REFERENCES

Aleksander, I. and Stonham, T.J. (1979) Guide to pattern recognition using random-access memories. *Computers and Digital Techniques,* 2: 29–40.

Aloni, D. (2004) Co-operative Linux. In *Proceedings of the Linux Symposium* (ed A.J. Hutton), Volume 1, Ottawa, Ontario, pp. 23–32.

Austin, J. (1998) *RAM-Based Neural Networks,* World Scientific, York, 200 pp.

Chesmore, D. (2004) Automated bioacoustic identification of species. *Anais da Academia Brasileirade Ciencias,* 76: 436–440.

Do, M.T., Harp, J.M. and Norris, K.C. (1999) A test of pattern recognition system for spiders. *Bulletin of Entomological Research,* 89: 217–224.

Gaston, K.J. and O'Neill, M.A. (2004) Automated image recognition: why not? *Philosophical Transactions of the Royal Society of London, Series B,* 359: 655–667.

Gauld, I.D., O'Neill, M.A. and Gaston, K.J. (2000) Driving Miss Daisy: the performance of an automated insect identification system. In *Hymenoptera: Evolution, Biodiversity and Biological Control* (eds A.D. Austin and M. Dowton), CSIRO, Canberra, pp. 303–312.

Janzen, D.H. (2004) Now is the time. *Philosophical Transactions of the Royal Society of London, Series B,* 359: 731–732.

Kohonen, T. (2001) *Self-Organising Maps.* Springer Series in Information Sciences, 3rd edition, volume 30, Springer, Berlin, 501 pp.

Lane, R.P. (1981) Allometry in size in three species of biting midges (Diptera, Cerotopogonidae, *Culicoides*). *Journal of Natural History,* 15: 775–788.

Lucas, S.M. (1997) Face recognition with the continuous *n*-tuple classifier. In *Proceedings of the British Machine Vision Conference, 1997* (ed A. Clark). Available online: http://www.bmva.ac.uk/1997/papers/113/paper.html.

MacLeod, N., O'Neill, M.A. and Walsh, S.A. (2003) PaleoDAISY: An integrated and adaptive system for the automated recognition of fossil species. *The Geological Society of America, Abstracts and Programs,* 35: 316.

MacLeod, N., O'Neill, M.A. and Walsh, S.A. (2004) A comparison between morphometric and unsupervised, artificial, neural-net approaches to automated species identification in foraminifera. In *Proceedings of the Micropalaeontological Society Calcareous Plankton Spring Meeting: Copenhagen, Denmark* (eds E. Sheldon, S. Stouge and A. Henderson), Geological Survey of Denmark and Greenland, Ministry of the Environment, GEUS, p. 21.

MacLeod, N., O'Neill, M.A. and Walsh. S.A. (in press) A comparison between morphometric and artificial neural net approaches to the automated species-recognition problem in systematics. In *Biodiversity Databases: From Cottage Industry to Industrial Network* (eds G. Curry and C. Humphries), Taylor & Francis, London.

O'Neill, M.A., Burns, G. and Hilgetag, C.-C. (2003) The PUPS-MOSIX environment: a homeostatic environment for neuro- and bio-informatic applications. In *Neuroscience Databases, a Practical Guide* (ed R. Kötter) Kluwer Academic Publishers, Amsterdam, pp. 187–202.

O'Neill, M.A. and Curtiss-Rouse, M. (2004) An engineering study for the design of a computational microbot (CMB), Rev. 1.6. *Digital Vision Internal Document.*

O'Neill, M.A. and Hilgetag, C.-C. (2001) The portable UNIX programming system [PUPS]: a computational environment for the dynamical representation and analysis of complex neurobiological data. *Philosophical Transactions of the Royal Society of London, Series B,* 356: 1259–1276.

Press, W.H., Tuekolsky, S.A., Vetterling, W.T. and Flannery B.P. (1992) *Numerical Recipes in C: The Art of Scientific Programming,* 2nd edition, Cambridge University Press, Cambridge, 1020 pp.

Richardson, T., Stafford-Fraser, Q., Wood, K.R. and Hopper, A. (1998) Virtual network computing. IEEE Internet Computing 2(1). Available online: http://www.cl.cam.ac.uk/Research/DTG/attarchive/pub/docs/att/tr.98.1.pdf.

Russell, K.N., Do, M.T. and Platnick, N.I. (2005) Introducing SPIDA-web: an automated identification system for biologcial species. Taxonomic Database Working Group Annual Meeting, Abstracts (http://www.tdwg.org/2005meet/paperabstracts/TDWG2005_Abstract_64.htm).

Schröder, S., Wittmann, D., Drescher, W., Roth, V., Steinhage, V. and Cremers, A.B. (2002) The new key to bees: automated identification by image analysis of wings. In *Pollinating Bees: The Conservation Link between Agriculture and Nature* (eds P.G. Kevan and V.L. Imperatriz-Fonseca), Proceedings of the Workshop on Conservation and Sustainable Use of Pollinators in Agriculture, with Emphasis on Bees, S. Paulo, October 1998. Ministry of Environment, Secretariat for Biodiversity and Forests, Brazil, pp. 209–218.

Turk, M. and Pentland, A. (1991) Eigenfaces for recognition. *Journal of Cognitive Neuroscience,* 3: 71–86.

Walsh, S.A., MacLeod, N. and O'Neill, M.A. (2004) Analysis of spheniscid tarsometatarsus and humerus morphological variability using DAISY automated digital image recognition. In *Abstracts of the 6th International Meeting of the Society of Avian Paleontology and Evolution*, Quillan, pp. 45–46.

Watson, A.T. (2005) Automated identification and network analyses in Acacia pollination ecology. Unpublished M.Phil thesis, University of St Andrews.

Watson, A.T., O'Neill, M.A. and Kitching, I.J. (2003) A qualitative study investigating automated identification of living macrolepidoptera using the Digital Automated Identification SYstem (DAISY). *Systematics & Biodiversity,* 1: 287–300.

Weeks P.J.D., Gauld, I.D., Gaston K.J. and O'Neill M.A. (1997) Automating the identification of insects: a new solution to an old problem. *Bulletin of Entomological Research,* 87: 203–211.

Weeks, P.J.D., O'Neill, M.A., Gaston K.J. and Gauld, I.D. (1999a) Automating insect identification: exploring the limitations of a prototype system. *Journal of Applied Entomology,* 123: 1–8.

Weeks, P.J.D., O'Neill, M.A., Gaston, K.J. and Gauld, I.D. (1999b) Species identification of wasps using principal component associative memories. *Image & Vision Computing,* 17: 861–866.

Yu, D.S., Kokko, E.G., Barron, J.R., Schaalje, G.B. and Gowen, B.E. (1992) Identification of Ichneumonid wasps using image analysis of wings. *Systematic Entomology,* 17: 389–395.

8 Automated Extraction and Analysis of Morphological Features for Species Identification

Volker Steinhage, Stefan Schröder, Karl-Heinz Lampe and Armin B. Cremers

CONTENTS

Introduction 115
Feature-Based Image Analysis 116
Automated Bee Identification System (ABIS) 116
  Image Acquisition in Collections and in the Field 117
  Automatic Extraction of Morphologic Wing Features 117
  Extraction of Edgels and Their Grouping into Cell Boundaries 117
  Genus Templates 118
  Genus Identification 119
  Knowledge-Based Processing with the Help of Genus Templates 120
  Automatic Extraction of Iconic Wing Features 122
  Species Identification by Non-linear Kernel Discriminant Analysis 122
  Encoded Knowledge 123
Results 123
Current Work 123
  Multiview Analysis 124
  Extracting Points of Interest 124
  Selection of Points of Taxonomic Interest 125
  Selection of Correspondences in Multiview Image Analysis 126
Conclusions 126
Acknowledgements 128
References 129

INTRODUCTION

Computer-aided systems based on image analysis have become popular for automated species identification. This contribution emphasizes that robust approaches to image-based automated species identification 'should know what they do'. This means these approaches should extract meaningful image features that depict corresponding object features (i.e. morphologic and morphometric features like veins, eye distances, lengths of antennas), instead of relying solely on pure image information (i.e. pixels, wavelets). To demonstrate the efficiency and reliability of a feature-based approach to image-based automated species identification, we first present a feature-based system for the automated identification of bee species that employs knowledge-based image analysis. The system has been successfully deployed in Germany, Brazil, and the USA. Second, we present a generalization of the feature-based approach to a multiview image analysis of different perspectives of specimens, which will be applicable also to different groups of insects.

FEATURE-BASED IMAGE ANALYSIS

The classical approach to species identification is that of qualitative morphological and quantitative morphometric analysis. Thus, taxonomists perform the identification task by visual observation and measurement of morphological and morphometric features (i.e. the colour of feathers, the shape of wings, the distance between the eyes, the angle between two veins). A natural way to automate the traditional taxonomic ways of species identification by visual observation and measurement is to employ methods of image analysis. Furthermore, since taxonomic descriptions rely on morphological and morphometric features, a feature-based approach seems appropriate.

This feature-based approach to image analysis considers any image as a collection of discrete image features rather than an array of pixels. While each pixel of an image encodes one intensity value in the case of a monochrome image, or several intensity values in the case of a colour image, an image feature represents the image of an object feature. An object feature, in turn, is a meaningful and characteristic part of the model of the object class, which the image-analysis system is aiming to recognize. Thus, the feature-based approach to image analysis is the key to knowledge-based image analysis, which employs explicit object models and allows explicit reasoning about the objects and their relations in the application domain (i.e. reasoning about the content of the images). This way of image analysis, which couples feature recognition and reasoning, is called image understanding and is part of the field of artificial intelligence (Shapiro and Eckroth, 1987).

The feature-based approach also represents the technical link to content-based image retrieval in large multimedia data-bases where content-based descriptions of the images, on the one hand, and the ontologies that describe relations between object and feature classes, on the other, help to retrieve images that show some specific content. Last, but not least, the feature-based approach supports explanation of the results of object identification or image retrieval. Comparing characteristic features allows the interpretation of results in terms of the application domain instead of raw statistics of pixel comparisons. In this way a feature-based approach helps to ensure better quality control. To summarize, the feature-based approach to image analysis enables users to:

- identify objects in terms of understanding the content of images;
- index images in large data-bases efficiently by content-based retrieval; and
- ensure quality control in terms of domain-specific criteria.

Obviously, the feature-based approach must rely on the presence of sufficient meaningful image features of the domain, where 'meaningful' is defined by the recognition purpose. This means that the image features must correspond to characteristic object features, which are sufficient to classify image content into the modelled object classes (i.e. identify the species). In the case where characteristic image features are not available for a desired classification result, a combination of feature-based and pixel-based image analysis is necessary. In the following sections we illustrate these principles of the feature-based approach to image analysis for species identification.

AUTOMATED BEE IDENTIFICATION SYSTEM (ABIS)

The automatic bee identification system (ABIS) was designed to identify species of bees. Bees play a key global role in the ecosystem and in agriculture because they are the most important pollinating insects. The annual value of the worldwide service of pollination is estimated to be

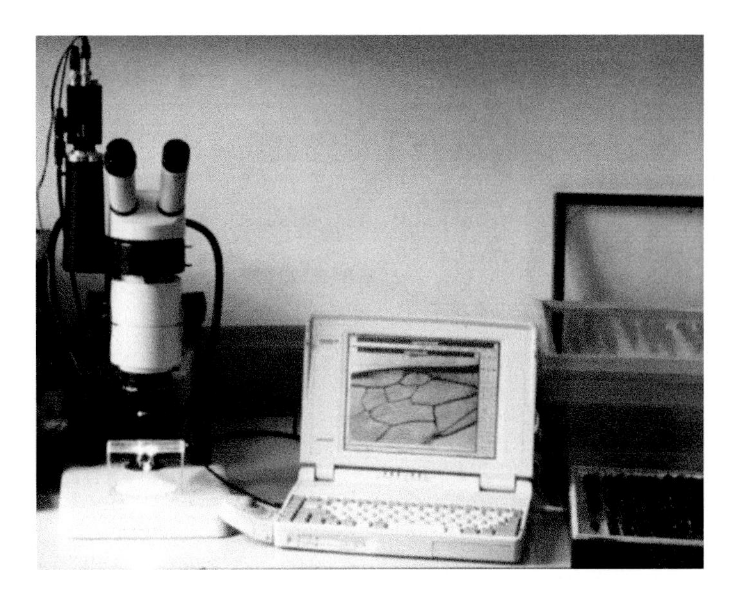

FIGURE 8.1 The setup of ABIS for work in a scientific collection. White LEDs illuminate the wing from below. The digital camera is mounted on the microscope. The ABIS software is running on a notebook computer.

around US $65 billion (Pimentel et al., 1997). But many of the bee species are currently threatened with extinction.

Efforts to preserve and protect bee species are severely hampered by their difficult taxonomy and the lack of entomological specialists who can identify bee specimens. The key idea behind ABIS is to identify bee species from images of their forewings. Using diffuse background illumination, bee wings show a clear venation within a transparent skin. The structure of this venation is genetically fixed and therefore suited to the identification of species (Schröder et al., 1999). Analysing each wing image gives rise to a set of well-defined, characteristic morphometric features. These features are used in a new knowledge-based classification approach to taxonomic identification.

Image Acquisition in Collections and in the Field

The first step in the recognition of a bee specimen is to acquire an image of its forewing. For this, ABIS requires only a microscope interfaced with a digital camera and a notebook computer. Live bees or collection specimens are mounted so as to permit images of their forewings to be made with the microscope (Figure 8.1). The equipment can also be used in the field (Figure 8.2). The wing images are stored as high-resolution, high-quality JPEG files for subsequent processing (Figure 8.3).

Automatic Extraction of Morphologic Wing Features

The ABIS system implements a fully automatic extraction algorithm for the image processing component of the identification system. The system automatically detects edges in the pattern of wing venation and, with the help of genera-specific wing templates, formulates hypotheses regarding the location of key cells within this pattern. Once these cells have been detected, numerical morphometric features can be generated that describe the cells and their topological relationships. These features are then input to a statistical classification process.

Extraction of Edgels and Their Grouping into Cell Boundaries

The first image processing step aims to extract edges of the wing venation by convolving the image with the derivates of the Gaussian kernel. Due to distortions (e.g. polls and hairs on the wings),

FIGURE 8.2 Employing ABIS in the field. By cooling live bees in a cooling box the wings of the sluggish bees can be clipped under the microscope slide to obtain wing images. Once released, the bees can fly off unharmed.

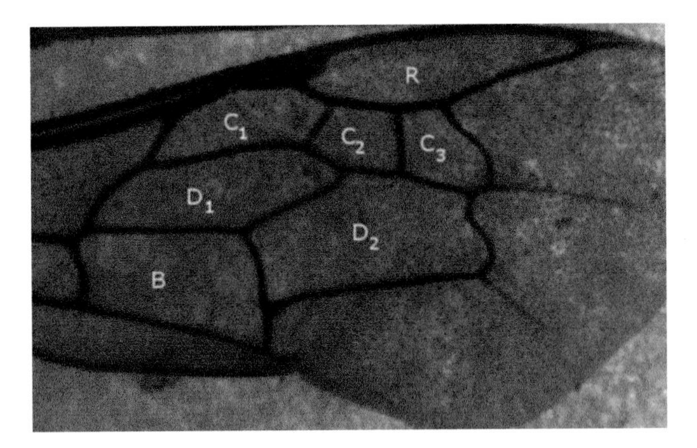

FIGURE 8.3 ABIS. The image of the forewing shows due to illumination from below nearly transparent cells enclosed by the venation. R, C_1, C_2, C_3, B, D_1 and D_2 depict the radial cell, up to three cubital cells, brachial cell and two discoidal cells, respectively.

the result of the convolution shows generally much positive false hypotheses of edge elements, the so-called edgels (Figure 8.4). To select and group the edgels into closed cell boundaries, ABIS uses *a priori* knowledge about the general shape of the cells. All cells are star shaped and every point of the cell boundary is visible from the centre of gravity of the cell. A boundary point is visible from the centre of gravity if the straight line segment between them lies completely within the cell area. Figure 8.5 depicts the result of the grouping process of edgels into closed cell boundaries.

Genus Templates

While the boundaries of the brachial cell B and the two discoidal cells D_1 and D_2 (c.f. Figure 8.3) can be detected easily and robustly, the boundaries of the other cells are generally difficult to detect. Therefore, ABIS allows users to employ for each bee genus a genus template. A genus template is derived in an automated way from a set of about 30 wing images of a genus by averaging the results of successful extractions of the cell boundaries. Thus, successful extractions of cell

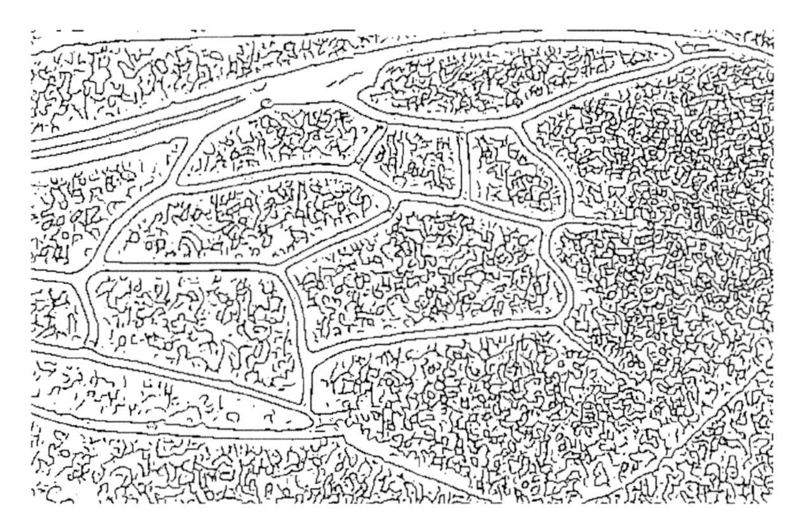

FIGURE 8.4 ABIS. Pure extraction of edge elements (i.e. pixels with high gradients of intensity) yields correct feature hypotheses, but also many positive false feature hypotheses due to hairs, polls, etc.

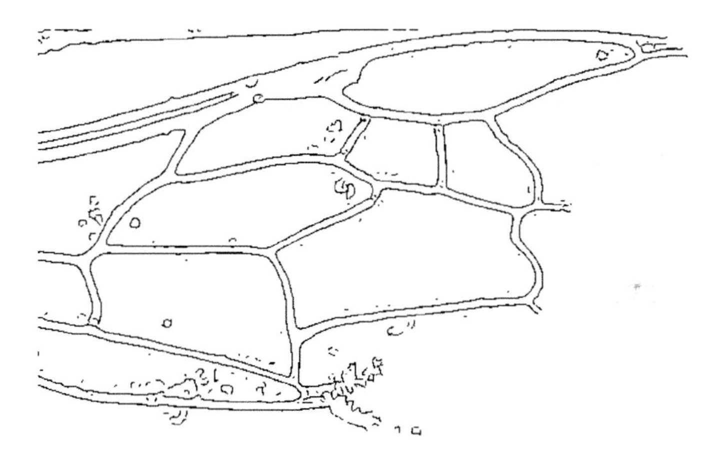

FIGURE 8.5 Because all cells are star shaped, we can select and group the edge elements to constitute cell boundaries of appropriate shape, closure and area.

boundaries in the wing images of each genus allows the software to support the extraction of cell boundaries for all other incoming wing images. This set of about 30 wing images of a genus to derive the genus template is called the training set of the genus.

Genus Identification

But how can the appropriate genus template be assigned to a new incoming wing image of unknown species? The answer is that ABIS must first identify the genus of the specimen under inspection. Identification of the genus can be done by detecting and analyzing only three cells of the wing image: the brachial cell, B, and two discoidal cells, D_1 and D_2 (c.f. Figure 8.3), which are – in contrast to the other cells – quite easily and robustly detected. A sufficient number of morphometric features (e.g. angles between axes of gravity (c.f. Figure 8.6) ratios of cell areas and perimeters) can be derived from these three cells to identify the genus by classical linear discriminant analysis (LDA). Furthermore, ABIS extracts the well-defined centres of gravity of the brachial cell and the two discoidal cells to compute the mapping of the genus template into the wing image under investigation (Figure 8.7).

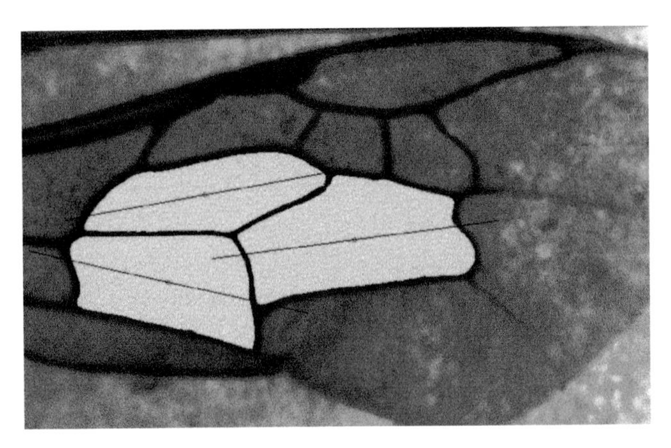

FIGURE 8.6 The successfully extracted cells B, D_1 and D_2 and their axes of gravity. The angles between these axes, the ratios of cell perimeters, cell areas, etc. are employed as morphometric features within classical LDA to derive the genus of the bee specimen.

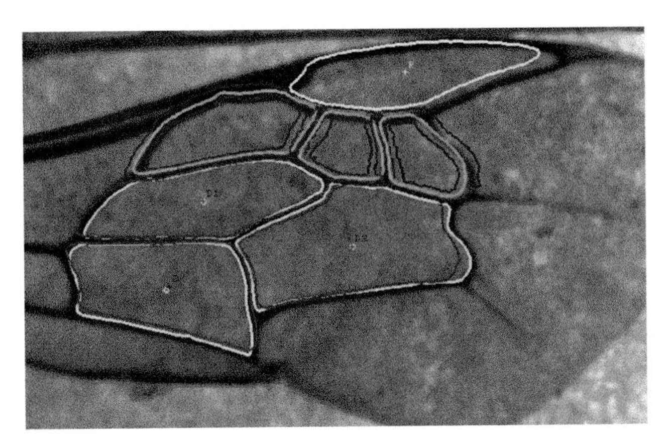

FIGURE 8.7 From successfully extracted cells B, D_1 and D_2 (light grey boundaries), the genus can be derived by LDA and the mapping of the appropriate genus template can be computed. The superimposed template predicts the expected locations of the boundaries (light medium or medium grey) of three cubital cells or C_1, C_2, C_3 (cf. Figure 8.3) to be extracted next, as well as the variances of these predictions (medium dark grey). The most expected locations and the variances are derived from the training set of each genus.

Knowledge-Based Processing with the Help of Genus Templates

The genus template superimposed on the wing image under investigation now guides the selection and grouping of edgels that constitute the boundaries of the cubital cells. Only those edgels within the variances of the predicted cell boundaries (dark grey in Figure 8.7) will be selected for estimating the cell boundaries. Once the boundaries of the cubital cells C_1, C_2 and C_3 have been extracted successfully (Figure 8.8), the mapping of the genus template is adjusted again, now using not only the centres of gravity of the cells B, D_1 and D_2, but also those of the newly derived cubital cells (Figure 8.9). Next, the genus template helps select and group the edgels that constitute the boundaries of the radial cell. Having detected the boundaries of all cells in this way, ABIS infers the venation enclosing all these cells and the venation junctions (Figure 8.10). From these morphological features – cells, veins and junctions – ABIS calculates a set of approximately 75 morphometric features of the wing. After a principal component analysis (PCA) for de-noising and dimensionality reduction, ABIS obtains a morphometric feature vector of 50 elements by choosing the first 50 principal components in the order of descending eigenvalues.

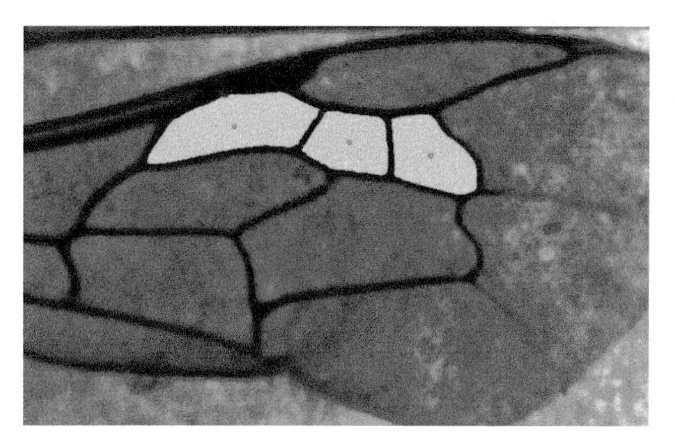

FIGURE 8.8 The successfully extracted cubital cells C_1, C_2 and C_3 derived with the help of the superimposed genus template.

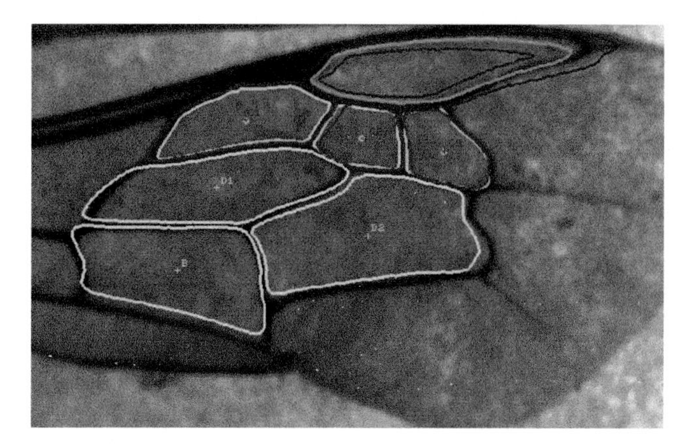

FIGURE 8.9 ABIS: the newly adjusted genus template now using the six centres of gravity of the cells B, D_1, D_2, C_1, C_2 and C_3. The prediction of the boundaries of the radial cell is now far better than it would have been in the first superposition based only on the centres of gravity of cells B, D_1 and D_2 (compare with Figure 8.7).

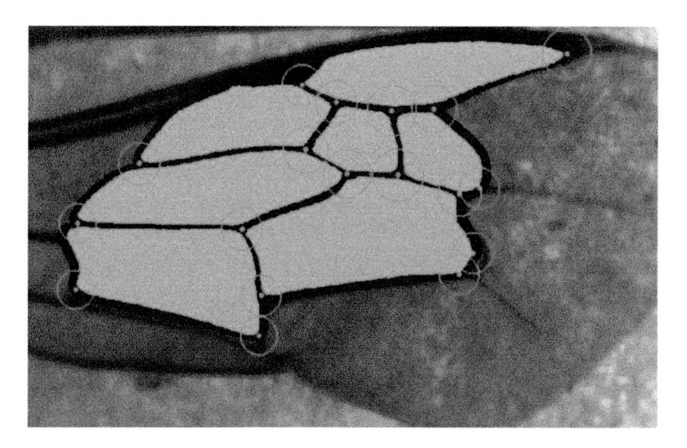

FIGURE 8.10 All successfully extracted morphological features: the cells (grey), the venation (dark) and the venation junctions (points).

Automatic Extraction of Iconic Wing Features

Using only morphometric wing features may neglect iconic information of the wings (i.e. the distribution of light and dark areas within the cells). The centres of the three initially extracted cells, B, D_1 and D_2, permit ABIS first to normalize the wing image by rotation and scaling. This is accomplished simply by deriving the parameters of the two-dimensional affine transformation given the coordinates of the centres of the extracted cells B, D_1 and D_2 on the one hand and their expected position in the normalized image on the other hand. Second, ABIS aligns a sampling window over the normalized wing image. The content of the sampling window is down-sampled by applying a second-order Gaussian filter three times. The resulting intensity matrix is of size 12×20 intensity values and thus forms an iconic feature vector of 240 elements.

Species Identification by Non-linear Kernel Discriminant Analysis

The resulting 290 dimensional feature vectors contain both the extracted 50 morphometric features and the 240 iconic features obtained by down-sampling. Performing classical LDA in such situations, where the input dimensionality is very high and only small training sets are available, will not be promising. To overcome this shortcoming, we have developed a non-linear generalization of LDA, the non-linear kernel discriminant analysis (NKDA) (see Roth and Steinhage, 1999). An NKDA classifier constructs non-linear decision functions that can be implemented efficiently using kernel functions. Compared with the support vector machine (SVM), the NKDA classifier has two advantages: (1) it allows data visualization (Figure 8.11) and (2) it is significantly faster due to simpler optimization and due to the possibility of handling multiclass problems directly.

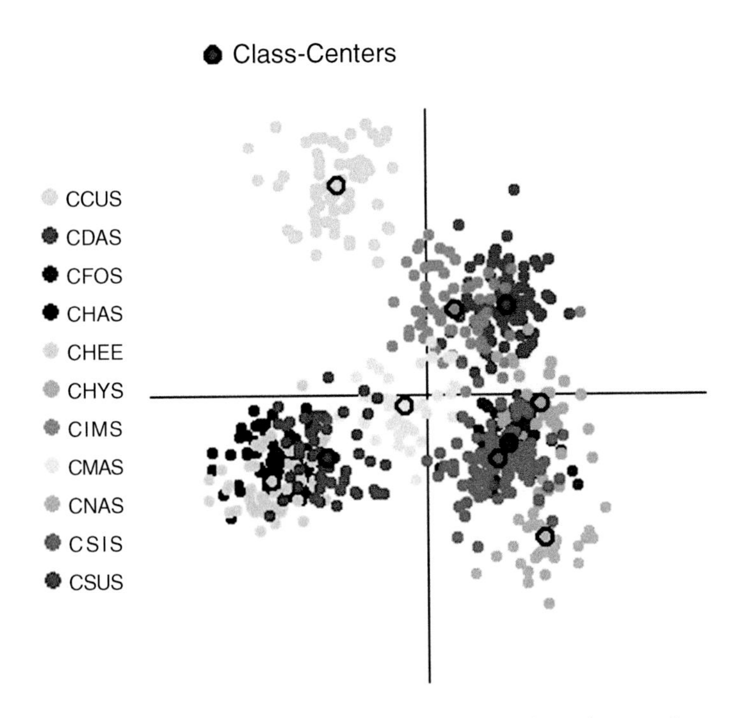

FIGURE 8.11 (Color Figure 8.11 follows page 110.) Visualization of species complexes as clusters in the feature space. Note: This is a two-dimensional projection of a high-dimensional feature space.

Encoded Knowledge

ABIS encodes different types of knowledge for identifying bee species from images of their forewings. General knowledge about the topology (structure) and geometry of the venation and cells enclosed by the venation of all bees is encoded within the basic wing template, which is fixed within the ABIS approach. The basic wing template encodes the star-shaped attitude of the cells, certain cell neighbourhoods, typical upper and lower bounds of cell areas, etc. This knowledge is valid for all those bee genera that ABIS 'knows' as a result of accessing the corresponding training sets.

Application of ABIS to identify species of wasps from images of their forewings, for example, requires users to change the basic wing template. The knowledge of the special topology and geometry of the venation and cells enclosed by the venation for all bees of one genus or a group of species is encoded within the genus templates (Figure 8.7). Each genus template predicts the number of cells (due to two or three cubital cells) and the genus-specific structure cell neighbourhoods, as well as the expected locations of the cell boundaries, including the variances of these predictions derived from the training set of each genus. Identification knowledge for genus diagnosis is derived by LDA during the training of the LDA step based on the training sets of all genera under consideration. The result of this training step is a set of linear decision functions that support the assignment of new specimens to the correct genus. Also, the NKDA learns and memorizes genus-specific non-linear decision functions that allow assigning new specimens of the trained species of each genus to the correct species.

In summary, the domain-specific knowledge for identifying bee species from images of their forewings consists of:

- the basic wing template: the common topology and geometry of the venation and cells enclosed by the venation for all bees;
- the genus templates: topology and geometry of the venation and cells enclosed by the venation for all bees of one genus or a group of species within one genus; and
- the linear decision functions of LDA and the non-linear decision functions of NKDA for genus and species identification in feature spaces, respectively. Each packet of knowledge is represented in specific data structures in the ABIS program and in specific relation tables in the associated data-base of ABIS.

RESULTS

In order to demonstrate the performance of the ABIS system, it has been tested with species even specialists find hard to identify. In such an example, we sought to separate the bees from a species complex comprising *Bombus lucorum, Bombus terrestris, Bombus cryptarum* and *Bombus magnus.* The ABIS system was trained with wing images of 70 individuals per species and successfully achieved an identification rate of far over 95 per cent by combining both types of features: the morphometric and the iconic (Table 8.1). Proving of the classification functions was done using 'leave-one-out' cross-validation. In further applications on the German *Colletes* and *Andrena* as well as with American *Osmia* species, ABIS again achieved identification rates of over 95 per cent.

CURRENT WORK

So far we have described ABIS as a feature-based approach for bee species identification by morphometric image analysis that works with great success on various bee genera. Our current work focuses on two topics. On the one hand, we aim to enhance the ABIS identification system into an ABIS identification framework, which includes database support, web interfaces and tools

TABLE 8.1
Recognition Rates with SVM and NKDA Classifiers and Different Sets of Features

	Morphometric features	Iconic features	Morphometric + iconic features
SVM	89.5%	95.8%	99.3%
NKDA	88.8%	95.1%	99.3%

for data visualization and project management. On the other, we aim to generalize our approach on image-based morphometric analysis in two ways. First, we want to combine the analysis of several images on different body parts (head, thorax, wings, abdomen, etc.) to obtain a multiview analysis. Second, new projects demand image-based species identification not only of bees but also for other insects. This section presents first results, which we obtained by a prototype implementation, which we called MorphoBox.

Multiview Analysis

Multiview analysis means (1) recognition of morphological features from different views of a single specimen, (2) derivation of several characteristic morphometric measurements (e.g. lengths, widths, surface areas) from these morphologic features and (3) combination of the morphometric measurements for identification using statistical approaches like LDA, NKDA or SVM. In contrast to ABIS (which works only on top views of bee forewings), multiview analysis allows information to be gained from different view angles of different morphological features. Therefore, multiview analysis requires the extraction of morphological features, or at least parts of features, that are invariant to image scale and rotation and provide robust identification and correspondence matching between different perspectives on the same feature across a substantial range of perspective distortion, change in viewpoint, addition of noise and change in illumination.

The last years show encouraging results for the extraction of those features, which are commonly referred to as salient points, key points or points of interest. The algorithms designed to extract these points of interest are commonly referred to as interest operators. Mikolajczyk and Schmid (2004) evaluated a variety of interest operators and identified the scale-invariant feature transform (SIFT) algorithm as being the most resistant to common image deformations. SIFT transforms image data into scale-invariant coordinates relative to local features.

Extracting Points of Interest

The best known SIFT operator was introduced by Lowe (2004) and involves the following major stages of computation used to generate the set of points of interest (PoIs):

1. Extraction of extrema of image intensity over all scales and image locations is implemented by using a difference-of-Gaussian function to identify potential candidates of PoIs.
2. At each candidate location, a three-dimensional quadratic function is fit to determine the interpolated location and scale of the extremum.
3. Orientation assignment is made to all PoIs based on local image gradient directions. All future operations are performed on image data that have been transformed relative to the assigned orientation, scale and location for each PoI, thereby providing invariance for these transformations.
4. Local image gradients are measured at the selected scale in the region around each PoI. These are transformed into a PoI descriptor of 128 elements that allows PoIs to be highly

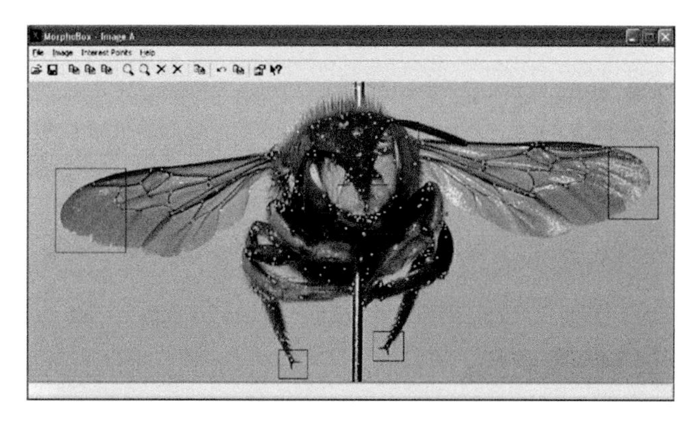

FIGURE 8.12 Automatically derived PoIs (points) and interactively defined window queries (rectangles).

distinctive, in the sense that a single PoI can be correctly matched with high probability against a large set of PoIs in other views.

Selection of Points of Taxonomic Interest

PoIs derived by SIFT operators are of interest due to their pure image appearance (i.e. local extrema of image intensity), but may not necessarily correspond to points of *taxonomic* interest (i.e. characteristic points of morphological features). Generally, the SIFT operator extracts a number of similar PoIs, where we expect only one point of taxonomic interest. Figure 8.12 shows the derived PoIs of a frontal view on a bee specimen. Thus, we have to select out of a set of PoIs the most appropriate PoI according to feature invariance or transformation constraints for matching across multiple views. Our current prototype system allows users to select PoIs and sets of PoIs interactively by simple 'point and window' queries.

A point query defines an image location (x,y) and selects the nearest neighboured PoI to (x,y) or the next n nearest neighboured PoIs within a user-defined radius around (x,y). The user asks a window query by defining interactively a rectangular window area in the image and selects thereby all PoIs within this window area. From a set of PoIs defined by a point or window query, it is possible to select that PoI by analysis of the PoI descriptors that shows a maximum of robustness in scale space and orientation invariance.

Figure 8.13 shows the selection of PoIs out of sets of PoIs defined by window queries. For selection of the PoIs, we use the Harris operator (Harris and Stephens, 1988). The Harris operator measures the likelihood of a PoI to be part of an edge or a distinct point (i.e. a corner or a centre

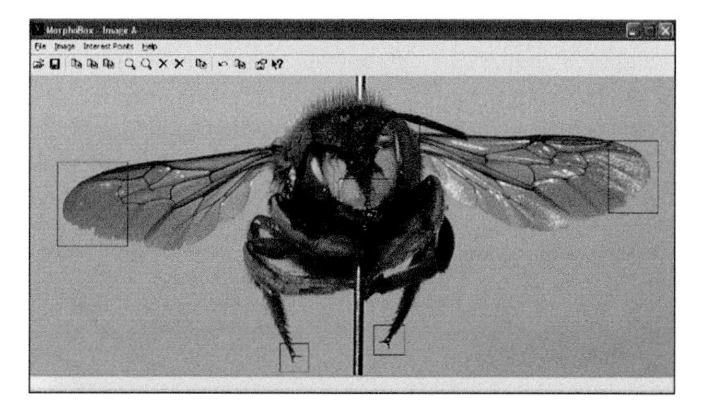

FIGURE 8.13 Automatically selected PoIs (points) within the image areas defined by the interactive window queries (rectangles).

of a circular feature). We select that PoI with the highest likelihood to be a distinct point. A more flexible and user-friendly kind of region query can be implemented by active contour approaches, which were introduced by Kass et al. (1987). Active contours – or snakes – are computer-generated curves that move within images to find object boundaries. Thus, active contours easily can be initially placed interactively by users near their region of interest and will automatically adapt their contour shape according to shape of the region.

Selection of Correspondences in Multiview Image Analysis

Having selected points of taxonomic interest in different perspectives of the same specimen, the aim of correspondence analysis is to match those points of taxonomic interest in two or more images that depict the same spatial morphological point of the specimen. Why is correspondence matching between different perspectives within the multiview analysis important? First, given an extracted and selected point p_1 of taxonomic interest in one image (e.g. p_1 depicting the top of the antenna), the established corresponding points $p_2,\ldots,p_n$ in the other $n - 1$ images, also depicting the top of the antenna, can be interpreted as support of the evidence of the correct identification of p_1.

Second, established correspondences allow reasoning *across images*. For example, one spatial feature point, p_1, might be involved in a trilateral relation with two other spatial feature points, p_2 and p_3, like the ratio of distances $|p_1\, p_2| / |p_1\, p_3|$, where p_1 and p_2 are only visible in image$_1$ and p_1 and p_3 are only visible in image$_2$, respectively. By establishing the correspondence between the images of p_1 in image$_1$ and image$_2$, this morphometric measurement is derivable by reasoning across image$_1$ and image$_2$.

Third, the correspondences established within the multiview image analysis are obviously the key for stereo analysis. This means that the spatial shape can be at least partially reconstructed to obtain spatial models and measurements of the observed specimen. The three-dimensional reconstruction can be enhanced by using parameterized shape models of insects and adjusting the shape parameters according to the spatial measurements derived by stereo analysis. A necessary requirement for stereo analysis is that the images are calibrated (i.e. the relative corresponding camera positions and the camera parameters like focal length are known). Such an approach is presented by Mortensen et al. (this volume) for the taxonomic identification of stonefly larvae.

To reduce positive false identification and matching between different perspectives, we employ geometric constraints that enforce consistent transformations for all matches between two images. This validation by correspondence, in turn, can be used to validate the PoI itself, because PoIs without support of a corresponding PoI in another image may be positive false hypotheses. Correspondence analysis can be enriched by employing relational matching (Vosselmann, 1992). Graph structures with PoIs as nodes will define topological relations of PoIs like neighbourhood or connectivity and will allow derivation of morphometric measurements like distances, angles, etc. Figure 8.14 shows a test on relational matching of the graph structure in a wing image with its rotated version. The graph is defined by the neighbourhoods of PoIs due to connecting veins. Employing these constraints in multiview analysis establishes successful correspondence matching between different perspectives on the same feature across a substantial range of change in viewpoint even up to 90° (c.f. Figure 8.15).

CONCLUSIONS

Acquisition and analysis of morphology is one of the fundamental issues in systematics, evolutionary biology and biodiversity. Of particular interest for all morphometric methods is the robust identification and extraction of characteristic morphological features. A successful example of a reliable feature-based approach to image-based species identification is given by ABIS. ABIS was designed to identify species of bees from images of their forewings by robust extraction of characteristic morphological features like veins, vein junctions and cells. Analysing all veins, vein

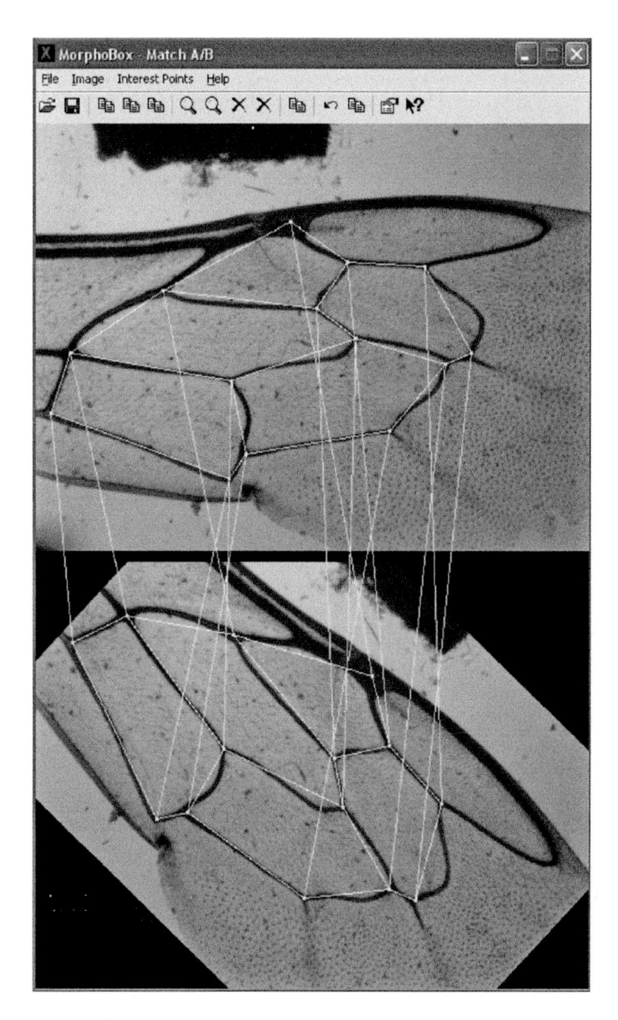

FIGURE 8.14 Test on relational matching of the graph structure in a wing image with its rotated version. The graph is defined by the neighbourhoods of PoIs due to connecting veins.

junctions and cells of each wing image gives rise to a set of about 300 well-defined characteristic features. These features are used in a knowledge-based classification approach to identify the bee species. This knowledge is encoded by domain-specific wing templates, genus templates and decision functions for genus identification and species identification. ABIS exhibits high correct identification rates (over 95%) for several bee genera and species groups in Germany, the USA and Brazil – among these, for example, the difficult to separate species complex comprising *Bombus lucorum, Bombus terrestris, Bombus cryptarum* and *Bombus magnus*.

Our current and future work aims to generalize the feature-based approach to image-based species identification demonstrated by ABIS by a multiview analysis that allows a combined morphological analysis of multiple images of a specimen taken from different perspectives and depicting different body parts. A necessary step in multiview analysis is the derivation of correspondences between the occurrences of the same morphological features of a specimen in different images. To achieve this correspondence matching, we detect so-called interest points that are invariant to image scale and rotation, and provide robust identification and correspondence matching across a substantial range of change in viewpoint. Furthermore, we employ geometric and topological constraints on the affine transformation associated with the correspondence matching. First results were derived on different animal groups and establish correspondences on images showing

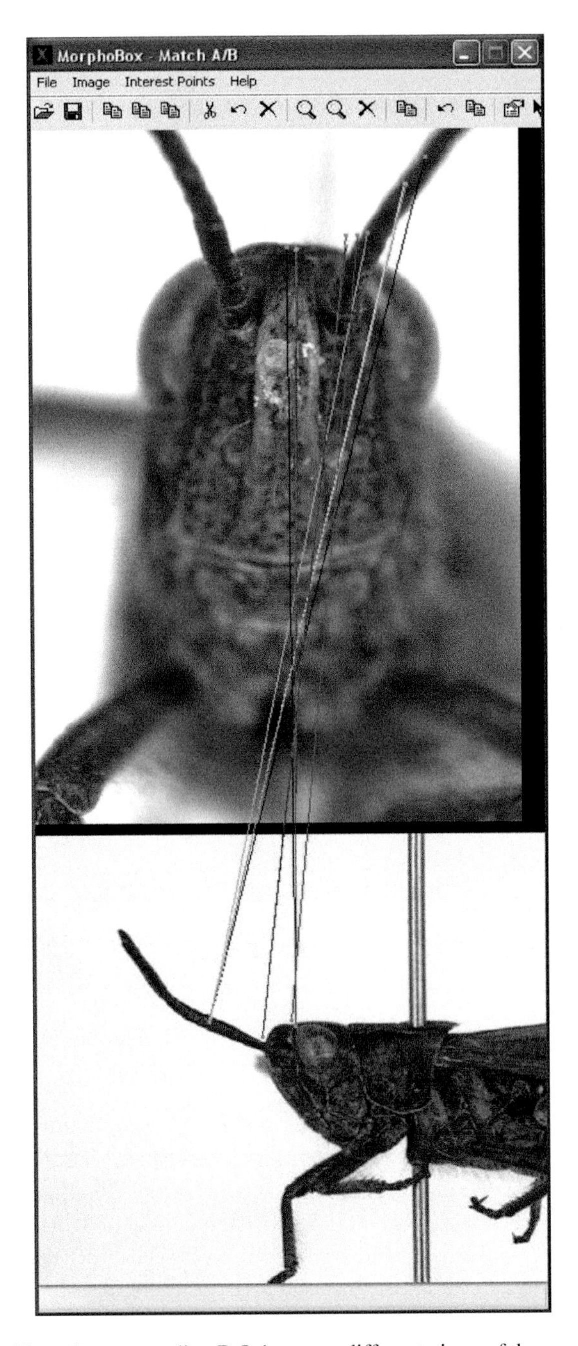

FIGURE 8.15 The matching of corresponding PoIs between different views of the same grasshopper specimen.

different perspectives up to 90°. Obviously, multiview analysis is also the key for three-dimensional reconstruction of spatial models of observed specimens by using calibrated images.

ACKNOWLEDGEMENTS

The ABIS project was funded by the German Research Council (Deutsche Forschungsgemeinschaft, DFG) and as part of the EDIS initiative (*Entomological Data Information System*) of the BIOLOG programme by the Ministry of Education and Research (Bundesministerium für Bildung und

Forschung, BMBF). The ABIS project is an interdisciplinary cooperation of the Institute of Applied Computer Science and the Institute of Agricultural Zoology and Bee Biology, both at the University of Bonn, Germany. The work on MorphoBox is funded as part of the REMO initiative (retrieval and monitoring in scientific applications) by the University of Bonn. The work on MorphoBox is an interdisciplinary cooperation of the Institute of Applied Computer Science and the Zoologisches Forschungsmuseum Alexander Koenig, both at the University of Bonn, Germany. We gratefully acknowledge the contributions on ABIS by Martin Drauschke, Artur Pogoda and Volker Roth as well as the contributions on MorphoBox by Markus Rall.

REFERENCES

Harris, C. and Stephens, M. (1988) A combined corner and edge detector. In *Proceedings of the Fourth Alvey Vision Conference* (ed M. Matthews), The University of Sheffield Printing Unit, Manchester, pp. 147–151.

Kass, M., Witkin, A. and Terzopoulos, D. (1987) Snakes: active contour models. *International Journal of Computer Vision,* 1: 321–331.

Lowe, D.G. (2004) Distinctive image features from scale-invariant keypoints. *International Journal of Computer Vision,* 60: 91–110.

Mikolajczyk, K. and Schmid, C. (2004) Scale and affine invariant interest point detectors. *International Journal of Computer Vision,* 60: 63–86.

Pimentel D., Wilson, C., McCullum, C., Huang, R., Dwen, P., Flack, J., Tran, Q., Saltman, T. and Cliff, B. (1997) Economics and environmental benefits of biodiversity. *Biological Sciences,* 47: 747–757.

Roth, V. and Steinhage, V. (1999) Nonlinear discriminant analysis using kernel functions. In *Advances on Neural Information Processing Systems* (eds S.A. Solla, T.K. Leen and K.-R. Mueller), MIT Press, Cambridge, MA, pp. 568–574.

Schröder, S., Wittmann, D., Roth, V. and Steinhage, V. (1999) Automated identification system for bees. In *Proceedings of the XIII International Conference of the International Union for the Study of Social Insects* (eds M.P. Schwarz and K. Hagendoorn), Dec. 29, 1998–Jan. 3, 1999, Flinders University Press, Adelaide, Australia, p. 427.

Shapiro, S. and Eckroth, D. (1987) *Encyclopedia of Artificial Intelligence,* Wiley Interscience, New York, 897 pp.

Vosselmann, G. (1992) *Relational Matching.* Volume 628 of Lecture Notes Computer Sciences, Springer, New York.

9 Introducing SPIDA-Web: Wavelets, Neural Networks and Internet Accessibility in an Image-Based Automated Identification System

Kimberly N. Russell, Martin T. Do, Jeremy C. Huff and Norman I. Platnick

CONTENTS

Introduction 132
Methods 134
    The Data 134
Imaging Protocol 137
Image Encoding 137
Construction of the Identification Engine 138
Artificial Neural Network Architecture 138
System Structure 140
Training 140
    Generalized Training 140
    SPIDA Training 141
Accessibility and SPIDA-Web 141
Results 142
Discussion 145
    Accuracy 146
    Accessibility 147
    Scalability 147
    Flexibility 148
Conclusions 149
Acknowledgements 150
References 150

INTRODUCTION

Attempts to understand how biodiversity originates and is maintained, and how it contributes to ecosystem functioning and human services, are hindered by lack of complete information. To understand the complexity of ecosystem function, and the likely impacts of human activities on these functions, ecologists and conservation scientists need to understand species interactions across multiple scales. Most studies to date have attempted to gain this understanding by looking at a very small subset of species, focusing primarily on vertebrates and other well-known or 'charismatic' groups. Unfortunately, recent syntheses (e.g. Goldwasser and Roughgarden, 1997; Platnick, 1999) suggest that such studies are not adequate in terms of predicting biodiversity patterns or signatures of disturbance. Additional information on lesser known groups is required to complete the picture. Yet, most studies avoid collecting data on diverse groups such as insects and arachnids precisely because they are less well known!

The inclusion of these groups in biodiversity studies has traditionally required both trained personnel who are able to identify known species correctly, and a systematist who can recognize and describe specimens new to science. Even when knowledgeable personnel can be found, the process of identification and description of new species takes time and money – assets in short supply for most ecologists, conservation biologists and wildlife managers. Non-specialists do not have the training or access to the materials necessary to produce accurate and consistent identifications on their own. The combined effect of this has heretofore led to the use and interpretation of questionable data or, more commonly, the complete abandonment of data from those taxonomic groups that comprise the bulk of biodiversity. We cannot hope to understand the complexity of ecosystem function and the relationship of human activities with ecosystem function without knowing how many, and what kinds of organisms are present.

Faced with these problems, and the increasing demand internationally for biodiversity research, some partial solutions have been pursued that attempt to delay or circumvent altogether the need for identifications. The use of technicians, or parataxonomists, to collect, sort and catalogue specimens prior to the input of a specialist has met with some success in Costa Rica (Instituto Nacional de Biodiversidad [INBio], 2001). The designation of RTUs (recognizable taxonomic units), or morphospecies, by non-specialists in order to obtain rapid richness estimates without requiring species-level identifications has proved reasonably accurate and useful in some cases (Oliver and Beattie, 1993, 1996). Certainly, the creation of biodiversity data-bases that catalogue collected specimens – particularly those that incorporate digital images of whole specimens and search procedures (similar to interactive keys) to help with identification (e.g. VirBas in Australia; Oliver et al., 2000) – will facilitate rapid, albeit cursory, biodiversity assessments. Although these methods provide a way to obtain quick species counts for initial richness comparisons, they do not provide enough information for in-depth biological or ecological studies. For serious analyses, identity is important. Therefore, tools must be developed to make routine identifications of specimens by non-experts both accurate and efficient.

An ideal identification system is one that encapsulates the knowledge of a systematist, requires little user input, and yields quick and accurate identifications. Some computer-aided identification systems such as interactive keys, multi-access keys, hypertext keys and expert systems are a significant improvement over the traditional, printed dichotomous key, but still require significant input from the user (and therefore require basic knowledge of the morphology and terminology of the target group; see Edwards and Morse, 1995; Dodd and Rosendahl, 1996; Rambold and Agerer, 1997). Methods that exhibit some level of automation are likely to be more accessible to non-specialists.

Many partly automated identification systems for multicellular organisms make use of digital imaging (e.g. Gerhards et al., 1993; Dietrich and Pooley, 1994; Chtioui et al., 1996; Weeks et al., 1997; Kwon and Cho, 1998; Do et al., 1999; Mancuso and Nicese, 1999; Weeks et al., 1999; Theodoropoulos et al., 2000). In very general terms, information is extracted from images in the form of specific measurements (taken manually or with the help of image tool programs), or the image itself is processed into a form that can be expressed numerically. The extracted observations are then subjected to statistical analysis (e.g. PCA, discriminant analysis), or submitted to some form of artificial neural network (ANN) in order to characterize and subsequently classify the species. Artificial neural networks are programming algorithms that simulate the structure of the brain and its processing of information (see Boddy et al., 1990, for an introduction). Species identification using ANNs, although similar in principle to statistical classification, relies on the ANN itself to create the group 'classifiers' by selectively weighting the input characters and adjusting its own internal configuration to maximize identification accuracy.

In the development of our identification system, we chose to focus on the ANN approach. This decision was based on a number of factors, including previous studies showing that in situations where both statistical and ANN-based approaches were tried using the same data as inputs, the ANNs almost always achieved equivalent or superior levels of accuracy (Chtioui et al., 1996; Goodacre et al., 1996; Wilkins et al., 1996; Parsons and Jones, 2000). The advantage of using ANNs is greatest when traditional identification procedures rely on somewhat subjective, qualitative characters that cannot be simply quantified (or even necessarily described). Qualitative features are subject to inter-and intra-observer variability arising from the user's level of knowledge, experience and frequency of use (Theodoropoulos et al., 2000).

There have already been many promising studies evaluating the potential of neural networks for the identification of cell types and organisms. ANNs have been used successfully in medical research to identify and classify cancer cells (Maollemi, 1991; Jiang et al., 1996; Hurst et al., 1997); to identify microorganisms of various kinds, including bacteria, yeasts and phytoplankton (Rataj and Schindler, 1991; Kennedy and Thakur, 1993; Goodacre et al., 1996; Wilkins et al., 1996; Goodacre et al., 1998; Wit and Busscher, 1998); and to identify macro-organisms, including plants of agricultural interest (Chtioui et al., 1996; Kwon and Cho, 1998; Mancuso and Nicese, 1999), parasitic larvae (Theodoropoulos et al., 2000), spiders (Do et al., 1999) and bats (from their echolocation signals – Parsons and Jones, 2000).

Of course, there are many different kinds of neural networks, ways of structuring an identification system and approaches to making such a system available to the public and there are many challenges to be faced when working with real data. Our system, SPIDA (*sp*ecies *id*entification, *a*utomated), or the web-accessible version, SPIDA-web, was created as a generalized identification system that can be tailored for virtually any group of organisms that can be distinguished visually (i.e. prior testing had demonstrated early versions' ability to distinguish five species of Ichneumonid wasp [unpublished data], six species of Lycosid spiders [Do et al., 1999] and twelve species of North American bees [Russell et al., in prep]). That said, by choosing to develop and refine our system using real data with which we have succeeded in creating a working prototype, we have of necessity had to face a number of challenges that will be common to most if not all automated identification systems.

Our test case, the Australasian ground spiders of the family Trochanteriidae, provided good examples of these challenges, including, among others, intraspecific variability (which itself varies in degree across species), variability in sample quality (due to debris or imaging techniques) and small sample sizes. In addition, we decided to tackle the problem of identifying all the closely related species included in a major taxon instead of the much simpler problem of distinguishing the species that happen to co-occur in a single area, most of which are only distantly related to each other and hence relatively easy to separate. Finally, spiders are considered by some to be one of the more difficult groups in terms of assigning species-level identifications, even compared with other arthropods. In the USA, only a tiny fraction of the roughly 3500 species are identifiable without the use of a microscope and the appropriate technical keys. Traditionally, one needs first to determine family membership with one key, genus membership with a different key (focusing on entirely different structures) and then, finally, species membership focusing on the complex structures of the genitalia, described in dizzying technical detail in published monographs. In sum, we have given ourselves a difficult task. But by doing so, we can more realistically assess the

TABLE 9.1A
Total Images Available for Training

Species	Pro		Anti	
	Unique	Total	Unique	Total
Desognaphosa kuranda	34	272	1171	13,393
Desognaphosa massey	20	160	1185	13,505
Desognaphosa millaa	26	208	1179	13,457
Desognaphosa yabbra	58	463	1147	13,202
Hemicloena julatten	20	80	1185	13,585
Longrita insidiosa	28	244	1177	13,441
Morebilus diversus	26	208	1179	13,457
Morebilus fumosus	25	100	1180	13,565
Morebilus plagusius	50	392	1155	13,273
Rebilus bulburin	24	192	1181	13,473
Trachytrema garnet	27	107	1178	13,558
Trachycosmus allyn	32	194	1173	13,471
Trachycosmus sculptilis	136	1055	1069	12,610

Notes: Statistics for the 13 species with 20 or more unique specimens. Total number of images available for training the species ANNs. 'Pro' images are from the species the ANN is being trained to recognize. 'Anti' images refer to all the images from the other species in the group (i.e. all the images that are not in the 'pro' set). Unique images are those taken from unique specimens, so this number reflects the number of individuals we had available for each species. Multiple images were taken of each specimen, so the 'total' number includes every image we have for a given species (pro) or all the remaining species (anti).

challenges of developing automated identification systems and the utility of our unique approach in meeting these challenges.

METHODS

The Data

We selected the recently revised (Platnick, 2002) Australasian ground spiders of the family Trochanteriidae as our prototype group. This decision was made primarily because of familiarity with the taxonomy of the group, and because specimens of all species were readily available and the size of the family – 121 species in 14 genera – seemed a reasonable and practical starting point for a practical identification system. Although some species in this family are relatively common, almost 80 per cent were represented by less than 10 individuals (of either sex); more than 50 per cent had fewer than 5. Thirteen species had 20 or more individuals (see Table 9.1A).

Species-level discrimination in spiders is based primarily on shape of the male and female genitalia. Anyone attempting identification to species, or a systematist describing new species, would need to examine these structures. Therefore, these are the structures we use for submission to SPIDA-web. This chapter will focus exclusively on the discrimination of the female specimens, as this work is entirely completed.

Female spider reproductive structures, known as epigyna, can be very complex or quite simple and so present a range of detail that will be useful in assessing the applicability of our system to other groups of organisms. The epigynum is found on the ventral side of an adult female and is

FIGURE 9.1 Adult female of *Desognaphosa bartle* Platnick. The female reproductive structure, the epigynum, is considered species diagnostic for most spiders. We used a single ventral view of the external structure visible here for input into SPIDA-web.

noticeable without dissection (see Figure 9.1). Although specialists often use structures of the internal epigynum (dorsal view) for species description and in constructing species-level keys, we chose to focus exclusively on the external features for the sake of simplicity and ease of use.

Once all the images were in place, it became apparent just how difficult a task we were about to put to a computer algorithm. Figure 9.2 provides illustrative examples of problems an automated ID system will need to overcome.

- Species similarity. Although the distinction between some species is easy to recognize, even between congeners (see Figure 9.2A), other groups appear to lack clear diagnostic structures in the ventral view (Figure 9.2B). Figure 9.2C illustrates the minor differences separating five species in the genus *Desognaphosa,* though in this case there is plenty of visible detail.
- Limited data. The third image in Figure 9.2C is from the single representative of the species *Desognaphosa karnak*. First, the structure is damaged. The human eye is able to compensate adequately, but getting the ANN to ignore this flaw is not so easy without any replicate specimens. But it is likely that even a human worker would have trouble separating this image from the previous image (of *Desognaphosa finnigan*) without additional specimens.
- Intraspecific variation. The relative degree of inter-and intraspecific variation in this group is not always predictable, as illustrated in Figures 9.2D, E and F. Figure 9.2D shows five

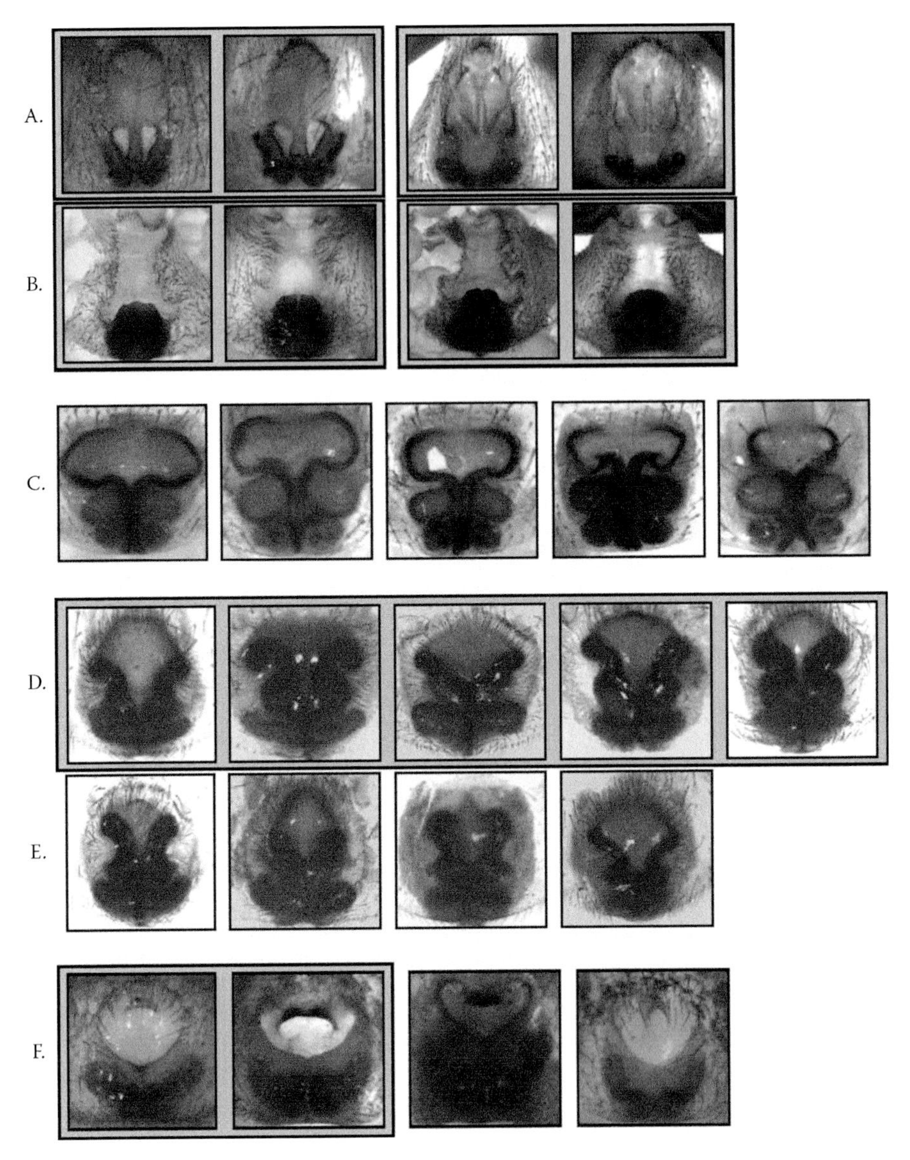

FIGURE 9.2 Illustrative examples of inter-and intraspecific variation in epigynal images among species in the spider family Trochanteriidae. Images enclosed in the same box are from the same species. (A) Two images each of two easy to distinguish congeners, *Trachyrema castaneum* and *T. garnet.* (B) Two images each of two congeners, *Longrita millewa* and *L. yuinmery,* with little information present in the epigynal images. (C) Five similar congeners in the genus *Desognaphosa*: *D. halcyon, D. finnigan, D. karnak, D. bartle* and *D. windsor.* (D) Five disparate individuals of the species *Rebilus bulburin.* (E) Representative images from four species in the same genus as (D), *R. lugubris, R. credition, R. brooklana* and *R. bilpin.* (F) The only two specimens from the species *Platorish nebo* and two related species, *P. churchillae* and *P. flavitarsus.*

images taken from individuals of the same species, *Rebilus bulburin.* This species exhibits one of the highest levels of intraspecific variation in this group. Other species in this genus, however, show much less intraspecific variation and the interspecific differences among some of these congeners (Figure 9.2E) is arguably similar in degree to the variation seen in *R. bulburin.* Finally, Figure 9.2F shows first epigynal pictures from the two individuals known of the species *Platorish nebo* and individual pictures of two other species in the genus, *P. churchillae* and *P. flavitarsus.* At first glance, it is difficult to see what the first two images have in common relative to the group. Without more unique specimens (or more information from other structures), it would be difficult to form a useful species image that would enable consistent species determinations for this group.

The point of this discussion is to emphasize the reality of the data used for our prototype in terms of the kinds of features available for species classification in spiders and to bring attention to the kinds of issues that are certainly *not* unique to spiders, such as intraspecific variation and specimen damage, with which any automated identification system will have to contend.

IMAGING PROTOCOL

All specimens were imaged using a Leica MZ 12.5 microscope fitted with a Q-imaging MicroPublisher CCD camera. Illumination was provided by an EIS fiber optic light source with a dual chrome gooseneck. The camera was connected to either a Dell Dimension 8200 Series or an Apple Titanium Powerbook G4 laptop. Images were converted to greyscale, cropped square, enhanced and resized as necessary in Adobe Photoshop.

Much like human students, ANNs distinguish objects by learning to focus their attention on particular aspects of an image, giving more weight to features that vary reliably between groups. If trained on only one image from each species, it is quite probable that a feature could be found to distinguish the two images that has nothing to do with the two species, but in reality is an artefact of the images themselves (e.g. presence of glare spots or background debris). Without multiple examples, the ANN can also form a much too specific 'vision' of a species, which could result in high numbers of false negatives when intraspecific variation is high. This is why the construction of an adequate training set is so important: the goal is to force the ANN to focus on the structures that are critical for distinguishing species, but also to encapsulate the likely variation both in the structures themselves and in the imaging of these structures (e.g. rotation, lighting, background, etc.).

Ideally, one would use many unique examples from every species to train the ANN, thereby encapsulating both types of likely variation. As previously stated, we did not have an adequate number of replicate specimens for most species. Hoping to compensate partially for this lack of unique samples, we collected either 4 or 12 images of every specimen, depending on how many specimens were available (greater than or less than 15, respectively). An attempt was made to introduce variation in the process by altering the lighting, repositioning the specimen and/or changing the rotation slightly between each picture. These replicate images were kept distinct from images taken of unique individuals. In a further attempt to generate more data for training, we created 'flipped' versions of each image using Adobe Photoshop. The female genitalia for this group are known to be bilaterally symmetrical. Flipping the images horizontally introduces some variation useful in training, particularly for species represented by fewer than three individuals.

IMAGE ENCODING

All automated identification systems face the task of reducing the feature space of the input data (i.e. reducing the total amount of information presented to the system), in order to minimize noise and facilitate more efficient classification criteria. Our approach to this is to use an encoding technique called wavelet transformation (Graps, 1995). Wavelet transforms are similar to the more commonly encountered Fourier transform. These are based on an iterative procedure in which an image is successively reduced to a coarser version of itself, through the removal of high-frequency information contained in wavelet coefficients (sometimes referred to as detail coefficients). These coefficients are parameters that modify the shape of a predetermined function, called a wavelet. Once the information in an image is parsed out into low- and high-frequency elements, the user can selectively eliminate the high-frequency information (usually noise, e.g. spines, hair, debris), keeping the more important shape information. Our previous work made use of the Daubechies 4 wavelet function, described in detail in Do et al. (1999).

The Daubechies 4 wavelet function requires that the input image be a square with a dimension of $2^j \times 2^j$, where j is an integer. We determined that an ANN with 4096 neurons in the input layer is the largest ANN that can be trained in a reasonable amount of time on a Sun Blade 100 or a Pentium 4 computer, which were the computers used in this part of the project. This means that wavelet coefficients in vector spaces V_0, V_1, V_2, V_3, V_4 and V_5 were used, producing an input matrix with a dimension of $2^6 \times 2^6$. This size input matrix can only be generated from an image scaled to 256 × 256 pixels ($2^8 \times 2^8$) prior to Daubechies 4 encoding.

For work with the Trochanteriidae, it was decided to investigate the Gabor wavelet function as well, as this type of filter had recently been applied to the problem of face recognition (e.g. Howell and Buxton, 1995; Krüger et al., 2000; Zhu et al., 2004; Bazanov et al., 2005) and had certain advantages, including being more robust to minor differences in lighting, orientation and scale.

The Gabor filter decomposes the image into data of varying resolutions by using banks of Gabor masks of different sizes to sample the image. The process is actually modeled after the receptive fields of the simple cells in the primary visual cortex of the mammalian eye (Pollen and Ronner, 1981). A set of image masks for different resolutions is available for this filter. At the first (coarsest) resolution, six masks are used. Each mask covers the entire image and represents real and imaginary components of the image with the orientations of 0, 120 and 240°. At the next level of resolution, 24 masks are used (four sets of six masks) each set of six masks covering one quadrant of the image and representing real and imaginary components of the image with the orientations of 0, 120 and 240°. The next level of resolution includes 16 sets of six masks, the next 64 sets of six masks, and the last (finest) level of resolution employs 256 sets of six masks (see Figure 9.3). Each mask yields a Gabor coefficient that was used as an input into the ANN, resulting in a total of 2046 inputs. Experiments indicated it was most efficient to use images scaled to the dimension of 51 × 51 pixels, as this limited computing time while providing acceptable accuracy (see following).

CONSTRUCTION OF THE IDENTIFICATION ENGINE

Although our proof-of-principle study (Do et al., 1999) indicated that back-propagation ANNs were an appropriate computing algorithm to use as an identification engine, we decided to investigate three other commonly used techniques: radial basis function ANNs (RBF), support vector machine (SVM) and the continuous *n*-tuple classifier (with log polar encoding). We wanted to be certain that we had chosen the identification engine that was most likely to succeed considering the organisms and data we had to work with. Do wrote the necessary software to test these other algorithms on a subset of the spider data. None of them performed as well as the back-propagation (specifically, cascade correlation) ANNs in our preliminary tests.

ARTIFICIAL NEURAL NETWORK ARCHITECTURE

Despite the decision to continue working with cascade correlation ANNs, there was a recognized need to change the way our ANNs were structured in order to address two separate issues: one a common criticism of automated identification systems in general and the other specific to back-propagation approaches:

- Classification of unknowns. One problem many automated identification systems face is the proper classification of unknowns (i.e. images from species the system was not trained to recognize; see Edwards and Morse, 1995; Morris and Boddy, 1995). Often these objects are forced into an erroneous classification. This was the case with our pilot study. The system was structured such that there was one ANN for the set of six species in the

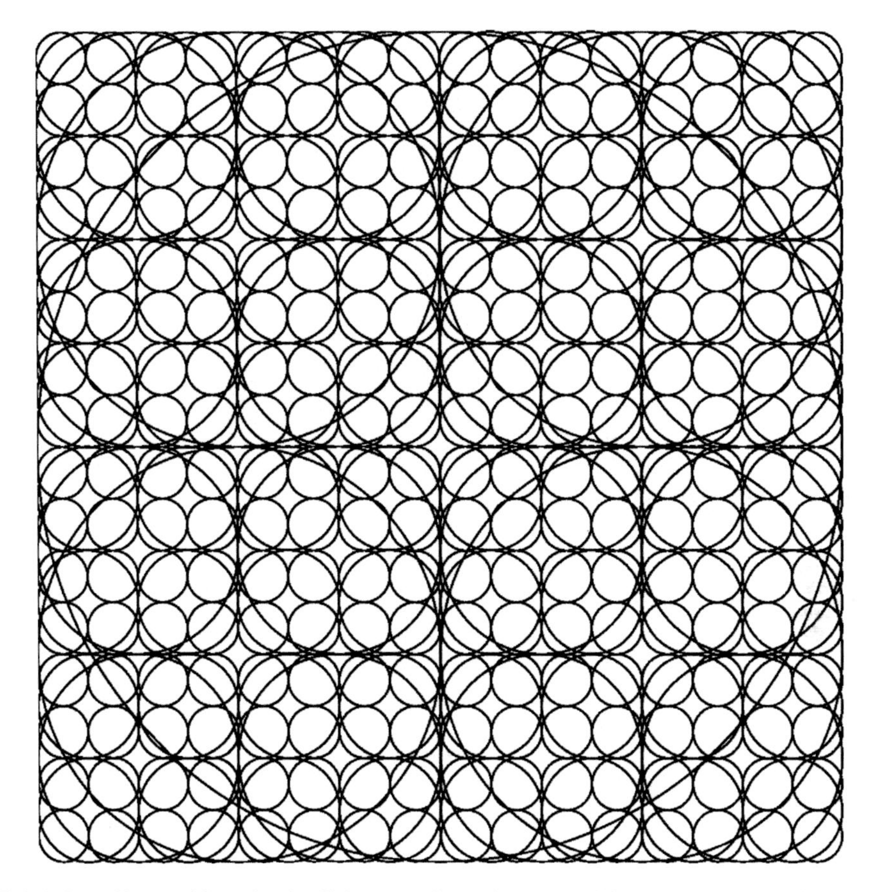

FIGURE 9.3 Sampling positions for the Gabor sampling scheme (modified from Howell and Buxton, 1995). Gabor filters decompose the image into data of varying resolutions by using banks of Gabor masks of different sizes to sample the image. Each circle on this figure represents a set of six Gabor masks representing real and imaginary components of the images with orientations of 0, 120 and 240°. Each mask yields a Gabor coefficient for use as an input into the ANN, resulting in a total of 2046 inputs.

trial, with an output node corresponding to each species. In this situation, an image submitted to the system would be forced through to one of the output nodes.

- Scalability. The issue of scalability is a common criticism of back-propagation ANNs. The traditional way of using such networks to classify species is to structure the ANN with an output node for each group the system attempts to distinguish. In this case, adding another species to the identification system would require the retraining of the entire ANN after adding another output node to represent the new group. This could be a very lengthy process, depending on the size of the group and the number of training images.

The solution to these problems eventually adopted for SPIDA was to create an ANN for each species in the group (Figure 9.4). Each of these ANNs has two output nodes, one positive and one negative, and is trained on images from the target species (the 'pro' training set) and a selection of images from other species in the group (the 'anti' training set). An image submitted for identification is presented to each ANN in the group. In this system, a true unknown can cycle through the group of ANNs without eliciting a positive response in any. If it is determined the unknown truly belongs in no group, adding the new species is as simple as training another relatively small ANN for the new species alone. One potential disadvantage to this approach is that, at some point, if the number of secondarily added ANNs exceeds some threshold, it is possible that accuracy may decrease, as

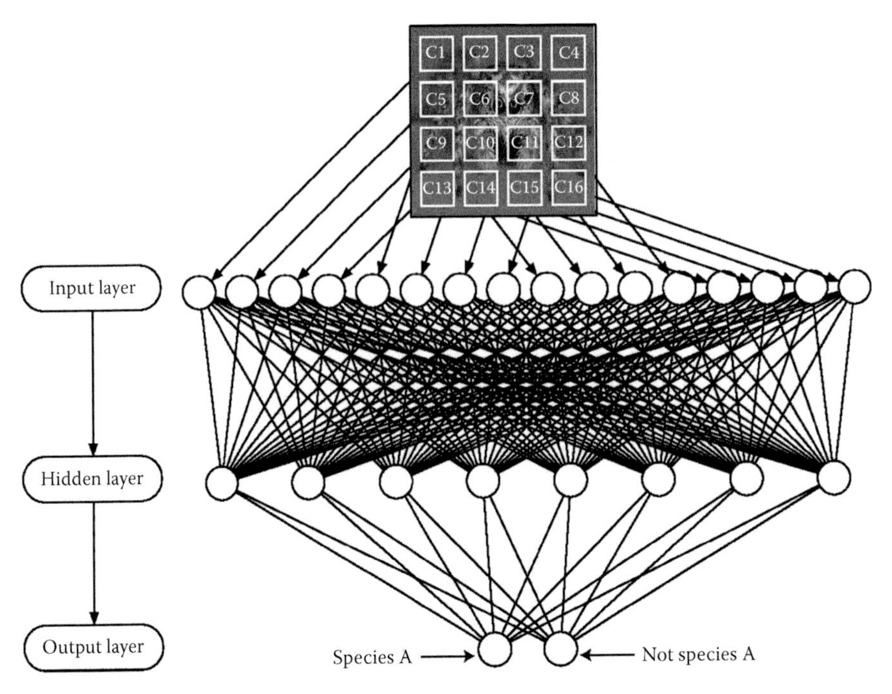

FIGURE 9.4 Simplified diagram of a back-propagation ANN, illustrating the basic structure of a typical species ANN in SPIDA-web. The output layer has only two nodes: species A and not species A (i.e. 'yes' or 'no'). Each species ANN is trained on information from members of that species ('pro' images) and from other species ('anti' images). Wavelet coefficients generated from the images of spider epigyna comprise the input data.

the newly added species will not be in the 'anti' training sets of the original set. In this case, more extensive retraining may be required, though it should be limited to closely related species only.

In order to test SPIDA's ability to classify previously unseen species correctly as unknowns, we randomly selected 20 images from species in related families and submitted them for identification.

SYSTEM STRUCTURE

As mentioned previously, SPIDA is structured as a collection of individually trained species-level ANNs. Images submitted to the system are cycled through all the ANNs in a predetermined grouping (e.g. family or genus). The positive output values for each ANN are saved to a file and ranked. The top three are selected and information on these species is retrieved from a data-base for presentation to the user. If the highest output value is above 0.59, then it is considered a positive ID and presented to the user as such. Although it is theoretically possible to have more than one ANN in the group return a value above 0.59, we did not experience this in the course of our study and in subsequent testing efforts. In fact, the difference between the first and second highest output values was usually very pronounced. However, we chose always to include information on the top three species in the event that a near tie were to occur, thus alerting the user to the fact that further scrutiny is required before a definitive ID can be assigned.

TRAINING

GENERALIZED TRAINING

Back-propagation ANNs consist of multiple layers of simple computing elements with many interconnections among the layers. The initial architecture of the ANN is established according to the amount of data making up each image (which determines the number of input neurons) and the number of groups that the system is designed to distinguish. The initial ANN consists of a layer of input neurons and a layer of output neurons fully interconnected between layers by random initial weights. Through a process of supervised learning, the network essentially enhances some features highly while diminishing the influence of others, using a complex method of averaging input parameters. The training process establishes additional neurons in a hidden layer between the input and output layers. In some cases, the number of hidden layers is fixed prior to training.

Other training algorithms, such as cascade correlation, allow neurons to be added one at a time as necessary, thereby minimizing the size of the ANN. These hidden neurons act as feature detectors that respond to specific patterns (e.g. a pattern unique to a given genus or species). The idea is to 'teach' the ANN to set the output neuron assigned to a given genus or species to its maximum value of 1.0 whenever a pattern indicative of that genus or species is presented, and set all other output neurons to their minimum value of 0.0. In practice, the ANN sets the output neurons to an intermediate value depending on the certainty of its identification (e.g. an output of 0.9999 indicates virtual certainty, whereas 0.6000 indicates lower confidence). The resulting output vector is then evaluated against the target function to compute an error. This error is then used to modify the weights in the connections. Training continues until the desired level of accuracy is attained. Once trained, the network is tested with previously unseen individuals to assess its ability to classify them into the correct groups (i.e. the network's ability to generalize from the training set to unknowns).

SPIDA Training

We used cascade correlation in conjunction with quick propagation (Fahlman, 1988; Fahlman and Lebiere, 1991) to train SPIDA ANNs using the Stuttgart Neural Network Simulator version 4.2 (SNNSv4.2). The training procedure was optimized for development of the final prototype system, which made use of all available data. In order to maximize the accuracy and future ability of the system to generalize, training of each ANN was a highly iterative and closely supervised procedure. For each species, a random set of images was selected for the training set, usually from a single individual. All other images were used for testing the network after it had been trained on this limited data-set. Then, the image that gave the most incorrect identification was added into the training set and the ANN was retrained and tested. This continued until all images were identified correctly. Finally, to ensure the system would likely be able to generalize appropriately (i.e. give accurate identifications to newly submitted images), images were sequentially removed from the training set to determine whether the ANN could still accurately identify all other images in the testing set. This process continued until the smallest possible training set that could accurately identify all the remaining images in the testing set had been defined.

In addition, for species that seemed to require the largest training sets, it was important to review the log files of this process to pick out any potentially contaminating images and remove them from the training sets. Contaminating images were defined as those with large amounts of debris, damage or occasionally questionable species designation (i.e. due to human error). These were noticed only when the identification logs indicated a persistent misclassification. Sometimes, when such an image was then added to the training set, the accuracy of subsequent identifications actually decreased after retraining. In other situations, it was merely a matter of examining the images in the 'pro' training set and picking out the oddball image if the ANN appeared to have trouble generalizing appropriately.

ACCESSIBILITY AND SPIDA-WEB

As important as creating a system with the ability to discriminate species using minimal data and computing power is making such a system adequately accessible to those most able to benefit from the technology. The overall goal was to create a user-friendly system, requiring a minimum of taxonomic knowledge and specialized equipment, that could be accessed from anywhere in the world. SPIDA-web is essentially SPIDA with an Internet interface, allowing users to have access to previously trained ANNs housed on a server, designed to distinguish specific organism groups.

The website consists partly of static html pages designed to give users some basic information about the project, such as how the system works and what its limitations are, as well as information on how to prepare and submit images. The remainder of the website is constructed with Java server pages, which essentially allow the software to display dynamic data in response to the user's input. These Java server pages are supported by a number of Java servlets served up by the open source Tomcat server software. Servlets are essentially computer codes written in the Java computer language designed to respond to a specific query submitted by the users via the Web. The servlets are capable of obtaining data from a data-base in order to process the user's request and send the results back to the Java server pages.

SPIDA-web supports a number of functions. First, all users are required to log in before accessing the site. This allows SPIDA-web managers to communicate with users to obtain more inputs for the data-base as well as more data to improve its ANNs. Once logged in, users can then access the Java server page that will allow them to select images for submission by browsing a local hard drive. Once selected, the images are submitted to a servlet that uses Sun's Java Advance Imaging (JAI) library to convert them to a usable format and scale them to a proper size. These images then undergo wavelet transformation and the resulting information is submitted to the trained ANNs obtained from the data-base via another servlet. The system then forwards the resulting identification to a Java server page that displays the identification (along with a second and third choice based on the ANN confidence values) and information from the data-base to the user (Figure 9.5). This information includes distribution maps, line drawings of genitalia, whole-body images, technical species descriptions and training images.

The entire process requires only a few seconds to complete on a local machine, but the ultimate speed of SPIDA-web will be dependent on the server and connection speed. SPIDA-web also has a number of administrative functions, such as adding and deleting user information; adding, deleting and editing genus and species information; and viewing user activity on the site. At this point, SPIDA-web is not set up to make automatic use of new data, such as images from new or under-represented species. These data will, however, be saved to a data-base, and reviewed and integrated by the SPIDA-web managers as needed.

RESULTS

SPIDA-web ANNs were trained successfully (i.e. convergence was attained) for each of the 121 species in the Australasian ground spider family Trochanteriidae, including those represented by only one or two individuals. Because so many species were data poor, the training method was optimized (as described previously) to produce the best result possible with the available information. In many cases, this meant that SPIDA-web had to use at least one of all the unique images available for a species in the training process. Also, the iterative training method, though leading to the best use of the data available, left any calculations of accuracy suspect as each ANN was trained intentionally until it was able to identify accurately all the images in the testing set. Until new users begin submitting images to SPIDA-web, there will be no truly new data from the Trochanteriidae on which to test the system.

One way to evaluate performance, however, is to look at the number of images required to train each ANN. Table 9.1B gives this information for the 13 species that had at least 20 unique individuals for the Gabor and Daubechies 4 wavelet encoding scheme. Networks trained with the Gabor-encoded images required a smaller percentage of images to achieve 100 per cent accuracy than the networks trained with the Daubechies 4 encoded images, though the difference was not great (49 vs. 53% unique for the 'pro' set and 3 vs. 6% for the 'anti' set, respectively). This result indicates that Gabor encoding may lead to ANNs better able to generalize, so all of the ANNs

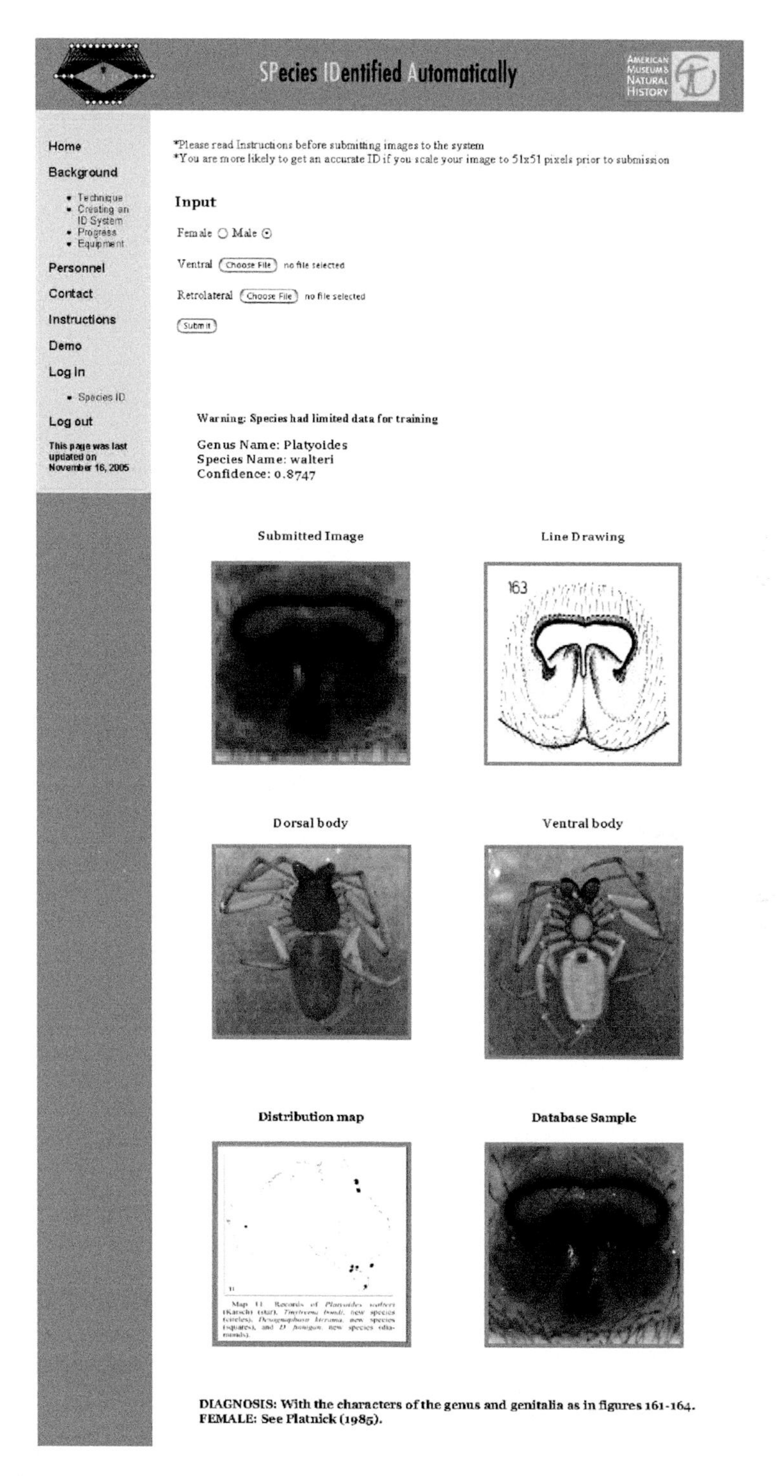

FIGURE 9.5 Screen shot of SPIDA-web. This is an example of a positive identification output after a user has logged in and submitted an image from his or her hard drive. In addition to the top ID, SPIDA-web is designed to show the second and third highest matches along with accompanying data-base information, which appears directly below the positive match, requiring the user only to scroll down.

TABLE 9.1B
Percent of Images Used for Training

	Gabor training				Daubechies training			
	Pro		Anti		Pro		Anti	
Species	**% of unique**	**% of total**	**% of unique**	**% of total**	**% of unique**	**% of total**	**% of unique**	**% of total**
Desognaphosa kuranda	62	14	5	0.53	41	7	3	0.36
Desognaphosa massey	75	21	4	0.39	75	14	4	0.44
Desognaphosa millaa	62	13	4	0.46	65	15	6	0.66
Desognaphosa yabbra	31	6	3	0.33	24	5	5	0.48
Hemicloena julatten	55	16	3	0.26	N/A	N/A	N/A	N/A
Longrita insidiosa	54	9	3	0.30	64	20	6	0.70
Morebilus diversus	65	15	6	0.56	69	30	14	1.54
Morebilus fumosus	52	22	4	0.34	68	37	6	0.67
Morebilus plagusius	20	3	1	0.12	30	7	5	0.47
Rebilus bulburin	63	10	4	0.32	71	20	7	0.83
Trachytrema garnet	44	15	2	0.22	N/A	N/A	N/A	N/A
Trachycosmus allyn	50	12	3	0.31	53	12	5	0.59
Trachycosmus sculptilis	10	2	3	0.21	18	4	6	0.67
Average	49	12	3	0.33	53	16	6	0.67

Notes: Statistics for the 13 species with 20 or more unique specimens. Percentage of unique and total images used in training each species ANN. Images were added to the training set iteratively until all the remaining images in the testing set were accurately identified. Gabor and Daubechies are the two different wavelet encoding techniques used in this project.

currently in place for SPIDA-web were produced using this technique. However, these same data also hinted that Daubechies 4 may be useful for some groups. Therefore, both algorithms are still being tested. On average, the number of unique images required to train an ANN to distinguish a given Trochanteriidae species adequately from all the rest was 16.

The SPIDA-web site is located at http://research.amnh.org/invertzoo/spida. Users have successfully logged on and submitted images (though disappointingly few from the Trochanteriidae).

Though extensive testing of SPIDA-web with new data from the Trochanteriidae was not possible at present, its ability to correctly classify species from other, related families as unknowns could be examined. Of the 20 randomly selected outgroup images, SPIDA-web correctly classified 19, or 95 per cent, as unknowns. The one image it missed was matched with a species that had limited data available for training. The lack of data may have created an ANN focused on inappropriate features that happened to be shared with the image from the outgroup species.

In lieu of new unique data, a handful of quick tests were performed to determine how SPIDA-web would react to reprocessed images. The system should be able to handle variation in user preprocessing techniques such as cropping and rotation. In the website documentation, users were instructed (with illustrations) how to crop and orient images, but minor variation is inevitable. A sample of 15 images that SPIDA-web had previously identified correctly was selected. Ten of these images we recropped, either zooming in or zooming out prior to resizing. Five images were deliberately cropped poorly, with the epigyna clearly not centred. Results are shown in Table 9.2A. All of the zoomed images and three of the five off-centre images were identified correctly. The two errors, though producing a 'no match found' output, had the correct species as the closest match found. A different selection of 10 images was also used to test the effect of rotation. Images were first rotated 2°, then 4° clockwise before being resized and submitted to SPIDA-web. The 2° rotation

TABLE 9.2A
Effects of Cropping

Image	Confidence		Type
	Original	Recropped	
mbpf2	0.8919	0.885	Zoom out
mbpf8	0.6554	0.7756	Zoom out
dgyff8a	0.9903	0.9934	Zoom out
dgyff11c	0.976	0.9743	Zoom out
dgyff55a	0.7972	0.741	Zoom out
dgyff6b	0.9812	0.9378	Zoom in
mbpff9	0.9561	0.9733	Zoom in
mbpf6	0.7807	0.8157	Zoom in
falf1	0.8793	0.8076	Zoom in
pdw1a	0.8183	0.9312	Zoom in
pdw1a	0.8183	0.3887	Off centre
mbpf8	0.6554	0.5439	Off centre
dgyff13a	0.9925	0.9876	Off centre
dgyff57c	0.8193	0.4503	Off centre
mbpff2	0.9239	0.9137	Off centre

TABLE 9.2B
Effects of Rotation

Image	Confidence		
	Original	Rotate 2°	Rotate 4°
mbpf1	0.7526	0.8648	0.3746
mbpf2	0.8919	0.8679	0.6472
mbpf3	0.427	0.3813	nmf
mbpf4	0.6887	0.6071	nmf
mbpf5	0.7355	0.8136	0.6159
mbpf6	0.7598	0.7845	Wrong ID
mbpf7	0.9694	0.9662	0.9113
mbpf8	0.7473	0.6019	nmf
mbpf9	0.874	0.8917	0.5371
mbpf10	0.6712	0.7065	0.2905

Notes: Effects of user variability in image cropping and rotation. The identification confidence from the species ANNs is reported before and after recropping and rotation. 'Off centre' was a deliberate attempt to badly crop an image. Numbers greater than 0.5 indicated a correct identification. Numbers less than 0.5 are reported when the system returned 'no match found', but the first choice was the correct species. 'nmf' refers to situations when there was no positive identification and the first choice was not the correct species.

produced no changes in the outcome of the identifications (Table 9.2B). A 4° rotation, however, reduced identification accuracy to 40 per cent.

DISCUSSION

Taking steps toward practical implementation is absolutely critical to the advancement of the field of automated object recognition, as it must be proven to be more than just a 'pie-in-the-sky' idea that works only in the abstract. This investigation was given the task of putting the reasonably established reality of automated species identification into practice in the form of a usable, accessible system. The goal, therefore, was not necessarily to create the *most* accurate or *most* flexible or *most* easily used system, but rather to design and implement an ID system from beginning ('John Smith', a collector who finds himself with a spider needing an ID) to end (technician recording the scientific name and confidence on a data-sheet). That goal was met. A prototype was successfully developed to identify the 121 Australasian species in the spider family Trochanteriidae.

The base system, SPIDA-web, receives digital images of unidentified specimens via the Internet, encodes the information in these images using wavelet transformation, circulates this information through a set of ANNs trained on sets of identified images, and returns identifications to the user – all in a matter of seconds. Output is structured to give users basic information on each species, including distribution maps, drawings, pictures and technical descriptions as well as an indication of confidence and alternative choices.

But how do we evaluate the true success of an automated object recognition system like this? Four criteria to determine the utility of such a system might be: (1) accuracy, (2) accessibility, (3) scalability and (4) flexibility. How does our prototype of SPIDA-web stack up?

Accuracy

Many potential users would rank this at the top of the list in terms of importance. That said, results of informal surveys of arachnologists suggest that acceptable cutoffs for accuracy vary widely and often depend on the background of the respondents. Systematists or taxonomic specialists demand the highest accuracy levels – 95 per cent minimum for such a system to be useful for them. Ecologists and conservationists would be happy with 85–90 per cent if it meant they could have a species list to work with. Certainly, automated ID systems should do the same or better than untrained or quickly trained novices (e.g. technicians, students). Ideally, they would do the same or better than those workers with moderate amounts of training (e.g. entomologists or PhD students after taking one or more courses in spider taxonomy).

As stated previously, we did not structure our training protocol in such a way that would allow us to measure accuracy in the true sense. We trained the ANNs to be able to identify all the images in the testing set accurately, but only after first expanding and then reducing the size of the training set. We can only offer that the accuracy of identification for well represented species (15–20 unique samples available for training) was consistently high – in the range of 90–96 per cent – in the present study when tested on subsets of data prior to final training. We suspect that accuracy levels for under-represented groups (<10 unique specimens) are much lower, perhaps below 75 per cent based on a very small test set. Unfortunately, we do not currently have access to more data on which to further test the prototype, as all known specimens from the Trochanteriidae are already in our possession.

Use of replicate, processed and flipped images certainly helps the system be more robust against variability due to lighting, image preprocessing and minor damage or debris. However, for species with very few individuals, these do not adequately replicate the information necessary to force ANNs to converge on the most useful and appropriate features in the images. This is absolutely necessary if an ANN system is going to be able to generalize and recognize what constitutes intraspecific variation versus interspecific variation.

The problem of small training set size will certainly not be limited to spiders. Most invertebrate communities consist of a few common species and many rare ones. One reason these organisms are often so difficult to identify is the fact that so many were described based on just a handful of individuals. Lack of data is likely going to be a common problem for all automated identification systems. One way to ameliorate this problem partially is to design ID systems to be evolving such that, as they are used, they improve. We have designed SPIDA-web to store all submitted images so that new images can be incorporated into the appropriate training sets as needed to improve accuracy. We would suggest that all systems be designed to have this capability.

Of course, we can measure the prototype's accuracy in terms of its ability to identify unknown, or 'new' species by testing it with out-group images. Having a system that is perhaps overly sensitive to the detection of unknowns is more useful than a system that errs on the side of misclassification. A non-identification forces the user to re-evaluate the specimen, perhaps setting it aside for a specialist to review. An incorrect, but positive identification is much more likely to be overlooked, as a technician will be less likely to question it. Though not perfect (only 95% accuracy from our small test set), SPIDA-web is much better at detecting unknowns than our previous system, which employed a more traditional ANN architecture (Do, 1996). For the data-set described earlier, most misclassifications (of true unknowns) were limited to species that were trained on inadequate data.

In most respects ANNs are unpredictable, as there is no way of knowing what information from the training images they weight highly as features. For example, if an ANN is trained on images from a single individual, and that individual had a tear or large piece of debris attached, it is possible that the ANN would consider those anomalies to be features. If an unknown happened to have a similar 'feature', real or not, it could result in a false positive identification. That is why, once again, it is always better to have a multitude (>15) of training specimens.

Accessibility

Accessibility refers to both the ease with which non-specialists can navigate through the identification process as well as the ability of users to gain access to the ID system. SPIDA-web ranks high in both categories. Because the input to SPIDA-web is the whole image of a structure (in the case of the prototype, a picture of the external genitalia of spiders), there is no need for users to measure or dissect or even know the name of what they are taking a picture of. It is as simple as finding the structure, centering it and snapping an image. Instructions on how to find the structure are included in the introductory pages of the website. Aside from rudimentary image processing, such as converting the image to greyscale and cropping it square, users can submit an image without having any technical software or technical knowledge.

What *do* they need? They need to have a digital camera (or access to a scanner) and any computer with Internet access. In the case of our prototype, they need to know they have a spider and what family that spider is in. One could imagine an auto-ID system that does not even require the user to know they have a spider (vs. a mite vs. a harvestman), but it seems unlikely that a single image could be used to identify everything from order to species (as unlikely, perhaps as a single portion of DNA that could be use to distinguish all species in the animal kingdom). Therefore, it is likely that all practical auto-ID systems will need to be hierarchical to some degree (see following discussion).

Scalability

Scalability is certainly an important issue, as any relevant auto-ID system will need to distinguish large numbers of species. As the taxonomy of most difficult to recognize groups (e.g. insects, arachnids, etc.) is relatively fluid, auto-ID systems must have the capacity to expand and/or be modified without requiring an excessive amount of computing time. As stated previously, traditionally structured back-propagation ANNs require full retraining each time a new species is either added or removed. Some forms of ANNs (e.g. plastic self-organizing maps, or PSOMs) can accept new species almost indefinitely without major adjustments.

SPIDA, with its collection of individually trained species-level ANNs, falls somewhere in between. Species can be added without affecting the rest of the species' ANNs in the group. The relatively small number of 'anti' images (images not belonging to the species an ANN is being trained on) required in the training process (Table 9.1B) supports the ease with which new species could be added to established systems. That said, it might be necessary to retrain some of the other ANNs if the addition of the new species caused a decrease in accuracy. There is a limit beyond which it would be advisable to retrain the whole set, if too many species are added. However, we do not see this as major limitation since any sensible identification system must be structured hierarchically.

It is unrealistic to think that any one morphological structure would be universally applicable to all groups of organisms. Even among spiders, certain families will likely require either different or supplemental information, not just genitalic structures, for accurate species designations. In fact, SPIDA was originally set up hierarchically, with a set of genus ANNs trained first to classify an image to genus, then to circulate the same image through only the set of species ANNs within that genus (instead of all 121). Though initial results gave very high accuracy for genus-level identifications (99%), for some genera, the genus ANNs failed to converge (i.e. during training, a consistent solution was never found), making them essentially useless as a discriminatory tool.

This happened with very large, very diverse genera. This was not surprising since, in order for a hierarchical system to work, the same characters used for species determination must also be used

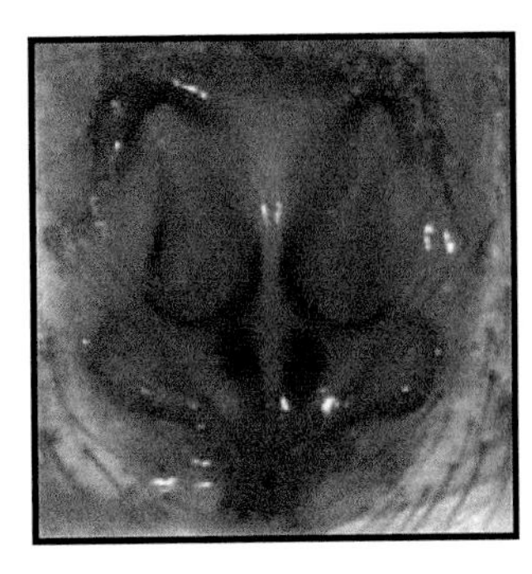
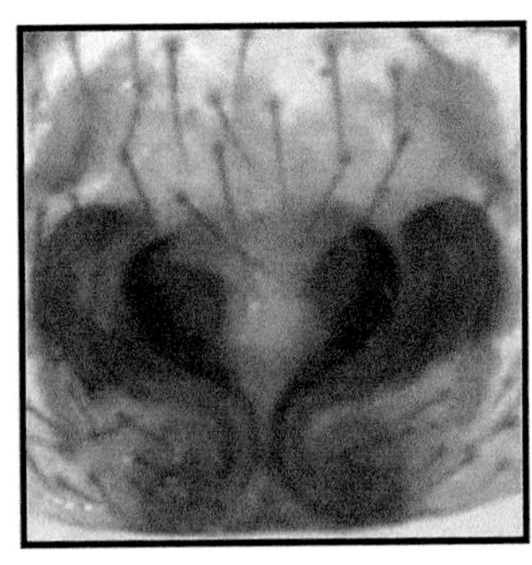
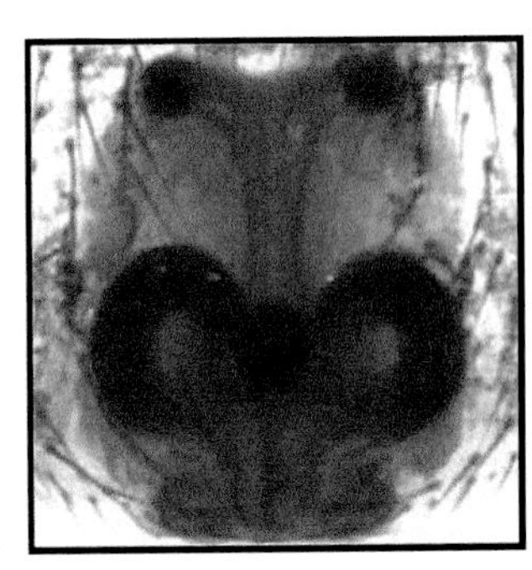

FIGURE 9.6 Epigyna of three species in the genus *Desognaphosa*: *D. massey, D. kroombit* and *D. bulburin*. These structures appear to share very few discernable features, if any. This illustrates the intrageneric diversity in many of the larger genera in this group, prohibiting the convergence of genus-level ANNs.

at higher levels of classification. For spiders, we know this is only sometimes the case. Typically, specialists do not use genitalic characters for genus classification. Despite this, for most genera, variation of epigynal structures within the genus was significantly less than variation between genera. In some cases, particularly in the larger genera (e.g. Desognaphosa with 26 species), the variation within a genus was extreme (Figure 9.6). Time was spent investigating potential solutions to this problem, but results proved inconclusive. Therefore, the final system was structured non-hierarchically, with each submitted image being circulated through all 121 species ANNs.

This simply illustrates the point that different information will be needed to distinguish different groups of organisms, so it makes more sense to design ID systems that operate within information type. Otherwise, one runs the risk of accidental similarities between processed images of very different structures. Perhaps the most efficient approach is to create systems tailored to manageable groups, adding a semi- or fully automated top-level system designed to shuttle the user to the appropriate subgroup. This could be accomplished with multiple-access keys or a separate ANN system based on different data. For example, family classification in spiders might be possible using images of eye-pattern or carapace shape (Roberts et al., in review). Finally, perhaps the most immediately useful application of this technology will be the identification of collections of species from a particular region or habitat, of which there will be a limited number of species in the pool. In either case, SPIDA is adequately scalable for most reasonable applications.

Flexibility

Flexibility is a measure of how easily an ID system can be applied to different groups of organisms. Some systems are amazingly accurate at distinguishing certain sets of organisms based on characters specific to that group – for example, ABIS (see Steinhag et al., this volume) using wing cell shape characters to identify bee species. Others have been shown to be amazingly flexible when tested on many different types of organisms and objects – for example, DAISY (see O'Neill, this volume). SPIDA, as defined by the combination of a wavelet encoding scheme and sets of individually trained ANNs, is not tied to any specific organism group, as its input is simply an image. This *is* a mild constraint, as at the moment, SPIDA can only be trained to identify organisms where the relevant characters can be imaged rather easily. Still, it has been tested with single images of wasp wings and bee wings and multiple images of male spider genitalia, all with successful results.

In summary, SPIDA-web is accurate (with adequate training data), highly accessible, reasonably scalable and quite flexible. There is plenty of room for improvement; as previously stated, our goal was to go from theory to practice, beginning to end. As that is now accomplished, the focus will shift to finding ways to improve (and better measure) accuracy, streamline the training process, explore the limits of SPIDA's scalability and further test its flexibility with new organism groups.

CONCLUSIONS

Automating the identification of specimens to species is a difficult task. There is no reason to believe that teaching a computer to identify species will be any easier than teaching a person to do so. In fact, it is likely a trickier process altogether, considering the amazing ability of the human mind to compensate for missing information and recognize the similarity in objects. The advantage, however, is that computers are fast, consistent and, once taught, do not forget. We chose to test our system on the most difficult of tasks: distinguishing individuals from closely related species. We also chose a challenging group, with species diagnostic characters that are difficult to quantify or even describe, making the use of traditional taxonomic keys problematic for the (relative) novice. Though sometimes strikingly similar within genera, these structures often vary widely in shape, size and dimensionality across genera, making even the basic description of differences complicated (if not impossible) for anyone but specialists in the group. Despite this, our results have been promising – even more so when you consider the growing evidence that specialists may not be as accurate or consistent as they think they are when assigning species names (e.g. Culverhouse, this volume). More to the point, we suspect that higher accuracy (~95–100%) will be attainable for ecological samples, as species will be more disparate.

As mentioned previously, we envision the most useful application of this technology to be in ecological and/or conservation studies. The majority of studies to date looking for ecological patterns in diversity, distribution, response to disturbance, etc. have relied on only a handful of surveys, often only one (Spellerberg, 1991; Green et al., 2005). This 'snapshot' approach limits our understanding of the processes governing the dynamic nature of species and communities and can often yield misleading data. There has been much discussion in the literature of promoting more multisurvey, long-term studies of biological communities to ameliorate this problem; there is widespread agreement about the need for repeated surveys to help expand our understanding of ecological phenomena, especially in the face of increasing human impacts (landscape alteration, global climate change, etc.; see Balmford et al., 2005; Green et al., 2005).

That said, most funding agencies require results in 2–3 years and often conservation efforts face even more pressing deadlines in terms of averting ecological disasters. When working with arthropods in particular, though they are relatively easy to sample, the identification process can often take years. This time delay is often used as an argument to not include arthropods in conservation studies and/or biological monitoring efforts. Our development of SPIDA-web targeted this type of need. We foresee identification modules being developed on data from the first set of surveys conducted at a site. Once trained, these modules could then be used to identify all subsequent surveys, leading to quick analysis of community dynamics. Technicians could be easily taught to image the specimens and submit them for identification via SPIDA-web; then, only the few individuals not recognized by SPIDA-web need be examined by a specialist, thus saving vast amounts of time.

In addition, there is no need to wait until newly collected species are given a proper scientific name – they could be added to the module based on a morphospecies designation in the short term. This has the added advantage of guaranteeing consistency in morphospecies classification throughout the monitoring period. This is relevant, as we anticipate SPIDA-web being most useful when trained on collections of species from a particular region, thereby being applied to pressing problems of biodiversity and conservation.

But what of the future beyond SPIDA-web? At a minimum, there needs to be cooperation and data sharing between groups working in the field of automated object recognition. We are at the stage where alternate approaches can be tested and evaluated based on the criteria outlined previously using real data. There will certainly be no single solution and no single approach that can be labeled as 'the best' for all tasks and/or organism groups. Yet seeing where each succeeds (or fails) will yield practical data and can only propel the field forward. To make automated object identification practically useful, enough infrastructure must be built up to make the creation and maintenance of identification modules suitably efficient and relatively independent of the whims of short-term funding agencies. Perhaps the creation of a research centre for automated identification or, on a smaller scale, the establishment of permanent research positions in this field at major research museums would provide adequate stability and resources. Of course, this will only happen if the powers that be are convinced that automated object recognition is an integral part of the future of taxonomy.

ACKNOWLEDGEMENTS

Funding for this project was provided by the National Science Foundation, grant #0119578. We thank the editor of this volume, Professor Norman MacLeod, and two anonymous reviewers for their helpful comments on the manuscript.

REFERENCES

Balmford, A., Crane, P., Dobson, A., Green, R.E. and Mace, G.M. (2005) The 2010 challenge: data availability, information needs and extraterrestrial insights. *Philosophical Transactions of the Royal Society, Series B,* 360: 221–228.

Bazanov, P.V., Buryak, D.Y., Mun, W.J., Murynin, A.B. and Yang, H.K. (2005) Comparison of Gabor wavelet and neural network-based face detection algorithms. In *Signal and Image Processing* (ed M.W. Marcellin), ACTA Press, Honolulu, Hawaii, pp. 200–208.

Boddy, L., Morris, C.W. and Wimpenny, J.W.T. (1990) Introduction to neural networks. *Binary,* 2: 179–185.

Chtioui, Y., Bertrand, D., Dattee, Y. and Devaux, M.-F. (1996) Identification of seeds by colour imaging: comparison of discriminant analysis and artificial neural network. *Journal of the Science of Food and Agriculture,* 71: 433–441.

Dietrich, C.H. and Pooley, C.D. (1994) Automated identification of leafhoppers (Homoptera: Cicadellidae: Draeculacephala Ball). *Annals of the Entomological Society of America,* 87: 412–423.

Do, M.T. (1996) *Pattern Recognition System Using Artificial Neural Networks and Wavelets for Taxonomic Identifications.* Master's thesis. University of Tennessee, Knoxville.

Do, M.T., Harp, J.M. and Norris, K.C. (1999) A test of a pattern recognition system for identification of spiders. *Bulletin of Entomological Research,* 89: 217–224.

Dodd, J.C. and Rosendahl, S. (1996) The BEG expert system – A multimedia identification system for arbuscular mycorrhizal fungi. *Mycorrhiza,* 6: 275–278.

Edwards, M. and Morse, D.R. (1995) The potential for computer-aided identification in biodiversity research. *Trends in Ecology and Evolution,* 10: 153–158.

Fahlman, S.E. (1988) An empirical study of learning speed in back propagation network. Technical Report (CMU-CS-88-162). Pittsburgh, PA, Carnegie Mellon University.

Fahlman, S.E. and Lebiere, C. (1991) The cascade-correlation learning architecture. Technical Report (CMU-CS-90-106). Pittsburgh, PA, Carnegie Mellon University.

Gerhards, R., Nabout, A., Sokefeld, M., Kuhbauch, W. and Nour Eldin, H.A. (1993) Automatic identification of 10 weed species in digital images using Fourier descriptors and shape parameters. *Journal of Agronomy and Crop Science,* 171: 321–328.

Graps, A. (1995) An introduction to wavelets. *IEEE Computational Sciences and Engineering,* 2: 50–61.

Green, R.E., Balmford, A., Crane, P.R., Mace, G.M., Reynolds, J.D. and Turner, R.K. (2005) A framework for improved monitoring of biodiversity: responses to the World Summit on Sustainable Development. *Conservation Biology,* 19(1): 56–65.

Goldwasser, L. and Roughgarden, J. (1997) Sampling effects and the estimation of food-web properties. *Ecology,* 78: 41–54.

Goodacre, R., Timmins, E.M., Burton, R., Kaderbhai, N., Woodward, A.M., Kell, D.B. and Rooney, P.J. (1998) Rapid identification of urinary tract infection bacteria using hyperspectral whole-organism fingerprinting and artificial neural networks. *Microbiology-UK,* 144: 1157–1170.

Goodacre, R., Timmins, E.M., Rooney, P.J., Rowland, J.J. and Kell, D.B. (1996) Rapid identification of *Streptococcus* and *Enterococcus* species using diffuse reflectance-absorbance Fourier transform infrared spectroscopy and artificial neural networks. *FEMS Microbiology Letters,* 140: 233–239.
Howell, A.J. and Buxton, H. (1995) Receptive field functions for face recognition. In *Proceedings of the 2nd International Workshop on Parallel Modelling of Neural Operators for Pattern Recognition (PAM-ONOP),* Faro, Portugal, pp. 83–92.
Hurst, R.E., Bonner, R.B., Ashenayi, K., Veltri, R.W. and Hemstreet, G.P.I. (1997) Neural net-based identification of cells expressing the p300 tumor-related antigen using fluorescence image analysis. *Cytometry,* 27: 36–42.
Instituto Nacional de Biodiversidad. (2001) National Biodiversity Inventory. <www.inbio.ac.cr/en/invn/Invent.html> (March 11).
Jiang, Y., Nishikawa, R.M., Schmidt, R.A., Wolverton, D.E. and Comstock, C.E. (1996) Evaluation of a computerized classification scheme for clustered microcalcifications for computer-aided diagnosis. *Radiology,* 201: 1304–1304.
Kennedy, M.J. and Thakur, M.S. (1993) The use of neural networks to aid in microorganism identification: a case study of *Haemophilus* species identification. *Antonie van Leeuwenhoek,* 63: 35–38.
Krüger, V., Bruns, S. and Sommer, G. (2000) Efficient head pose estimation with Gabor wavelet networks. In *Proceedings of the 11th British Machine Vision Conference,* University of Bristol. UK, pp. 72–81.
Kwon, Y.K. and Cho, R.K. (1998) Development of identification method of rice varieties using image processing technique. *Agricultural Chemistry and Biotechnology,* 41: 160–165.
Mancuso, S. and Nicese, F.P. (1999) Identifying olive (*Olea europaea*) cultivars using artificial neural networks. *Journal of the American Society for Horticultural Science,* 124: 527–531.
Maollemi, C. (1991) Classifying cells for cancer diagnosis using neural networks. *IEEE Expert,* 6(6): 8–12.
Morris, C.W. and Boddy. L. (1995) Artificial neural network identification and systematics of eukaryotic microorganisms. *Binary,* 7: 70–76.
Oliver, I. and Beattie, A.J. (1993) A possible method for the rapid assessment of biodiversity. *Conservation Biology,* 7: 562–568.
Oliver, I. and Beattie, A.J. (1996) Invertebrate morphospecies as surrogates for species: a case study. *Conservation Biology,* 10: 99–109.
Oliver, I., Pik, S., Britton, D., Dangerfield, J.M., Colwell, R.K. and Beattie, A.J. (2000) Virtual biodiversity assessment systems. *Bioscience,* 50: 441–450.
Parsons, S. and Jones G. (2000) Acoustic identification of twelve species of echolocating bat by discriminant function analysis and artificial neural networks. *Journal of Experimental Biology,* 203: 2641–2656.
Platnick, N.I. (1999) Dimensions of biodiversity: Targeting megadiverse groups. In *The Living Planet: Biodiversity Science and Policy* (eds J. Cracraft and F.T. Grifo), Columbia University Press, New York, pp. 33–52.
Platnick, N.I. (2002) A revision of the Australasian ground spiders of the families Ammoxenidae, Cithaeronidae, Gallieniellidae, and Trochanteriidae (Araneae: Gnaphosoidea). *Bulletin of the American Museum of Natural History,* 271: 1–243.
Pollen, D.A. and Ronner, S.F. (1981) Phase relationships between adjacent simple cells in the visual cortex. *Science,* 212: 1409–1411.
Rambold, G. and Agerer, R. (1997) DEEMY – The concept of a characterization and determination system for ectomycorrhizae. *Mycorrhiza,* 7: 113–116.
Rataj, T. and Schindler, J. (1991) Identification of bacteria by multilayer neural networks. *Binary,* 3: 159–164.
Roberts, A.K., Smith, D.M., Guralnick, R.P., Cushing, P.E. and Krieger, J. (in review). Quantitative prediction of fossil biodiversity: eocene spiders from Florissant, Colorado. *Palaios.*
Spellerberg, I.F. (1991) *Monitoring Ecological Change.* Cambridge University Press, Cambridge, 350 pp.
Theodoropoulos, G., Loumos, V., Anagnostopoulos, C., Kayafas, E. and Martinez-Gonzales, B. (2000) A digital image analysis and neural network based system for identification of third-stage parasitic strongyle larvae from domestic animals. *Computer Methods and Programs in Biomedicine,* 62: 69–76.
Weeks, P.J.D., Gauld, I.D., Gaston, K.J. and O'Neill, M.A. (1997) Automating the identification of insects: a new solution to an old problem. *Bulletin of Entomological Research,* 87: 203–211.
Weeks, P.J.D., O'Neill, M.A., Gaston, K.J. and Gauld, I.D. (1999) Automating insect identification: exploring the limitations of a prototype system. *Journal of Applied Entomology,* 123: 1–8.

Wilkins, M.F., Boddy, L., Morris, C.W. and Jonker, R. (1996) A comparison of some neural and non-neural methods for identification and phytoplankton from flow cytometry data. *Cabios,* 12: 9–18.

Wit, P. and Busscher, H.J. (1998) Application of an artificial neural network in the enumeration of yeasts and bacteria adhering to solid substrata. *Journal of Microbiological Methods,* 32: 281–290.

Zhu, J., Vai, M.I. and Mak, P.U. (2004) Gabor wavelets transform and extended nearest feature space classifier for face recognition. In *Proceedings of the Third International Conference on Image and Graphics,* Hong Kong, pp. 246–249.

10 Automated Tools for the Identification of Taxa from Morphological Data: Face Recognition in Wasps

Norman MacLeod, Mark A. O'Neill and Stig A. Walsh

CONTENTS

Introduction 153
Methods Review 154
Multivariate Methods 156
Discriminant Analysis (DA) and Canonical Variates Analysis (CVA) 156
Principal Components Analysis (PCA) 157
Neural Nets 161
An Example 165
Two-Group Results 170
Three-Group Results 172
Improving Neural Net Approaches to Automated Taxon Recognition 175
Discussion and Summary 180
Prospectus 183
Acknowledgements 184
Notes 184
References 184

INTRODUCTION

Morphological data are so ubiquitous in systematics they are often taken for granted. At a time when many students of modern organisms have turned their attention to molecular methods in order to solve problems of phylogenetic inference (Felsenstein, 2003; Salemi and Vandamme, 2003) and acknowledging that many see a bright future for DNA barcoding (e.g. Hebert et al., 2005; Tautz et al., 2003; Hebert et al., 2005), it is important to recall the multiple roles of morphological data in systematic contexts. These include a realization that the production of cladograms and biodiversity assays does not exhaust the scope of contemporary systematics (e.g. Bello et al., 1995; Cracraft, 2002) and an understanding that an analysis of morphological data will be crucial to evaluating the success of non-morphological data for reconstructing systematic relations among taxa (Wheeler, 2003; Ebach and Holdrege, 2005; Wheeler, 2005 and references therein). Moreover, it needs to be acknowledged that over 99 per cent of all life forms that have existed on Earth have left no discernable molecular record, yet all are crucially important for understanding relations between both ancient and *modern* taxa (Gauthier et al., 1988; Donoghue et al., 1989). For these reasons, the study of morphological data will continue and, where possible, should proceed in tandem with the study of other data types useful in systematics. Molecular and morphological sources of information should play their own unique roles in contributing to the discovery of systematic relations among taxa, with each used to evaluate, probe and test inferences derived from the other.

There is, however, a practical problem in converting this sentiment into a coherent and productive research programme – a problem that, we believe, lies at the heart of much of the criticisms traditionally made of morphological data. That problem, put badly, is that while some newer data types (e.g. codon sequences) can be delivered for systematic analysis rapidly, reproducibly and in easily analyzed quantitative form, morphological data are delivered slowly and are based on qualitative assessments of complex variation patterns made by an ever dwindling number of taxonomic specialists. Moreover, these specialists often disagree among themselves as to taxon identifications and use criteria for making those identifications that often seem difficult to communicate (see Zachariasse et al., 1978; MacLeod, 1998 and references therein; Culverhouse et al., 2003, this volume and references therein). In order for morphological data to play the roles they should in systematic investigations, strategies for their rapid collection and automated analysis must be developed.

Granted, morphological data are complex. The seemingly inexhaustible variety of organic forms is a constant source of delight and, if truth be told, accounts for much of the attraction most people feel for living things (e.g. Wilson, 1986). This complexity is much richer than the simple molecular codes of proteins or DNA, and rapidly exhausts the ability of spoken or written language to capture in any but the most superficial detail. But is such complexity really inexpressible? Surely over 2000 years of successful results from qualitative morphological analyses are sufficient to answer this question in the negative.

Given patience, adequate samples and cooperative working relationships, systematists can come to agreements on general questions of morphological diagnoses (e.g. Culverhouse, 2007). What is presently lacking is not a common language for expressing the boundaries of morphological variation with adequate scope to represent organic complexity. That language – geometry – has been available for, literally, millennia. Rather, what has been lacking is the technological toolkit necessary to collect geometric data, segment these data into characteristic patterns, perform the computations required to make comparisons across specimens and summarize these comparisons into representations of similarity relations meaningful in various systematic contexts (e.g. taxonomic, phylogenetic, biogeographical, palaeontological, ecological).

Fortunately, owing to ongoing developments in multivariate analysis, artificial intelligence, computer vision and allied fields, sophisticated tools continue to be developed that provide systematists with an ever increasing ability to extract systematic information from morphological data. Over the next decade, integration of these technological innovations is set to revolutionize the manner in which morphological data are collected, analyzed and interpreted. These coming improvements will likely reinvigorate the study of comparative morphology by placing it on a firm mathematical–geometric footing. The purpose of this chapter is to sketch out the basic approaches now being used to forge a re-engagement between systematics and 'numerical morphology', especially as these relate to the questions of automated identification. This will be done through the discussion of concepts and via application of exemplar methods to a single data-set. The intention is to provide readers with a basic, qualitative understanding of some of the newer numerical approaches to these perennial systematic problems that have proven successful, illustrate the advantages and disadvantages of these approaches, and suggest areas where future development is needed.

METHODS REVIEW

All methods employed in morphological approaches to automated taxonomic identification can be classified generically as 'pattern recognition' techniques, though this classification places emphasis on ends rather than means. Traditionally, group-characterization and specimen-identification problems in systematics have been viewed as being analogous to either ordination or discrimination problems in multivariate data analysis (e.g. Blackith and Reyment, 1971; Mardia et al., 1979; Pimentel, 1979; Reyment et al., 1984; Reyment, 1991). With important changes in our understanding of shape theory, these approaches have, for the most part, carried through the reformulation of morphometrics during the late 1980s and early 1990s that produced 'geometric morphometrics' – also known as the 'morphometric synthesis' (see Bookstein, 1993) – though the assignment of unknown specimens to classes or groups has only recently been taken up as an explicit topic by this school (e.g. Polly and Head, 2004; Zelditch et al., 2004; Klingenberg and Monteiro, 2005).

Accordingly, geometric morphometric applications exist that make use of discriminant approaches. These explicitly recognize and take advantage of differences in the within-groups and between-groups covariance structure of measurements across a sample of objects, as well as simple distance and likelihood approaches to the characterization of group boundaries. Use of neural nets of various sorts as generalized tools for characterizing taxonomic groups and assigning unknown specimens to groups so characterized is also becoming increasingly popular and successful (e.g. Gaston and O'Neill, 2004; MacLeod et al., 2007 and references therein).

Before moving on to a discussion of particular approaches, though, it is necessary to review briefly the practical differences between group-characterization and identification tasks and how these affect the choice of analytic approach. Both tasks assume that the existence of groups is a feature of nature and that these groups can be objectively recognized via reference to the data at hand. Demonstrating the ontological status of such groups constitutes a separate – though related – issue that is beyond the scope of this discussion. Here we will only be concerned with characterizing and identifying groups whose existence has been established by some means *a priori*.

The group-characterization problem involves an assessment of what subsets of information are involved in distinguishing one group from another. This is analogous to the phylogenetic data-analysis task of recognizing a group's autapomorphies. In order to accomplish this task, information is needed about the degree to which variables are distributed continuously or discontinuously across an optimized taxonomic space and an assessment of where any distributional discontinuities fall (see MacLeod, 2002a, for examples). This task is most closely aligned with that of the systematist whose primary tasks are to discern the broad distribution of groups (e.g. species, genera, families) in nature based on commonalities in the patterning of their characteristics, recognize new groups and infer the processes responsible for creating/maintaining those patterns.

The group-identification problem is more limited. It involves only the assignment of unknown specimens to previously recognized groups (e.g. species, genera, families). This is analogous to a major aspect of the taxonomist's task. Here, complete information about all the ways a group might be characterized is often unnecessary. The only information required is a rule-based system capable of assigning unknown specimens to the correct group. In a sense, the differences between group characterization and group identification are analogous to the differences between the systematic description of a group (which should include a discussion of all characterizing attributes) and its diagnosis (which usually consists of rules of thumb applied to a readily accessible subset of the former).

The fact that these two tasks are interlinked should come as no surprise. Together, they form the corpus of systematic biology (see Simpson, 1961; Mayr, 1969, 1982) and, indeed, they have traditionally have been carried out more or less simultaneously by the same person. Nevertheless, there is an important practical distinction between them. In many systematic biodiversity and ecological applications, the primary focus resides in placing specimens into established groupings or taxonomies – in identifying them as belonging to this or that *a priori* group. The rules used to place the specimens into these groupings are not as important as the correctness of the placement (i.e. identification) itself and are rarely the subject of much description in technical reports that deal with biodiversity, ecological and/or conservation topics. The important thing is that these rules be (1) appropriate (i.e. capable of assigning specimens to their correct groups), (2) able to be applied objectively and (3) consistently applied in practice. Both morphometric and neural-net approaches can be used to automate the group-identification task. How they accomplish this, the degree of success that can be expected from their application and the means by which their performance in systematic contexts can be improved are the subjects of this chapter.

Multivariate Methods

Discriminant Analysis (DA) and Canonical Variates Analysis (CVA)

The standard multivariate solution to the group-identification problem is through discriminant analysis (e.g. Sneath and Sokal, 1973). Classical linear DA was developed by Fisher (1936). Typically, it begins with a set of specimens, features of whose morphologies have been quantified by measurements. These have traditionally consisted of distances between topologically corresponding landmark points, but more recently have been based on the coordinate positions of the landmarks themselves. Once such data have been secured for a collection of specimens from different, authoritatively identified groups or taxa, classical DA amounts to a multivariable regression of the differences between the mean vectors of two groups and the pooled covariance matrix (see Reyment et al., 1984; Davis, 2002). When more than two groups are present, the related method of canonical variates analysis (CVA) is usually preferred (see Blackith and Reyment, 1971; Reyment, 1971; Sneath and Sokal, 1973; Mardia et al., 1979; Campbell and Atchley, 1981; Reyment et al., 1984; Davis, 2002).

Canonical variates analysis – in the sense used here – was first described by Rao (1952) and is based on the principle that the total covariance matrix of a data-set consisting of k groups can be partitioned into within-groups (i.e. W, the sum of products of deviations of each specimen from its group mean over all variables) and between-groups (i.e. B, the sum of products of each group mean from the grand mean over all variables) components. A representation of the geometry of a canonical variates analysis is presented in Figure 10.1. Note the directionality differences between the orientation of the within-groups and between-groups trends (Figure 10.1A), which in this example are almost at right angles to one another. Canonical variates analysis, in effect, solves the problem of

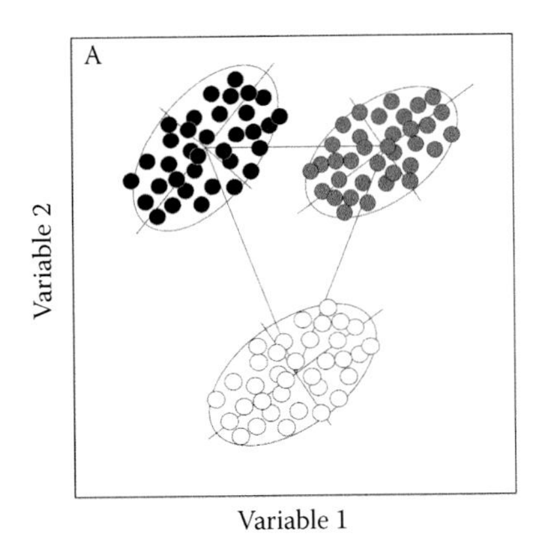

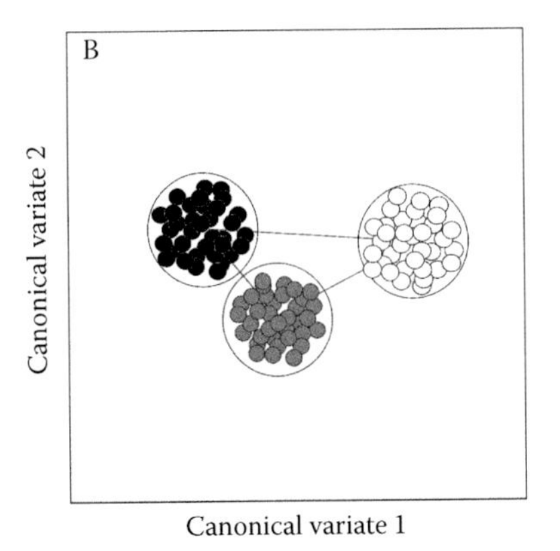

FIGURE 10.1 Fundamental geometry of canonical variates analysis (CVA). A. Scatterplot of two variables for a data-set subdivided *a priori* into three groups. Group axes indicate major directions of within-groups covariation. Note that the triangle joining the group centroids is isosceles and that the covariance matrix of the black group differs slightly from those of the other two groups. B. Canonical variates transform of the data represented in A. This transform adjusts scaling relations within the canonical variates space such that within-groups variation is rendered isotropic and between-groups variation maximized to reflect separation between group centroids and the similarities in the group-specific covariance patterns. In this case the black and grey groups are represented as being more similar to each other than either is to the white group (reflecting patterns of separation of group centroids, primarily) and the grey group represented as being slightly more similar to the white group (reflecting similarities in the structure of these groups' covariance matrices).

finding a set of multivariate axes that maximize the *B/W* ratio. High values of this ratio denote a system in which groups are well separated from one another, with each clustered tightly about its group mean; low ratios denote geometries in which group means and group variances are subequal.

An important aspect of CVA (also present in classical discriminant analysis) is the desire to scale the discriminant space to reflect not only differences between the group means (*B*), but also differences in the within-groups covariance structure (*W*). In most applications, this is accomplished by premultiplying *B* by the inverse of *W* (= $W^{-1}B$). Such a transformation deforms the scale of the discriminant space so that all groups have isotropic variance. Once this operation is performed, the distances within this transformed space reflect differences in the group-specific covariance structure. Thus, in Figure 10.1B, group 3 (white symbols) occupies a position further from group 1 (black symbols) in the canonical variates space because its covariance structure is more similar to that of group 2 (grey symbols) (see also Campbell and Atchley, 1981; Klingenberg and Montiero, 2005).

Canonical variates analysis has a long history of use in group-identification contexts. More recently, MacLeod et al. (2007) used CVA to achieve excellent group-identification results for Procrustes-registered landmark data from modern planktonic foraminiferan tests. Nevertheless, unlike the space formed by principal components (see later discussion), the canonical variates space is complexly deformed relative to the original variables. Like principal components spaces, though, CVA axes form a highly group-dependent space. This is because the inclusion of new groups will invalidate the optimization of the entire discriminant space for all groups. The only way to re-establish this optimization is to recompute the discriminant system for all the old groups and the new groups together. Of course, only when the morphology of the new group is sufficiently similar to those of the original groups can all variables employed in a previous analysis be used and be appropriate to use (see MacLeod et al., 2007, for additional discussion). As a result, CVA-based discriminant systems are limited in their morphological–taxonomic scope as well as in their computational stability. Also, owing to the space-deformation issues described above, CVA does not support direct theoretical modelling of shape geometries (see MacLeod, 1999, 2002b, 2005a).[1]

Principal Components Analysis (PCA)

Principal components analysis (PCA) is based on methods first described by Pearson (1901) in terms of the overall theory, and Hotelling (1933) in terms of computational approaches. As with DA and CVA, PCA regards individual specimens as being characterized by a series of measurements (usually linear distances between topologically corresponding landmarks) and attempts to re-express the variance of these as linear series of composite variables that summarize the covariance structure and are themselves uncorrelated with one another (Manley, 1994; MacLeod, 2005b). Principal components analysis is a fundamental tool in multivariate data analysis, where it is used to (1) reduce the dimensionality of multivariate data-sets by transforming the original variable set to a smaller number of composite variables, the first few of which usually represent most of the sample variance; (2) examine the structure of covariances or correlations among the original variables; and (3) achieve a low-dimensional representation of overall or phenetic similarity relations among the specimens comprising the sample (Figure 10.2). In terms of automating the specimen-identification procedure, it is this last attribute on which we will focus.

Although PCA is a sample-dependent method in the sense that its results are strictly a function of the sample being analyzed – and can be expected to change if the composition of the sample changes – this is not necessarily as problematic for the purpose of group identification as might first appear. Indeed, most descriptive statistics used routinely throughout data analysis are similarly sample dependent (e.g. mean, variance, standard deviation, coefficient of variation; see Simpson et al., 2003). The feature that makes these statistics useful for characterizing underlying features of biological reality is the manner in which the sample is obtained.

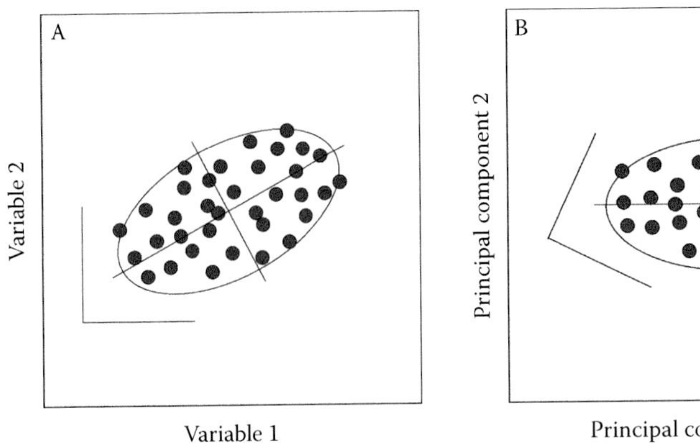

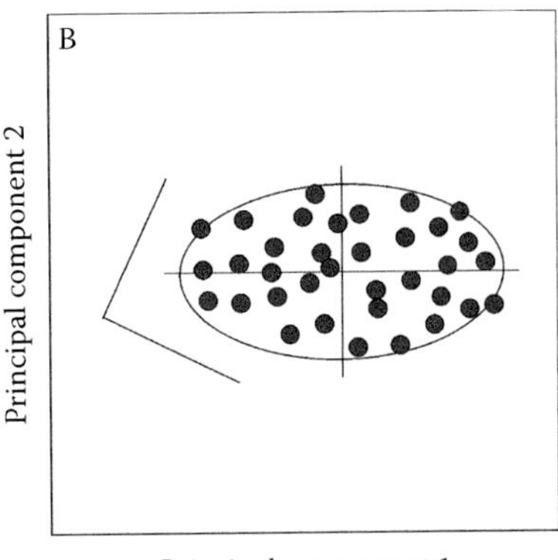

FIGURE 10.2 Fundamental geometry of principal components analysis (PCA). A. Scatterplot of two variables from a single data-set. Enclosing confidence ellipse indicates major directions of group-specific covariation with axes representing the major and minor directions of within-groups covariance. Angle figure in the lower left corner of the variable space represents the variable axes. B. Principal components transform of the data represented in A. This transform rigidly rotates the variable space to a position that corresponds to the major directions of covariation within the data. Note that, unlike canonical variates analysis (see Figure 10.1), there is no internal deformation of the space. This aids interpretation of the space in that the geometric relations between the original variables – represented in B by the angular figure – and the principal components bear a simple and consistent relation to one another.

In order to make descriptive statistics useful for estimating features of the populations from which they were collected, it is of the utmost importance that the sample be drawn randomly from the population and be sufficiently large to estimate the structure and range of variation exhibited by that population accurately. Otherwise, the sample will characterize only a subset the population, which, depending on the hypotheses under consideration, may or may not be useful. Many data analysts fall into the trap of mistaking an inappropriately constructed sample for a valid population proxy or, even worse, for an entire complexly structured population. By the same token, results obtained through the analysis of a sample pertain only to the variables actually measured and cannot be used to infer the character, variation structure or relations of other variables.

Provided a sample is statistically representative of a population of interest, there is no reason to suppose results of a PCA (or DA, or CVA) are inherently non-representative of a population. Indeed, the degree of confidence one can place in the values of any descriptive statistic is an issue that can be evaluated empirically for populations that can be resampled or, for samples, via bootstrap, jacknife or other Monte Carlo simulations (see Manley, 1997). This observation, of course, does not mean one is free from concern regarding the degree to which the sample is representative of the underlying population or from the obligation of testing this where possible. Rather, it is mentioned as an antidote for statements alluding to the instability of PCA results – and morphometric variables generally – that litter the systematics literature (e.g. Pimentel and Riggins, 1987, Cranston and Humphries, 1988; Chapill, 1989; Crowe, 1994; see MacLeod 2002a for additional discussion and examples). Any and every correctly calculated PCA result is a valid description of the sample used in its determination and may be valid beyond the limits of that sample depending on the sample's statistical representativeness in the same way that sample means, standard deviations, etc. are used routinely to make population-level inferences.

The other commonly alluded to theoretical concern with using PCA for specimen identification is the fact that PCA is predicated on a characterization of the covariance structure among variables across the entire sample taken as a whole. This differs from the typical discriminant analysis situation where the purpose is to determine the character of structural relations between different

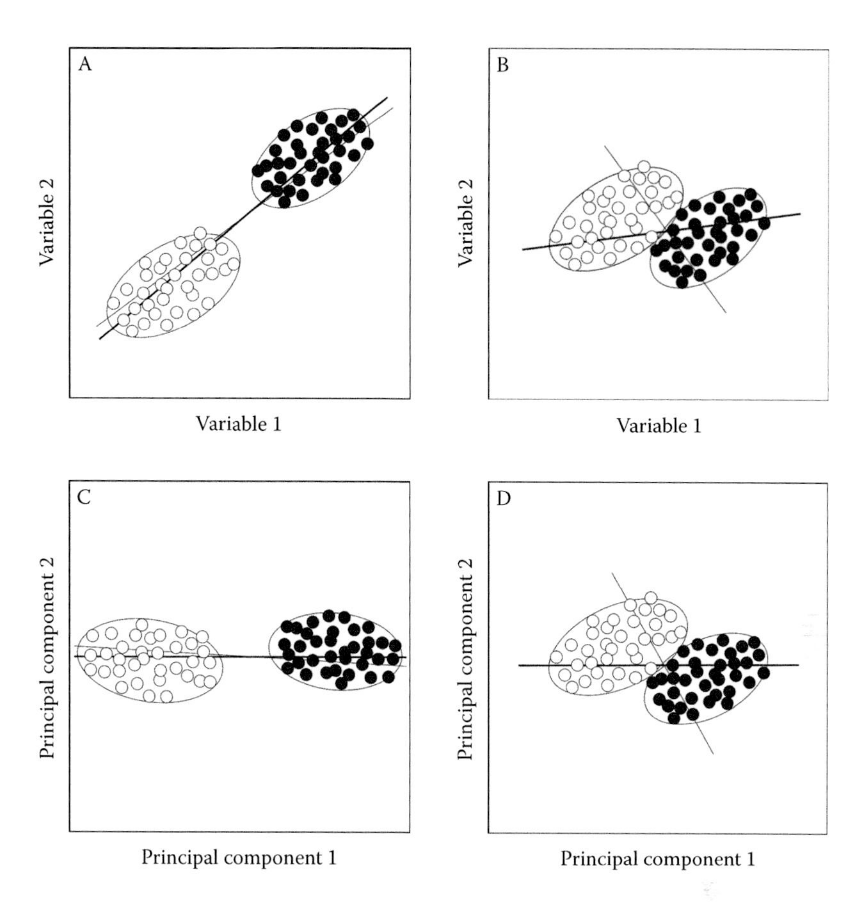

FIGURE 10.3 Behavior of PCA when data are partitioned into groups. A and B represent two different group geometries plotted in the space of the original variables. C and D represent PCA transforms of the data shown in A and B, respectively. A and C represent the case in which the best, linear, group-discrimination axis (light) and major axis of the pooled-sample covariation (bold) are well aligned. B and D represent the case in which these axes lie at a high angle to one another. Note the difference in the ability of the resulting first principal component axes to approximate a group-discrimination axis. Since, for complex, multivariate data-sets, these geometries can rarely be assessed prior to analysis, standard PCA is of limited use in addressing the group identification problem.

samples (see earlier discussion). In this sense a distinction is usually made between data-sets that exhibit no subordinate internal structure (Figure 10.2) and those that do (Figure 10.3).

For example, if the major dimensions of variation within a sample arise as a result of within-sample groupings (Figure 10.3), orientation of the principal component axes will reflect this distinction. In such cases very little information about the nature of variation within the subsidiary groups can be predicted *a priori.* This obtains because there is no obvious way to predict the position of the group discrimination axis (thin axes in Figure 10.3) relative to the major axis of pooled-group variance (thick axis in Figure 10.3) at the outset of an analysis. If these axes are approximately aligned with one another (Figure 10.3A), the PCA result (Figure 10.3B) will reflect this fact by causing the groups to separate more or less cleanly along PC-1. If these axes exhibit a high angular relation to one another (Figure 10.3C), the PCA result (Figure 10.3D) will not specify an axis on which group distinctions are well represented.

Other geometries are also possible (e.g. axis of group discrimination lies at right angles to the axis of pooled-groups variance, in which case the higher PCA axes may capture between-groups distinctions). Regardless, whenever subsidiary groupings are present in the sample, placement of

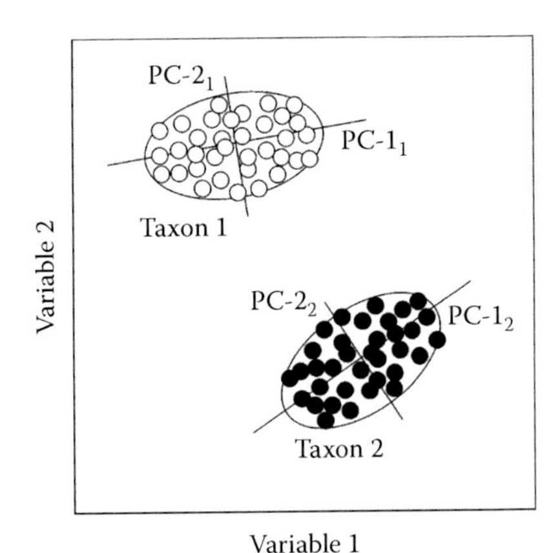

FIGURE 10.4 The method of disjoint PCA models to address the group-identification problem. Here, rather than attempting to represent the problem in the space of the pooled-sample principal components, PCA models are calculated separately for each *a priori* group. In this sense, PCA is used to perform a major axis, multivariate linear regression of individual data-sets with model boundaries being defined by the amount of scatter about the regressions exhibited by a training set of authoritatively identified individuals. Unknown specimens can then be fit to each regression-based model and appropriate identification decisions made on the basis of which model exhibits the best fit. Note also that this approach does not require that an unknown specimen be assigned to any known model if its fit does not meet objectively and empirically defined criteria.

PC-1 (and, by extension, all subsequent PCA axes) will represent an unpredictable amalgam of all variation sources.[2]

Because PCA (also DA/CVA) represents a rigid rotation that preserves the inherent linear geometry of the data, within-sample group boundaries may be rotated, compressed or rarefied (depending on scaling conventions), but they will not be mixed. As a result, clouds of data that interdigitate prior to PCA will do so when re-expressed as PCA ordinations and data representing distinct and mutually exclusive domains within the space of the original variables will continue to do so within these ordination spaces. Nevertheless, trying to encompass all types of variation in a single PCA analysis is impractical and violates the 'single sample' heuristic upon which the technique was first formulated. The basic PCA model does suggest a simple alternative, however, through which the single-sample limitations of PCA can be turned into strengths for the purpose of generalized group identification.

Consider the situation in which a data-set contains individuals drawn from different taxa. Rather than utilizing an exploratory PCA approach and then partitioning the resultant space using *ad hoc* methods, one could confirm the characteristics of the groups (i.e. test to determine whether the taxon-specific samples exhibited subsidiary groupings) and then utilize separate PCA analyses to create different quantitative models of variation for each taxon (Figure 10.4). In this context, PCA is perhaps best regarded as a regression-based, generalized modelling method capable of summarizing the linear variation of any data-set, no matter how complex. Moreover, once a PCA-based model has been determined for a statistically representative sample, it can be compared to similar models constructed from other samples. Such models can be cross-validated against each other, subjected to permutation tests to assess their distinctiveness and, if these tests are passed, used to allocate unknown specimens to appropriate groups or recognize individuals that do not fit any established model (e.g. possibly a new group).

This approach was originally developed by Wold (1976) under the name 'disjoint principal components models', later termed 'simple modelling of class analogy' (SIMCA) (see also Wold and Sjöström, 1977; Wold et al., 1983). While biological applications of SIMCA have been limited (e.g. Wold, 1976; Dahl et al., 1984), the technique exhibits some of the attributes of much more advanced neural-net architectures (see following discussion). Moreover, because of its basis in eigenanalysis, SIMCA can readily be adapted to take advantage of geometric morphometric data and concepts – both landmarks and outlines, or combinations of the two (e.g. Bookstein, 1991; Bookstein and Green, 1993; MacLeod, 1999), thus further extending its applicability.

Neural Nets

Despite being a relative newcomer on the biological object-identification scene, the concept behind neural nets actually dates back to the early days of electronic computing. The computers most people are familiar with today are based on symbolic logic. Such systems were developed originally by Turing, von Neumann and others as machines that accept data in the form of symbol strings (usually ones and zeros), operate on those strings using a formalized set of rules (a program usually coded as sets of simple rules or algorithms stored in the system's memory) and output another string of symbols deterministically related to the input string. The rules specified by the computer programmes for CVA and PCA are examples of this approach to computer design. Using this generalized approach, complex shapes can be reduced to simple symbols (the data matrix of measurements), submitted to the DA/CVA or PCA algorithms, and the results – consisting of another symbol string (the eigenvalue, eigenvector and score matrices) – output.

Neural nets originally represented an alternative approach to computer design. Here, a single processor addressing a linear memory is replaced (at least in a logical sense) by an interconnected assembly of simple processors that store numerical weights configured into networks by interprocessor pathways, one weight per interprocessor pathway (see Lang, this volume, for a more technical discussion). Instead of submitting a programme to this network-based computer architecture, one 'trains' the network to perform a task such as characterizing a group by submitting a set of generalized input data to the network's input nodes and comparing its output to an expected or 'correct' answer. The match between the observed and expected output is then assessed, areas of agreement and disagreement summarized, and this information fed back into the network. This feedback causes the interprocessor, or internode weights, to be adjusted, and a new output, hopefully closer to the expected result, produced. This process is then repeated until the correct answer is obtained.

Once the network computer has 'learned' its groups using a training data-set to 'tune' the network recursively, it is ready to accept new data and perform the identification operation. The advantages of this approach to computing include: (1) lack of any need to devise specific algorithms and write computer programs for complex operations (neural nets train themselves), (2) the same net architecture can be used on very different data types (neural nets can also cope with substantial proportions of missing data – unlike morphometric approaches – and will work successful on many holistic features that are difficult to characterize using simple morphometric descriptors), (3) neural nets are robust to degradation (indeed, they can reprogramme themselves if necessary) and (4) incorporation of an ability to get better at tasks the more such networks are used; to take advantage of 'on-the-job' training.

Any similarity of this process to human learning is entirely deliberate. The neural network concept is based on the architecture of biological neural systems. Whereas 'von Neumann machines' are particularly good at performing simple, repetitive tasks quickly and accurately on data adequately represented by simple symbols – precisely the sorts of operations humans are very bad at performing, neural nets often have the edge in performing complex, subtle tasks on generalized data.

Unfortunately, because of their non-deterministic character, the performance of neural nets in handling any given task is difficult to predict in advance. Also, neural nets can take long time periods to train and many neural net architectures are sample dependent in the sense that the addition of a new class requires the entire weight system to be recomputed from its initial (usually random) configuration. In addition, the search methodologies neural nets use to adjust the weights during training may converge on a suboptimal result (e.g. local minima) from which there is no way to further improve the net without re-initialization.[3] Note that these last two aspects of neural network calculations are also characteristic of many complex algorithm-based systems (e.g. inference of the structure of maximum-parsimony phylogenetic networks).

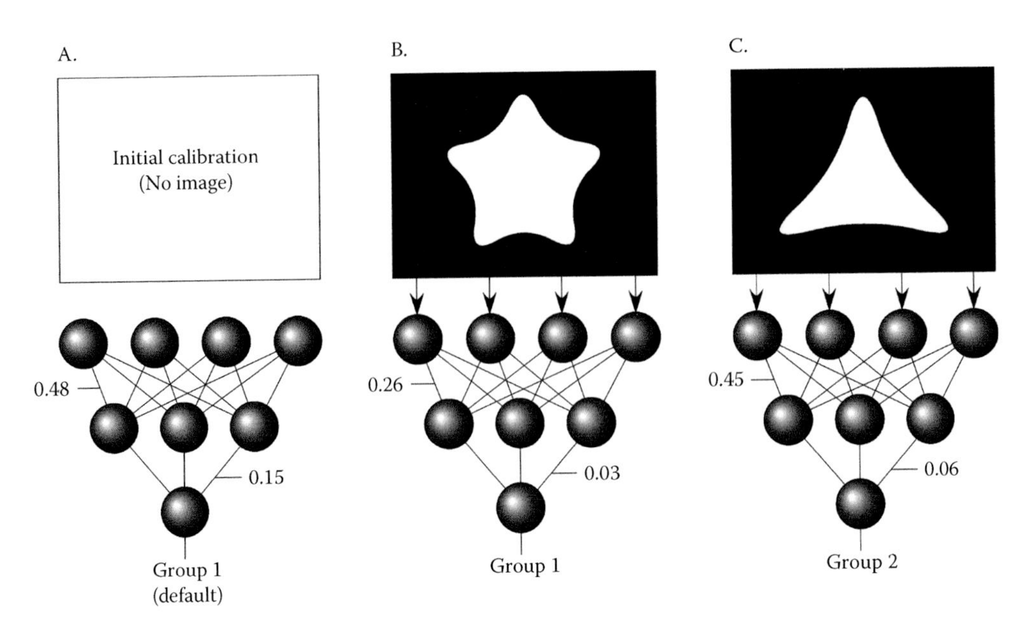

FIGURE 10.5 Design of a standard, back-propagation neural network (e.g. multilayer perceptron). Artificial neurons are arranged in layered banks connected by interneuron paths that can be assigned numerical weights. It is a requirement of such designs that neurons within the same bank are not connected to each other. A. Interneuron weights are assigned randomly during initialization of the net. B. Presentation of data from a training-set object (arrows) to the net causes the weight system to change. C. Presentation of data from another training-set object (arrows) to the net causes the weight system to undergo further changes in an attempt to express the difference between the objects. This procedure is repeated recursively for all objects in all training sets. The result is a set of interneuron weights that express the observed between-groups differences. Note that, because this design requires global optimization across all nodes, introduction of a new group requires that the entire weight scheme be recalculated.

In order to improve the performance of neural networks, a number of variant designs have been developed. Traditional or classical neural nets (formally called multilayer perceptrons, MLPs) consist of a fixed, layered structure that remains static during the training process (Figure 10.5). Traditional MLP nets are trained using supervised learning, with weights adjusted during training iterations by a process termed backpropagation (see Bishop, 1996; Lang, this volume).

Once trained, backpropagation nets can be used to identify unknowns (which belong to one of the classes the net was trained on) with typical identification accuracies in excess of 90 per cent, though this, of course, is data set dependent. Each time a new image is submitted, a probability estimate is obtained for the fit between the pattern and the various group models (represented by the net's weight system) resulting in a decision being made regarding group affiliation. Data that do not conform to any of the net's group models to a sufficient degree may be rejected as unidentifiable. Addition of a new group to the net requires complete recalculation of the weight system. Nevertheless, supervised backpropagation networks have been the sort used most extensively for image identification (Bishop, 1996; Haykin, 1999; Duda et al., 2000).

Neural nets can also be based on unsupervised learning strategies. To date these nets have been employed primarily to support data visualization, but their flexibility is such that they are becoming more common in a wide variety of applications. A simple version of an unsupervised neural net is the Kohonen self-organizing map (SOM) (Kohonen, 1982, 1984; Lang, this volume). These nets also use a set number of nodes, but operate according to different principles.

The SOM net is better thought of as a lattice, or map in which the dimensions of the grid correspond to the number of input observations or variables. The idea here is to adjust the structure of the map dynamically, based on the training-set data, until it forms an accurate representation of

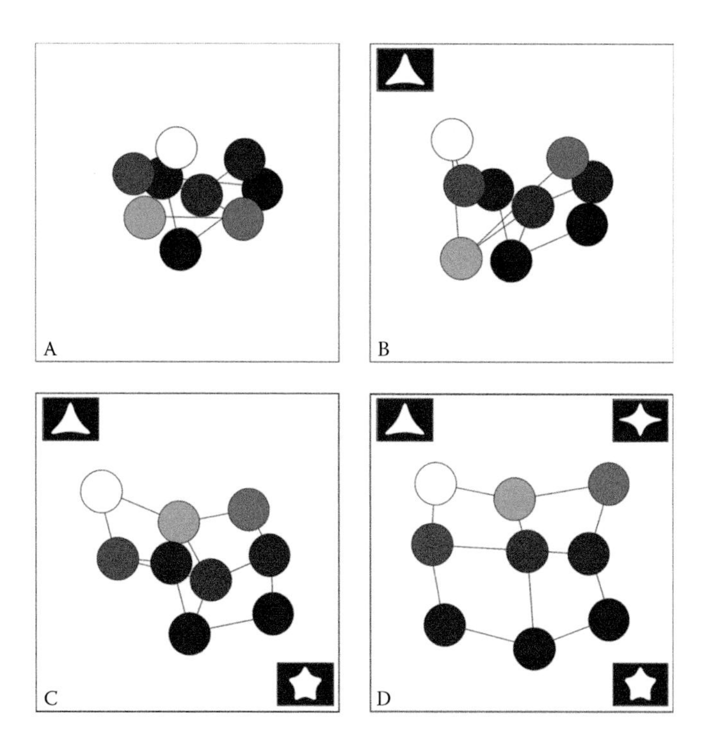

FIGURE 10.6 Design and development of a self-organizing map (Kohonen) neural network. A. Neurons assigned random positions in the weight space during net initialization of the net. This is a two-dimensional map and so represents relations among only two variables (say) circularity (x-axis) and triangularity (y-axis). B. Presentation of data from a training-set object with high triangularity and low circularity to the net causes a node to be assigned as the focus and the positions of all other nodes to be recomputed based on proximity to the focus. C. Presentation of data from a training-set object with high circularity and low triangularity to the net causes the weight system to undergo further changes in an attempt organize the expression of the difference between the objects. In a self-organizing map or Kohonen net, changes in the weight system reflect the spatial location of the group-specific models within the set of possible relations between input data-sets. Also note the net knows nothing about 'triangularity' or 'circularity' at the outset. It develops a model of these relations as data are presented. This procedure is repeated recursively for all groups present in all training sets (D). The result is a set of interneuron weights that reflect the observed relations between input variables. The ability to draw distinctions between a large number of groups will be dependent on the size of the network (map).

similarity relations among the different groups. As before, the net is initialized using random values. When the first object of the first group is presented to the net, a node whose initialized values are close to that of the object is selected as the focus of the group (Figure 10.6A) and the values assigned to the other nodes modified to reflect their similarity (as assessed by a distance calculation) relative to the focus. Thus, the values for nodes immediately adjacent to the focus are changed more than the values of nodes located at a distance from the focus. As examples of other groups are added, they form their own foci within the SOM structure and, as more examples of these groups are added, the system of values coalesces around these foci (Figure 10.6B–D). With training, the SOM system comes to resemble a topographic surface that reflects phenetic similarity relations within the original data.

Interestingly, Kohonen (1984) advocated use of a supervisory computational engine to perform the training calculations quickly. His method also specified a decrease in the size of the regionalized neighbourhoods over time (to allow the net to encompass finer differences between groups) and a gradual lowering of the learning rate over time (to promote overall net stability). While Kohonen SOMs represent a better data-analysis strategy for understanding the relations between groups, they are more difficult to construct and are limited by their need to specify the size of a SOM at the outset of an analysis. This makes SOMs ideal for dealing with well-constrained problems (e.g. identification systems for well-known groups whose taxonomy is stable), but limits their overall applicability in open-ended situations.

The most advanced neural net designs incorporate elements of dynamic learning strategies in non-stationary network design environments. An example of this neural net type is the plastic self-organizing map (PSOM) (Lang and Warwick, 2002; Lang, this volume). A PSOM net begins with a few nodes and randomly initialized links (Figure 10.7A). When data from an initial group are presented to the net, more nodes are created and the existing internode weights adjusted so that they 'move' closer to the established group (Figure 10.7B). As more examples of the initial group are added, high-value internode weights are further strengthened and low-value weights diminished, changing the structure of the net (Figure 10.7C). When data from a new group are introduced, a new set of nodes is created and placed in an appropriate position relative to the established topology (Figure 10.7C). As more data from the new group are evaluated the weights of high-value internodes are adjusted, eventually to the point where most or all internodal connections between established groups are lost (Figure 10.7D). This process is repeated until all groups are located relative to one another in the dynamic net topology.

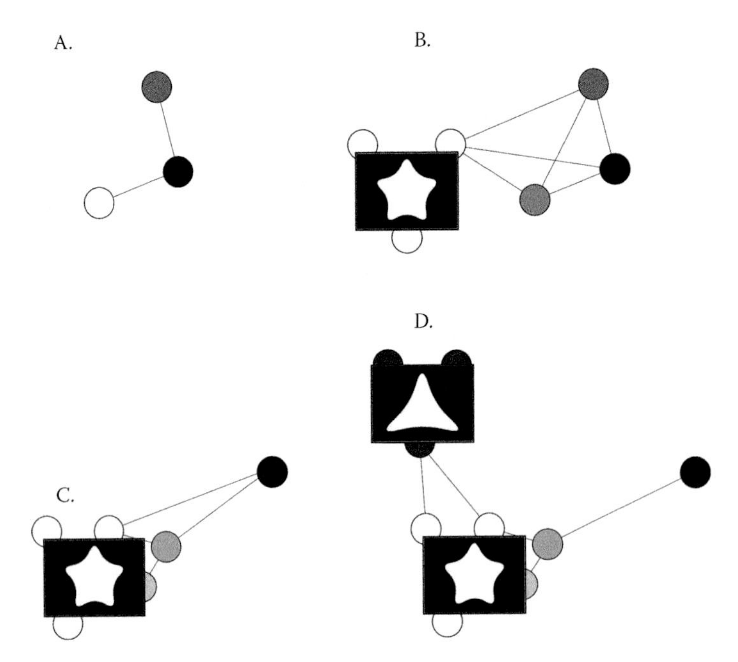

FIGURE 10.7 Design and development of a plastic self-organizing map (PSOM) neural network. This is similar to a Kohonen net (map, see Figure 10.6). In this case an original (primitive) neuron configuration (A) is specified and interneuron weights assigned randomly. B. Presentation of data from a training-set object to the net causes the weight space to change (symbolized by shade of grey, as in Figure 10.6, and by the length of the interneuron paths) and new neurons to be added to the net. C. As more objects from the same training set are added, the weight space continues to change, consolidating certain neurons into a tight cluster representing the group and driving others away (black neuron). D. Presentation of data from another training-set object to the net causes a cluster of neurons to appear in a region adjacent to a similar existing cluster. As more objects from the second training set are added, the weight space continues to change, further consolidating group-specific neurons into tighter clusters and driving the existing clusters away from each other (black neuron), possibly to the point at which intercluster links are broken. The primary advantage of this design is its speed (once links have been broken, only weights linked with active regions of the net need be recalculated during calibration), scalability (the number of clusters that can be managed by the net is limited only by hardware) and incorporation of dynamic learning into the net's fabric (discreteness of the net will improve with use).

Important features of dynamic neural nets are their ability to accommodate new groups (and 'forget' disused ones), without having to recompute the entire weight system, and the ability of some designs (e.g. PSOM) to increase in scope and complexity without bound (at least theoretically). These attributes will likely be very important for creating practical neural nets for a variety of routine biological and systematic applications – especially identification. Such nets will need to be able to accommodate at least hundreds and likely thousands of groups simultaneously.

In addition, practical neural nets for many (if not most) systematic applications will need to be able to be updated quickly as new groups are discovered, or discovered to be important in resolving a particular problem. Preliminary comparisons suggest dynamic nets do not offer great advantages over traditional designs in terms of accuracy or identification speed (though truly large-scale nets have yet to be populated and tested with biological data). What is unequivocal, however, are the advantages of the PSOM and related designs in terms of scalability, overall net maintenance, and flexibility. One sign of the success of dynamic nets is the fact that hybrid designs are beginning to appear that combine the advantages of both by embedding a dynamically structured set of nodes within an overall static framework (see SPIDA and SPIDA-web, Russell et al., this volume)

AN EXAMPLE

In order to illustrate the advantages and outstanding issues needing further development when these approaches are applied to the group-identification problem, it is useful to compare and contrast results obtained from a specific exemplar data-set. Since human face characterization is a classic problem in the general field of pattern recognition (e.g. Gong et al., 2000), the example selected for this study represents a systematic view of the 'face recognition' problem. Thus, our example compares and contrasts the application of morphometric and PSOM neural net approaches to the design of automated systems to separate two wasp families based solely on female facial characteristics. In addition, we will use these same facial characters to distinguish the parthogenetic and sexually reproducing forms of the one of the families. Facial features are obvious, contain a wide variety of characters (see later discussion), and good images of wasp faces could be collected by relatively untrained personnel. Nevertheless, identification of these groups based on these characters alone has, so far as we are aware, never been attempted before.

Our subjects all belong to the wasp superfamily Cynipoidea (suborder Apocrita), which represents over 3000 species and over 200 genera; all are commonly referred to as 'gall wasps', though only one family (Cynipidae) actually contains species that induce plant galls. Other families within the superfamily (e.g. Figitidae) are insect parasites, developing initially as koinobiont endoparasitoids but spending the last instars feeding on the host. These latter species usually attack endopterygote larvae. Reproductively, cynipid wasps (approx. 1300 species) alternate between sexual and parthenogenic modes. In some cases these alternative sexual modes are expressed as characteristically different variants in facial morphology (Figure 10.8). Other cynipid taxa do not undergo this alternation of generations but produce all progeny sexually.

The cynipoid wasp faces used in this example analysis were all female and came from a collection of scanning electron microscope (SEM) images made available on the *MorphBank* website (http://www.morphbank.com/) as part of the *MorphBank Project*. Each image in this database is associated with fully searchable text information, and images can be downloaded at different resolutions and in different formats (jpeg, tiff). As work in numerical morphology progresses, it is expected that image-data repositories such as *MorphBank* (see also *MorphoBank*, http://www.morphobank.org/) will be set up, contributed to and used on a routine basis. A complete list of all species and thumbnail representations of the images used in this investigation are presented in the appendix.

This example will focus on a higher taxonomic-level characterization of facial morphology among sexually reproduced species of the Cynipidae (cynipid-s), parthenogenically reproduced

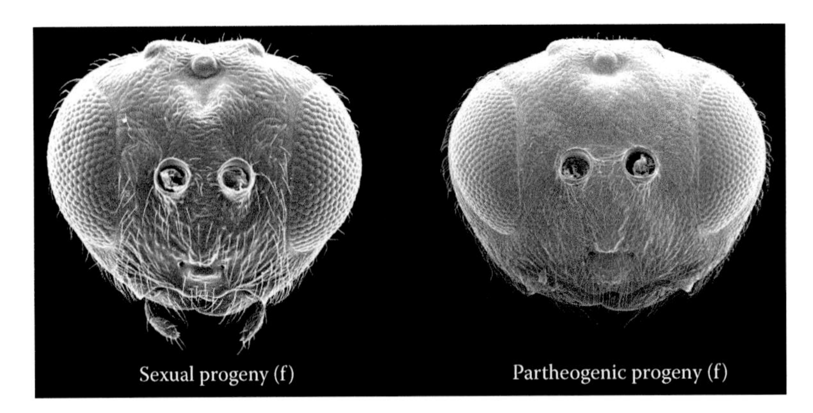

FIGURE 10.8 Head-shields of cynipid wasps from the species *Andricus gallaeurnaeformis* (Cynipidae) showing facial dimorphism related to reproductive mode: sexually produced female (left) and parthenogenically produced female (right).

species of the Cynipidae (cynipid-p), and members of the Figitidae (figitid). Generally speaking, cynipids have wide faces with very large, laterally placed eyes, whereas figitids have long, narrow faces with relatively smaller eyes placed well up on the face (Figure 10.9). These taxa, however, vary widely (see Appendix 1) and face morphology *per se* has not been regarded as a group-diagnostic character (e.g. Ritchie, 1993).

Original sample sizes for cynipid-s, cynipid-p and figitid data-sets were 32, 32 and 35, respectively. Of the 44 genera and 59 species represented in the combined cynipid data-base, 4 genera and 5 species are shared between cynipid-s and cynipid-p data-sets (see Appendix 2). This represents significant taxonomic overlap at the genus level (21%), less so at the species level (7%). Owing to this discrepancy, there is a good chance that at least some of the differences between cynipid-s and cynipid-p groups are the result of taxonomic differentiation rather that sexual-morph differences. However, there is nothing in the *Morphbank* data to indicate that the species comprising these groups were selected because they represent different modes of morphological variation or that they are not representative samples of the sexual groups from which they were drawn. Accordingly, they will be treated in this study as quasi-random samples drawn from the cynipid taxonomic and morphological space.

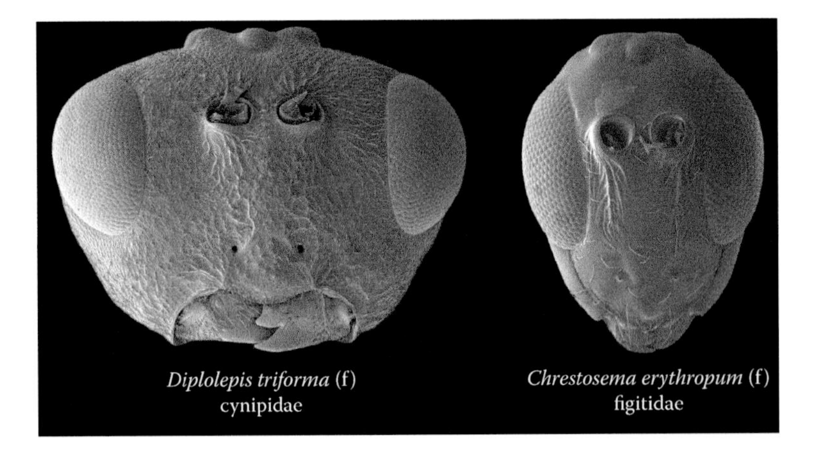

FIGURE 10.9 Head-shields of cynipid wasps from the species showing facial dimorphism related to taxonomic group: sexually produced *Diplolepis triforma* (Cynpidae) female (left) and sexually produced *Chrestosema erythropum* (Figitidae) female (right).

These considerations present no obstacle to the primary goal of this chapter, which is to use a variety of quantitative methods to assess whether differences can be detected between *a priori* defined morphological groups. Nevertheless, if consistent differences between sexual and parthenogenic cynipid taxa are found, they will need to be treated as provisional indications of possible sex-linked morphological differentiation that should be confirmed (or refuted) with subsequent analysis of data collected for that purpose.

Morphological characters and the positions of landmarks used to quantify face shape variation for this study are presented in Figure 10.10. For both the discriminant and Procrustes distance analysis (see later discussion), landmark locations were hand-digitized (using *ImageJ* software, http://rsb.info.nih.gov/ij/), assembled into a data matrix and transformed into Procrustes-registered shape coordinates (Rohlf, 1990), which were then used as input to the aforementioned procedures. This operation took approximately three hours to complete, but had to be repeated several times before the untrained operator (NM) became sufficiently familiar with the facial morphology of these wasps to achieve an acceptable level of landmark-placement consistency. The neural net analysis used raw tiff images that had been processed only insofar as blanking out the backgrounds surrounding the face, standardizing the face orientations in the frame and standardizing the image resolution and size (150 dpi, 600 × 450 pixels). This operation was automated and so required essentially no additional analysis time. No scaling information was included in any analysis because the purpose was to compare the objects on the basis of shape alone.

In terms of sample groupings, the analysis was subdivided to produce two different test situations. The first of these focused on recognizing gross distinctions between the cynipid ('sexual' + 'parthenogenic') and figitid groups. This is the more standard form of group-identification test in which both groups encompass broadly comparable levels of taxonomic distinction. The second test attempts the more difficult problem of distinguishing between the two cynipid sexual groups as well as the generally more distinct figitid taxa. The latter problem is especially challenging in that, owing to the taxonomic level at which these comparisons are being made, within-groups variation is likely to be quite high relative to between-groups variation, especially for the two cynipid groups. Nevertheless, this is exactly the sort of problem quantitative identification tools will need to handle successfully if they are to be useful in systematic applications. Both the two-group and three-group data-sets were submitted to CVA, a new method (described later) based on an assessment of Procrustes distance, and an implementation of the PSOM neural net.

The CVA method employed here was a stepwise variant of the classical CVA technique (see previous discussion) in which the set of variables is evaluated (using an F-test) to order the variables in terms of their relative contribution to group discrimination. This is a standard model for yielding efficient, stable and optimally interpretable discriminant results. In both two-group and three-group analyses a forward stepwise strategy was employed with the critical F-value set to 0.10. Raw data constituted the 22 shape variables specified by the x and y coordinates of the 11 landmarks shown in Figure 10.10B after being transformed to their generalized least-squares Procrustes shape coordinates about the pooled-sample consensus shape.

Efficiency of the shape variation models determined by the CVA was assessed by performing a cross-validation analysis of the specimen set used to create the model. This is, admittedly, not as useful as knowing how well the models perform using a completely unknown set of specimens, but adequate numbers of specimens from the same localities as the training set were not available and the purpose of this analysis is to illustrate use of the methods and compare their results, not to create a final gall wasp identification engine. During the cross-tabulation, specimens were assigned to groups by using the linear-discriminant equations to project the measurement set into the discriminant space, calculating the Mahalanobis distance between the projected specimens and each group centroid, and assigning the specimen to the closest group.

The Procrustes distance method employed in this study is a variant of the SIMCA disjoint linear pattern recognition method (Wold, 1976). A generalized least-squares Procrustes consensus shape was determined for each group and used as the best single representation of overall group

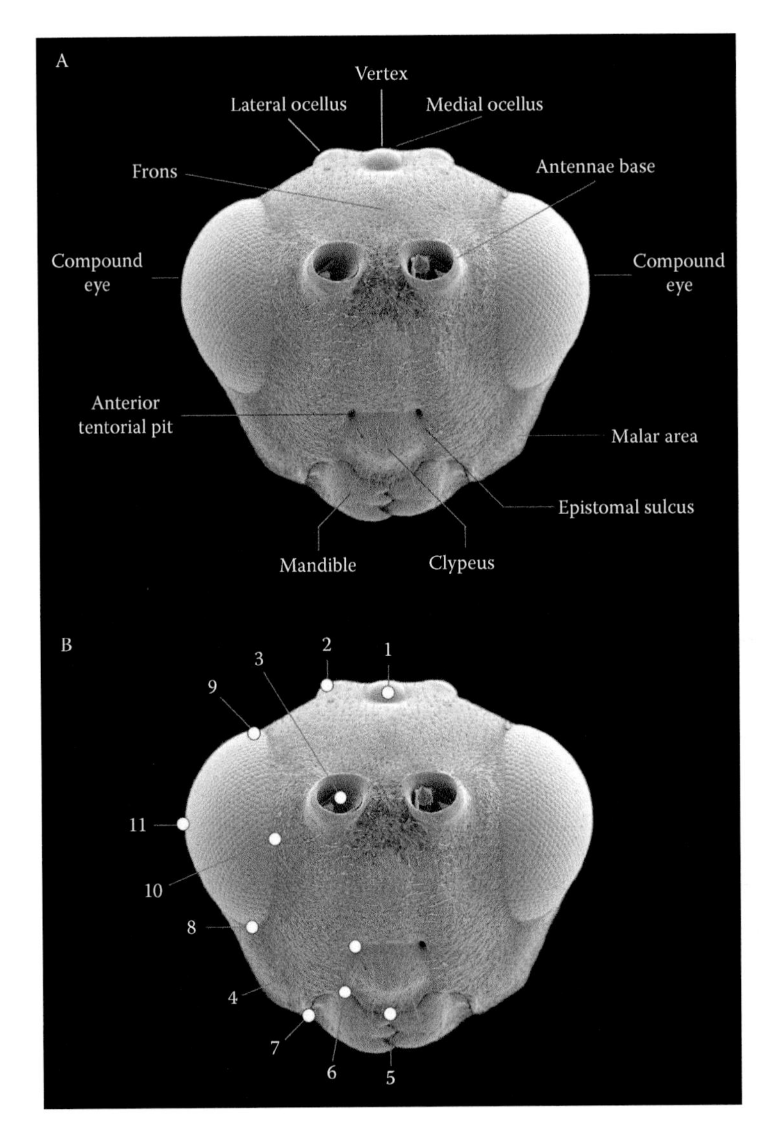

FIGURE 10.10 Morphological features of cynipid wasp faces. A. Labelled scanning electron micrograph of *Melanips opacus* (Figitidae). B. Landmarks used to quantify facial shape variation. 1: position of the medial ocellus; 2: position of the lateral ocellus; 3: position of the antenna base; 4: position of the anterior tentorial pit; 5: position of the ventral margin of the clypeus along the facial midline; 6: position of the ventral terminus of the epistomal sulcus; 7: position of the ventral margin of the malar area; 8: position of the ventral margin of the compound eye; 9: position of the dorsal margin of the compound eye; 10: position of the proximal inflection point in the compound eye outline; 11: position of the distal inflection point in the compound eye outline.

shape variation after confirming that no significant intragroup clustering was present at each landmark vertex for the training set. Once these consensus shapes were calculated, simple 2σ group-specific tolerance envelopes were estimated from the distribution of Procrustes distances from consensus shape exhibited by the training set.[4] These consensus shapes and distance envelopes then constitute generalized models of shape variation characteristics of each group.

Using the same cross-validation strategy described earlier, specimens were identified by calculating the Procrustes distance between their landmark configurations and each group-specific consensus configuration and determining whether overall Procrustes distance lies within the 2σ tolerance envelope of one or more groups. In the case where a specimen lay within the tolerance

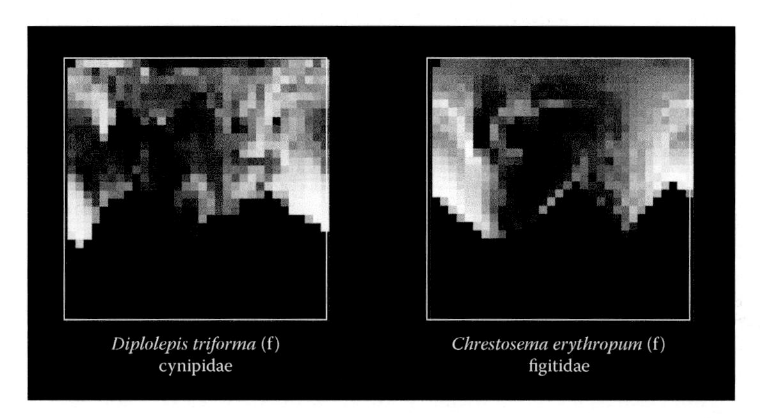

FIGURE 10.11 Examples of the image data used by the DAISY implementation of a PSOM, *n*-tuple neural net for taxonomic identification. Sexually produced *Diplolepis triforma* (Cynpidae) female (left) and sexually produced *Chrestosema erythropum* (Figitidae) female (right). See Figure 10.9 for original images. Both images represent 32- × 32-pixel subsamples of 600- × 450-pixel processed images that have been adjusted for equalization of the image histogram and transformed to a polar coordinate pixel sequence.

envelopes of more than a single group (i.e. group-specific shape models overlap), it was assigned to the group whose consensus configuration was closest to that of the specimen.

Overall, this Procrustes-distance method has the advantage of being much more easily calculated than either standard SIMCA or CVA models; it more faithfully represents the geometric information encoded in the group-level training set data and uses those data to inform the analysis, yielding a test statistic that is similarly quick and easy to calculate as well as being free of group dependence (unlike the CVA approach discussed earlier). Polly and Head (2004) employed a more sophisticated version of this approach to the analysis of marmot and pipesnake morphology in which specimen identification was referred to a maximum likelihood model. Interestingly, their approach did not perform as well for the characterization of these groups as the simpler approach described before did for the wasp data (see later discussion).

The PSOM neural-net approach to species identification was implemented by the digital automated identification system (DAISY; see Weeks et al., 1997, 1999a, 1999b). DAISY was chosen as the best currently available implementation of a PSOM-like neural net that has been programmed for use in systematic applications. The DAISY system accepts training sets in the form of standard-format images (e.g. tiff) of authoritatively identified specimens. Each image is processed by (1) reducing its spatial resolution (via subsampling) to a 32- × 32-pixel grid, (2) adjusting its pixel-level spectrum to achieve brightness equalization and (3) transforming each pixel grid from a Cartesian to a polar format. The first step in this process represents an empirically determined optimum resolution needed to maximize the signal-to-noise ratio and quantify topological correspondences. The second reduces interimage variations due to lighting/exposure artefacts. The third allows the analysis to utilize spatially irregular 'regions of interest' as well as the more traditional rectilinear image boundaries. Results of these transformations for characteristic cynipid and figitid images are presented in Figure 10.11.

Once DAISY had processed all images in the training set, a non-linear discriminant space was calculated based on the *n*-tuple classifier. The proximate basis for this classification is a pairwise comparison between RGB brightness values for each of the 32- × 32-pixel locations. The result allows each object in each training set to be placed into a multidimensional, distance-based, ordination space whose character can be varied based on the estimated affinity (via NVD correlation) between similarly processed images of unknown specimens and the training set array. It is this ability to modify the character of the base training set ordination that gives the DAISY implementation of the PSOM concept its adaptive quality.

To achieve specimen identification, DAISY uses the computed affinity vector to project objects into a non-linear discriminant space. The position of an unknown object in this space can be compared to the locations of training set objects using simple distance metrics. Cross-validation identifications were achieved by a multilayer strategy based on this discriminant space. The primary DAISY discriminator converts metric distances to a more robust rank measure and assesses the eight nearest training set neighbors to the unknown object. Under this 'coordination' metric, the strength of group affiliation is measured by the number of neighbors belonging to the same group (e.g. all eight nearest neighbors in same group represent a coordination value of 1.00; six out of the eight neighbors represent a value of 0.75). In the current DAISY implementation, a coordination value of three or more is regarded as sufficient to associate the specimen with a group to a high confidence level.

If an object cannot be identified by the coordination metric, DAISY concludes the unknown object is not embedded within a group cluster. In that case a 'SILL' metric is then used to determine whether an unknown object lies at the edge of a group cluster. Similarly, a 'vote' metric is used to determine the probable identity of objects in regions of the space containing members of more than a single group. If these metrics fail to attribute the unknown to a group, an identification is made based on using a 'first part the post' (FPTP) metric that assigns the unknown object to the group of its nearest neighbor, equivalent to a coordination of 1.00.

Two-Group Results

Stepwise discriminant analysis of the combined cynipid (sexual and parthenogenic progeny) and figitid data-sets identified 15 of the 24 Procrustes-registered shape-coordinate variables as being important for group discrimination. Three full landmark locations were eliminated from further consideration by this process (distal terminus of the epistomal sulcus, distal terminus of the malar area, distal inflection point of the compound eye) along with x-coordinate of the lateral ocellus. Of the remaining variables, the discriminant axis was aligned with a contrast between the proximal inflection point and ventral terminus of the compound eye (x-coordinates) and the antennae base and anterior tentorial pit (see Figure 10.10). These results suggest that relative size and placement of the eye, along with the relative width of the face, are the best means for discriminating between cynipid and figitid wasps. This axis enabled 97 per cent of the training set to be identified correctly; only three figitid species (3.0%) were assigned incorrectly to the cynipid group. Visual inspection confirmed that these three species images (*Phaenoglyphis villosa*, *Euceroptres montanus* and *Melanips* sp.; see Appendix 3) did indeed represent figitid species characterized by atypically large eyes and broad faces.

Results of the simple Procrustes distance models were comparable to those of the stepwise CVA analysis. For these data, identification of which landmarks are more important for group characterization can be gained from comparisons of differences in the group-specific consensus configuration of landmarks and the standard deviations for each shape coordinate. Consensus-image comparison indicated that landmarks specifying the antenna base (3), anterior tentorial pit (4), ventral margin of the clypeus along the facial midline (5), ventral terminus of the epistomal sulcus (6) and lateral position of the compound eye (10,11) were primarily responsible for inter-group differences (Figure 10.12). When these locations were combined with an assessment of the group-specific variability in landmark placement, it was apparent that location of the antenna base, anterior tentorial pit and ventral terminus of the epistomal sulcus are the primary between-groups discriminating features.

As before, these data suggest cynipids have characteristically broader faces than figitids, but they also suggest that these differences are most reliably assessed in the central portion of the face than at its margins. Cross-validation results indicate the Procrustes distance method performs virtually as well as discriminant analysis; only four figitid taxa were incorrectly assigned to the cynipid group (96% correct). Interestingly, only one of these four taxa (*Phaenoglyphis villosa*) was also

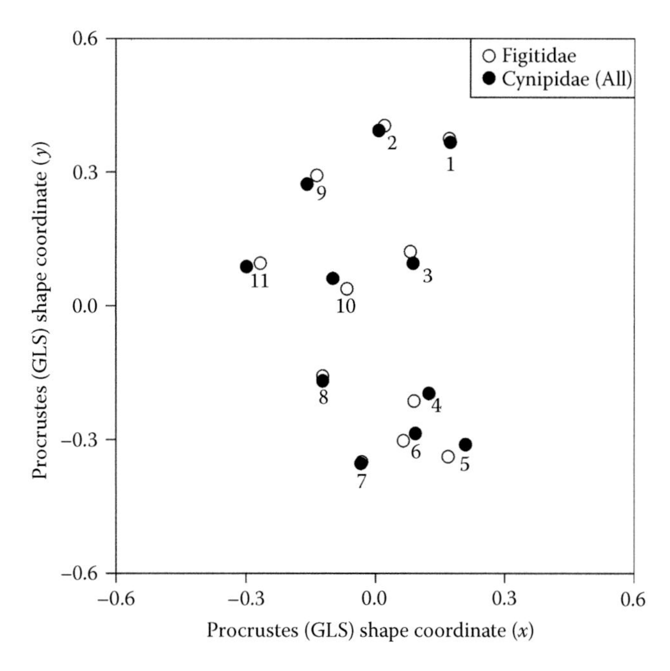

FIGURE 10.12 Generalized least squares (GLS) consensus landmark configurations for cynipid (closed circles, sexual and parthenogenic progeny) and figitid landmark data-sets (open circles). Landmark numbers as in Figure 10.10B. Use of consensus models mimics the disjoint shape model approach of Wold (1976), described in Figure 10.4, in which the different group-specific ranges of variation can be thought of as representing a single landmark location (e.g. compare geometry of Figure 10.4 with that of landmark 10) or the distribution of Procrustes-registered shapes on the Kendall shape manifold (Kendall, 1984). A complete analogue to the Wold (1976) method could be realized by computing disjoint relative warp shape models for each landmark data-set. However, results obtained by this investigation (see text) suggest disjoint models based on the simple Procrustes metric can facilitate remarkably consistent and subtle discriminations between higher taxonomic groups.

misassigned by discriminant analysis. This suggests there is a broader degree of similarity between the two groups than is indicated by the results of either the CVA or Procrustes distance methods taken alone. Inspection of the Procrustes distance results confirms this suspicion. An additional three figitid specimens are within 0.02 units of being attributable to the cynipid group (i.e. subequally distant from the consensus configuration of both groups) including *Euceroptres montanus.*

The DAISY two-group results (cross-validation) indicated that 88 of the objects were assigned to the correct groups. Of these, 74 passed the coordination test with an average confidence value of 0.97, two passed the SILL test with certainty values of 1.00, six passed the vote test with a value of 0.75, and six passed the FPTP test. While this overall result (90% correct) is marginally lower than the morphometric results obtained, it must be remembered that DAISY was required to sort through much more information extraneous to group identification than the morphometric methods, did so quickly (less than 3 minutes required to construct all models and perform the cross-validation analyses), and without any human intervention to guide its comparisons.

All taxa misidentified by both morphometric methods were also listed among the set of DAISY's failed identifications suggesting its results represent a more consistently conservative estimate of identification certainty than either of the morphometric approaches. If the question being asked was whether these two groups were present in the sample, the quality of DAISY's answer would be indistinguishable from those of the much more labor-intensive morphometric methods. Even in terms of estimating relative abundances of these taxa, the DAISY estimates do not differ markedly from the morphometric results (see Table 10.1). These results are not unusual, but rather typical of performance levels that can be expected of neural net approaches to the group

TABLE 10.1
Percentage Elative Abundance Estimates Based on Two-Group Cross-Validation

	Actual	CVA	Procrustes distance	DAISY (PSOM, *n*-tuple)
Cynipidae	64.65	67.68	68.69	61.62
Figitidae	35.35	32.32	31.31	38.38
Total	100.00	100.00	100.00	100.00

identification problem using current technology (see Gaston and O'Neill, 2004; MacLeod et al., in press and references therein).

Three-Group Results

Discrimination between taxa belonging to different families of organisms is a routine and useful task that empirical results suggest can be handled effectively by either morphometric or neural net methods. Possibly the most challenging situation for automated group-identification systems arises from the simultaneous discrimination of taxa at supraspecific and intraspecific levels. The wasp data-set facilitates examination of this case in that, alongside the figitid data used before, the cynipid data were now subdivided into sexual (cynipid-s) and parthenogenic (cynipid-p) groups.

A larger suite of discriminant variables was needed to resolve the three-group problem. In this case the medial ocellus (*y*-coordinate), ventral margin of the clypeus along the facial midline (*y*-coordinate), as well as the lower (*x*- and *y*-coordinates) and upper (*x*-coordinate) boundaries of the compound eye were judged not to be significant contributors to group discrimination. Among the remaining variables, those most highly aligned with the major discriminant axis represent a contrast between the positions of the tentorial pit, antennae base and medial ocellus with the proximal and distal inflection points of the compound eye (all *x*-coordinates). This axis accounts for 83.64 per cent of the between-groups variation and primarily separates the cynipid and figitid species (Figure 10.13).

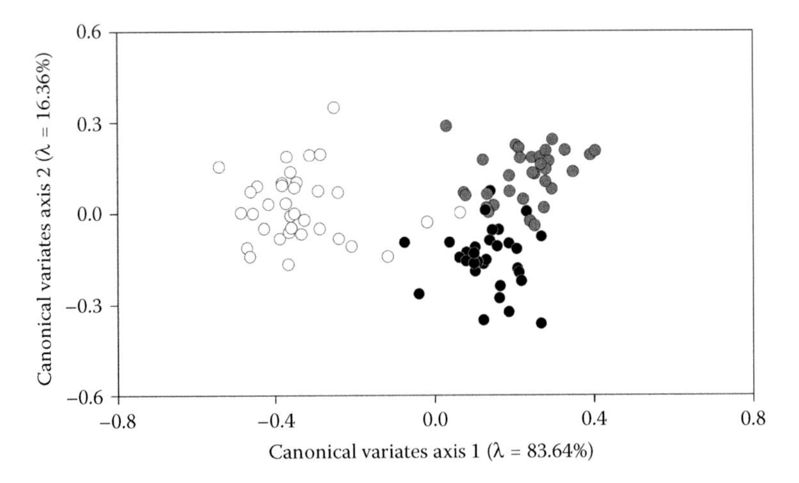

FIGURE 10.13 Scatterplot of Procrustes-registered training-set data along the two canonical variates axes for the discrimination between cynipid sexual progeny (black symbols), cynyipid parthenogenic progeny (grey symbols), and figitid (open symbols) taxa. See text for discussion.

TABLE 10.2
Cross-Validation Results for Three-Group Analyses

	Cynipid-p	Cynipid-s	Figitid	Total
		CVA		
Cynipid-p	30	2	—	32
Cynipid-s	3	29	—	32
Figitid	3	—	32	35
		Procrustes distance		
Cynipid-p	27	5	—	32
Cynipid-s	5	27	—	32
Figitid	—	5	30	35
		DAISY (PSOM, *n*-tuple)		
Cynipid-p	22	9	1	32
Cynipid-s	14	16	2	32
Figitid	2	1	32	35

Note: Numbers refer to specimens.

As was the case in the previous two-group discriminant analyses, these patterns suggest that width of the face and placement of the eyes are the primary dimensions of difference between cynipid and figitid wasp faces. The second discriminant axis subsumes the remaining 16.36 per cent of between-groups variation and represents a contrast between the distal inflection point of the compound eye (*x*- and *y*-coordinates) with the antenna base and ventral terminus of the malar area (both *x*-coordinates). This pattern indicates that cynipid wasps in this dataset that were produced parthenogenically exhibit faces that are characteristically broader and lower, with narrower eyes than sexually reproduced cynipid progeny.

Cross-validation tests for the discriminant analysis (Table 10.2) indicate these models placed 91 (91.92%) of the specimens into the correct group; two parthenogenic cynipids were misassigned to the sexually reproduced cynipid group (*Biorhiza pallida, Neuroterus numismalis*), three sexually reproduced cynipid individuals to the parthenogenic group (*Gonaspis potentillae, Phanacis centaureae, Synergus crassicornis*) and three figitid species to the parthenogenic group (*Aspicera scutellata, Zaeucoila* sp., *Melanips opacus*). Given the high within-groups variability overall and subtlety of the distinction between the two cynipid groups, this is an unexpectedly encouraging result, especially for a character complex that has not heretofore been used to distinguish between these groups taxonomically.

Procrustes distance results were, once again, similar to those for CVA. For this method, the major directions of difference between cynipid and figitid species were focused on the relative placement of landmarks specifying positions of the antennae base, and lateral inflection points of the compound eye (Figure 10.14). As with the three-group CVA results, the primary discriminating factors specify characteristic differences in the width and length of the face, especially those centering on the distance between the antenna base and proximal margin of the compound eye.

Distinctions between the two cynipid groups, however, were captured a bit differently in the Procrustes distance analysis. Here, the primary landmarks differentiating these groups are the positions of the lateral ocellus, ventral terminus of the clypeus along the facial midline and the ventral terminus of the compound eye. These data indicate the dorsal and lateral portions of the face are characteristically wider in the parthenogenic group and the tentorial pits, clypeus and distal terminus of the epistomal sulcus shifted slightly to a more central position. This differs from

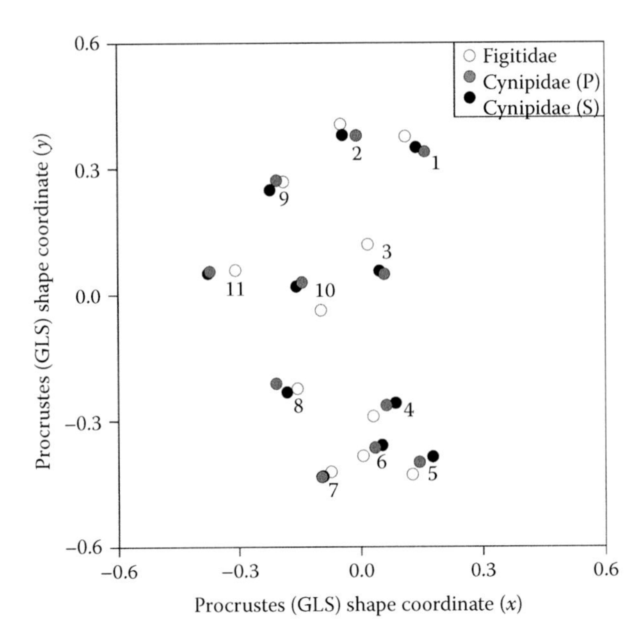

FIGURE 10.14 Generalized least squares (GLS) consensus landmark configurations for cynipid sexual progeny (black symbols), cynipid parthenogenic progeny (grey symbols) and figitid (open symbols) taxa. Landmark numbers as in Figure 10.10B. Use of consensus models mimics the disjoint shape-model approach of Wold (1976), described in Figure 10.4, in which the different group-specific ranges of variation can be thought of as representing a single landmark location (e.g. compare geometry of Figure 10.4 with that of landmark 10) or the distribution of Procrustes-registered shapes on the Kendall shape manifold (Kendall, 1986). A complete analogue to the Wold (1976) method could be realized by computing disjoint relative warp shape models for each landmark data-set. However, results obtained by this investigation (see text) suggest disjoint models based on the simple Procrustes metric can facilitate remarkably consistent and subtle and simultaneous discriminations between intraspecific shape variation and variation between higher taxonomic groups.

the more generalized results provided by the discriminant analysis with respect to intraspecific variation among the cynipids and offers an intriguing new insight into the comparative morphology of these wasps.

In terms of group identification, cross-validation results suggest that the Procrustes distance approach performed nearly as well as discriminant functions with a total of 15 misassignments out of the 99 total specimens (84.85% correct assignments). Interestingly, these errors were not confined to the more difficult distinction between the two cynipid groups, as might be expected, but were spread evenly among all three groups; five figitid species were incorrectly assigned to the cynipid-s group, five cynipid-s species incorrectly assigned to the cynipid-p group and five cynipid-p species incorrectly assigned to the cynipid-s group.

DAISY performance for this most challenging analysis was much more variable than either CVA or Procrustes distance (Table 10.2). The group-identification models constructed by DAISY were able to assign only 70 (70.71%) of the specimens to the correct group. This is much lower than either of the morphometric methods, but still quite respectable given the nature of the problem, character of the data and complete automation of the system. DAISY's figitid model was the most successful, allocating 32 of the 35 (91.43%) true figitid specimens correctly, with the vast majority of those passing the most rigorous coordination test (overall certainty index of 0.96). For the cynipid models, 22 of the 32 (68.75%) cynipid-p species were correctly assigned, with misassigned species overwhelmingly placed in the cynipid-s group. However, performance for the cynipid-s identifications was much poorer, with only 16 of the 32 species (50.00%) correctly assigned. Of these incorrect assignments, the overwhelming majority were placed in the cynipid-p group.

In interpreting these results, it must be remembered that DAISY 'sees' much more variation than morphometric data analysis methods and has no recourse to human intelligence to guide its comparisons. Also, while noting the discrepancy between the morphometric and DAISY results it should also be kept in mind that all quantitative approaches yielded better results than had been obtained by any human systematist to date using just these data, as well as results that compare favorably to estimates of reproducibility in data generated by human experts (typically *c.*70%; see Zachariasse et al., 1978; MacLeod, 1998; Culverhouse, this volume). DAISY encountered difficulties distinguishing between the two cynipid groups because these are genuinely difficult to separate on the basis of an unweighted analysis of all morphological data present in the face. Both morphometric methods yielded better results when compared to DAISY, but relied on the *a priori* selection of a drastically reduced subset of the available information and were much more labor intensive to perform.

IMPROVING NEURAL NET APPROACHES TO AUTOMATED TAXON RECOGNITION

Given the results of the morphometric methods, it seems clear that the subset of possible landmarks chosen for those analyses did a remarkably good job of capturing primary morphological distinctions that can be used to separate these groups to a high degree of reliability, at least insofar as these samples are representative of variation within their parent populations. Such detailed analyses are undoubtedly appropriate when a complete understanding of the geometries involved in group distinction is required. An example of the power of this approach is the rather clear result obtained when consensus configurations of landmarks were compared for the two cynipid groups (Figure 10.14).

However, the large amount of work required to set up morphometric analysis strategies could prove a severe limiting factor. Such systems exist (e.g. Steinhage et al., this volume) but have, to date, been developed for commercially important species (e.g. bees) whose taxonomy is based on characters that are relatively easy to characterize using morphometric approaches (e.g. wing venation patterns). Since there are comparatively few species in which morphometric characters are used routinely to make identifications, if extensive research and technological development need to go into tuning image segmentation algorithms for use in obtaining data for taxonomic group discriminations based on morphometric approaches, this will likely prohibit development of automated taxonomic group identification systems for generalized use. Accordingly, it is important to explore the methods for improving DAISY's performance on such difficult data-sets.

There are two obvious hypotheses that need to be tested to better understand the DAISY results. First, is it the case that variation in these wasp faces is copious and unstructured other than in the regions identified by the morphometric analysis? If so, generalized neural net-based systems may need to be equipped with image preprocessors that identify regions of variation across the sample on which the neural nets can be focused. Such preprocessors would have the advantage of making identification systems not only more efficient, but also more interpretable insofar as alternative geometries of data points or regions of interest could be explored for their ability to contribute to group characterization. Alternatively, it may be the case that the amount of data available to DAISY – rather than its inherent structure – was the reason for the DAISY's lower level of performance relative to the morphometric methods. It has long been known that neural nets work best on large data-sets (see Bishop, 1996; Haykin, 1999; Duda et al., 2000; Lang, this volume), but few recommendations are available regarding how large training sets need to be, and what aspects of the data are relatively important to include for biological training sets.

To evaluate the first hypothesis it is necessary to devise a method to represent the spatial information encoded by localized features in an image-based format acceptable to DAISY. This step is not strictly necessary in terms of rendering data compatible with neural-net processing insofar as standard alpha–numeric symbols can be input directly into any net algorithm. Nevertheless, these morphometric data need to be analyzed by DAISY in order to test its ability to characterize these groups from generalized pixel-based data. Since DAISY is, in effect, an artificially intelligent image-recognition engine, what is needed is a method to represent the morphometric data as an image in order to render the DAISY results more strictly comparable to those obtained from the CVA and Procrustes distance analyses.

To accomplish this transformation, we used a simple scaling function that matches spatial position to an 8-bit greyscale to create a 'landmark image':

$$b_i = (1-(s_{max} - s_i/s_{max} - s_{min}))\cdot 255 \tag{10.1}$$

where b_i = brightness value of the ith shape coordinate, s_i = value of the ith shape coordinate, s_{max} = maximum value of the shape coordinate, and s_{min} = minimum value of the shape coordinate.

The 22-pixel brightness values resulting from this calculation for each specimen were then inserted into a 5- × 5-pixel array (Figure 10.15). Once the morphometric data had been converted into this 'grid-image' form, the three image sets were submitted to DAISY for processing.

Results of this test (Table 10.3) showed a slight increase in the accuracy of the DAISY identifications. Using the same data submitted to the morphometric methods, DAISY assigned 79 of the 99 (79.80%) to the correct group in the cross-tabulation test. Moreover, the coordination statistics indicate that over 60 per cent of these identifications are high quality (>95% certainty). This places the DAISY results within just a few percentage points of matching the morphometric results.[5] On a group-specific basis, all three groups exhibited subequal numbers of misclassifications.

Thus, the raw information used in the morphometric analyses was sufficient to enable the PSOM neural net to characterize these groups adequately and recognize group-level distinctions. Based on these results, it seems clear that – for this data-set at least – focusing the neural net on particular types of data (landmark coordinates) representing particular subregions of the image represents a viable strategy for improving its accuracy. Of course, this was an extreme test of the data-focusing hypotheses. A variety of intermediate image sampling strategies are available (see Figure 10.16 for an example) that would combine heightened sensitivity to the spatial locations of various subregions of interest in a manner that would preserve aspects of the overall spatial information content provided by the image frame.

Regardless, this experiment showed that highly detailed information on the relative locations of taxonomic characters can cause neural nets to yield increased group-characterization/identification power. This increased power, however, comes at a steep price as these operations remove the inherent advantages of raw image-based neural net systems in terms of their scalability and generalizability. In effect, this approach turns neural nets into an alternative form of non-linear multivariate discriminant analysis, with all the limitations associated with such methods (see earlier discussion).

What about sample size? Sample sizes on the order of 30 individuals are usually considered adequate for the characterization of most populations whose measurements that exhibit a normal distribution. Unfortunately, very few morphological features used in systematic research exhibit such well-behaved distributions at low sample sizes. Figure 10.17 gives an indication of the extent of this common problem for the figitid Procrustes (GLS)-registered shape coordinate data.

As indicated before, the traditional method of addressing this issue (collecting more data from the population in question) is not available because the wasp face data were taken from a static, historical image collection. However, a simulation of the effect additional individuals would have on the DAISY results is possible. This can be accomplished by randomly deforming the original wasp image sets to mimic the effect of obtaining new individuals. Because of the linear character of the CVA and Procrustes distance methods, such simulated data-sets will not improve their results. Indeed, those results would be expected to be degraded since the image distortions would inevitably

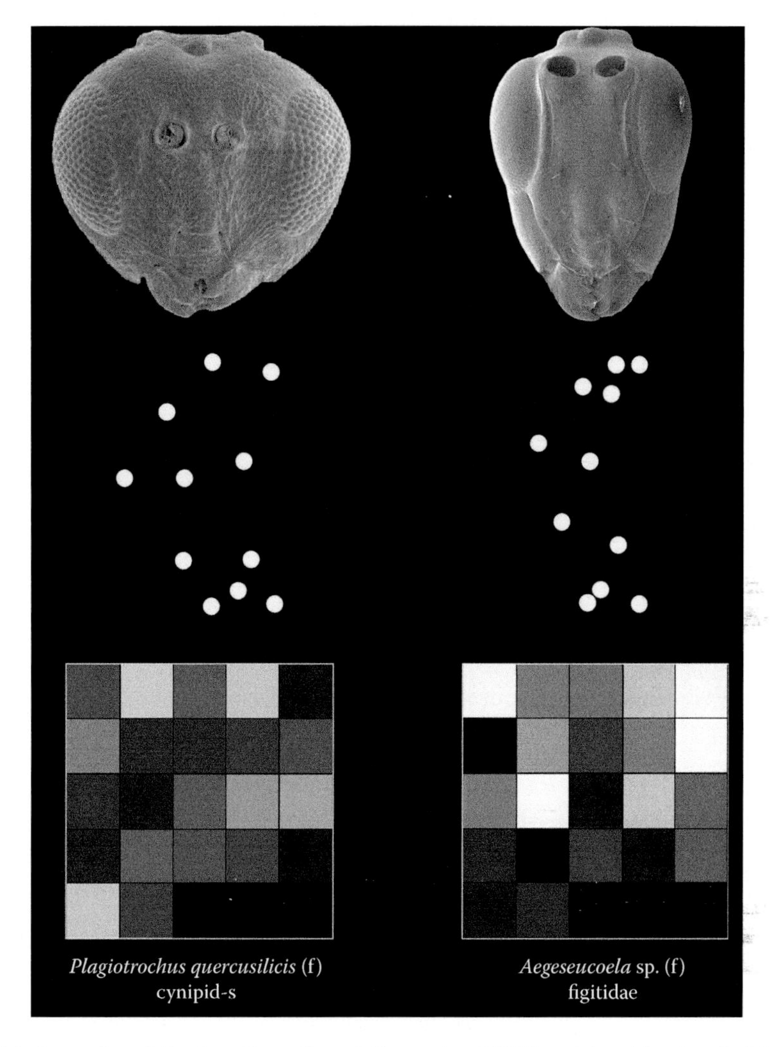

FIGURE 10.15 Examples of the transformation of Procrustes (GLS) registered landmark data into images that can be analyzed by a pixel-based neural net. Upper row: example images of typical cynipid-s (*Plagiotrochus quercusilicis,* left) and figitid (*Aegeseucoela* sp., right) species. Middle row: shape coordinates for the 11 landmarks were used to represent this morphology in the CVA and Procrustes distant analyses. Bottom row: image-based representations of the 11 Procrustes registered landmarks. Note greyscale values are used to represent shape coordinate positions along both x and y shape axes. Images like these were used as input to DAISY to test that system's ability to characterize all three wasp groups. See text for discussion of results.

TABLE 10.3
Cross-Validation Results for DAISY-Grid Analyses

	DAISY (PSOM, *n*-tuple)			
	Cynipid-p	Cynipid-s	Figitid	Total
Cynipid-p	26	4	2	32
Cynipid-s	4	25	3	32
Figitid	2	5	28	35

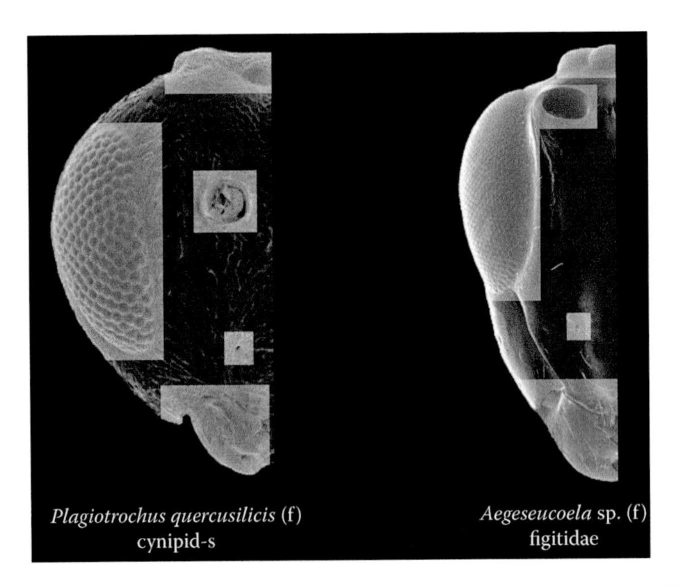

FIGURE 10.16 Use of image masking to highlight corresponding regions of interest within an image. Use of this technique can effectively force neural nets programmed for image processing to focus on particular parts of an image, thus differentially weighting their information content.

lead to placement of a greater number of simulated individuals into the regions of overlap between those models. The non-linear characteristics of neural nets (including DAISY), however, allow additional data to help these systems devise better non-linear rules for group identification. The more subtle and robust nature of the identification statistics also enables DAISY to take advantage of a conservative, but nuanced, assessment of group membership.

To simulate increased sample size, each image from the cynipid-p, cynipid-s and figitid image sets was randomly perturbed or 'morphed' using a distortion factor of ≤10 per cent of the original image's height and/or width dimensions (see Figure 10.18) to form four or five additional images each of the same species. This operation was used to simulate a four- to fivefold increase in overall sample size, with totals of 189, 192 and 216 simulated images for the cynipid-s, cynipid-p and figitid data-sets, respectively (including the original images). The 10 per cent figure represents a liberal, but qualitative, estimate of the amount of within-species morphological variation that can be expected for this group based on comparative inspection of examples of within-species morphological variation. This is, of course, no substitute for the actual acquisition of new specimens and such will be necessary before a system to recognize these wasp groups from facial morphology is finalized. However, these randomly morphed images are sufficient to test the effect of including additional unique images in the training sets. This 'morphed' data-set was then submitted to DAISY analysis in the manner described previously.

Results of the DAISY analysis of this expanded data-set are dramatic (Table 10.4). In the total image set (597 individuals), only one was inappropriately assigned in the cross-tabulation test. This was a morphed cynipid-s image that was assigned to the cynipid-p group. Thus, under increased (simulated) sample size conditions, the DAISY system was able to identify all the images with absolute accuracy. Moreover, 580 (97.15%) of these correctly identified images were assigned a high coordination value (≥3), indicating high-confidence identifications. This result suggests that the non-linear point clouds representing group-specific morphological variation all exhibit good convexity. Accordingly, even in cases where there are very subtle variations in group-specific morphologies with broad areas of overlap between group boundaries, neural nets can likely meet or beat the best results produced by linear morphometric approaches, provided a sufficiently large number of specimens are used to model group-specific patterns of variation.

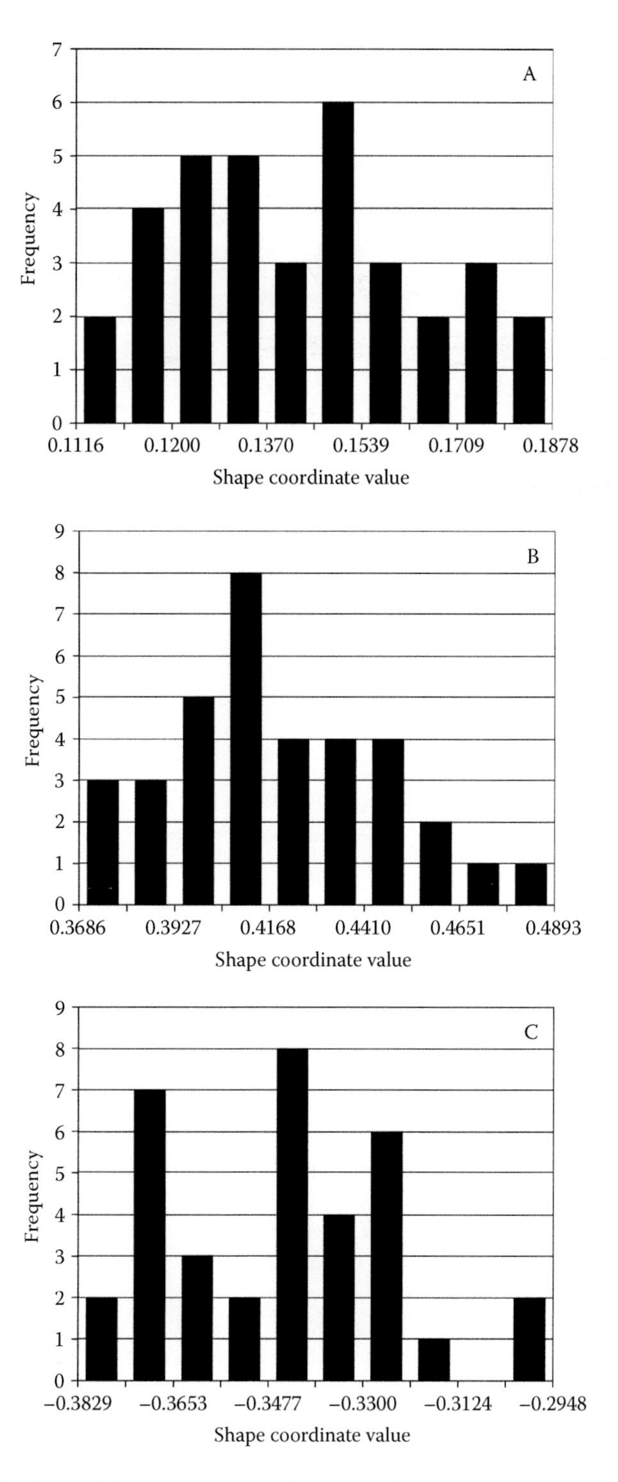

FIGURE 10.17 Distributions of shape coordinate values for various figitid landmarks. A. Medial ocellus (*x*-coordinate). B. Lateral ocellus (*y*-coordinate). C. Distal clypeus terminus along mid-line (*y*-coordinate). Note variety of distribution geometries with only B approximating a normal shape.

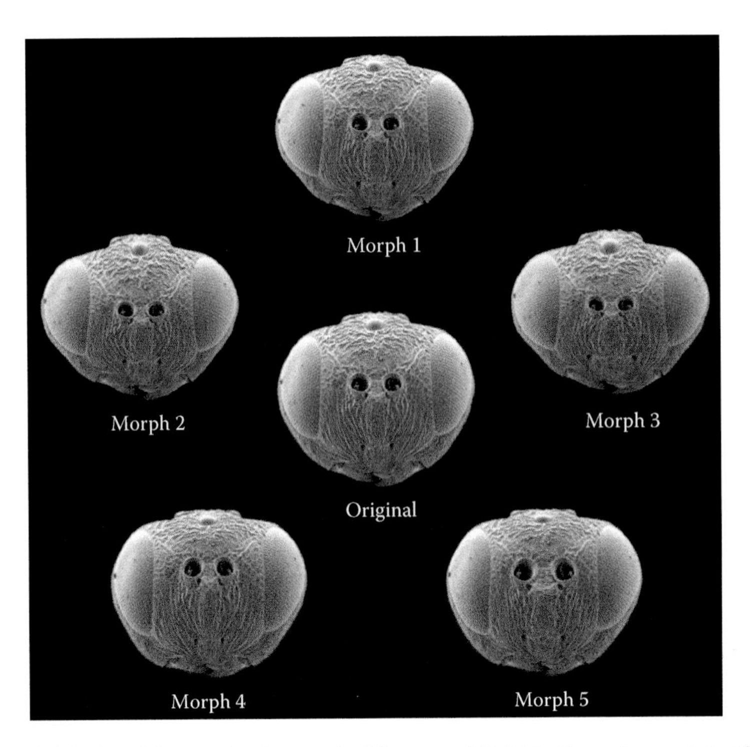

FIGURE 10.18 Original and five randomly morphed images of *Callirhytis erythrocephala* (female, parthenogenic form).

TABLE 10.4
Cross-Validation Results for DAISY-Simulation Analyses

	DAISY (PSOM, *n*-tuple)			
	Cynipid-p	Cynipid-s	Figitid	Total
Cynipid-p	192	—	—	192
Cynipid-s	1	188	—	189
Figitid	—	—	216	216

The implications of this result are clear. Given alternative strategies to focus neural nets on specific aspects of morphological variation or increasing the sample size in order to improve accuracy, the latter will likely be far more effective. This improvement is much simpler to implement and can be accomplished while preserving the scalability and generalizability attributes that made neural nets such an attractive approach for solving the automated object identification problem in the first place. Indeed, given adequate sample sizes, generalized, dynamic neural nets will likely be more effective at the group discrimination task than any alternative approach currently available, as well as being far more rapid, consistent and reliable than either human experts or linear morphometric analysis.

DISCUSSION AND SUMMARY

Because the need for group identification in systematics is so broad, it seems obvious that there is a place for a diversity of different algorithmic approaches. Each taxonomic group and each application of taxonomic data to 'real-world' problems will have their own unique aspects. Consequently, no 'one system fits all' approach is either advocated or anticipated here.

In this study, geometric morphometric approaches to group-identification problems were represented by the linear discriminant analysis of Procrustes-registered shape coordinates and by a new method that made identifications based on the shortest overall Procrustes distance between an unknown object and a set of consensus shape coordinate configurations determined previously from training sets. Both methods performed well in two-group (interspecific) and three-group (interspecific and intraspecific) trials based on an assessment of cynipid and figitid wasp faces; cross-validation results ranged from 85 to 97 per cent correct identifications.

While these are very encouraging results, they remain, in essence, 'benchtop' studies, valid certainly for these samples, but of a presently unknown degree of generalizability. Additional research will be needed to determine whether these methods can be employed on randomly selected samples of real wasp populations – preferably using reflected-light imaging instead of expensive and time-consuming scanning electron photomicrography – to characterize group differences and achieve the same high levels of correct identifications for unknown specimens that are not part of the training sets. This having been said, our results strongly suggest there is no reason to reject the null hypothesis that no shape difference exists between cynipid and figitid wasp faces. Moreover, these results support the tantalizingly strong suspicion that previously unidentified, characteristic facial differences exist, not only between gall wasp families, but also, possibly, between the sexual and parthenogenic morphs of cynipid wasp species.

With respect to comparisons between the methods themselves, discriminant analysis exhibited slightly better overall performance than the new Procrustes distance strategy, though it is debatable whether the difference is significant. [Note: for the two-group test these methods differed in terms of correct identifications by a single specimen.] What is significant, though, is the fact that the classic discriminant strategy carries a substantial computational overhead and cannot be extended to the inclusion of new groups without recomputing the entire discriminant space. Discriminant analyses are also heir to the 'curse of dimensionality' (the fact that sample sizes need to increase exponentially to provide the same level of characterization information to discrimination procedures as the number of variables increases; see Belman, 1961) and, in standard modes of operation, cannot recognize specimens that do not fit into any of the predefined groupings. These factors explain the tendency of many discriminant approaches to deliver false positive identifications, especially as the number of possible groups increases (e.g. Duda et al., 2000). For these reasons the classical discriminant approach will likely find its best use in the context of group discriminations of limited scope and where absolute accuracy in both species richness and overall diversity is critical.

The Procrustes distance method is more generalized, arguably more conservative, and more flexible than the classical discriminant analysis approach. Based on Wold's (1976) idea of calculating disjoint models of shape variation for each group, this approach is eminently scalable in terms of adding groups to the system with the (important) limitation that the same set of landmarks needs to be able to be recognized in all species included in an analysis (a constraint that also limits discriminant analysis). In addition, the Procrustes distance method is very easy to apply and provides direct information about the geometric nature of between-groups differences. This is not the case for discriminant analysis, where the geometry of the discriminant space is distorted to achieve maximal distinction with regard to within-groups and between-groups variation.

As pointed out in Wold's (1976) original discussion of the disjoint model concept, the Procrustes distance method can be set up to impose limits automatically on the range of shape variation allowable under each group-specific model, assessed, of course, via reference to statistical samples of the variation characteristic of each group. Such limits can be used as a guide for the interpretation of an overall Procrustes distance or on a landmark-by-landmark basis. Use of this empirically determined 'tolerance envelope' approach would allow the Procrustes distance method to recognize the presence of specimens that do not adequately conform to any group model (e.g. aberrant specimens, members of new species). In terms of the purposes of systematic investigations, this is a crucial advantage.

As with the CVA approach, the 'curse of dimensionality' would operate on Procrustes distance methods as a result of the expected exponential increase in sample sizes needed to specify the disjoint models required to characterize large numbers of groups, but this would be mitigated somewhat as a result of there being progressively fewer common measurements that could be collected in order to characterize morphologically dissimilar groups – especially if strict topological correspondence were desired. Of course, this latter tendency would place a severe restriction on the taxonomic scope of problems that can be considered using morphometric approaches.

Neural nets (at least those based on PSOM strategies) represent a radically different and fully automated approach to group identification that is similar in many ways to the Procrustes distance method. Like Procrustes distance, PSOM-based neural nets such as DAISY base their characterization metrics on the matchings of within-groups features without trying to achieve any optimization of the between-groups space. As the PSOM within-groups disjoint models become more distinct they cause other models within the space to 'drift away', but no attempt is made to enforce any directional optimization of this drift globally (as is the case with discriminant analysis). These features account for the scalability of PSOM-neural nets and Procrustes distance approaches. Unlike both morphometric methods, though, PSOM-based neural nets are able to process generalized morphological data – indeed, digital data of any type. This is a very substantial advantage overall because it frees the net to compare morphologies of any type with one another, not just those on whose surfaces the same subset of landmarks can be located.

For most interspecific studies this latter feature works to the net's – and the systematist's – advantage. DAISY's performance in the two-group analysis, while less precise than those of the morphometric methods under the same sample size constraints, was well within the range of acceptability for routine assessments of both richness and diversity. Given the speed with which DAISY was able to complete its analyses and its fully automated nature, PSOM neural nets have clear and practical advantages over both qualitative and morphometric approaches for most routine group-identification needs.

DAISY performed as well as could be expected of human experts in terms of making fine distinctions between all three cynipid groups, though less well than the morphometric methods. This was undoubtedly due to the fact that the groups being characterized embodied both superspecific and subspecific levels of morphological distinction. Such data-sets will always include more conflicts among the features used to summarize between-groups distinctions than would be the case if the between-groups morphological distinctions were more clearly drawn. Presentation of only the landmark data used by the morphometric methods to recognize between-groups distinctions improved the PSOM results to the level of those delivered by the morphometric methods. However, adopting this strategy as a way of improving the performance of neural-net applications in systematics forces the latter down the self-limiting pathways trod by the morphometric approaches.

Certainly there is scope for adding image-processing capabilities to group-identification systems in order to better focus their attention on aspects of the morphology already known to be important for group identification (see Steinhag et al., this volume, for an advanced example of the morphometric approach). Such system designs can and will be developed for specialized needs where optimal accuracy is required (e.g. medical and forensic applications) and where the characters being assessed lend themselves to study using simple image-segmentation algorithms (e.g. insect wing venation patterns). Nevertheless, we very much doubt such approaches will be able to serve as generalized platforms for the routine recognition of thousands of species, few of which will have any characteristics in common. For these situations, a more general-purpose solution is needed.

The DAISY results described earlier suggest that PSOM-like, n-tuple neural nets can provide the platform needed for construction of generalized group-recognition engines, provided sample sizes are large enough to permit robust, non-linear models of within-groups morphological variation to be developed. Indeed, DAISY's group recognition results for the simulated large-size data-set were the best of any method applied to these data. This result is all the more remarkable when it is realized that DAISY required *no guidance* regarding which aspects of the morphology were important for drawing group-level distinctions.

PROSPECTUS

Properly designed, flexible and robust automated identification systems organized around distributed computing architectures and referenced to authoritatively identified collections of training set data (e.g. images, gene sequences) can provide all systematists with the analytic tools necessary to handle routine identifications of common taxa. The systems available today, such as DAISY, SPIDA, ABIS and others, represent excellent starting points that can be used to test concepts and build even better systems. This study points up some exciting new directions that can be explored to help resolve long-standing issues with automated taxon recognition systems, especially through development of a dialogue between neural net and morphometric data-analysis strategies.

Ironically, the preprocessing of image data-sets based on generalized input (e.g. pixels) for the purpose of reducing within-groups variation is a classic neural-net problem and can be introduced without sacrificing the automated nature of the net's core function *or* its generalizability. The generalized nature of the neural net's input data makes the space within which its discriminations are made inherently high dimensional (but whose dimensionality, once optimized, can effectively remain static irrespective of the number of groups or distinctiveness of the morphologies analysed). Nevertheless, when data sets that have high dimensionality are submitted to a neural net, most of the net's resources go into characterizing parts of the space that are irrelevant to the central problem of intergroup discrimination. If subsets of the generalized information can be identified by preprocessing (see Bookstein, this volume, for a proposed technique whereby this could be accomplished), the net can focus on those aspects of the variation pertinent to the discrimination problem. However, even in the absence of advanced processing features such as this, the present generation of generalized neural net-based systems are eminently practical for making richness assessments of biota containing both obviously and subtly distinct taxa.

Despite decades of research into the general issue of automated species identification, and the existence of a significant literature on this topic, most systematists continue to believe that construction of a machine that can identify taxa quickly, consistently, on demand, in large quantities and with higher accuracy than human experts remains a science-fiction fantasy. We respectfully disagree. There is already much solid evidence that, for particular species groups, such systems not only can be constructed, but have been constructed. These systems do not rely on problematic exotica for their observations (e.g. DNA sequences), but rather the common observation of morphology that systematists have been using to make taxonomic identifications for millennia. This remarkable research has been done with scant publicity and little direct funding from the major governmental and private supporters of biological research even though most organism-based research projects are absolutely dependent on access to high-quality, consistent taxon identifications.

The challenge to the systematics community no longer lies in the need to turn the science-fiction fantasy into reality (e.g. Janzen, 2004). Rather, it lies in (1) encouraging (and funding) the further development and testing of systems that already exist and techniques that have been proven to work, (2) populating available systems with sufficient numbers of groups (as well as a sufficient number of examples of each group, even in the case of rare species) to make them useful, (3) encouraging new systematists to participate in the interdisciplinary teams that will be required to extend this research programme and (4) embedding current and future taxon identification systems in an electronic infrastructure that will deliver their benefits to those who need them most: practicing taxonomists and parataxonomists, especially those working in the third world where biodiversity assessment problems are most acute.

But even more than this, there is a need to change the mindset of the systematics community, many of whose members currently see research into such systems as an idiosyncratic diversion from mainstream systematics or as a threat to the employment prospects of future taxonomists. These are not credible issues. Systematics has suffered, and continues to suffer, from an image of being out of step with a postmodern world that expects such systems to be developed, as they have in many other fields. The ability of systematics and taxonomy to confront and successfully contribute to the resolution of many of the major biological and ecological questions of the twenty-first century would not only benefit from the existence of such systems, *but also requires that they exist.* The alternative of simply training more systematists to provide the required number of rapid, high-quality and consistent taxonomic identifications is not a realistic option either in principle or in practice.

ACKNOWLEDGEMENTS

A preliminary version of this chapter was commented on by two anonymous reviewers whose efforts are acknowledged here and greatly appreciated. Development of the DAISY system has benefited from the input of many students and colleagues over the years. We thank them here as a group for their work on behalf of the goal – and the vision – of achieving reliable, automated, taxonomic identifications by means of artificial intelligence and machine vision.

NOTES

1. See Klingenberg and Montiero (2005) for a CVA variant that circumvents this problem to some extent; experiments with partial least squares comparisons of CVA scores to shape data suggest this approach may also be used to obtain heuristic linear models of the CVA space.
2. Common principle component analysis (cPCA) (see Flury, 1987, 1988) is a multigroup variant of classical PCA that seeks to estimate the major axes of variation that multiple groups have in common with one another, but is not an approach to the group characterization and/or specimen-identification problems *per se*.
3. If neural nets are trained using multimodal methods – for example, simulated annealing or genetic algorithm – there is a far better chance of finding a global minimum.
4. An alternative, and possibly more conservative, formulation of this test would be to construct 2σ tolerance envelopes for each landmark.
5. A separate implementation of the DAISY algorithm designed to accept numeric (rather than image) input was also run and yielded virtually identical results.

REFERENCES

Bello, E., Becerra, J.M. and Valdecasas, A. (1995) The burden is description not identification. *Trends in Ecology and Evolution,* 10: 416–417.

Belman, R. (1981) *Adaptive Control Processes: A Guided Tour,* Princeton University Press, Princeton, NJ, 255 pp.

Bishop, C.M. (1996) *Neural Networks for Pattern Recognition,* Oxford University Press, Oxford, 504 pp.

Blackith, R.E. and Reyment, R.A. (1991) *Multivariate Morphometrics,* Academic Press, London, 412 pp.

Bookstein, F.L. (1991) *Morphometric Tools for Landmark Data: Geometry and Biology,* Cambridge University Press, Cambridge, 435 pp.

Bookstein, F.L. (1993) A brief history of the morphometric synthesis. In *Contributions to Morphometrics* (eds L.F. Marcus, E. Bello and A. García-Valdecasas), Museo Nacional de Ciendcias Naturales 8, Madrid, pp. 18–40.

Bookstein, F.L. and Green, W.D.K. (1993) A feature space for edgels in images with landmarks. *Journal of Mathematical Imaging and Vision,* 3: 213–261.

Campbell, N.A. and Atchley, W.R. (1981) The geometry of canonical variate analysis. *Systematic Zoology,* 30: 268–280.

Chapill, J.A. (1989) Quantitative characters in phylogenetic analysis. *Cladistics,* 5: 217–234.

Cracraft, J. (2002) The seven great questions of systematic biology: an essential foundation for conservation and sustainable use of biodiversity. *Annals of the Missouri Botanical Garden,* 89: 303–304.

Cranston, P.S. and Humphries, C.J. (1988) Cladistics and computers: a chironomid conundrum. *Cladistics,* 4: 72–92.

Crowe, T.M. (1994) Morphometrics, phylogenetic models and cladistics: means to an end or much ado about nothing? *Cladistics,* 10: 77–84.

Culverhouse, P.F., Williams, R., Reguera, B., Herry, V. and González-Gils, S. (2003) Do experts make mistakes? *Marine Ecology Progress Series,* 247: 17–25.

Dahl, C., Wold, S., Nielsen, L.T. and Nilsson, C. (1984) Simca pattern recognition study in taxonomy: claw shape in mosquitoes (Culicidae, Insecta). *Systematic Zoology,* 33: 355–369.

Davis, J.C. (2002) *Statistics and Data Analysis in Geology,* 3rd edition, John Wiley & Sons, New York, 638 pp.

Donoghue, M.J., Doyle, J.A., Gauthier, J., Kluge, A. and Rowe, T. (1989) The importance of fossils in phylogeny reconstruction. *Annual Review of Ecology and Systematics,* 20: 431–460.

Duda, R.O., Hart, P.E. and Stork, D.G. (2000) *Pattern Classification,* Wiley-Interscience, New York, 654 pp.

Ebach, M.C. and Holdrege, C. (2005) DNA barcoding is no substitute for taxonomy. *Nature,* 434: 697.

Eldredge, N. and Cracraft, J. (1980) *Phylogenetic Patterns and the Evolutionary Process,* Columbia University Press, New York, 349 pp.

Felsenstein, J. (2003) *Inferring Phylogenies,* Sinauer Associates, Sunderland, MA, 664 pp.

Fisher, R.A. (1936) The utilization of multiple measurements in taxonomic problems. *Annals of Eugenics,* 7: 179–188.

Flury, B.K. (1987) A hierarchy of relationships between covariance matrices. In *Advances in Multivariate Statistical Analyses* (ed A.K. Gupta), Reidel, Dordrecht, the Netherlands, pp. 31–43.

Flury, B.K. (1988) *Common Principal Components and Related Multivariate Models,* Wiley, New York, 258 pp.

Gaston, K.J. and O'Neill, M.A. (2004) Automated species identification – Why not? *Philosophical Transactions of the Royal Society of London, Series B,* 359: 655–667.

Gauthier, J., Kluge, A.G. and Rowe, T. (1988) Amniote phylogeny and the importance of fossils. *Cladistics,* 4: 105–209.

Godfray, H.C.J. (2002) Challenges for taxonomy. *Nature,* 417: 17–19.

Gong, S., McKenna, S.J. and Psarrou, A. (2000) *Dynamic Vision: From Images to Face Recognition,* World Scientific Publishing Co., Hackensack, NJ, 344 pp.

Haykin, S. (1999) *Neural Networks, a Comprehensive Foundation,* 2nd edition, Prentice Hall, New York, 852 pp.

Hebert, P.D.N., Penton, E.H., Burns, J.M., Janzen, D.H. and Hallwacks, W. (2005) Ten species in one: DNA barcoding reveals cryptic species in the neotropical skipper butterfly *Astaptes fulgerator. Proceedings of the National Academy of Sciences,* 101: 14812–14817.

Hotelling, H. (1933) Analysis of a complex of statistical variables into principal components. *Journal of Educational Psychology,* 24: 417–441.

Janzen, D.H. (2004) Now is the time. *Philosophical Transactions of the Royal Society of London, Series B,* 359: 731–732.

Kaesler, R.L. (1993) A window of opportunity: peering into a new century of paleontology. *Journal of Paleontology,* 67: 329–333.

Kendall, D.G. (1984) Shape manifolds, procrustean metrics and complex projective spaces. *Bulletin of the London Mathematical Society,* 16: 81–121.

Klingenberg, C.P. and Monteiro, A.R. (2005) Distances and directions in multidimensional shape spaces: implications for morphometric applications. *Systematic Biology,* 54: 678–688.

Kohonen, T. (1982) Self-organized formation of topologically correct feature maps. *Biological Cybernetics,* 43: 59–69.

Kohonen, T. (1984) *Self-Organization and Associative Memory,* Springer–Verlag, New York, 312 pp.

Lang, R. and Warwick, K. (2002) The plastic self-organizing map. In *Proceedings of the International Joint Conference on Neural Networks,* Society, I.C. Institute of Electrical & Electronics Engineers, Honolulu, HI, pp. 727–732.

MacLeod, N. (1998) Impacts and marine invertebrate extinctions. In *Meteorites: Flux with Time and Impact Effects* (eds M.M. Grady, R. Hutchinson, G.J.H. McCall and D.A. Rotherby), Geological Society of London, London, pp. 217–246.

MacLeod, N. (1999) Generalizing and extending the eigenshape method of shape visualization and analysis. *Paleobiology,* 25: 107–138.

MacLeod, N. (2002a) Phylogenetic signals in morphometric data. In *Morphology, Shape and Phylogeny* (eds N. MacLeod and P.L. Forey), Taylor & Francis, London, pp. 100–138.

MacLeod, N. (2002b) Geometric morphometrics and geological form-classification systems. *Earth-Science Reviews,* 59: 27–47.

MacLeod, N. (2005a). Shape models as a basis for morphological analysis in paleobiological systematics: dicotyledenous leaf physiography. *Bulletins of American Paleontology,* 369: 219–238.

MacLeod, N. (2005b). Principal components analysis (eigenanalysis & regression 5). *Palaeontological Association Newsletter,* 59: 42–54.

MacLeod, N., O'Neill, M.A. and Walsh, S.A. (2007) A comparison between morphometric and artificial neural net approaches to the automated species-recognition problem in systematics. In *Biodiversity Databases: From Cottage Industry to Industrial Network* (eds G. Curry and C. Humphries), Taylor & Francis, London, pp. 37–62.

Manley, B.F.J. (1994) *Multivariate Statistical Methods: A Primer,* Chapman & Hall, London, 215 pp.

Manley, B.F.J. (1997) *Randomization, Bootstrap and Monte Carlo Methods in Biology,* Chapman & Hall, London, 399 pp.

Mardia, K.V., Kent, J.T. and Bibby, J.M. (1979) *Multivariate Analysis,* Academic Press, Inc., New York, 521 pp.

Mayr, E. (1969) *Principles of Systematic Zoology,* McGraw–Hill, New York, 234 pp.

Mayr, E. (1982) *The Growth of Biological Thought: Diversity, Evolution, and Inheritance,* Harvard University Press, Cambridge, MA, 974 pp.

Pearson, K. (1901) On lines and planes of closest fit to a system of points in space. *Philosophical Magazine,* 2: 557–572.

Pimentel, R.A. (1979) *Morphometrics: The Multivariate Analysis of Biological Data,* Kendall/Hunt, Dubuque, IA, 275 pp.

Pimentel, R.A. and Riggins, R. (1987) The nature of cladistic data. *Cladistics,* 3: 201–209.

Polly, P.D. and Head, J.J. (2004) Maximum likelihood identification of fossils: taxonomic identification of quaternary marmots (Rodentia, Mammalia) and identification of vertebral position in the pipesnake *Cylindrophis* (Serpentes, Reptilia). In *Morphometrics – Applications in Biology and Paleontology* (ed A.M.T. Elewa), Springer–Verlag, Heidelberg, pp. 197–222.

Rao, C.R. (1952) *Advanced Statistical Methods in Biometric Research,* Wiley, New York, 390 pp.

Reyment, R.A. (1971) *Quantitative Palaeoecology,* Elsevier, Amsterdam, 226 pp.

Reyment, R.A. (1991) *Multidimensional Paleobiology,* Pergamon Press, Oxford, 539 pp.

Reyment, R.A., Blackith, R.E. and Campbell, N.A. (1984) *Multivariate Morphometrics,* 2nd edition, Academic Press, London, 231 pp.

Ritchie, A.J. (1993) Superfamily Cynipoidea. Pp. 521–530 In *Hymenoptera of the World: An Identification Guide to Families* (eds H. Goulet and J.T. Huber), Canada Communications Group, Ottawa, Canada.

Rohlf, F.J. (1990) Rotational fit (Procrustes) methods. In *Proceedings of the Michigan Morphometrics Workshop* (eds F.J. Rohlf and F.L. Bookstein), The University of Michigan Museum of Zoology, Special Publication No. 2, Ann Arbor, pp. 227–236.

Salemi, M. and Vandamme, A.-M. (2003) *Phylogenetic Handbook: A Practical Approach to DNA and Protein Phylogeny,* Cambridge University Press, Cambridge, 430 pp.

Simpson, G.G. (1961) *Principles of Animal Taxonomy,* Columbia University Press, New York, 247 pp.

Simpson, G.G., Roe, A. and Lewontin, R.C. (2003) *Quantitative Zoology,* revised edition, Dover Publications, New York, 454 pp.

Sneath, P.H.A. and Sokal, R.R. (1973) *Numerical Taxonomy: The Principles and Practice of Numerical Classification,* W. H. Freeman, San Francisco, 573 pp.

Sokal, R.R. and Sneath, P.A. (1963) *Principles of Numerical Taxonomy,* W. H. Freeman, San Francisco, 359 pp.

Tautz, D., Arctander, P., Minelli, A., Thomas, R.H. and Vogler, A.P. (2003) A plea for DNA taxonomy. *Trends in Ecology and Evolution,* 18: 70–74.

Weeks, P.J.D., Gauld, I.D., Gaston, K.J. and O'Neill, M.A. (1997) Automating the identification of insects: a new solution to an old problem. *Bulletin of Entomological Research,* 87: 203–211.

Weeks, P.J.D., O'Neill, M.A., Gaston, K.J., and Gauld, I.D. (1999a) Automating insect identification: exploring limitations of a prototype system. *Journal of Applied Entomology,* 123: 1–8.

Weeks, P.J.D., O'Neill, M.A., Gaston, K.J. and Gauld, I.D. (1999b) Species identification of wasps using principle component associative memories. *Image and Vision Computing,* 17: 861–866.

Wheeler, Q.D. (2003) Transforming taxonomy. *The Systematist,* 22: 3–5.

Wheeler, Q.D. (2005) Losing the plot: DNA 'barcodes' and taxonomy. *Cladistics,* 21: 405–407.

Wilson, E.O. (1986) *Biophilia,* Harvard University Press, Cambridge, MA, 111 pp.

Wold, S. (1976) Pattern recognition by means of disjoint principal component models. *Pattern Recognition,* 8: 127–138.

Wold, S., Albano, C., Dunn, W.J., III, Esbensen, K., Hellberg, S., Hohansson, E. and Sjöström, M. (1983) Pattern recognition: finding and using regularities in multivariate data. In *Food Research and Data Analysis* (eds H. Martens and M. Ruswurm), Applied Science Publications, London, pp. 147–188.

Wold, S. and Sjöström, M. (1977) SIMCA: a method for analyzing chemical data in terms of similarity and analogy. In *Chemometrics: Theory and Application* (ed. B.R. Kowalski), American Chemical Society, Washington, DC, pp. 243–262.

Zachariasse, W.J., Riedel, W.R., Sanfilippo, A., Schmidt, R.R., Brolsma, M.J., Schrader, H.J., Gersonde, R., Drooger, M.M. and Brokeman, J.A. (1978) Micropaleontological counting methods and techniques – an exercise on an eight meters section of the Lower Pliocene of Capo Rossello, Sicily. *Utrecht Micropaleontological Bulletins,* 17: 1–265.

Zelditch, M.L., Swiderski, D.L., Sheets, H.D. and Fink, W.L. (2004) *Geometric Morphometrics for Biologists: A Primer,* Elsevier/Academic Press, Amsterdam, 443 pp.

11 Pattern Recognition for Ecological Science and Environmental Monitoring: An Initial Report

Eric N. Mortensen, Enrique L. Delgado, Hongli Deng, David Lytle, Andrew Moldenke, Robert Paasch, Linda Shapiro, Pengcheng Wu, Wei Zhang and Thomas G. Dietterich

CONTENTS

Introduction 190
Project Overview 190
Recognizing Stonefly Larvae 191
Mechanical Manipulation and Imaging 192
Soil Mesofauna 192
Mechanical Manipulation and Imaging 195
Classification Framework 195
Segmentation 195
Coarse Classification 196
Fine Classification 196
Recent Methods for Object Recognition 196
Step 1: Apply a 'Region Detector' to the Input Image 197
Step 2: Represent Each Region by a Fixed-Length 'Descriptor' 197
Step 3: Apply a Classifier to the Bag of Descriptor Vectors 197
Classification Methods 198
Region Detectors 198
Region Descriptors 198
Defining Object Parts by Clustering Descriptor Vectors 200
Converting Sets of SIFT Vectors into Standard Feature Vectors 200
Training the Classifier 201
Results and Discussion 201
Concluding Remarks and Future Work 203
Acknowledgements 204
Notes 204
References 204

INTRODUCTION

Progress in the ecological sciences is limited by the lack of high-resolution sensing of either the abiotic or the biotic biosphere. While remote sensing from satellites or aircraft provides valuable information about gross spatial distribution by organism type (e.g. broadleaf vs. needle-leaf trees), high-resolution measurement of organism population sizes and spatiotemporal distribution is virtually impossible to acquire. Existing methods for obtaining population counts involve the manual collection and identification of specimens by human experts, which is too costly to provide ongoing high-resolution data. Among the many technologies being developed to address this problem, pattern recognition from image data is one of the most promising.

This chapter presents the results to date in an ongoing project by a multidisciplinary team of computer scientists, entomologists and mechanical engineers to develop high-throughput methodologies for the identification and classification of insects. Mechanical devices for automatically photographing insect specimens have been developed along with general-purpose pattern-recognition algorithms for classifying these specimens to genus or species levels. These methods and devices are applied to two important scientific, environmental and agricultural problems: (1) water quality monitoring in streams (by recognizing and counting stonefly larvae) and (2) measurement and characterization of soil biodiversity (by recognizing and counting soil mesofauna).

A fundamental scientific challenge for computer science research is to develop general-purpose pattern-recognition methods that can be applied to many different classification problems without requiring manual redesign for each new task. Some existing pattern-recognition methods in systematics require carefully designed feature extraction and/or classification algorithms for each new application (Roth et al., 1999; Steinhage et al., 2001; Jalba et al., 2005). Consequently, each new application requires substantial time and expertise to construct. Our work addresses this challenge by developing a robust pattern-recognition system that can be applied without modification to various situations.

A second fundamental challenge is to develop pattern-recognition methods that can handle highly articulated three-dimensional objects. Many existing pattern-recognition methods are largely limited to either objects or object parts that are roughly two dimensional (e.g. insect wings; see Roth et al., 1999; Steinhage et al., 2001) or to specific views of semirigid three-dimensional object parts (e.g. human faces, spider genitalia, etc.; see Turk and Pentland, 1991). The insects studied in this project are three-dimensional objects with many articulated parts (legs, antennae, abdomen, tails, etc.) that cannot be reliably placed into consistent poses. To address this challenge, we are applying recently developed computer vision techniques that detect distinctive image regions and represent them in ways that capture important invariants that are then combined to classify the specimens.

PROJECT OVERVIEW

While the environmental monitoring tasks of identifying stonefly larvae and counting soil mesofauna populations have quite different characteristics, our goal is to develop a robust pattern-recognition system that can adapt to new identification tasks simply by relearning each new domain from a set of training data. Here, we present an overview of the two tasks and our approach to classification. Note that this overview discusses both completed and in-progress work.

Figure 11.1 illustrates the entire classification system from imaging of specimens to taxonomic identification. The stonefly larvae and soil mesofauna are prepared and mechanically manipulated for imaging using different methods and mechanical hardware. Likewise, the software control of the mechanical apparatus is, by necessity, performed by different modules within the integrated imaging software. Images captured by the digital camera are first segmented to identify the image regions belonging to specimens and to separate the specimens from the background. Each foreground region in the segmented images (ideally corresponding to individual specimens) is then categorized by means of a coarse classification that groups specimens using simple-to-compute

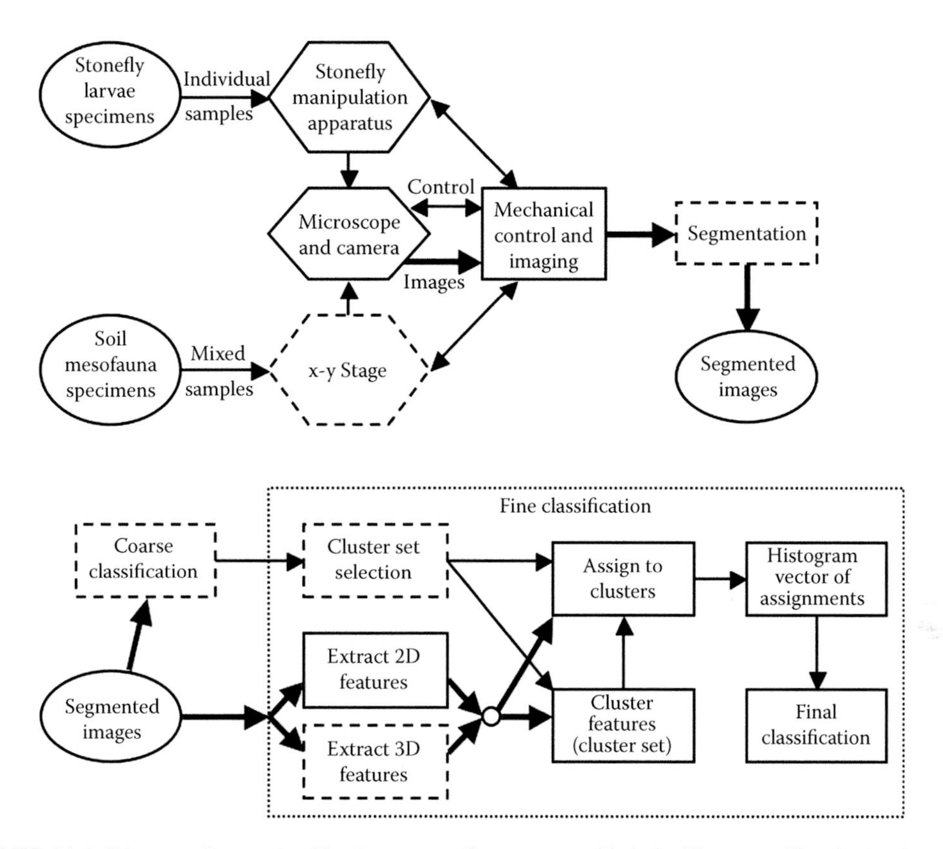

FIGURE 11.1 Diagram of insect classification system. Components with dashed lines are still under development.

object properties (e.g. eccentricity, color histograms, compactness). The segmented images and the specimen's coarse grouping are then employed for fine classification, where the goal is to identify each specimen to the species level, though in some cases classification to genus or even just to family is beneficial. As this project is still a work in progress, system components that are under development have dashed outlines.

The remainder of this section discusses the two application problems: recognizing stonefly larvae and classifying soil mesofauna. For each application, we describe the motivation for choosing the taxonomic identification task and the methods for manipulating specimens and capturing images.

RECOGNIZING STONEFLY LARVAE

Stream water quality measurement could be revolutionized if an economically practical method were available for monitoring aquatic insect populations. Since species differ in their water quality requirements, population counts of stonefly (*Plecoptera*) larvae and other aquatic insects inhabiting stream substrates are known to be a sensitive and robust indicator of stream health and water quality (Johnson et al., 1993; Resh and Jackson, 1993). Consequently, changes in water quality can be tracked by monitoring changes in aquatic insect community composition. Because aquatic insects integrate stream water quality over time, they provide a more reliable measure of water quality than single-time-point chemical measurements.

Aquatic insects are especially useful as biomonitors because (1) they are found in nearly all running-water habitats, (2) their large species diversity offers a wide range of responses to water quality change, (3) the taxonomy of most groups is well known and identification keys are available, (4) responses of many species to different types of pollution have been established and (5) data analysis methods for aquatic insect communities are available (Resh et al., 1996). Because of these advantages, biomonitoring using aquatic insects is routinely employed by federal, state, local, tribal and private resource managers to track changes in river and stream health and to establish baseline criteria for water quality standards. Collection of aquatic insect samples for biomonitoring is inexpensive and requires relatively little technical training. However, the sorting and identification of insect specimens can be extremely time consuming and requires substantial technical expertise. As a result, aquatic insect identification is a major technical bottleneck for large-scale implementation of biomonitoring.

Larval stoneflies are especially important for biomonitoring because they are sensitive to reductions in water quality caused by thermal pollution, eutrophication, sedimentation and chemical pollution. On a scale of organic pollution tolerance from 0 to 10, with 10 being the most tolerant, most stonefly taxa have a value of 0, 1 or 2 (Hilsenhoff, 1988). Because of their low tolerance to pollution, change in stonefly abundance or taxonomic composition is often the first indication of water quality degradation. Most biomonitoring programs identify stoneflies to the taxonomic resolution of family, although when expertise is available, genus-level (and occasionally species-level) identification is possible. Unfortunately, because of constraints on time, budgets and availability of expertise, some biomonitoring programs fail to resolve stoneflies (as well as other taxa) below the level of order. This results in a considerable loss of information and, potentially, in the failure to detect changes in water quality.

Although automated identification of all types of aquatic insects is a long-term goal of this research agenda, stoneflies are an ideal model group for the development of these methods for several reasons. First, they encompass a wide range of identification challenges, from very easy (highly patterned, distinctive species) to very difficult (species complexes of nearly indistinguishable taxa). Second, ontological changes in body size, patterning and allometry present an identification challenge that must be overcome for any automated technique to be viable. Third, local and regional variability within species provides even further challenges for training identification algorithms. Finally, automated identification of stoneflies will be immediately useful to biomonitoring programs even before the technique is available for other aquatic insect orders.

Mechanical Manipulation and Imaging

A fundamental problem in pattern recognition is to exploit variability between categories (e.g. taxa) while eliminating variability within categories. One important source of variability that can be eliminated is variation during image capture. To achieve consistent, repeatable image capture, we have designed and constructed a software-controlled mechanical stonefly larval transport and imaging apparatus that positions specimens under a microscope, rotates them (to obtain views from various angles) and photographs them with a digital camera. Using this apparatus, imaging rates of a few tens of specimens per hour can be achieved. A minimum of eight images (from different viewing angles) is taken of each specimen. The imaging apparatus has a series of mirrors so that each image acquires two simultaneous views of a specimen from approximately 90° apart. Light diffusers reduce glare and eliminate hard shadows. In summary, the apparatus can quickly acquire several images of a specimen from various angles with consistent imaging conditions across specimens and species. Figure 11.2 shows the imaging apparatus, including the mirror setup used for acquiring two simultaneous images of each specimen. Figure 11.3 shows example images obtained using the imaging assembly.

SOIL MESOFAUNA

Agricultural and forest management is hampered by a lack of cost-effective methods for measuring insect populations and insect biodiversity. Such measurements can help society understand the impact of various forest and agricultural management practices on ecosystem health. Agricultural

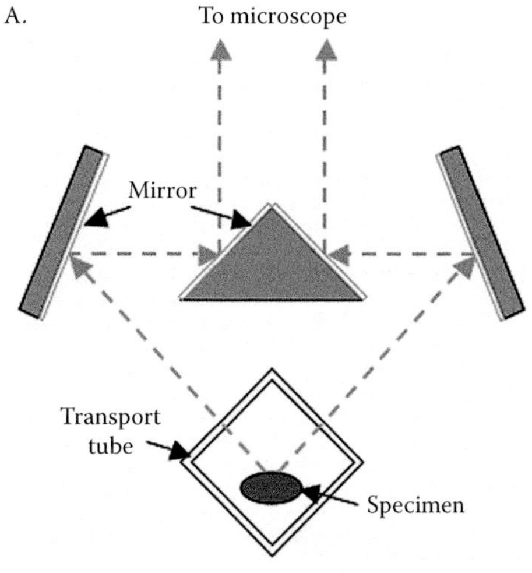

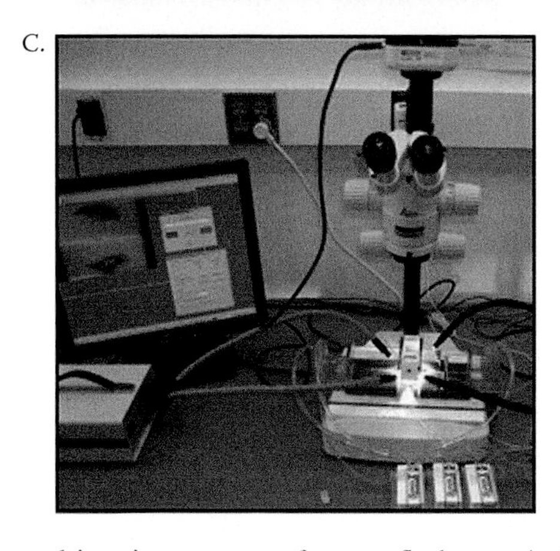

FIGURE 11.2 Transportation and imaging apparatus for stonefly larvae. A. Diagram of mirror system for obtaining two simultaneous views of a specimen (from approximately 90° apart) in a single image. B. Image of prototype mirror and transportation apparatus. C. Image of entire stonefly transportation and imaging setup (with microscope and attached digital camera, light boxes and computer controlled pumps for transporting and rotating the specimen.

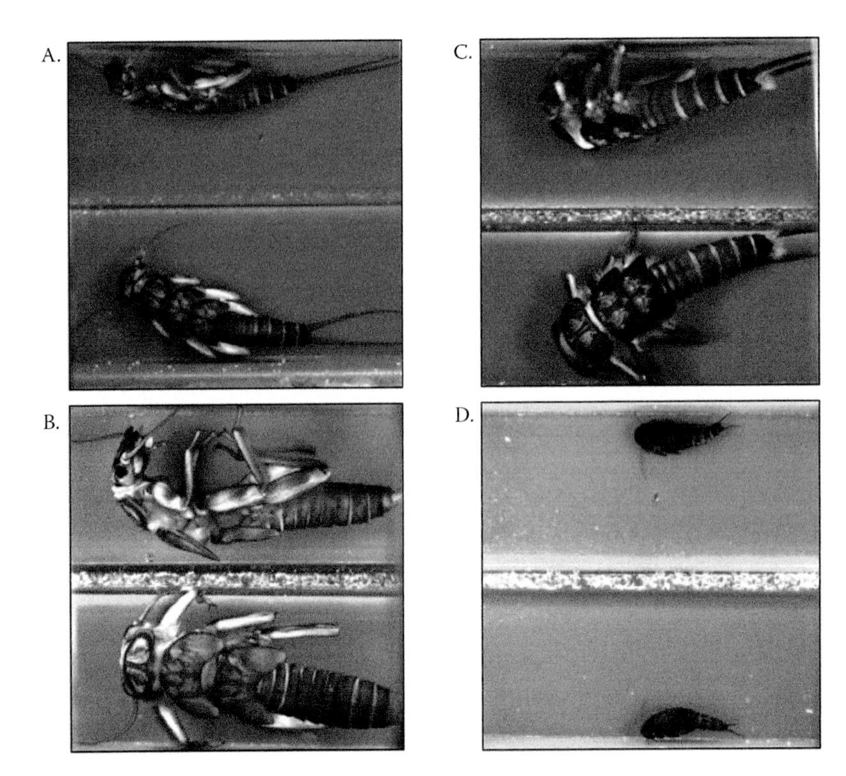

FIGURE 11.3 Example images of stonefly specimens taken with our imaging apparatus. A: *Calinueria californica*; B: *Doroneuria baumanni*; C: *Hesperoperla pacifica*; D: *Yoraperla* sp.

soils have been reduced in organic content worldwide, leading to a large portion of the carbon being lost to the atmosphere as CO_2 (Adl, 2003; van der Putten, 2004; Wall, 2004). Numerous attempts to reverse this process have been tried in different cropping systems (Coleman and Hendrix, 2000). Tests for organic content in the soil are destructive and do not reveal anything about the biological diversity present. Population counts of soil mesofauna are recognized as one of the most sensitive, cheapest and least destructive assays of soil biodiversity and nutrient cycling functions (Wardle, 2002).

Biodiversity can be measured in many ways – in particular by assessing the population size, variety and geographical distribution of key species. These species range in size from micro-organisms (bacteria and fungi) to megafauna (mammals and birds). For practical and theoretical reasons, arthropods (insects and their allies) are generally regarded as the best potential biodiversity indicators: their high species richness allows for fine taxonomic resolution, their ease of capture results in low-cost assays and their widespread distribution provides good generalizability and applicability (Szaro and Johnston, 1996; Brown, 1997; McGeoch, 1998; Niemelä et al., 2002). Historically, the practical limitations facing arthropod soil ecologists have been high densities (100–500,000/m^2) and high species diversity (a dozen in a 3-inch diameter soil core to hundreds in a metre squared). As a practical compromise, sample volume is always drastically reduced and the decreased number of replicate samples is often insufficient with respect to the environmental heterogeneity inherent in soil samples.

Many of the most significant characteristics of soil mesofaunal samples are changes that take place in repeated subsequent sampling under changing ecological conditions. The ability to scan over enough samples (or large enough samples) and to detect changes in the species present or their relative abundances are challenges that perhaps only automated identification systems are uniquely capable of solving. Coarse identification (to the family-level resolution) is often sufficient to assign ecological function and is attainable with a modest image data-base. Species-level identification within a functional group is not necessarily as important as being able to distinguish the presence of multiple species; that is, correctly identifying five species of entomobryid springtails, which necessitates an immense archive of pictures (an ultimate goal), is not as significant as recognizing that there are five different 'morphospecies' (a practical goal).

Mechanical Manipulation and Imaging

Soil mesofauna specimens are routinely processed by Berlese funnel extraction. Berlese extraction involves slow drying of a soil sample, which causes the soil arthropods to crawl out of the soil and into a collecting vessel. The specimens are then separated from the associated debris by addition of an organic liquid followed by agitation, settling and decanting (Moldenke, 1994). Decanted samples are then placed in petri plates and diluted sufficiently such that the probability of one specimen occluding another is low.

The mechanical manipulation of the soil mesofauna specimens is still under development (as indicated in Figure 11.1); however, our current design is as follows. The petri plate containing soil mesofauna specimens will be placed on a computer-controlled, motorized *x–y* stage that allows the computer to scan the plate systematically to find each specimen. The specimens are so small (ranging from 50 to 2500 μ) that only a portion of each is in focus at any one time. This problem will be addressed by a montage process in which the focal plane of the microscope scans vertically through the field of view (via computer-control focus column) while a series of 10–20 images are taken. The images will then be combined to produce a synthetic image in which all parts of the specimens are simultaneously in focus. An additional benefit of this process is that it can produce an approximate three-dimensional map of the height of each specimen, which may be useful for reconstructing the specimen's three-dimensional shape. After a specimen has been photographed, it can be removed mechanically from the petri plate and placed in an appropriate receptacle to ensure that no specimen is counted twice and to provide a means of manual performance auditing.

Many of the soil arthropods are partially transparent or translucent. While specimen translucency may be an important cue for identification, it poses difficulties for automatically separating the specimens from the image background. Image matting methods (Chuang et al., 2001; Ruzon and Tomasi, 2001; Wang and Cohen, 2005; Levin et al., 2006) can be applied to determine the amount of transparency (on a per-pixel basis) of a specimen if it is photographed with different background color patterns (Smith and Blinn, 1996). Hence, an LCD display panel will be placed between the petri plate and the *x–y* stage to provide automatic control over the background color pattern.

CLASSIFICATION FRAMEWORK

Having motivated the classifications tasks and discussed the image acquisition methodology, we now address the problem of automatically identifying the family, genus and species of each specimen. This proceeds in three steps: segmentation, coarse classification and fine classification.

Segmentation

The first step in classifying a specimen is to segment the image to separate a specimen from the background and, possibly, from other specimens in the frame. While segmentation is not required for identification, it does simplify many aspects of both learning and classifying objects. Without segmentation, the classification algorithm may confuse irrelevant background features (such as bubbles, dirt, etc.) with desired specimen features.

Automatic segmentation of a general class of images is a very difficult problem that remains unsolved. However, one advantage of the imaging mechanisms described previously is that we can control the background, which greatly simplifies segmentation. Part of the image capture protocol for stoneflies is to acquire a background image (without a specimen) prior to positioning each new specimen under the microscope. This background image allows for a simple background subtraction process to provide a majority of the segmentation. For soil mesofauna, known background images are provided by the LCD display. In both cases, a Bayesian matting process (Chuang et al., 2001) can be applied to improve segmentation and provide partial transparency for pixels near the specimen's boundary or for specimens that are translucent (as are some soil mesofauna species). Our current automatic segmentation algorithm performs most, but not all, of the work of segmentation. It is often still necessary to post-process the images manually to further refine the segmentation (e.g. remove bubbles).

Coarse Classification

Our work to date has focused primarily on stoneflies. Since they share the same body plan and many have a similar general shape, there has been little need for performing coarse classification of stonefly larvae beyond manually identifying them as *Plecoptera* during specimen collection. However, we anticipate that the identification of soil mesofauna will greatly benefit from a coarse classification step that automatically groups the specimens with similar body plans, sizes, shapes and colors. Our plan is to extract, for each specimen, a set of shape cues such as eccentricity, compactness and Fourier shape coefficients. We will also examine more detailed shape models that capture local curvatures of the specimen's silhouette, which may allow us to count protuberances such as legs, antennae, tails and so on (Wu and Dietterich, 2004). Additionally, a specimen's color histogram information can also provide an indicator for separating specimens into groups. Using the shape and color information, each specimen will be classified into a general morphological group, which will then provide the basis for fine classification within each such group.

Fine Classification

Identification to finer taxonomic levels will utilize both two-dimensional image information and three-dimensional reconstructions of the specimens. Two-dimensional methods operate directly on one or more images of the specimen, typically taken from preferred views. Our two-dimensional approach is to extract various interest regions and construct invariant descriptors for each region. The descriptors are then clustered and a training algorithm learns feature-cluster associations. Details of two-dimensional classification are provided in Classification Methods (below).

Three-dimensional methods apply various techniques to construct a three-dimensional model of the specimen from two-dimensional images. One technique is to fit a parameterized three-dimensional model so that, when projected onto the two-dimensional image plane, it produces an image that matches the image of the specimen. Another technique is to combine multiple two-dimensional images to construct a three-dimensional model of the specimen, which is then matched to three-dimensional models. Our current approach is to create a three-dimensional depth map of the specimens from multiple images with different focal planes, as mentioned previously in relation to creating images of the soil mesofauna specimens with all their body parts in focus. We believe that the resulting depth map will then allow us to extract three-dimensional features that can be classified in a manner similar to the two-dimensional features as discussed later.

RECENT METHODS FOR OBJECT RECOGNITION

Recent work in computer vision has led to the development of a new family of methods for object recognition. In this project, we are refining and extending these techniques so that they can be applied to recognize insects and soil arthropods. Hence, before describing our specific methods, we first review this family of modern object recognition methods. These methods are based on the following general approach.

Step 1: Apply a 'Region Detector' to the Input Image

Many recent object-recognition methods classify objects by identifying and describing a collection of small regions or 'patches' in an image. These patch-based approaches work even when objects are partially occluded or when they are photographed in front of complex background settings. The first step in object recognition is to apply region detectors to identify and extract a set of 'interesting regions' from the image. A *region detector* identifies a location (i.e. a pixel or region), a scale (i.e. a circle of specified radius or an ellipse with specified major and minor axes) and an orientation for each region. A good region detector should be reliable and informative. A detector is reliable if it is robust to changes in viewpoint, scale, illumination and noise. That is, given two images of the same object taken with different viewpoints, illumination and noise levels, the region detector should still detect the same regions. A detector is informative if the detected regions are useful for discriminating among the different object classes. Some popular region detectors include the Harris (Harris and Stephens, 1988), Harris-affine (Mikolajczyk and Schmid, 2004), maximal difference-of-Gaussian (DOG) (see Lowe, 1999), maximally stable extremal regions (MSER) (see Matas et al., 2002) and entropy (or Kadir) (Kadir and Brady, 1996) detectors. Region detectors are also known as interest operators because they find interesting (i.e. locally unique) points or regions within the image.

Step 2: Represent Each Region by a Fixed-Length 'Descriptor'

After detection, each interest region is represented by a *descriptor* (i.e. a real-valued vector of features) that succinctly and discriminatively characterizes the local image properties. By constructing each region's descriptor relative to the local coordinate frame determined by the interest operator (i.e. the position, orientation and scale), the descriptors are translation-, rotation- and scale invariant. Additionally, a descriptor should be insensitive to changes in illumination and, especially for object class recognition (as required in this project), non-rigid transformations. Recent descriptors include spin images (Lazebnik et al., 2003), shape context (Belongie et al., 2002), SIFT (Lowe, 2004) and PCA-SIFT (Ke and Sukthankar, 2004).

In this work we employ the SIFT (scale invariant-feature transform) descriptor, which has been shown to perform better than other local descriptors (Mikolajczyk and Schmid, 2003). The SIFT descriptor is invariant to scale, rotation, intensity and contrast changes and, to a small degree, affine transformations. SIFT divides the region into a set of bins. For each bin, it computes a histogram of the intensity gradient orientation at each pixel. The result is a 128-dimensional real-valued vector. Once each detected region has been converted to a SIFT vector, the input image is discarded, and only the 'bag' of SIFT vectors is retained for further analysis.

Step 3: Apply a Classifier to the Bag of Descriptor Vectors

Several classifiers have been developed that can analyze the bag of descriptor vectors and predict the class of the object. The simplest kind of classifier represents each object as a collection of 'parts'.[1] Each descriptor is classified according to which part it represents. If a sufficient number of matching parts is detected, the object is assigned to the corresponding class. More sophisticated classifiers (see later discussion) compute a weighted sum of the detected parts. The most complex methods take into consideration the spatial relationships among the detected parts (Agarwal and Roth, 2002; Fergus et al., 2003), although subsequent work (Opelt et al., 2004; Dorkó and Schmid, 2005) has obtained better results without including this information.

Representing an object by local salient regions has many advantages that have made region-based recognition very popular in recent years. Region-based representations cope better with images that have cluttered backgrounds and objects that are partially occluded. This is because classifiers can be trained to make a decision even if not all parts are detected. In cases where images need to be compared to one another, interest operators reduce the matching task from comparing hundreds of thousands (or millions) of pixels to comparing just a few hundred highly salient regions. Another major benefit of using local regions for object class recognition is that they provide a degree of object pose invariance. In particular, affine-invariant detectors and descriptors allow for small out-of-plane object rotations (up to approximately 30°).

CLASSIFICATION METHODS

As noted before, we are pursuing two strategies for insect classification: (1) recognition using features extracted from the two-dimensional image and (2) recognition based on three-dimensional reconstruction of the specimen from multiple images. As our work on three-dimensional reconstruction is still preliminary; this chapter reports only on the methods and results to date of our two-dimensional approach.

Our two-dimensional approach follows the general region-based methodology introduced previously. This section presents the details of the approach. To date, we have focused only on discriminating between pairs of taxa (i.e. binary classifiers). Hence, we refer to one as the positive class and the other as the negative class. If we attain high accuracy on pair-wise discrimination, there are many machine learning methods for extending this to discriminate among tens or hundreds of taxa (Allwein et al., 2000).

REGION DETECTORS

We have experimented with many of the region detectors discussed previously, and we have chosen two: Harris-affine (Mikolajczyk and Schmid, 2003) and Kadir (Kadir and Brady, 1996). Our system constrains these detectors to consider only points that lie within the specimen so that spurious detections in the background are not a problem. This is possible because the specimens have been segmented from the background.

REGION DESCRIPTORS

For every interest region detected, we construct a 128-element SIFT vector that describes each region's local neighborhood. The SIFT descriptor computes a 16 × 16 neighborhood centred on the detected region, normalized to scale, rotation and (for Harris-affine) the affine parameters produced by the detector. This neighborhood is partitioned into 16 subregions of 4 × 4 pixels each. For each pixel within a subregion, SIFT adds the pixel's gradient vector to a histogram of gradient directions by quantizing each orientation to one of eight directions and weighting the contribution of each vector by its magnitude. Each gradient direction is further weighted by a Gaussian of scale $s = n/2$, where n is the neighborhood size, and the values are distributed to neighboring bins using trilinear interpolation to reduce boundary effects that occur when pixels move across bin boundaries.

The final descriptor is a 128-dimensional real vector representing the 4 × 4 grid of eight-bin orientation histograms. Figure 11.4 shows a graphical representation of the SIFT descriptors created for three regions detected in two stonefly images. The SIFT vectors in Figures 11.4C and 11.4D are very similar, and these points are in corresponding positions on the two specimens. In contrast, the SIFT vector shown in Figure 11.4E is quite distinct, and it does not correspond to the other two points. This shows the ability of SIFT to capture distinctive characteristics of the detected regions.

The SIFT descriptor is invariant to scale, rotation, contrast and intensity changes and small out-of-plane rotations. Scale invariance is achieved by describing the local neighborhood around each feature point at that feature's characteristic scale (as computed by the Harris or Kadir detectors). To achieve rotation invariance, the descriptor bins and gradient directions are defined relative to the dominant gradient orientation in the neighborhood. Invariance to intensity and contrast changes results from normalizing the 128 vector to unit magnitude. The 4 × 4 bin size

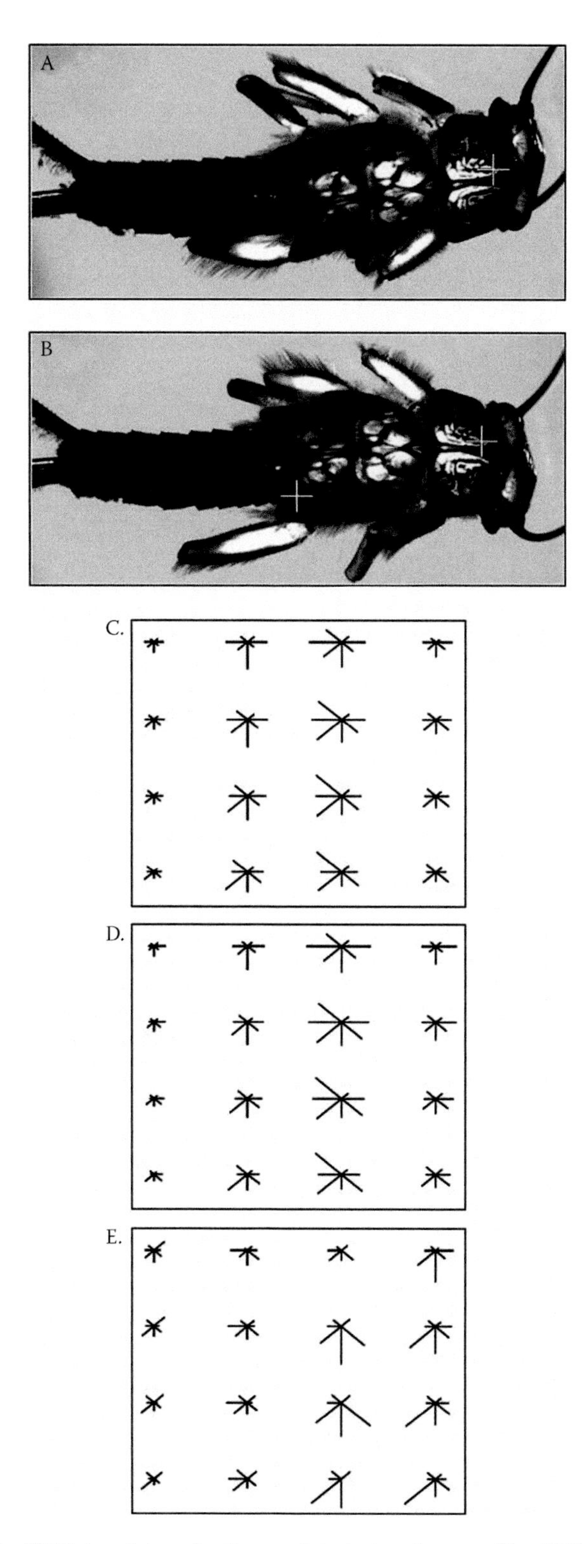

FIGURE 11.4 Example SIFT descriptors for three points in two images. The SIFT histograms for (C) the detected region in image (A), the matching region in image (B) and (E) a random region in image (B).

(relative to the detected feature's characteristic scale) provides some invariance to minor affine and perspective transformations as well as some non-rigid distortion (such as typically occurs from interclass variability).

DEFINING OBJECT PARTS BY CLUSTERING DESCRIPTOR VECTORS

The first two steps of detecting interesting regions and representing them by SIFT descriptor vectors are performed both during training of the classifier and during classification of new specimens. The third step – defining object parts – is only performed during training.

Parts are defined by performing a cluster analysis on the SIFT descriptor vectors extracted from the input images of the positive specimens. As do Dorkó and Schmid (2005), we cluster the vectors by fitting a Gaussian mixture model (GMM). A GMM assumes that each SIFT vector is generated from one of K clusters according to a probability $P(C_i)$, where $i = 1, \ldots, K$ denotes the cluster number. The values of the 128-element descriptor vectors are assumed to have a Gaussian distribution within each cluster with mean vector $\boldsymbol{\mu}_i$ and diagonal covariance matrix $\boldsymbol{\Sigma}_i$. According to this model, the probability assigned to any particular SIFT vector x can be computed as

$$p(x) = \sum_{i=1}^{K} p(x \mid C_i) P(C_i) \tag{11.1}$$

where $p(x \mid C_i) = gauss(x \mid \boldsymbol{\mu}_i, \boldsymbol{\Sigma}_i)$ is the multivariate Gaussian probability density function.

The adjustable parameters in the model are $\boldsymbol{\mu}_i$, $\boldsymbol{\Sigma}_i$ and $P(C_i)$ for $i = 1, \ldots, K$. These parameters are fit by maximum likelihood (i.e. to maximize $p(x)$ on the training data vectors x) via the well-known expectation-maximization (EM) algorithm (McLachlan and Krishnan, 1997). The EM algorithm is initialized using the very efficient K-means algorithm, and it typically converges in 20–100 iterations.

To apply the GMM, we must choose a value for K. If K is too small, then the clusters found by EM will be very broad (with large variances in $\boldsymbol{\Sigma}_i$). This will cause them to fail to be distinctive or informative. On the other hand, if K is too large, then the clusters will over-fit the training data and not generalize well. Values within the range $50 \leq K \leq 100$ give the best results.

CONVERTING SETS OF SIFT VECTORS INTO STANDARD FEATURE VECTORS

Once the 'parts' have been defined, we then convert each set of SIFT vectors (extracted from one training image) into a single feature vector of length K as follows:

1. Initialize a histogram vector: $hist[i] = 0$ for $1 \leq i \leq K$.
2. For each SIFT descriptor vector x extracted from the image:
   a. Let $i^* = \arg\max_i P(C_i \mid x) = \arg\max_i p(x \mid C_i)\, P(C_i)$
   b. $hist[i^*] = hist[i^*] + 1$.
3. Normalize the histogram to unit magnitude.

Step 2a computes the 'part' cluster i^* that is most likely to have generated the observed SIFT vector x according to the GMM. Consequently, the ith entry in the histogram vector is proportional to the number of SIFT vectors that 'belong' to cluster i.

This conversion gives us training data in a format suitable for analysis by standard supervised learning algorithms. Let us denote the normalized histogram for the jth input image by h_j and denote the corresponding class label (taxon) by y_j. Our training data thus consist of pairs (h_j, y_j) for $j = 1, \ldots, N$, where N is the total number of training images.

TRAINING THE CLASSIFIER

After the SIFT vector sets have been converted to standard feature vectors, they are used to train a classifier. In this chapter, we report results of a 'bagged' decision-tree classifier. A decision-tree classifier has the form of a nested set of *if–then–else* statements, where each statement has the form

If $h[i] > \theta_n$

then statement

else statement.

Statement can either be a nested *if–then–else* or else a predicted class label. A new specimen is assigned a predicted class by executing this tree of *if–then–else* statements until a predicted class label is reached.

Decision tree classifiers are learned top-down by first selecting a feature i and a threshold θ_i for the outermost *if–then–else* and then splitting the training data according to the results of this test. If all of the examples that reach a *statement* belong to a single class, then the recursion halts and that class is predicted. Similarly, if a new *if–then–else* would result in sending two or fewer examples down either the *then* or the *else* branches, then the recursion halts, and the class belonging to a majority of the data points is assigned at that *statement*. We have employed the J48 decision tree learning algorithm, which is part of the WEKA machine learning system (Witten and Frank, 2005).

A single decision tree is typically not a very good classifier. However, very high performance can often be obtained by constructing an ensemble of decision trees through a method known as 'bagging' (Breiman, 1996). In bagging, the decision tree learning algorithm is applied to L different training sets. Each training set is constructed by drawing N examples uniformly, with replacement, from the original training data-set. Such a training set is known as a bootstrap replicate. Each bootstrap replicate may contain multiple copies of some of the original training examples, and it may be missing other components of the original examples. On average, a bootstrap replicate contains approximately 62 per cent of the original training examples (but with enough copies so that there are still N total examples). In the results reported next, bagging has been applied to construct 15 decision trees. To classify a new example, the predictions of these 15 decision trees are computed and the class (taxon) with the largest number of predictions is chosen as the overall bagged prediction.

RESULTS AND DISCUSSION

We report the results of three experiments on stonefly classification. In each experiment, we train our method to discriminate between two species. The three experiments present progressively increasing levels of difficulty:

1. The somewhat easy task of discriminating between the very distinctive *Calineuria californica* and *Yoraperla* sp.
2. A moderately difficult task of discriminating between the *Hesperoperla pacifica* and *Doroneuria baumanni.*
3. A hard task of discriminating between the very similar *Calineuria californica* and *Doroneuria baumanni.*

TABLE 11.1
Classification Results Using Decision Trees with Bagging

Experiment	Classification accuracy (%)
1: *Calineuria* vs. *Yoraperla*	94.40 ± 2.66
2: *Hesperoperla* vs. *Doroneuria*	90.47 ± 3.36
3: *Calineuria* vs. *Doroneuria*	73.33 ± 6.88

These experiments are performed as follows: the image data-set is randomly divided into three completely disjoint sets of equal size. To avoid any kind of influence in the test results, different images of the same insect instance are placed in the same set. The first set is used as the 'clustering set' to create the GMM clusters for each object class (as detailed before); the second set is used to train the 15 decision trees that comprise the final classifier (described earlier), while the third is used to measure the classification accuracy of the classifier. We use separate clustering and training sets to reduce over-fitting of the classifier to the training data.

Table 11.1 presents the resulting classification accuracy rates for our three experiments (along with 95% confidence intervals). For experiment 1 (*Calineuria* vs. *Yoraperla*), the method achieved 94 per cent accuracy; for experiment 2 (*Hesperoperla* vs. *Doroneuria*), the accuracy was 90 per cent; and, for the very difficult experiment 3 (*Calineuria* vs. *Doroneuria*), the method attained only 73 per cent correct classifications. These results represent the best matching rates achieved thus far in differentiating between pairs of insect species. However, because some of our design decisions (number of clusters, choice of species for clustering, number of decision trees to combine) were made to optimize these numbers, they are probably optimistically high.

Because these are binary classification experiments, either species' clustering set can be used to learn the GMM part clusters. In our experiments, the GMM clusters of both species were tried and the one that rendered the best results is presented. *Calineuria*'s part clusters were employed in the first experiment, *Hesperoperla*'s in the second experiment, and *Doroneuria*'s in the third. The choice of clustering set affects the accuracy, and there are two possible reasons for this: (1) some species have more clustering examples in the data-set than others and (2) some species exhibit more distinctive visual cues that help with recognition. In the first experiment, the data-set contains many more images of *Calineuria* than of *Yoroperla*; this difference probably accounts for the GMM clustering algorithm producing better *Calineuria* part clusters, which in turn helped increase the classification rate. By contrast, the data-set in experiment three contains similar numbers of *Calineuria* and *Doroneuria* images, and both clustering sets give similar results. In the second experiment, the high classification accuracy is surprising. It is possible that the *Hesperoperla* clustering set has certain visual cues that produce very distinctive GMM clusters.

Tables 11.2, 11.3 and 11.4 present the confusion matrices of experiments 1, 2 and 3, respectively. Table 11.2 shows that all of the 16 misclassification errors involved misclassifying a *Calineuria* as a *Yoraperla,* which is quite surprising because, normally, misclassification favors the class with the larger number of training examples (in this case, *Calineuria*). Table 11.3 shows a similar pattern: most of the errors misclassify *Hesperoperla* as *Doroneuria*. Finally, in Table 11.4, we see a more balanced ratio of errors. The poor results in experiment 3 accord with our own informal experience: both *Calineuria* and *Doroneuria* are from the Perlidae family and look very similar – so much so that the non-entomologists on our team cannot generally distinguish them using only the collected image data. While *Hesperoperla* is also in the Perlidae family, it is more visually distinct from *Calineuria* and *Doroneuria*.

TABLE 11.2
Confusion Matrix: *Calinueria* vs. *Yoraperla*

Taxa	Classified as *Calineuria*	Classified as *Yoraperla*
Calineuria	171	16
Yoraperla	0	99

TABLE 11.3
Confusion Matrix: *Hesperoperla* vs. *Doroneuria*

Taxa	Classified as *Hesperoperla*	Classified as *Doroneuria*
Hesperoperla	171	24
Doroneuria	4	95

TABLE 11.4
Confusion Matrix: *Calineuria* vs. *Doroneuria*

Taxa	Classified as *Calineuria*	Classified as *Doroneuria*
Calineuria	126	66
Doroneuria	30	138

CONCLUDING REMARKS AND FUTURE WORK

This chapter describes our ongoing work on developing robust pattern recognition methods for automatically classifying insects. Although this project focuses on identifying stonefly larvae for stream water quality monitoring and soil mesofauna for soil biodiversity assessment, the goal is to create general-purpose techniques that can be applied to many automated systematics tasks by simply retraining the system for each application. In addition, we have designed and developed (or are developing) computer-controlled mechanical systems to facilitate automated imaging of specimens. Our insect classification system uses recent advances in object-class recognition based on interest operators and bagged local-region descriptors combined with boosted decision trees.

While we are pleased with the classification accuracy achieved thus far on stonefly larvae, there is still significant room for improvement. First and foremost, we continue to add and refine components that, as indicated in Figure 11.1, are still under development. These components include the control and imaging software for the soil mesofauna, fully automated segmentation by incorporating Bayesian matting (or a similar method), coarse classification, and the extraction and inclusion of three-dimensional features from the depth map constructed during the focal-montage process. In addition to these necessary components, we continue to make improvements to the existing mechanical specimen manipulation and imaging apparatus with the ultimate goal of mechanically delivering each identified specimen to a separate bin based on its classification.

We are also developing a new watershed-based region detector (Zhang et al., 2004) that appears to be better suited for insect classification and we continue to explore additional region and feature descriptors to supplement the SIFT feature vectors. Furthermore, we are experimenting with several additional classifiers, including those that work directly on the bags of descriptor vectors instead of converting them to standard feature vectors. We are also developing a multiclass system that will classify a specimen into one of many taxonomic categories rather than our current binary classifiers that distinguish only between two species at a time.

Our current effort has focused on classifying each specimen to the species level (though in the experiments presented here, genus-level classification implies species-level identification as well). However, in cases where the classifier is uncertain about the species, it is often useful (both for stream monitoring and for soil biodiversity studies) to make a coarser classification at the level of genus or family. For example, while it is difficult to achieve high confidence in automatically distinguishing between the very similar *Calineuria californica* and *Doroneuria baumanni* species of stonefly larvae, high-confidence classification to the family level is much more feasible and still very valuable for stream health assessment. Hence, we plan to explore methods for trading off the benefit of fine classification against the risk of making an error in order to automatically choose the best level of the taxonomic hierarchy for classifying each specimen.

We would also like to explore additional systematics applications to test the robustness of our system in new classification domains, such as recognition of plant species. Our long-term goal is to develop commercial products to make this emerging technology available to the environmental monitoring and research communities.

ACKNOWLEDGEMENTS

The authors gratefully acknowledge the support of the US National Science Foundation under grant number IIS-0326052. The views expressed are those of the authors and do not represent the views of the US government or the National Science Foundation.

NOTES

1. We write 'parts' in quotes because these 'parts' do not necessarily correspond to real parts of the object; rather, they correspond to interest regions in the object, which may or may not correspond to meaningful physical parts of the object to be classified.

REFERENCES

Adl, S.M. (2003) *The Ecology of Soil Decomposition,* CABI Publishing, Oxforshire, UK, 352 pp.

Agarwal, S. and Roth, D. (2002) Learning a sparse representation for object detection. *Proceedings of the European Conference Computer Vision (ECCV),* 113–128.

Allwein, E.L., Schapire, R.E. and Singer, Y. (2000) Reducing multiclass to binary: a unifying approach for margin classifiers. *Journal of Machine Learning Research,* 1: 113–141.

Belongie, S., Malik, J. and Puzicha, J. (2002) Shape matching and object recognition using shape contexts. *IEEE Transactions Pattern Analysis Machine Intelligence,* 24(4): 509–522.

Breiman, L. (1996) Bagging predictors. *Machine Learning,* 24(2): 123–140.

Brown, K.S. (1997) Diversity, disturbance and sustainable use of neotropical forests: insects as indicators for conservation monitoring. *Journal of Insect Conservation,* 1(1): 25–42.

Chuang, Y.-Y. Curless, B., Salesin, D.H. and Szeliski, R. (2001) A Bayesian approach to digital matting. *Proceedings of Computer Vision and Pattern Recognition* (CVPR), 2: 264–271.

Coleman, D.C. and Hendrix, P.F. (2000) *Invertebrates as Webmasters in Ecosystems,* CABI Publishing, Oxfordshire, UK, 352 pp.

Dorkó, G. and Schmid, C. (2005) Object class recognition using discriminative local features, Technical Report RR-5497, INRIA–Rhone-Alpes, 2005.

Fergus, R., Perona, P. and Zisserman, A. (2003) Object class recognition by unsupervised scale-invariant learning. *Proceedings of Computer Vision and Pattern Recognition* (CVPR), 2: 264–271.

Harris, C. and Stephens, M. (1988) A combined corner and edge detector. *Proceedings of the Fourth Alvey Vision Conference,* pp. 147–151.

Hilsenhoff, W.L. (1988) Rapid field assessment of organic pollution with a family level biotic index. *Journal of the North American Benthological Society,* 7(1): 65–68.

Jalba, A.C. Wilkinson, M.H.F., Roerdink, J.B.T.M., Bayer, M.M. and Juggins, S. (2005) Automatic diatom identification using contour analysis by morphological curvature scale spaces. *Machine Vision and Applications,* 16(4): 217–228.

Johnson, R.K., Wiederholm, T. and Rosenberg, D.M. (1993) Freshwater biomonitoring using individual organisms, populations, and species assemblages of benthic macroinvertebrates. In *Freshwater Biomonitoring and Benthic Macroinvertebrates* (eds D.M. Rosenberg and V.H. Resh), Chapman & Hall, London, pp. 40–158.

Kadir, T. and Brady, M. (1996) Saliency, scale and image description. *International Journal of Computer Vision,* 24(2): 83–105.

Ke, Y. and Sukthankar, R. (2004) PCA-SIFT: a more distinctive representation for local image descriptors. *Proceedings of Computer Vision and Pattern Recognition* (CVPR), 2: 511–517.

Lazebnik, S., Schmid, C. and Ponce, J. (2003) Sparse texture representation using affine-invariant neighbourhoods. *Proceedings of Computer Vision and Pattern Recognition* (CVPR), 2: 319–324.

Levin, A., Lischinski, D. and Weiss, Y. (2006) A closed form solution to natural image matting. *Proceedings of Computer Vision and Pattern Recognition* (CVPR), 1: 61–68.

Lowe, D.G. (1999) Object recognition from local scale-invariant features. *Proceedings of Computer Vision and Pattern Recognition* (CVPR), 2: 1150–1157.

Lowe, D.G. (2004) Distinctive image features from scale-invariant keypoints. *International Journal of Computer Vision,* 60(2): 91–110.

Matas, J., Chum, O., Urban, M. and Pajdla, T. (2002) Robust wide-baseline stereo from maximally stable extremal regions. *Proceedings of the British Machine Vision Conference* (BMVC), 384–393.

McGeoch, M.A. (1998) The selection, testing and application of terrestrial insects as bioindicators. *Biological Reviews of the Cambridge Philosophical Society,* 73: 181–201.

McLachlan, G. and Krishnan, T. (1997) *The EM Algorithm and Extensions*, Wiley, New York, 304 pp.

Mikolajczyk, K. and Schmid, C. (2003) A performance evaluation of local descriptors. *Proceedings of Computer Vision and Pattern Recognition* (CVPR), 2: 257–264.

Mikolajczyk, K. and Schmid, C. (2004) Scale and affine invariant interest point detectors. *International Journal of Computer Vision,* 60(1): 63–83.

Moldenke, A.R. (1994) Arthropods. In *Methods of Soil Analysis, Part 2* (eds R.W. Weaver, S. Angle and P. Bottomley), Microbiological and Biochemical Properties, SSSA Book Series, No. 5, Soil Science Society of America, Madison, WI, pp. 517–541.

Niemelä, J., Kotze, D.J., Venn, S., Penev, L., Stoyanov, I., Spence, J., Hartley, D. and de Oca, E.M. (2002) Carabid beetle assemblages (Coleoptera, Carabidae) across urban–rural gradients: an international comparison. *Landscape Ecology,* 17(5): 387–401.

Opelt, A., Fussenegger, M., Pinz, A. and Auer, P. (2004) Weak hypotheses and boosting for generic object detection and recognition. *Proceedings of Computer Vision and Pattern Recognition* (CVPR), 2: 71–84.

van der Putten, W.H., Anderson, J.M., Bardgett, R.D., Behan-Pelletier, V., Bignell, D.E., Brown, C.G., Brown, V.K., Brussaard, L., Hunt, H.W., Ineson, P., Jones, T.H., Lavelle, P., Paul, E.A., St. John, M., Wardle, D.A., Wojtowicz, T. and Wall, D.H. (2004) The sustainable delivery of goods and services provided by soil biota. In *Sustaining Biodiversity and Ecosystem Services in Soil and Sediments* (ed D.H. Wall), Island Press, Washington, DC, pp. 15–44.

Resh, V.H. and Jackson, J.K. (1993) Rapid assessment approaches to biomonitoring using benthic macroinvertebrates. In *Freshwater Biomonitoring and Benthic Macroinvertebrates* (eds D.M. Rosenberg and V.H. Resh), Chapman & Hall, London, pp. 195–233.

Resh, V.H., Myers, M.J. and Hannaford, M.J. (1996) Macroinvertebrates as biotic indicators of environmental quality. In *Methods in Stream Ecology* (eds F.R. Hauer and G.A. Lamberti), Academic Press, London, pp. 647–667.

Roth, V., Pogoda, A., Steinhage, V. and Schröder, S. (1999) Integrating feature-based and pixel-based classification for the automated identification of solitary bees. Jahrestagung der Deutschen Gesellschaft für Mustererkennung (Annual Convention of the German Society for Pattern Recognition), pp. 120–129.

Ruzon, M. and Tomasi, C. (2001) Alpha estimation in natural images. *Proceedings of Computer Vision and Pattern Recognition* (CVPR), 1: 18–25.

Smith, A.R. and Blinn, J.F. (1996) Blue screen matting. *Proceedings of Computer Graphics (SIGGRAPH)*, pp. 259–268.

Steinhage, V., Arbuckle, T., Schröder, S., Cremers, A.B. and Wittmann, D. (2001) ABIS: automated identification of bee species. *BIOLOG Workshop, German Programme on Biodivesity and Global Change, Status Report*, pp. 194–195.

Szaro, R.C. and Johnston, D.W. (1996) *Biodiversity in Managed Landscapes,* Oxford University Press, Oxford, UK, 808 pp.

Turk, M. and Pentland, A. (1991) Eigenfaces for recognition. *Journal of Cognitive Neuroscience,* 3(1): 71–86.

Wall, D.H. (2004) *Sustaining Biodiversity and Ecosystem Services in Soils and Sediments,* Island Press, Washington, DC, 275 pp.

Wang, J. and Cohen, M. (2005) An iterative optimization approach for unified image segmentation and matting. *Proceedings of Computer Vision and Pattern Recognition* (CVPR), 2: 936–943.

Wardle, D.A. (2002) *Communities and Ecosystems: Linking the Aboveground and Belowground Components,* Princeton University Press, Princeton, NJ, 400 pp.

Witten, I.H. and Frank, E. (2005) *Data Mining: Practical Machine Learning Tools and Techniques,* 2nd edition, Morgan Kaufmann, San Francisco, 416 pp.

Wu, P. and Dietterich, T.G. (2004) Improving SVM accuracy by training on auxiliary data sources. *International Conference on Machine Learning* (ICML), pp. 871–878.

Zhang, W., Deng, H., Dietterich, T.G. and Mortensen, E.N. (2004) A hierarchical object recognition system based on multiscale principal curvature regions. *Proceedings of the International Conference on Pattern Recognition* (ICPR), 1: 778–782.

12 Plant Identification from Characters and Measurements Using Artificial Neural Networks

Jonathan Y. Clark

CONTENTS

Introduction 207
Case Study 209
Materials and Methods 210
Training Data-Set 210
Validation Data-Set and Data Partitions 211
Test Data-Set 212
Neural Network 212
Tests Performed 213
Performance Assessment 215
Results 215
Tests with Amended Classification Only 215
Tests with Amended Classification plus Geographic Information 216
Discussion 218
Summary 220
Acknowledgements 223
References 224

INTRODUCTION

Identification of taxa, usually to species level, is one of the main activities of professional botanists. It is still mostly performed with the help of paper-based taxonomic keys, although there are now some computer-based identification guides (e.g. DELTA, INTKEY; see Dallwitz, 1974, 1980; Watson et al., 1989; Partridge et al., 1993; Dallwitz et al., 1997). The validity and accuracy of such an identification procedure relies heavily on the underlying classification, the experience of the botanist who compiled the key and the user's interpretation and experience. Interactive, computer-based systems are often extremely useful, because characters can usually be chosen in any order. However, this very feature makes it difficult to compare their performance with some other methods. As has already been demonstrated (Clark, 2004), the DELTA key generator (Dallwitz, 1974; Dallwitz et al., 1997) can be used to generate a conventional printable key, whose performance can then be compared to that achieved by other methods. A good account of computer-based identification methods is provided by Pankhurst (1991).

Artificial neural networks (ANNs) are computer-based systems that can learn from previously classified known examples and can perform generalized recognition – that is, identification – of previously unseen patterns. Multilayer perceptrons (MLPs) are supervised neural networks and, as such, can be trained to model the mapping of input data (e.g. in this study, morphological character states of individual specimens) to known, previously defined classes.

A distinct advantage of these kinds of systems is that a human expert is only needed to define and name the original taxa and to provide the appropriate training samples. Although they might require some help with choice of characters and definitions of character states, non-experts can perform actual data collection, and the generation of the identification model is relatively straightforward. However, the process involves the setting of a number of system parameters that are not easy to determine *a priori*. Here, a methodology is proposed that can enable this kind of artificial intelligence system to be used by systematists without requiring specialized knowledge of ANNs. Although the example shown here is botanical, the principles and methodology can be applied equally well to zoological identification problems.

Conceptually, an MLP (Figure 12.1) contains a number of layers, usually three: the input layer (which serves merely to distribute the input data to the next layer), the hidden layer (which is the first layer that performs functions on the data) and the output layer (which receives as inputs the outputs of the hidden layer). The number of input nodes is equal to the number of characters, and the number of output nodes is equal to the number of classes being identified (in this study, the number of species). The number of nodes in the hidden layer depends on the complexity of the identification model and is often (as here) determined by experiment. Training an MLP to produce an identification system is carried out by presenting a succession of data records to the input nodes of the network. These data records comprise the training set; each record contains character states derived from a specimen of known taxonomic identity.

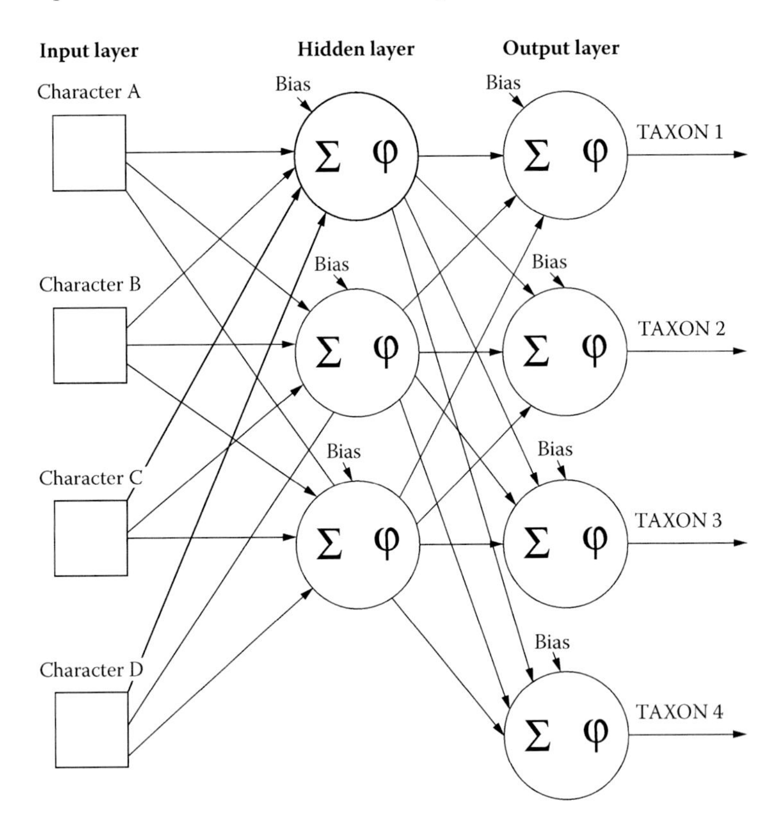

FIGURE 12.1 Multilayer perceptron architecture.

At each node in the hidden layer, numerically coded character states from each record are multiplied by (initially random) weights, summed and then passed through a 'squashing function' – in this case, the sigmoid logistic function (Haykin, 1994, p. 138). The resultant numeric values are then used as inputs to the nodes in the output layer, where similar processing occurs. The output node with the highest final output value is regarded as the 'winner' and corresponds to a particular target class (in this case, species). During training, the internal weights are adjusted so that the forward pass produces a better result, using a learning method called backpropagation (Rumelhart and McClelland, 1986). A better result is one where the target output node (the one corresponding to the correct species class) gives a higher output and the other nodes (corresponding to other species classes) give a lower output.

Periodically, the ability of the network to recognize previously unseen patterns needs to be tested using a totally independent *validation* data-set, which also contains data whose taxonomic class is known. By testing the network performance with this validation set, training can be stopped before over-training occurs. Over-training can result in the model representing the training data very thoroughly, but having a poor ability to generalize when presented with other data. The goal of any practical identification system is to enable recognition of data records that have not been used in the production of the system itself. Therefore, a completely independent *test* data-set containing data records to be identified is presented to the network in order to test the practical effectiveness of the model. Information derived from this test set must not be used to optimize network parameters. The interested reader is referred to Freeman and Skapura (1992), Haykin (1994) and Looney (1997) for further information about ANNs.

There is, of course, a wealth of knowledge already available in the form of published descriptions that could be used as the basis for training artificial neural network keys (ANNKEYs). Earlier studies have been carried out (Clark and Warwick, 1998; Clark, 2000, 2003) in which it was found that suitable training data could be prepared from published descriptions in the form of virtual data records by adding low levels of Gaussian noise to data derived from general descriptions of taxa. For example, descriptions drawn from an existing generic monograph (Cole, 1988) of the genus *Lithops* N.E. Brown were used to provide the original character states for infraspecific taxa, and the MLP-ANN trained to classify these into the 35 species to which they belonged. The trained system was tested using data derived from living plants in the author's collection. Preserved or pressed specimens were not used because these plants are succulent xerophytes and most of the available diagnostic characters are only apparent in living plants. Although it is likely that further work could result in improved results using descriptions as a starting point, this chapter focuses on the use of data derived from real specimens because earlier studies suggested better identification systems would result (Clark, 2000, 2004).

CASE STUDY

The case study presented next relates to cultivated species of the genus *Tilia*, a group of about 30 species of deciduous north temperate woody trees. In the UK, *Tilia* are usually referred to as lime trees, although they are not related to the citrus tree of that name. In other countries, they are often called lindens or basswoods. Their leaves are mostly heart shaped, with an acuminate tip. This can provide a number of readily measurable characters such as length and width, together with various details of hairs. Other morphological characters can be easily extracted from the inflorescence (cluster of flowers), which is subtended by a distinctive bract, characteristic of the genus.

This project was restricted to 19 species grown in European gardens (see Table 12.1) and included in the taxonomy of the genus reported in the *European Garden Flora* (Pigott, 1997). The exception to this is that *T. neglecta* is now included in *T. americana* (Pigott, 2002). Cultivated species were chosen for this study because both living trees and herbarium specimens (see Figure 12.2) were readily available. Furthermore, cultivated *Tilia* trees can present a challenge to identify because many species hybridize.

TABLE 12.1
Acronyms and *Tilia* Species

Acronym	Species	Acronym	Species
AME	*americana*	KIU	*kiusiana*
AMU	*amurensis*	MAN	*mandshurica*
CAR	*caroliniana*	MAX	*maximowicziana*
CHI	*chinensis*	MIQ	*miqueliana*
COR	*cordata*	MON	*mongolica*
DAS	*dasystyla*	OLI	*oliveri*
HEN	*henryana*	PLA	*platyphyllos*
HET	*heterophylla*	TOM	*tomentosa*
INS	*insularis*	TUA	*tuan*
JAP	*japonica*		

Although printed identification keys have already been used for the identification of cultivated *Tilia* species (Schneider, 1912; Pigott, 1997), so far the only computer-based identification systems are those of Rath (1996) and Clark (2000, 2004) relating to these 19 species. Rath, however, only separated 13 species of rather different woody trees (in 12 genera) using a neural network trained on leaf image data; the only lime tree included was *T. cordata*. The work presented here builds on earlier work (Clark, 2004) and determines whether improvements in performance can be obtained by using suggestions made by the ANN to make changes in the classification or by the addition of minimal geographic information.

MATERIALS AND METHODS

TRAINING DATA-SET

The training data-set potentially contained data derived from three examples of each species – that is, comprising three different data records for each of 19 cultivated species, each record containing data derived from a different (mostly field collected) herbarium specimen, including type material where practical. There were thus 57 potential training records, each containing data from one specimen, although some of these were used for the creation of a validation data-set (see following discussion). These data records were derived from specimens in the herbarium of the Royal Botanic Gardens, Kew (K), and that of the Natural History Museum, London (BM). Each specimen provided the states of 22 morphological characters and three geographic characters (binary coding for continent of origin: Europe, W. and C. Asia, E. Asia and N. America).

The selection of characters deserves some discussion. The identification system described and developed here is intended for identification of (and derived from) mature flowering specimens taken from the crown of the tree. Botanical specimens are usually collected from the canopy, and the leaf morphology can vary considerably on the same tree. Leaves sprouting from the base are called 'sprout leaves' and are usually not suitable for identification, being somewhat different in character from the normal crown leaves. Although subjective descriptions of characters have been widely used in classical botany, it was decided to concentrate here on measurements, since they can be more objectively evaluated by non-specialists. Furthermore, such continuous characters provide data that can be directly input to an ANN without further coding. Some character states (e.g. presence or absence of staminodes) are usually constant within a *Tilia* species, whereas others (e.g. leaf length) are variable. In such cases, three or four separate measurements were taken and the mean value recorded. Much of the leaf measurement variability is caused by different ages of leaves, so measurements from particularly immature leaves were not included.

FIGURE 12.2 *Henry 7452*: Holotype of *Tilia tuan* Szyszylowicz (herbarium of the Royal Botanic Gardens, Kew).

Although fruit characters are often of importance (C.D. Pigott, personal comm., 1999), from a practical point of view it was decided to concentrate on flowering specimens to avoid damaging important specimens such as nomenclatural types and because additional characters (e.g. number of flowers per inflorescence, presence or absence of staminodes) could then be used. Also, since fruits often drop off herbarium specimens, the number of flowers in the inflorescence is a more reliable character – essentially the same as the number of fruits in the infructescence (cluster of fruits). A full character list is given by Clark (2000).

Validation Data-Set and Data Partitions

During training, it is necessary to test the generalization ability of the network by means of a validation data-set to enable early stopping and avoid over-training. This data-set consisted of data from one specimen of each species from the potential training set described before; the other two remained in the training set itself. In order to reduce the effect of the choice of validation set records, three different partition pairs (A, B and C) of training and validation sets were produced, where the one record of each species to be transferred to the validation set was chosen randomly. The ANN tests described here were carried out using each partition pair.

Test Data-Set

The test set data records were extracted from herbarium specimens of cultivated trees of known taxonomic identity, but where that identity was not known to the ANN during training. Thus, final testing was performed using an independent dataset derived from 30 herbarium specimens of trees cultivated at Kew, the specimens themselves being held in the cultivated folders of the Kew Herbarium. In a few rare instances, some character states were not readily visible on the specimen and it was necessary to collect further material from the original tree. This material was pressed, dried and mounted before examination. The final evaluation with the test set, therefore, represents an independent test of the effectiveness of the identification system and its ability to generalize – a test that is only rarely performed with conventional taxonomic keys.

These data were presented to the MLP-ANN in ASCII tabulated numeric format; each record for each specimen consisted of a single line, starting with a short acronym representing the specimen number (including an acronym for the species) followed by space-delimited character states, with each record terminated by a class number corresponding to the species. Extreme values of numeric ranges were replaced by their mean value, as when a key is generated using DELTA. A list of sources of material is available in Tables 12.11 and 12.12, with the species names and acronyms used there given in Table 12.1.

NEURAL NETWORK

A simple three-layer MLP, consisting of one input layer, one hidden layer and one output layer, was used in this investigation. The number of input nodes corresponds to the number of characters (22, or 25 with geographic data), each input node accepting the numeric character state of one character. The number of hidden nodes was variable; their number was determined by experiment. Each output node corresponded to one class – in this case, one species (theoretically a taxon at any rank) to be identified. There are no connections between nodes in the same layer, and no recurrent (backward) connections. This network architecture is shown in Figure 12.1, although the actual number of nodes differs from that shown in the diagram.

The input vectors were normalized (scaled) to the range $\pm$ 0.9 to reduce the training time required for inputs to the hidden nodes to reach the domain of the sigmoid activation function. Normalization was carried out for each character independently over all training records to prevent unintended character weighting. The maximum and minimum values for each character determined during the normalization were retained for use during similar normalization of the validation and test data-sets to ensure comparable scaling. The initial network weights were set to small random values in the range of ± 0.5 (see Freeman and Skapura, 1992), the exact pseudo-random sequence of values used being dependent on an arbitrarily chosen random seed. The presentation order of input vectors (training data records) was randomized between epochs and a bias input of 1.0 was used. [Note: An epoch is one complete run through the training data-set.] The interested reader is referred to Clark (2003, 2004) for more details on neural network parameters and the training.

The error value used in this study to evaluate training, validation and testing was the squared error percentage (E; see Prechelt, 1994), with corrections (Clark, 2000, 2003), given in Equation 12.1:

$$E = 100 \cdot \frac{1}{NP(o_{max} - o_{min})^2} \sum_{n=1}^{P} \sum_{i=1}^{N} (o_{pi} - t_{pi})^2 \tag{12.1}$$

where o_{max} and o_{min} = maximum and minimum desired target output values used during training (here 0.9 and 0.1, respectively), N = number of output nodes, P = number of patterns, or examples (records), in the data-set under consideration, o_{pi} = actual output value at output node i when input pattern p is presented, and t_{pi} = desired target output at output node i, when pattern p is presented.

Training was first carried out with a constant learning rate (gain) set to 0.1 and one fixed random number seed, changing only the number of nodes in the single hidden layer in order to determine an optimized size of the hidden layer. Momentum, a technique often used with neural networks to increase the speed of convergence (see Haykin, 1994, p. 149), was not used here, because an effort was being made to simplify the use of the methodology by non-experts and thus to minimize the number of variable parameters. The optimized number of hidden nodes was chosen by determining the lowest mean E_{val} (error on the validation data-set) over all the partition pairs, A, B and C. After determining a sensible number of hidden nodes in this way, that number was then fixed, and the learning rate varied to find an optimized value. After the network parameters were set to these experimentally determined optimized values, the tests were run again 10 times, each time using a different random number seed. These runs were repeated, using the same set of 10 random number seeds, for each of the training/validation partition data-set pairs (A, B and C). Thus, a collection of parameterized networks and their results was achieved. The overall neural network results were obtained by collating these sets of results.

TESTS PERFORMED

Results of the initial tests performed as part of earlier studies (Clark, 2000, 2004), using 22 morphological characters of the 19 cultivated species, are shown here in Table 12.2 (confusion matrix) and compared with the overall results from this study in Table 12.10. From Table 12.2, it can be seen that there are three major anomalies. *T. amurensis* (TAMU) is usually misidentified as *T. insularis* (INS), *T. henryana* (THEN) is usually misidentified as *T. mongolica* (MON) and *T. heterophylla* (THET) is usually misidentified as *T. tomentosa* (TOM). It has already been suggested that the specimen of *T. heterophylla* (THET) may be wrongly named and could be a hybrid (Clark, 2004). Further investigations will be needed to resolve this matter. Regarding *T. henryana* and *T. mongolica,* since it does not seem reasonable to consider them conspecific, we should probably accept that this is a limitation of the choice of characters (and suggest that fruit characters might need to be included in a future study). However, it is much more reasonable to include *T. insularis* in a wider (i.e. amplified) concept of *T. amurensis,* as suggested by Pigott (2000).

The further tests performed in this study were, therefore, carried out in the light of additional changes to the underlying classification. Clark (2004) suggested that the system would perform even better after such reclassification. To examine this, similar tests were performed in which all specimens and data relating to *T. insularis* were renamed as *T. amurensis,* with the number of species classes and the number of ANN output nodes reduced accordingly. It is also suggested here that *T. japonica* (Miquel) Simonkai, a very similar taxon, should also be included within an amplified (expanded) *T. amurensis* concept. Thus, the new tests presented here were conducted using a hypothetically amplified *T. amurensis,* including both *T. insularis* and *T. japonica.*

To evaluate the effect on performance by adding geographical data, further tests were performed with the amplified concept of *T. amurensis,* but with the addition of three extra binary (Boolean) characters referring to the continent of origin. This might seem an artificial concept when applied to the cultivated test specimens considered here, but it provides an illustration of the effectiveness

TABLE 12.2
***Tilia* Confusion Matrix: 19 Species – Classification Unchanged**[a]

Species	AME	AMU	CAR	CHI	COR	DAS	HEN	HET	INS	JAP	KIU	MAN	MAX	MIQ	MON	OLI	PLA	TOM	TUA
TAME	**100**																		
TAMU		**0.0**		6.7		3.3			80.0	10.0									
TCOR		15.0			**47.5**	0.8			15.8								20.8		
TDAS	3.3	3.3				**6.7**			33.3				23.3				30.0		
THEN			6.7				**13.3**								80.0				
THET								**0.0**				3.3				3.3		93.3	
TINS	11.1			1.1	2.2	32.2			**33.3**	14.4							5.6		
TKIU											**100**								
TMAX				6.7									**93.3**						
TMIQ														**98.3**		1.7			
TMON	3.3									3.3					**93.3**				
TOLI									10.0				2.2	44.4		**37.8**		5.6	
TPLA						0.7											**99.3**		
TTOM				1.7								22.5	7.5	1.7		35.8		**30.8**	
%Conf	*84.9*	*0.0*	—	—	*95.5*	*15.2*	*100*	*0.0*	*19.3*	—	*100*	—	*73.8*	*68.1*	*53.8*	*48.1*	*63.8*	*23.8*	—

[a] AMU, INS and JAP treated as distinct.

of the system for identification of wild specimens and cultivated specimens grown from wild-collected seed or cuttings where the continent of origin would be known.

PERFORMANCE ASSESSMENT

In these ANNKEY trials, a confusion matrix was produced showing the species identifications. This lists the percentage of identifications referred to each species by the system and is similar in concept to the misclassification and misidentification matrices of other authors (Boddy et al., 1998, 2000). All 30 final identification attempts by the system when performing the random seed tests against the final test data-set were collated and summed to produce the pooled species-based results in Tables 12.5 and 12.8.

When a test specimen was available for a species, the bottom row of the matrix shows the confidence of correct identification (%Conf). This is identical to the confidence of correct classification index used by Morgan et al. (1998) and shows the probability that a given taxonomic identification is correct. This is calculated by expressing the proportion of correct identifications as a percentage of the total number of identifications (including incorrect identifications).

$$\%Conf = \frac{correct}{correct + incorrect} \times 100 \tag{12.2}$$

Each row of the confusion matrix refers to a species in the test set (shown as T...). Columns represent the taxon (in this case, species) to which the test specimens were referred by the ANN. The percentages shown refer to the proportion of the total specimens of each test (row) species identified as belonging to the target (column) species class. Ideal correct identifications are shown in bold.

RESULTS

TESTS WITH AMENDED CLASSIFICATION ONLY

Table 12.3 shows results for different numbers of hidden nodes, ranged between 64 and 104. At the point of training termination, the errors (E_{val}) produced by the network on presentation of the validation set are as shown. A hidden-node number of 104 produced the lowest mean validation error, so the number of hidden nodes was then fixed to that optimized number. Table 12.4 shows

TABLE 12.3
***Tilia* with Amended Classification[a]: Optimization of Number of H Nodes Showing E_{val} at Point of Training Termination**

	Hidden nodes					
Data-set	**64**	**72**	**80**	**88**	**96**	**104**
A	2.94	2.84	2.71	2.99	3.05	2.65
B	3.34	3.24	3.01	3.21	2.93	3.10
C	3.01	2.88	2.99	2.79	3.12	2.88
Mean	3.09	2.99	2.90	3.00	3.03	2.88

[a] INS and JAP included within AMU.

TABLE 12.4
***Tilia* 22 with Amended Classification[a]: Optimization of Learning Rate Showing E_{val} at Point of Training Termination**

	Learning rate					
Data-set	**0.05**	**0.10**	**0.15**	**0.20**	**0.25**	**0.30**
A	2.66	2.65	2.77	3.34	3.21	3.44
B	3.01	3.10	3.00	3.24	3.60	3.55
C	2.89	2.88	2.82	3.33	3.20	3.12
Mean	2.86	2.88	2.86	3.30	3.34	3.37

[a] INS and JAP included within AMU.

TABLE 12.5
***Tilia* with Amended Classification[a]: Overall Results**

	Data-set	E_{val}	R_{val}	R_{test}
A	Mean of 10 runs	2.73	74.71	64.67
	Lowest E_{val}	2.49	70.59	66.67
B	Mean of 10 runs	2.91	81.18	67.67
	Lowest E_{val}	2.68	88.24	70.00
C	Mean of 10 runs	2.81	79.41	66.33
	Lowest E_{val}	2.56	82.35	66.67
	Overall mean (30 runs)	2.82	78.43	66.22
	Mean lowest E_{val}	2.58	80.39	67.78

[a] INS and JAP included within AMU.

the results produced after fixing the number of hidden nodes to 104, but with the learning rate (gain) varied in the range 0.05–0.30. The optimized learning rate was determined to be 0.05 (when E_{val} is considered to three decimal places), so the learning rate was then fixed to that optimized value.

Table 12.5 presents a summary of the final results from runs with the network parameters fixed to the preceding optimized values, but now run with 10 different random seeds (the earlier tests were run with the same random seed). For each training/validation pair (A, B and C), the mean E_{val} errors over the 10 runs are shown, together with the lowest value. The recognition accuracy (R_{val}) is also shown in a similar manner, and shows the percentage of correct species identification for data records in the validation set. The recognition accuracy (Rtest) resulting from presentation of the independent test set to the trained network (saved at the point of minimum validation error) is also shown. A good result of 66.22 per cent recognition accuracy on the test set was achieved and, when compared to previous results (Clark, 2004), represented an 8.90 per cent improvement after changing the classification.

Tests with Amended Classification plus Geographic Information

Table 12.6 shows results for different numbers of hidden nodes ranging between 64 and 104. At the point of training termination, the error (E_{val}) and recognition accuracy (R_{val}) produced by the

TABLE 12.6
***Tilia* with Amended Classification[a] and Continent of Origin: Optimization of Number of H Nodes Showing E_{val} at Point of Training Termination**

	Hidden nodes					
Data-set	**64**	**72**	**80**	**88**	**96**	**104**
A	1.87	1.74	1.86	1.58	1.88	1.86
B	2.53	2.19	2.36	2.17	2.09	1.96
C	2.34	1.97	2.20	2.03	2.13	2.05
Mean	2.25	1.97	2.14	1.93	2.03	1.96

[a] INS and JAP included within AMU.

TABLE 12.7
***Tilia* with Amended Classification[a] and Continent of Origin: Optimization of Learning Rate Showing E_{val} at Point of Training Termination**

	Learning rate					
Data-set	**0.05**	**0.10**	**0.15**	**0.20**	**0.25**	**0.30**
A	1.59	1.58	1.57	2.04	1.61	1.64
B	2.19	2.17	2.22	2.17	2.17	2.19
C	2.05	2.03	2.22	2.09	2.27	2.21
Mean	1.94	1.93	2.00	2.10	2.02	2.01

[a] INS and JAP included within AMU.

network on presentation of the validation set are as shown. A hidden node number of 88 produced the lowest mean validation error (E_{val}), so the number of hidden nodes was then fixed to that optimized number.

Table 12.7 shows the results produced after fixing the number of hidden nodes to 88, but with the learning rate (gain) varied in the range of 0.05–0.30. In this case, the optimized learning rate was determined to be 0.1, so the learning rate was then fixed to that optimized value. Table 12.8 presents a summary of the final results from runs with the network parameters fixed to the previous optimized values, but now run with 10 different random seeds (the earlier tests were run with the same random number seed). For each training/validation pair (A, B and C), the mean E_{val} errors over the 10 runs are shown, together with the lowest value. The recognition accuracy (R_{val}) is also shown similarly and shows the percentage of correct species identification for data records in the validation set. The recognition accuracy (Rtest) resulting from presentation of the independent test set to the trained network (saved at the point of minimum validation error) is also shown.

An improved result of 82.45 per cent recognition accuracy on the test set was achieved, representing a 16.23 per cent improvement after including information regarding the continent of origin, when compared to the results obtained without geographic information. Table 12.9 shows the new misidentification matrix created from the ANN runs with revised classification and with addition of geographical data. A comparison between the effective performances of all the tests, including those from earlier studies (Clark, 2004), is shown in Table 12.10.

TABLE 12.8
***Tilia* with Amended Classification[a] and Continent of Origin: Overall Results**

Tilia25

	Data-set	E_{val}	R_{val}	R_{test}
A	Mean of 10 runs	1.73	93.53	86.00
	Lowest E_{val}	1.48	94.12	86.67
B	Mean of 10 runs	2.18	90.59	79.67
	Lowest E_{val}	1.84	94.12	83.33
C	Mean of 10 runs	2.09	89.42	81.67
	Lowest E_{val}	1.92	94.12	83.33
	Overall mean (30 runs)	2.00	91.18	82.45
	Mean lowest E_{val}	1.75	94.12	84.44

[a] INS and JAP included within AMU.

DISCUSSION

It is clear from Table 12.10 that the MLP-ANN provides a better identification system than the *DELTA* system and that improvements to the underlying classification are helpful. Clearly, however, the greatest improvement was obtained by the inclusion of minimal geographic information. Any identification system is of use in suggesting inadequacies with respect to the underlying existing classification of taxa, and neural networks are no exception. In the original results (Table 12.2), the *T. amurensis* test specimen (TAMU) was always identified as its close relation, *Tilla insularis* (INS). However, now it is believed that these species should be considered conspecific, as *T. amurensis* Rupr. (Pigott, 2000). The case study presented here supports this view. The suggestion was also made here that *T. japonica* (Miquel) Simonkai should be included within a wider concept (i.e. an amplified view) of *T. amurensis* Rupr. Although this work adds weight to this hypothesis, further botanical studies regarding character variation in both taxa are needed before a firm conclusion can be made, because this study was based on relatively few specimens and therefore the model of the species variation and limits may be incomplete.

Note also that, in earlier work (Clark, 2004), it was suggested that the tree from which the *T. heterophylla* test specimen (THET) was taken was incorrectly named and should perhaps have been *T.* × *moltkei,* a probable hybrid between *T. americana* and *T. tomentosa* 'Petiolaris'. This test specimen closely matches specimens labelled *T.* × *moltkei* in the Kew Herbarium, and it is also identified as this species using Pigott's (1997) key. However, in this study (see Table 12.9), this specimen is variously referred to other American species, and sometimes to the Chinese *T. oliveri.* Certainly this specimen warrants further study – perhaps the indecision of the ANN provides further suggestions of its possible parents. Other anomalies in the confusion matrix include the various misidentifications of *T. dasystyla,* which might be a further indication of the presence of hybrids in the test set. It would certainly be useful in further studies to deliberately include hybrid specimens to investigate how a network trained on species would behave.

Although the anomalous identification of *T. henryana* as *T. mongolica* is still present in the latest results (Table 12.9), the most likely explanation is that they share relative absence of characters, which is affecting the model. Further studies could usefully explore this area, with regard to character coding before training.

It is particularly interesting that the identification performance of the ANN system was greatly improved (by over 16%) by including a small amount of geographical information, providing only

TABLE 12.9
***Tilia* Confusion Matrix: 17 Species with Amended Classification[a] plus Continent of Origin**

Species	AME	AMU	CAR	CHI	COR	DAS	HEN	HET	KIU	MAN	MAX	MIQ	MON	OLI	PLA	TOM	TUA
TAME	**100**																
TAMU		**99.2**		0.8													
TCOR					**78.3**										21.7		
TDAS					46.7	**36.7**					3.3				13.3		
THEN							**20**						80				
THET	13.3		36.7					**20**						20		10	
TKIU									**93.3**					6.7			
TMAX				10							**90**						
TMIQ												**100**					
TMON													**100**				
TOLI											1	44		**45**			
TPLA															**100**		
TTOM					8.3					2.5						**89.2**	
%Conf	*88.2*	*100.0*	—	—	*79.7*	*100.0*	*100.0*	*100.0*	*100.0*	—	*93.1*	*57.7*	*71.4*	*84.9*	*83.3*	*97.3*	—

TABLE 12.10
***Tilia* Overall Results and Comparison of Methods and Performance**

Identification method	% Test specimens correct
DELTA TILIA 22 chars	42.5
MLP TILIA 22 chars	57.3
MLP TILIA 22 chars + revised classification	66.2
MLP TILIA 25 chars + revised classification including geography	82.5

the continent of origin. This is not surprising given that most herbaria arrange specimens of any given genus geographically before actually reaching the species folders. This is, of course, in part because specimens are first collected from a given geographical location, and their first priority is to be filed according to that since the identity of the specimen may only be determined at a later date. However, a primary function of herbaria is identification, and even simple geographic information reduces the possible taxa considerably.

Indeed, a human expert, unable to identify a specimen immediately, often first asks questions about geographic origin to reduce the possibilities. It is therefore considered valid in a practical identification study such as this to include geographical data, as typically might be available with an unidentified collected specimen. In the case of *Tilia,* this is particularly relevant because each species is restricted in the wild to a particular continental area: namely, (1) Europe and western and central Asia, (2) Eastern Asia or (3) North America (Pigott, 2002). Thus, although species themselves are not primarily defined by geography, it is possible to provide easier identification, given that a specimen is from a known continent of origin.

SUMMARY

A simple methodology for the practical application of supervised ANNs for botanical identification has been presented. Here, the main parameters of the ANN are chosen by systematic trial and error – a methodology that can be followed when considering any taxonomic group. Clearly, the main limitations of such a system are those of any identification system: namely, that performance depends on the quantity, accuracy and validity of the available specimens, and the nature of the underlying classification system. It has been shown that a multilayer perceptron (MLP) can be used for an effective identification system based on character and measurement data, provided a systematic methodology, such as that demonstrated here, is employed to tune the system parameters effectively. Neural networks require this fine-tuning for effective performance, and such a simple strategy is needed in order that these kinds of systems can be used in a practical sense by non-experts.

In conclusion, the results presented here (see Tables 12.9 and 12.10) demonstrate that the MLP neural network performance can be improved considerably by the addition of geographical information and a re-evaluation of the underlying classification system. Although encouraging results were produced using only 22 morphological characters and amendments to the existing species concepts, considerable improvements were achieved by including extra information relating merely to the continent of origin, along with a few minor changes in the taxonomy.

Artificial neural networks train best and learn to generalize best if they are presented with good examples of the classes that they are trying to model, especially if based on many examples showing variations representative of those classes the net is attempting to discriminate. Herbarium specimens can provide much data of this kind and are also a primary source of information for taxonomists. The use of neural networks as tools for herbarium systematics is, therefore, to be

TABLE 12.11
Sources of Material for Training and Validation Data-Sets

Specimen	Collection/herbarium no./data on sheet	Specimen location
AME1	B.F. Bush No. 61	K
AME2	H.H. ILTIS 22539, Wisconsin, July 4, 1964	BM
AME3	C.C. Deam 16305, Clarke Co, Indiana, June 23, 1915	BM
AMU1	F. Karo 1080 (Komarov, Fl. Man) isotype	K
AMU2	Wilson 8871 (Arnold Arboretum Expedition) v. *glabrata* Nakai	K
AMU3	U. Faurie 489, July 1906, Korea	BM
CAR1	T.G. Harbison 3 (or sn.), June 14, Lake City, Florida	K
CAR2	(TYPE) Miller's Dictionary No. 4, see Aiton Hort. Kewensis (*pubescens*) (4)	BM
CAR3	(ISOTYPE) A.H. Curtiss No. 401* (TYPE) *floridana*	BM
CHI1	(TYPE) Potanin (1885) Herb Hort. Pet.	K
CHI2	Forrest 6130 (Yunnan)	K
CHI3	Forrest 12765 (Yunnan)	K
COR1	Lindberg 1237	K
COR2	Petunnikow 1012	K
COR3	Pigott 94-31	BM (British herb)
DAS1	(TYPE) Steven sn. (flowering sheet of 2)	K
DAS2	ex Herb. R.J. Shuttleworth, received 1877; 1 of 2, Kastel Dag (locus typus)	BM
DAS3	C.R. Fraser Jenkins 2928, June 29, 1971, Abkhazskaya, ASSR ssp. *caucasica*	BM
HEN1	Wilson 597, W. Hupeh 8/07 and 11/07	BM
HEN2	Wilson 414 6/07 and 10/07	BM
HEN3	(TYPE) *Henry 7452*A	K
HET1	Biltmore Herb. 1030b	K
HET2	Palmer 15471 (v. *michauxii*)	K
HET3	Harper 1302	BM
INS1	(TYPE) Wilson 8532 Ulreung-do [Ullung-do], May 31, 1917	BM
INS2	C6926 Cult. Edinburgh. August 1970 B.02	BM
INS3	#10859 Cult. Arnold Arboretum, Clausen et al.	BM
JAP1	E.H. Wilson 7305, Arnold Arboretum Expedition, Hokkaido 1914	K
JAP2	Shirasawa sn., received at Kew: April 24, 2002 (as *cordata* v. *japonica*)	K
JAP3	(ISOTYPE) Herb. Hort. Bot. Petrop. 1 of 2 (HongK3910), Maxim., Japan: Hakodate 1861	K
KIU1	HerbHongK3916, Hayakawa sn.	K
KIU2	H. Shirasawa sn. (Middle Japan) syntype? (in Type folder) [not proposed lectotype]	K
KIU3	S. Matusima, July 5,1952. TSM861	BM
MAN1	V. Komarov1081, Ninguta valley, 1896 (year), ex Herb. Petrop.	BM
MAN2	Wilson 10400, Arnold Arboretum Korea, Prov. Kogen: Kongo-san, June 30, 1917	K
MAN3	Ex Herb. Hort. Petro (received at Kew September 4, 1890)	K
MAX1	(TYPE) Shiras sn, Middle Japan	K
MAX2	Wilson 7129	K
MAX3	Wilson 7387, August 16, 1914, Hokkaido, Japan	BM
MIQ1	(ISOTYPE) Maximowicz, sn., Yokohama, Japan (12/85a)	K
MIQ2	Homi Shirasawa sn. Middle Japan	K
MIQ3	H.E. Fox, sn. July 26, 1912	BM
MON1	(ISOTYPE) Przewalski [3F]	K
MON2	A. David No. 1923. (Herb. Mus. Paris), June + August 1866, Mt. Ta-chan-Kou, Mong.	K
MON3	Hsia 2700, Herb. Inst. Bot. Peiping, China	K
OLI1	Wilson 2274 (705) (flowering sheet of 2)	K

Continued.

TABLE 12.11 ***(Continued)***
Sources of Material for Training and Validation Data-Sets

Specimen	Collection/herbarium no./data on sheet	Specimen location
OLI2	Wilson 2332, W. Hupeh, June 1907	BM
OLI3	Wilson 2335 (specimen with damaged bracts)	BM
PLA1	Fraser 109	K
PLA2	Schreiber 2480 (as *sphaerocarpa*)	K
PLA3	Pigott 006 (cult) Sheet 2, seedling, Slovenija, 46.20′N, 14.04′E	BM
TOM1	J.S. Mill, sn. (Herb. 1862 [as *argentea*])	K
TOM2	(TYPE of petiolaris) Hort. K 1893, t. BotMag.	K
TOM3	Wierzbicki 1600/Herb J. Gay (as *argentea*)	K
TUA1	(HOLOTYPE) *Henry 7452*	K
TUA2	Wilson 2316 (7/07!) a.	K
TUA3	Wilson 2334, July 1907, W. Hupeh	BM

encouraged, as through this use more can be made of these still rather underused and extremely valuable data repositories.

ACKNOWLEDGEMENTS

Grateful thanks are due to Donald Pigott for many helpful discussions and advice. Gratitude is also due to Simon Owens and Martin Cheek for permission to study specimens in the Herbarium of the Royal Botanic Gardens, Kew, UK. Use of the graphic in Figure 12.1 is given by kind permission of the Trustees of RBG Kew. Also, thanks are due to Nigel Taylor for permission to collect specimens from trees growing in the gardens, as well as to Roy Vickery of the Natural History Museum, South Kensington, London, for access to the herbarium.

TABLE 12.12
Sources of Material for Test Data-Set

Specimen	Collection/herbarium no./data on sheet	Specimen location
TAME1	000-73-11560 463 (Dot19) GDNS No. 62: Summer 1980 [+JYC301]	K (Cult.)
TAMU1	1982-8500 JYC320	K (Tree)
TCOR1	000-73.11561 463 (Dot13) Summer 1980 [+JYC319]	K (Cult.)
TCOR2	044-06.04401 461-06 (Dot2) Summer 1980	K (Cult.)
TCOR3	1972-12982	K (Tree)
TCOR4	000-69.16042 124 (Dot39) Summer 1980	K (Cult.)
TDAS1	14/7/61 H/2762/61 near *Camellia cuspidata*	K (Cult.)
THEN1	462-34.46201 2 sheets September 26, 1980: sheet 1 only	K (Cult.)
THET1	000-72.10878 461-06 (Dot1): Summer 1980	K (Cult.)
TINS1	(notNPT/JYC) 000-72-10804 462 GNDS No. 229 (best specimen)	K (Cult.)
TINS2	000-73.11580 463 (Dot11)	K (Cult.)
TINS3	000-73-18796	K (Tree)
TKIU1	1959-77603	K (Tree)
TMAX1	191015602 as R. Melville sn. 29/6/1943 (not NPT/JYC)	K (Cult.)
TMIQ1	463-00.46301 462 (Dot 63) GDNS No. 77 Summer 1980	K (Cult.)
TMIQ2	000-69-16884 GDNS No. 187 Arboretum loc 454 8 Aug 1977	K (Cult.)
TMON1	238-05.23801 462 (Dot 56) Summer 1980	K (Cult.)
TMON2	238-05.23802 462 (Dot 37) Summer 1980	K (Cult.)
TOLI1	134-11.13424 142 (Dot 9) Summer 1980	K (Cult.)
TOLI2	000-69.16887 454 (Dot 3) Summer 1980	K (Cult.)
TOLI3	000-73.18799 462 (Dot 44) Summer 1980	K (Cult.)
TPLA1	207-89.20711 462 (Dot 34) Summer 1980	K (Cult.)
TPLA3	401-71.40116 482 (Dot 18) Summer 1980	K (Cult.)
TPLA4	468-04.46809 462 (Dot 84) Summer 1980	K (Cult.)
TPLA5	624-23.62405 462 (Dot 45) Summer 1980	K (Cult.)
TPLA6	000-73.18793 462 (Dot 71) Summer 1980	K (Cult.)
TTOM1	401-71.40117 482 (Dot 39) Summer 1980	K (Cult.)
TTOM2	470-00.47018 323 GDNS No. 210 Summer 1980	K (Cult.)
TTOM3	1972-12978	K (Tree)
TTOM4	254-36.25403 323 (Dot 48) Summer 1980	K (Cult.)

Notes: Cult. = in cultivated herbarium folders; tree = new specimen collected from living tree in the RBG Kew Gardens.

REFERENCES

Boddy, L., Morris, C.W. and Morgan, A. (1998) Development of artificial neural networks for identification. In *Information Technology, Plant Pathology & Biodiversity* (eds P. Bridge, P Jeffries, D.R. Morse and P.R. Scott), CAB International, Wallingford, UK, pp. 221–231.

Boddy, L., Morris, C.W., Wilkins, M.F., Al-Haddad, L., Tarran, G.A., Jonker, R.R. and Burkill, P.H. (2000) Identification of 72 phytoplankton species by radial basis function neural network analysis of flow cytometric data. *Marine Ecology Progress Series,* 195: 47–59.

Clark, J.Y. (2000) *Botanical Identification and Classification Using Artificial Neural Networks.* PhD thesis, University of Reading, UK.

Clark, J.Y. (2003) Artificial neural networks for species identification by taxonomists. *BioSystems,* 72: 131–147.

Clark, J.Y. (2004) Identification of botanical specimens using artificial neural networks. In *Proceedings of IEEE Symposium on Computational Intelligence in Bioinformatics and Computational Biology* (CIBCB'04), La Jolla, CA, IEEE Press, San Diego, pp. 87–94.

Clark, J.Y. and Warwick, K. (1998) Artificial keys for botanical identification using a multilayer perceptron neural network (MLP). *Artificial Intelligence Review,* 12: 85–115.

Cole, D.T. (1988) *Lithops – Flowering Stones,* Acorn Books, Randburg, South Africa, pp. 99–213.

Dallwitz, M.J. (1974) A flexible computer program for generating identification keys. *Systematic Zoology,* 23: 50–57.

Dallwitz, M.J. (1980) A general system for coding taxonomic descriptions. *Taxon,* 29: 41–46.

Dallwitz, M.J., Paine, T.A. and Zurcher, E.J. (1997) *Users Guide to the DELTA System – A General System for Processing Taxonomic Descriptions*, Edition 4.07, CSIRO Division of Entomology: Canberra, Australia, 136 pp.

Freeman, J.A. and Skapura, D.M. (1992) *Neural Networks: Algorithms, Applications, and Programming Techniques,* Addison–Wesley, Reading, MA, 401 pp.

Haykin, S. (1994) *Neural Networks – A Comprehensive Foundation,* Macmillan College Publishing Company, Inc., New York, 853 pp.

Looney, C.G. (1997) *Pattern Recognition Using Neural Networks,* Oxford University Press, New York, 480 pp.

Morgan, A., Boddy, L., Mordue, J.E.M. and Morris, C.W. (1998) Evaluation of artificial neural networks for fungal identification, employing morphometric data from spores of *Pestalotiopsis* species. *Mycological Research,* 102: 975–984.

Pankhurst, R.J. (1991) *Practical Taxonomic Computing,* Cambridge University Press, Cambridge, 214 pp.

Partridge, T.R., Dallwitz, M.J. and Watson, L. (1993) *A Primer for the DELTA System.* 3rd edition, CSIRO Division of Entomology: Canberra, Australia, 17 pp.

Pigott, C.D. (1997) *Tilia.* In *European Garden Flora* (eds The European Garden Flora Editorial Committee), Cambridge University Press, Cambridge, pp. 205–212.

Pigott, C.D. (2000) The taxonomic status of *Tilia insularis. The New Plantsman,* 7: 178–183.

Pigott, C.D. (2002) A review of chromosome numbers in the genus *Tilia* (Tiliaceae). *Edinburgh Journal of Botany,* 59: 239–246.

Prechelt, L. (1994) Proben1 – A set of neural network benchmark problems and benchmarking rules. Technical Report 21/94, Universität Karlrühe, Germany.

Rath, T. (1996) Klassifikation und Identifikation gartenbaulicher Objekte mit könstlichen neuronalen Netzwerken. *Gartenbauwissenschaft,* 61: 153–159.

Rumelhart, D.E. and McClelland, J.L. (1986) *Parallel Distributed Processing,* Vols. 1 and 2, MIT Press, Cambridge, MA, 1784 pp.

Schneider, C. (1912) *Handbuch der Laubholzkunde,* Vol. 1, Verlag Gustav Fischer, Jena, Germany, pp. 367–389.

Watson, L., Gibbs Russell, G.E. and Dallwitz, M.J. (1989) Grass genera of southern Africa: Interactive identification and information retrieval from an automated data bank. *South African Journal of Botany,* 55: 452–463.

13 Spot the Penguin: Can Reliable Taxonomic Identifications Be Made Using Isolated Foot Bones?

Stig A. Walsh, Norman MacLeod and Mark A. O'Neill

CONTENTS

Introduction 225
Methods and Materials 226
Results 228
Discussion 232
Conclusions 236
Acknowledgements 236
References 236

INTRODUCTION

With a fossil record known to extend as far back as the Early Palaeocene (61 Ma; Slack et al., 2006), Sphenisciformes (penguins) is among the oldest of extant avian clades. Penguins exhibit a remarkable range of adaptations to life in the marine environment, including flattened limb bones, feathers that are unique among birds and solid bones (Davis and Renner, 2003). This latter characteristic has been a feature of the group from earliest times, and to some extent explains their abundance in the fossil record compared with some other avian clades (Fordyce and Jones, 1990). Despite this, the origin, early evolution and systematics of penguins remain poorly understood.

The two most commonly recovered fossil penguin elements are the humerus and tarsometatarsus (a fusion of the three primitive theropod metatarsals common to all modern birds). Most described species are based on either one or the other of these distinctive elements (Fordyce and Jones, 1990). This presents a problem for penguin systematics because neither bone exhibits a wealth of obvious synapomorphies (Simpson, 1946; Zusi, 1975; Fordyce and Jones, 1990). As a result, until very recently (e.g. Slack et al., 2006; Walsh and Suárez, 2006), no cladistic analysis of the group attempted to include fossil taxa.

Examination of skeletons of modern penguin species held in any collection will reveal – even to a non-expert – a high degree of intraspecific variation in postcranial bones (Simpson, 1946; Zusi, 1975). Character polymorphism is also common (Walsh and Suárez, 2006), making character-based taxonomic identification of isolated material (e.g. from abandoned rookeries; Emslie and Woehler, 2005) problematic. This observation does not bode well for the reliability of identifications of fossil material, which on the whole has been (and continues to be) diagnosed on the basis of aspects of overall morphology rather than by clear apomorphies. Comparison of new fossil penguin material with described species is also often problematic because holotype material is held in widely separated institutions worldwide. Many comparisons of fossil humeri and tarsometatarsi are, by necessity, made using published monochrome images. Consequently, determination of differences in overall morphology is generally made solely on patterns of light and shade contained within a photograph. Considering these factors, can reliable identifications of fossil or even extant penguin species ever be made?

In an attempt to answer this question we conducted a preliminary investigation of shape variability in a selection of modern penguin species using the DAISY unsupervised artificial neural net (uANN) (Walsh et al., in press). In this earlier study we focused on two standard views of the penguin humerus (cranial and caudal) and tarsometatarsus (dorsal and plantar). Our rationale was that if an experienced palaeo-ornithologist could recognize species from published images (as is the common practice), then an automated image-recognition system should also be able to provide similar results given that they are both using the same tonal pattern data. We have previously found DAISY image libraries with around 30 images per class to provide optimal results for most object types (MacLeod et al., 2007).

However, due to a lack of suitable specimens, it was only possible to acquire a handful of images for most of the species, and it was only possible to include 10 of the 17 or so living penguin species. Results indicated that, despite image libraries with low numbers of images, DAISY was able to detect taxonomic groupings within all of the training sets. The plantar view of the tarsometatarsus proved to be the more useful of the two elements for identification purposes, although for some taxa (e.g. species of *Spheniscus*) the humerus had more utility. Overall these results indicated that, given sufficient images, DAISY might be capable of accurately identifying all living species of penguin solely on the basis of an image of either the tarsometatarsus or humerus.

These encouraging findings were even more surprising given that the current version of DAISY does not take specimen size into account. This factor is one of the main identification criteria used by human experts. Furthermore, palaeo-ornithologists are able to integrate multiple specimen views and written descriptions in their comparisons. By restricting our previous investigation to object shape, we had effectively denied DAISY access to the most useful information employed by palaeo-ornithologists to make their identifications.

DAISY was clearly able to make its identifications solely on the basis of specimen shape and/or internal patterning. However, the relative importance of these factors is difficult to gauge. In theory, examination of the interneuron weighting scheme constructed when the system builds a new training set should provide some indication of what features DAISY was actually using to make its identifications. However, because the original concept of DAISY as an easily used identification system for biological species did not require this facility, such a tool is not included with the current system. This information is of considerable interest, however, because it could potentially be used to locate features that could be coded for phylogenetic analysis.

Here, we use the plantar view of the tarsometatarsus as a morphometric case study to attempt to discover whether the overall element shape or the patterning within its outline is likely to be of more importance to DAISY as an identification criterion. The selection of this element and view was based on its overall performance in our preliminary study. Although Myrcha et al. (2002) used simple linear distance measurements and ratios of tarsometatarsi to revise fossil Antarctic penguins, only Livezey (1989) has attempted to use true multivariate morphometric techniques on penguin postcrania. Unfortunately, that study was not concerned with intraspecific morphological variability, and this problem has remained uninvestigated.

METHODS AND MATERIALS

Landmark and outline coordinates were acquired from the same images used for the original DAISY training set of Walsh et al. (in press). Although each specimen was originally imaged in two views, only the plantar view was used for the present study. Results of the Walsh et al. investigation are reproduced here. Identification using the DAISY approach is dependent on the accuracy of the original identifications. Since the majority of spheniscid specimens held at The Natural History Museum (Tring) are from wild or captive individuals identified *in vivo,* the collection identifications were regarded as accurate.

For both the DAISY and morphometric analyses, RGB images of the dorsal view of 61 tarsometatarsi images were captured at 72 dpi using a handheld Fujifilm S3000 digital camera. All images were of extant taxa comprising 10 species in five genera from 59 individual skeletons held in the collections of the Natural History Museum, Tring, and were captured on site at Tring (Appendix 4). Number of individuals imaged for each taxon comprised: *Aptenodytes patagonicus* (6), *A. fosteri* (6), *Pygoscelis papua* (6), *P. antarctica* (2), *P. adeliae* (9), *Eudyptes crestatus* (9), *Eudyptula minor* (4), *Spheniscus demersus* (6), *S. humboldti* (3) and *S. magellanicus* (5). Although 56 skeletons were imaged, the extent of adherent connective tissue, mounting materials and damage meant that only 61 tarsometatarsus images were usable. Due to lack of suitable specimens, *Megadyptes antipodes, Eudyptes chrysocome, E. robustus, E. pachyrhynchus, E. schlegeli, E. sclateri* and *Spheniscus mendiculus* were not included in this study, and *Eudyptula minor albosignata* was included as *Eu. minor.*

Each specimen was imaged on a matte black background to maximize contrast and segmentation of the specimen in the frame. All specimens were illuminated consistently from the top left for right-hand specimens and from the top right for left-hand specimens in order to minimize variations in shadowing caused by lighting (necessary for the DAISY analysis). The acquired images were digitally balanced for brightness and contrast, and corrected for physical artefacts (e.g. accession numbering, wire connectors, small amounts of adherent soft tissue) using Corel Photo-Paint 11.0. For consistency, each specimen image was segmented from the original background, inserted onto a uniform black 72-dpi 500- × 500-pixel grid and, where necessary, re-orientated as accurately as possible to a standard north–south pose. Images of left-hand elements were digitally flipped for right-hand standardization. Each image was then resampled to an 8-bit greyscale to minimize colour pattern interference from variable post-mortem staining of the bone surface. Each image in the training set was then named according to taxon at species and genus levels, and assigned a sequential number for cross-reference to the raw image library.

Once prepared in the manner described earlier, the image library was uploaded and built as a training set using the DAISY BUILDTOOL algorithm, which performs three transformations on the images (Figure 13.1). Firstly, the images are histogram normalized to reduce variability between images caused by lighting and exposure. The images are also subsampled to a 32- × 32-pixel grid. This resolution has been found to maximize the signal-to-noise ratio, while reducing processing requirements. Lastly, the images are transformed from a standard *x,y* format to a polar format, which serves to emphasize internal patterns of pixel brightness values while retaining information about the object perimeter in the analysis.

Once built, the consistency of the training set was tested via cross-validation analysis provided by the system's own JACKTOOL algorithm. Identification results are reported at one of three levels. A coordination is a high-level pass that classifies the unknown image according to the number of its nearest neighbours of a given class. The higher the number of nearest neighbours is, the better the classification is, although the lower limit required for a coordination pass is set at three nearest neighbours. Sill identifications classify the unknown image based on its position in the ordination relative to the edge of a class cluster. Although there are a number of classification levels below the Sill level, these are classed as fails as the identifications lack the degree of confidence we require.

The images of the generic-level plantar view training set were renamed using random numbers and the training set was rebuilt. A randomized training set should contain no groupings, or very weak groupings where there is a strong numerical bias towards objects in a given class. This randomized training set was therefore used to test the validity of the groups recovered in the original analysis.

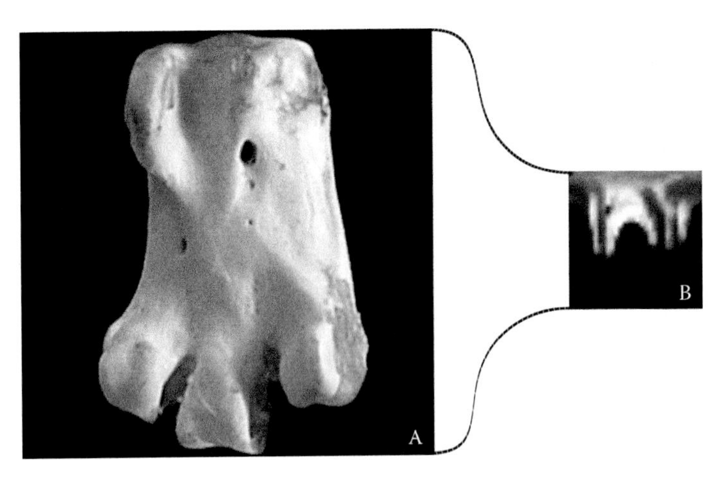

FIGURE 13.1 A 'before and after' example of the DAISY input, and output after the training set build transformations. A: a 500 × 500 image of the plantar view of a tarsometatarsus of *Spheniscus demersus*. B: same image after transformation; the image has been subsampled to 32 × 32 pixels and has been transformed from a Cartesian to a polar arrangement of pixels (note that this image is shown 125 per cent larger than the original for ease of viewing).

Landmark and outline coordinates were collected using tpsDIG 2.04 (Rohlf, 2005). A total of 10 landmarks were used (Figure 13.2A) that were chosen to specify the position of morphological features common to all specimens. Landmarks 1–3 represent the lateral medial and distal points that delimit the triangle of the medial hypotarsal crest; landmarks 4 and 5 delimit the proximal and distal extent of the lateral hypotarsal crest. Landmarks 6 and 7 mark the proximal and distal extent of the lateral proximal vascular foramen; the medial vascular foramen is not visible in some taxa (Zusi, 1975). Finally, landmarks 8–10 indicate the proximal extent of the articular surface of trocleae IV-II. These landmarks were chosen because they mark important segments of highlight and shadow that are likely to be important to the DAISY analysis. Captured landmark coordinate pairs were Procrustes registered in PAST 1.40 (Hammer et al., 2001) and the Procrustes residuals subjected to simultaneous entry canonical variates analysis in SPSS 13.0 at genus and species levels.

Outline data were analysed using the extended eigenshape approach (MacLeod, 1999) for closed curves. The outline was captured using 100 nodes beginning at the medial extremity of the medial cotyla, and registered using seven other landmarks selected at homologous points along the outline (see Figure 13.2B). Using MacLeod's extended eigenshape software (freely available at http://www.nhm.ac.uk/hosted_sites/paleonet/), the 100-node outline was interpolated to 89 nodes at the recommended accuracy level of 99 per cent. Eigenshape scores resulting from the analysis were subjected to simultaneous entry canonical variates analysis in SPSS 13.0 at generic and specific levels.

RESULTS

The DAISY analysis revealed groupings of taxa at both generic and specific levels (Figure 13.3) with an overall pass rate of 43 per cent at the specific and 70 per cent in the generic level. Although these rates may appear comparatively low, it must be remembered that the numbers of images for this test were insufficient to provide dense groupings. The higher results achieved for the generic training set indicate that this is most likely the case, since images for each of the specific classes were combined to form the generic classes. Because the randomized training set detected no clear groups, the recovered partitions were taken to be reliable.

At the specific level (Figure 13.3A) *Aptenodytes forsteri* (62.5%), *Pygoscelis adeliae* (60.0%) and *Spheniscus demersus* (62.5%) achieved the highest overall pass rates. *Pygoscelis antarctica*,

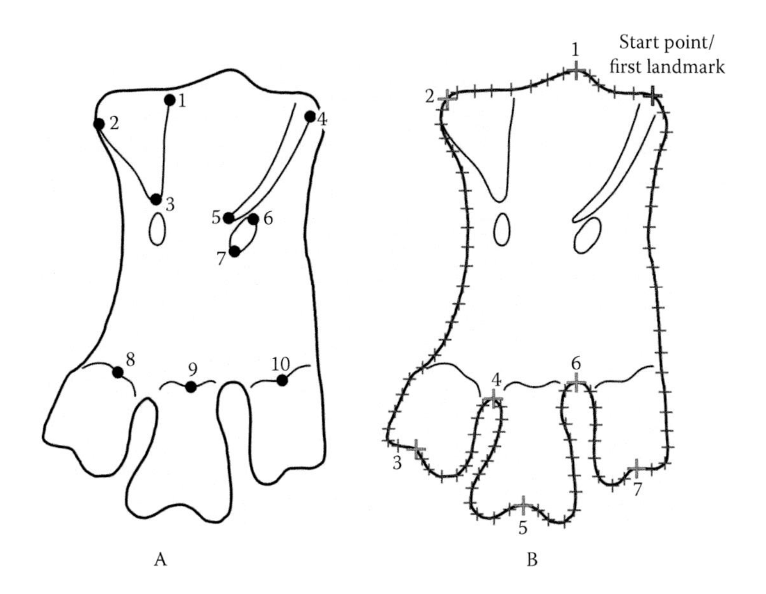

FIGURE 13.2 Morphometric data collection. A: ten landmarks that define broad areas of highlight and shadowing were collected: 1–3, medial, and 4 and 5 lateral hypotarsal crests; 6 and 7, lateral proximal foramen; 8 and 9, proximal limits of the metatarsal trochleae. Note that the medial proximal foramen is not visible in *Spheniscus, Eudyptes* or *Eudyptula* and was therefore not suitable for inclusion in the landmark constellation. B: 100-node outline with seven landmarks (excluding the start point): 1, intercotylar prominence; 2, medial extent of medial cotyla; 3, 5 and 7, groove of metatarsal trochleae; 4 and 6, proximal extent of the intertrochlear notches.

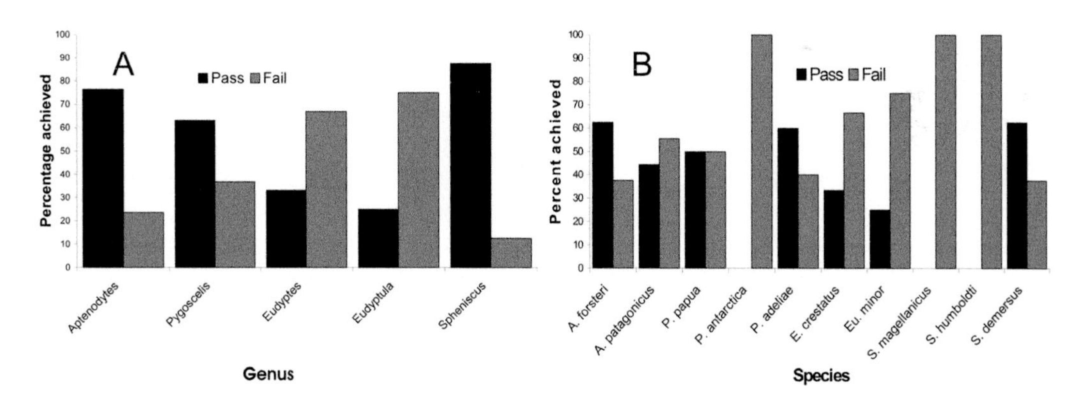

FIGURE 13.3 Results of cross-validation test of the DAISY training set. A: at the genus level all images pass, but only *Aptenodytes, Pygoscelis* and *Spheniscus* achieve coordination-level passes. B: the species level test also achieved coordination-level passes for *Aptenodytes* (*A. forsteri*), and two species of *Pygoscelis* (*P. adeliae, P. papua*); only *Spheniscus demersus* achieved coordination passes for *Spheniscus.*

Spheniscus magellanicus and *S. humboldti* achieved no correct identifications, a result probably explained by the low number of images (between three and five) in these classes. A list of failed matches is given in Table 13.1. At the generic level (Figure 13.3B) *Aptenodytes* (76.5%), *Pygoscelis* (63.0%) and *Spheniscus* (87.5%) achieved the highest pass rates. As the only monotypic genera (in this training set) and with only three and four images, respectively, *Eudyptes* (33.0%) and *Eudyptula* (25.0%) achieved the lowest number of correct identifications. A list of failed matches is given in Table 13.2.

Canonical variates analysis of the landmark data provided reasonable separation for species and genera (Figure 13.4A), with the first seven functions significant at the species level (function

TABLE 13.1
Accurate and Failed Matches for DAISY Analysis of the Specific Level Training Set

	forsteri	*patagonicus*	*crestatus*	*minor*	*adeliae*	*antarctica*	*papua*	*demersus*	*humboldti*	*magellanicus*
forsteri	62.5	37.5	0	0	0	0	0	0	0	0
patagonicus	11.1	44.4	22.2	0	0	0	11.1	0	0	0
crestatus	0	0	33.33	0	0	0	0	33.3	0	33.3
minor	0	0	0	25.0	0	0	0	25.0	0	50
adeliae	0	0	10.0	0	60.0	0	20	0	0	10.0
antartica	0	33.3	33.3	0	33.3	0	0	0	0	0
papua	0	50.0	0	0	0	0	50.0	0	0	0
demersus	0	0	0	0	0	0	0	62.5	12.5	25.0
humboldti	0	33.3	0	0	0	0	0	66.67	0	0
magellanicus	0	0	20	0	0	0	0	80	0	0

TABLE 13.2
Accurate and Failed Matches for DAISY Analysis of the Generic Level Training Set

	Aptenodytes	*Eudyptes*	*Eudyptula*	*Pygoscelis*	*Spheniscus*
Aptenodytes	76.5	11.75	0	11.75	0
Eudyptes	0	33.33	0	0	66.67
Eudyptula	0	0	25.0	0	75.0
Pygoscelis	21.0	10.5	0	63.0	5.5
Spheniscus	6.25	6.25	0	0	87.5

1: $\chi^2 = 497.216$, $p < 0.001$, Wilk's $\lambda = < 0.001$; function 2: $\chi^2 = 346.461$, $p < 0.001$, Wilk's $\lambda = 0.001$; function 3: $\chi^2 = 241.223$, $p = < 0.001$, Wilk's $\lambda = 0.006$; function 4: $\chi^2 = 165.809$, $p < 0.001$, Wilk's $\lambda = 0.023$; function 5: $\chi^2 = 117.532$, $p < 0.001$, Wilk's $\lambda = 0.080$; function 6: $\chi^2 = 77.698$, $p = 0.004$, Wilk's $\lambda = 0.188$; function 7: $\chi^2 = 48.964$, $p = 0.036$, Wilk's $\lambda = 0.349$). Although not identical to the DAISY statistical output, cross-validation (leave one out) was applied to the ordination to allow a better degree of comparability between the morphometric approaches and uANN (Table 13.3). In this test *P. adeliae* performed very poorly, with 66.7 per cent misidentified as *P. papua* and 33.3 per cent as *A. patagonicus*. *S. demersus* also performed poorly, achieving only 42.9 per cent correct identifications, with the rest misidentified as *S. humboldti*.

At the genus level, the landmark results were much improved, with good separations for *Aptenodytes* and *Pygoscelis* in particular. All four canonical functions were highly significant (function 1: $\chi^2 = 294.028$, $p < 0.001$, Wilk's $\lambda = 0.002$; function 2: $\chi^2 = 171.568$, $p < 0.001$, Wilk's $\lambda = 0.030$; function 3: $\chi^2 = 87.705$, $p = < 0.001$, Wilk's $\lambda = 0.167$; function 4: $\chi^2 = 38.153$, $p < 0.001$, Wilk's $\lambda = 0.459$). The two monotypic genera in this data-set, *Eudyptes* and *Eudyptula*, performed poorly in the cross-validation test (Table 13.4), although the direction of failure remained the same as that for the species level.

Extended eigenshape analysis of the element outline provided poorer separations than the landmark data. The first eigenshape axis accounted for 98.7 per cent of the shape variation, confirming the subjective visual impression that these elements are all very similar in outline shape. Canonical variates analysis (CVA) of the eigenscores for the species level (Figure 13.5) revealed obvious groupings, although there was a large degree of between-group overlap. Only the first three canonical functions were significant (function 1: $\chi^2 = 284.862$, $p < 0.001$, Wilk's $\lambda = 0.003$; function 2: $\chi^2 = 181.742$, $p < 0.001$, Wilk's $\lambda = 0.026$; function 3: $\chi^2 = 95.264$, $p = 0.001$, Wilk's $\lambda = 0.149$).

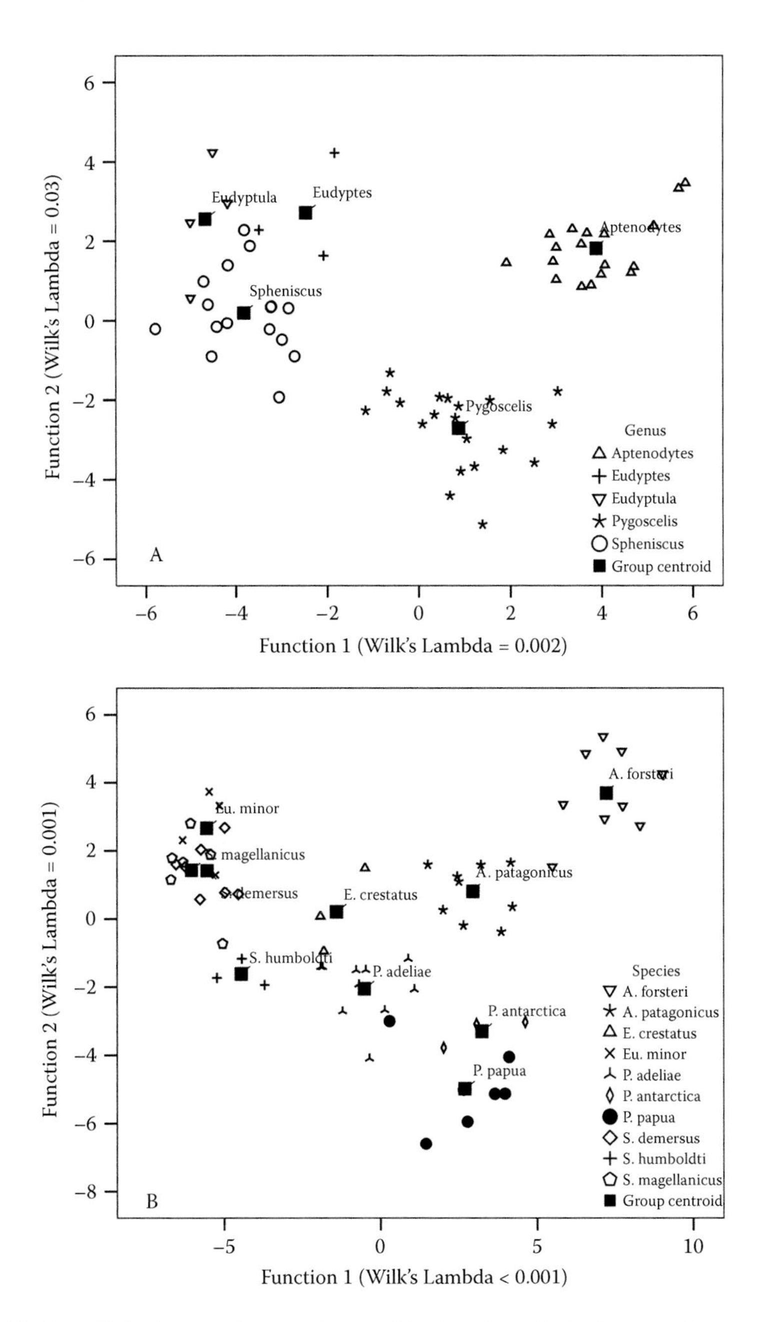

FIGURE 13.4 CVA plots for the superimposed landmark analysis. A: genus-level results; B: species-level results. See text for discussion.

A cross-validation test (Table 13.5) showed all *E. crestatus* were misidentified as *P. adeliae,* and all of *Eu. minor* as *E. crestatus, P. antarctica, S. demersus* or *S. humboldti.* No species of *Spheniscus* achieved more than 40.0 per cent correct identifications. These species were mostly mistaken for each other, although 37.5 per cent of *S. demersus* outlines are misidentified as *Eu. minor.*

At the genus level, eigenshape-based groupings were generally distinct, but *Eudyptula* clearly overlapped *Spheniscus,* and *Eudyptes* is widely spaced within *Pygoscelis.* Again, only the first three canonical functions were significant (function 1: $\chi^2 = 197.303$, $p < 0.001$, Wilk's $\lambda = 0.023$; function 2: $\chi^2 = 99.670$, $p < 0.001$, Wilk's $\lambda = 0.150$; function 3: $\chi^2 = 32.272$, $p = 0.009$, Wilk's $\lambda = 0.541$).

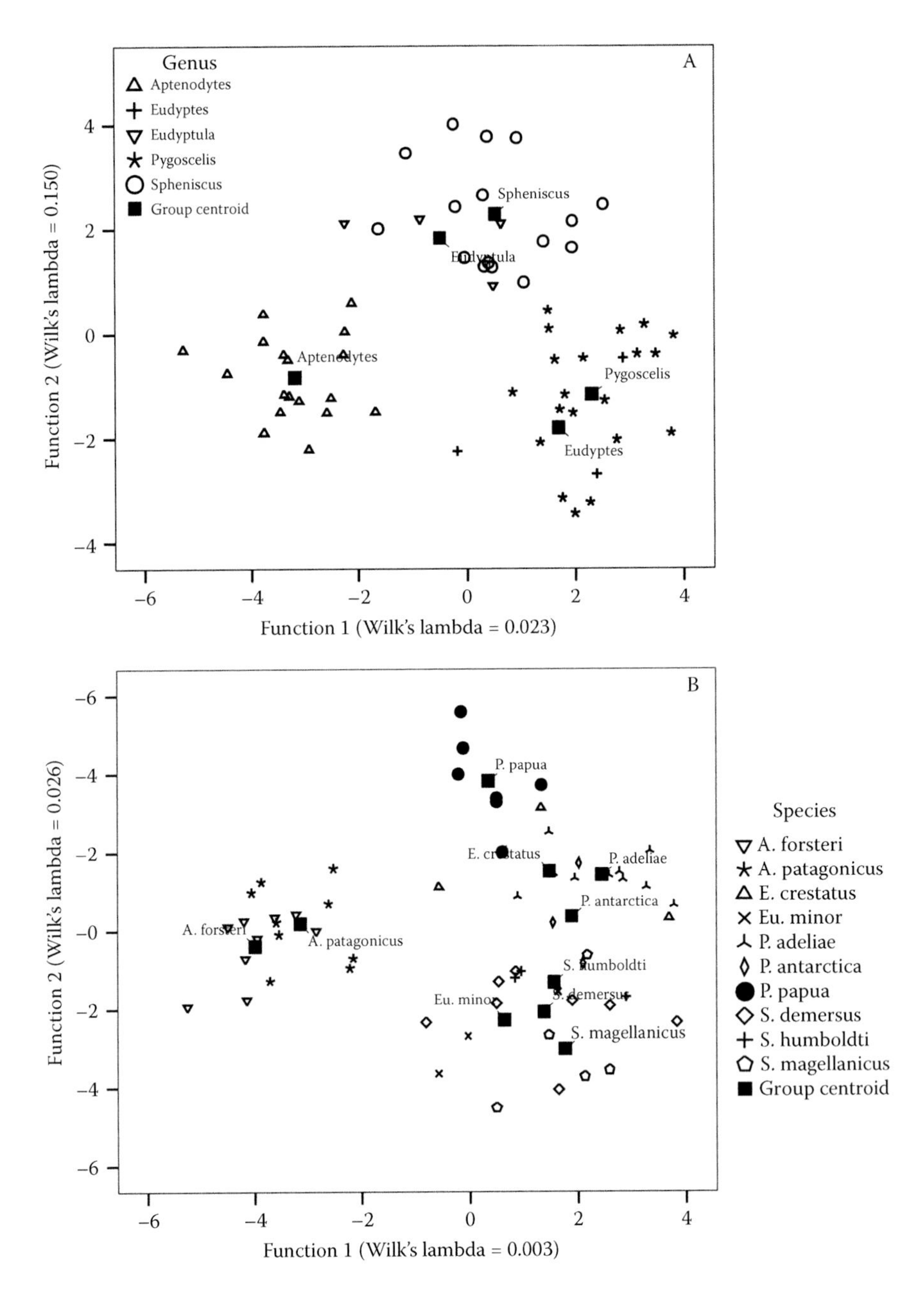

FIGURE 13.5 CVA plots for the extended eigenshape analysis. A: genus-level results; B: species-level results. See text for discussion.

Cross-validation results (Table 13.6) were much improved compared with those of the species level test. With 66.7 per cent of *Eudyptes* misidentified as *Pygoscelis,* results of the two cross-validation tests are consistent.

DISCUSSION

The original study using DAISY (Walsh et al., in press) was intended as an investigation of whether a human expert could realistically identify a penguin taxon from only a black-and-white image.

TABLE 13.3
Cross-Validation Results (Percentages) for CVA of Superimposed Landmark Data at Species Level

	forsteri	*patagonicus*	*crestatus*	*minor*	*adeliae*	*antarctica*	*papua*	*demersus*	*humboldti*	*magellanicus*
forsteri	88.9	11.1	0	0	0	0	0	0	0	0
patagonicus	0	100	0	0	0	0	0	0	0	0
crestatus	0	0	66.7	0	0	0	0	0	33.3	0
minor	0	0	25	50	0	0	0	25	0	0
adeliae	0	0	0	0	90	0	0	0	0	0
antartica	0	33.3	0	0	0	0	66.7	0	0	0
papua	0	0	0	0	14.3	14.3	71.4	0	0	0
demersus	0	0	0	0	0	0	0	42.9	0	57.1
humboldti	0	0	0	0	0	0	0	33.3	66.7	0
magellanicus	0	0	20	0	0	0	0	16.7	0	83.3

TABLE 13.4
Cross-Validation Results (Percentages) for CVA of Superimposed Landmark Data at Genus Level

	Aptenodytes	*Eudyptes*	*Eudyptula*	*Pygoscelis*	*Spheniscus*
Aptenodytes	100	0	0	0	0
Eudyptes	0	66.7	0	0	33.3
Eudyptula	0	25	50	0	25
Pygoscelis	10	5	0	80	5
Spheniscus	0	6.3	0	0	93.8

TABLE 13.5
Cross-Validation Results (Percentages) for CVA of Extended Eigenshape Data at Species Level

	forsteri	*patagonicus*	*crestatus*	*minor*	*adeliae*	*antarctica*	*papua*	*demersus*	*humboldti*	*magellanicus*
forsteri	88.9	11.1	0	0	0	0	0	0	0	0
patagonicus	33.3	66.7	0	0	0	0	0	0	0	0
crestatus	0	0	0	0	100	0	0	0	0	0
minor	0	0	25	0	0	25	0	25	25	0
adeliae	0	10	10	0	70	20	0	0	0	0
antarctica	0	0	0	0	33.3	66.7	0	0	0	0
papua	0	0	0	0	0	14.3	85.7	0	0	0
demersus	0	0	0	37.50	0	0	0	25	25	12.5
humboldti	0	0	0	0	0	0	0	33.3	33.3	33.3
magellanicus	0	0	0	0	0	0	0	40	20	40

TABLE 13.6
Cross-Validation Results (Percentages) for CVA of Extended Eigenshape Data at Genus Level

	Aptenodytes	*Eudyptes*	*Eudyptula*	*Pygoscelis*	*Spheniscus*
Aptenodytes	94.4	0	5.6	0	0
Eudyptes	0	33.3	0	66.7	0
Eudyptula	0	0	50	0	50
Pygoscelis	0	10	5	85	0
Spheniscus	0	0	18.8	0	81.3

The results of that study did indeed suggest that this is possible, particularly if the expert also has available extra information in the form of measurements and written descriptions. Outline shape and internal patterning in these images are obvious and potentially important sources of information for the DAISY uANN and human expert alike. The present study attempted to investigate the relative importance of each by extracting and analysing these data separately.

Each morphometric approach provided clear taxonomic groupings in both genus and species, with up to 100 per cent correct identifications achieved. Overall, superimposed landmarks provided the highest percentages of correct species identifications (mean 66%), with only two species achieving less than 50 per cent correct identifications. Eigenshape analysis was less effective at identifying these species, with a mean of 48 per cent correct identifications and five species achieving less than 50 per cent correct (two of which failed to achieve any correct identifications). Set against these results, the DAISY species analysis was less effective, with an average of 34 per cent correct identifications; three species achieved less than 50 per cent and three achieved no correct identifications. Consequently, for these objects, the information extraction and analysis strategy using these morphometric techniques is clearly much more effective than the whole image analysis employed by DAISY. The results achieved using DAISY for other object types (MacLeod et al., 2007) indicate the system is capable of far better performance, and it seems likely that the extra information contained within these particular images is impeding its effectiveness. Whether this problem is a result of the presence of conflicting morphological information or merely large amounts of uninformative tonal 'noise' is not possible to determine based on these tests. These possibilities are currently being investigated.

At the genus level it is obvious that *Aptenodytes* represents a distinct shape, with 100 per cent correct identifications in the landmark analysis, and only 5.6 per cent failing as *Eudyptula* in the outline test. The tarsometatarsus of this genus is perhaps the most easily identified by human experts, but this is unsurprising as it also happens to be by far the largest. *Aptenodytes*, *Pygoscelis* and *Spheniscus* returned the highest percentages of correct identifications in the DAISY analysis, although *Spheniscus* scored the highest at 87.5 per cent. *Spheniscus* was also 14 per cent higher than *Pygoscelis* in the landmark test, but four percent lower than that genus in the outline analysis. *Spheniscus* thus appears to be a slightly more robust genus than *Pygoscelis*. *Eudyptes* and *Eudyptula* apparently performed poorly in all analyses – particularly in the outline test. However it must be remembered that these taxa were represented by too few images to form well defined clusters (*Eudyptes* = 3, *Eudyptula* = 4).

The spread and direction of misidentifications for each analysis is informative with respect to group overlap. The DAISY analysis misidentified some *Aptenodytes* images as *Eudyptes* and *Pygoscelis,* whereas the outline analysis misidentified a small proportion (5.6%) as *Eudyptula*. The latter result seems somewhat surprising considering that the short and robust aspect of the *Aptenodytes* tarsometatarsus is almost the opposite of the narrow and gracile *Eudyptula* element. Despite being a relatively distinct taxon in each of the analyses, *Spheniscus* appears to be something of a wastebasket with regard to misidentifications; only *Aptenodytes* was never misidentified as *Spheniscus* across the analyses. However, in the outline analysis, only *Eudyptula* was misidentified as *Spheniscus*. *Spheniscus* itself was misidentified as *Eudyptula* in the outline analysis and as *Eudyptes* in the landmark test, but as *Eudyptes* and *Aptenodytes* in the DAISY analysis. In the DAISY and landmark analyses, most incorrect *Pygoscelis* images were misidentified as *Aptenodytes*, a tendency consistent with the close relationship between the genera found in most cladistic analyses of the Spheniscidae (see Walsh and Suárez, 2006, for a review). Overall, these results indicate that, apart from *Aptenodytes,* there is a considerable morphological overlap at the generic level, particularly with regard to the position of the *Eudyptes* and *Eudyptula* point clouds within *Spheniscus* and *Pygoscelis.*

The species-level results provided more insight into partitioning of the generic clusters. In the landmark analysis, *A. patagonicus* achieved 100 per cent correct identifications in the cross-validation test, probably largely accounting for the result at generic level. This finding is, however, at odds with both the eigenshape (67% correct) and the DAISY (44% correct) analyses. In the outline and landmark tests, the two *Aptenodytes* species are only mistaken for each other, further confirming the distinctiveness of this genus, though not the individual species. However, *P. antarctica* failed to achieve any correct identifications in the DAISY and landmark analyses, and *E. crestatus* and *Eu. minor* were entirely misidentified in the eigenshape analysis. This is again probably mostly a result of the low numbers of images (<4) for these species. In the landmark and outline analyses *Pygoscelis papua* achieved reasonably high proportions of correct identifications (86 and 71.5%, respectively), but was misidentified exclusively as other members of the same genus. A similar situation was observed in *P. adeliae* (although the misidentifications were spread between *A. patagonicus* and *E. crestatus*), suggesting that, notwithstanding the poor performance of *P. antarctica*, species of *Pygoscelis* are reasonably distinct.

The misidentification of *Eudyptula* as *Spheniscus* proved to be due to misidentification as *S. humboldti* and *S. demersus* in the eigenshape analysis, and *Eudyptula* was also mistaken for *S. demersus* in the landmark test. Landmark analysis proved to be the most effective strategy for separating species of *Spheniscus,* indicating that differences between those species relate to internal features rather than overall shape. This finding is interesting, since living species of *Spheniscus* are remarkably homogeneous and polymorphic in their postcranial osteology, a situation that may relate to a comparatively recent origination of the genus within Spheniscidae (Bertelli and Giannini, 2005; Walsh and Suárez, 2006). Otherwise, in both morphometric analyses the *Spheniscus* species are mostly misidentified as each other, with only *S. demersus* misidentified as *Eu. minor* in the outline analysis. By comparison, DAISY misidentified *S. humboldti* as *A. patagonicus* and *S. demersus* (the highest achiever at 62.5% correct) as *E. crestatus.* In both morphometric tests, *S. demersus* performed poorly, with a highest score of only 43 per cent correct identifications in the landmark analysis, and only 25 per cent in the eigenshape analysis. These results suggest that *S. demersus* occupies a generalized shape space within *Spheniscus.*

Can these results shed any light on the criteria DAISY uses to make its identifications? The results of these three analyses are neither strikingly similar to nor different from each other (a repeated measures ANOVA revealed no significant difference between them). However, there are some similarities between the DAISY list of failed matches and the CVA cross-validation results for the morphometric approaches. The DAISY results are similar to the eigenshape analysis in the direction of misidentifications of *Eudyptula* at generic level (both are misidentified exclusively as *Spheniscus*). However, DAISY more closely matches the landmark results in (1) the ranked order of correct identifications at generic level (differing only in the position of *Aptenodytes* and *Spheniscus* as first and second place), (2) the direction of misidentifications at generic level for *Eudyptes* (both exclusively misidentified as *Spheniscus*), (3) *Pygoscelis* (first = *Pygoscelis,* second = *Aptenodytes,* third = *Eudyptes,* fourth = *Spheniscus,* although *Eudyptes* and *Spheniscus* are equal third in the landmark results), (4) *Spheniscus* (both misidentified as *Eudyptes,* although DAISY also allocated an equal proportion of misidentifications to *Aptenodytes*) and (5) the total absence of correct identifications for *P. antarctica* at specific level.

These general similarities suggest that, for this training set, DAISY primarily used tonal patterns within the object outline to construct the ordination. Since the purpose of the polar transform in training set construction is partly to emphasize internal patterning, these results suggest that the use of polar thumbnails as training set objects is effective. It also suggests that enhancement of tonal patterns internal to object outlines will be necessary to boost the signal-to-noise ratio. For many identification problems, this may be unnecessary providing sufficiently large image libraries are available. However, pattern enhancement may allow smaller image libraries (such as the one tested here) to be effective as full training sets, and may also improve the performance of the system for larger libraries. We are currently investigating these possibilities.

To date only Livezey (1989) has attempted a morphometric analysis of penguin tarsometatarsi. That study used only a traditional interlandmark distance-based approach. Although we were not able to include all living species in our investigation, our results provide more information about the shape variability of this element in penguins than has been present in any systematic study of this group to date. However, it is also because our data-set is limited with regard to taxa and specimens that we have chosen not to take our analysis further to include modelling of shape change across spheniscid species. Our intention is to investigate this aspect when more material is available. New three-dimensional approaches to morphological analysis (see MacLeod et al., this volume) also offer powerful tools for this research and might help to characterize shape change with better resolution than two-dimensional approaches.

CONCLUSIONS

Our study indicates that the interspecific morphological variability of the tarsometatarsus of extant penguins exceeds the intraspecific morphological variability. This skeletal element is, therefore, useful for taxonomic identification of isolated remains, and probably also to serve as species-diagnostic fossil material. A landmark morphometric strategy offers the most effective approach to identification of the penguin tarsometatarus at both generic and specific levels. This result suggests that DAISY uses mostly tonal patterning internal to object outlines, indicating that enhancement of this information may be the key to improving the performance of the system. Although the morphometric approaches used here were more effective at species recognition, their application is time- and effort intensive compared with that required for a DAISY analysis. Providing a straightforward method of increasing the signal-to-noise ratio can be found, DAISY and similar uANN image recognition systems will be able to offer a fast and effective alternative to traditional two-dimensional morphometric approaches for the purpose of automated species and generic identifications.

ACKNOWLEDGEMENTS

We thank Monja Knoll (University of Portsmouth) for her help in improving this manuscript, and Jo Cooper and Robert Prys-Jones of the Natural History Museum, Tring, for access to osteological material and useful discussion.

REFERENCES

Bertelli, S. and Giannini, N.P. (2005) A phylogeny of extant penguins (Aves: Sphenisciformes) combining morphology and mitochondrial sequences. *Cladistics,* 21: 209–239.

Davis, L.S. and Renner, M. (2003) *Penguins,* T. and A.D. Poyser, London, 224 pp.

Emslie, S.D. and Woehler, E.J. (2005) A 9000-year record of Adelie penguin occupation and diet in the Windmill Islands, East Antarctica. *Antarctic-Science,* 17: 57–66.

Fordyce, R.E. and Jones, C.M. (1990) Penguin history and new fossil material from New Zealand. In *Penguin Biology,* Academic Press (eds L.S. Davis and J.T. Darby), San Diego, pp. 419–446.

Hammer, Ø., Harper, D.A.T. and Ryan, P.D. (2001) PAST: paleontological statistics software package for education and data analysis. *Palaeontologia Electronica,* 4(1): http://palaeoelectronica.org/2001_1/past/issue1_01.htm.

Livezey, B.C. (1989) Morphometric patterns in recent and fossil penguins (Aves, Sphenisciformes). *Journal of the Zoological Society of London,* 219: 269–307.

MacLeod, N.M. (1999) Generalizing and extending the eigenshape method of shape space visualization and analysis. *Paleobiology,* 25: 107–138.

MacLeod, N., O'Neill, M.A. and Walsh, S.A. (2007) Comparison between morphometric and artificial neural net approaches to the automated species-recognition problem in systematics. In *Biodiversity Databases: From Cottage Industry to Industrial Network* (eds G. Curry and C. Humphries), Taylor & Francis, London, pp. 37–62.

Myrcha, A., Jadwiszczak, P., Tambussi, C.P., Noriega, J.I., Gadzicki, A., Tatur, A. and del Valle, R. (2002) Taxonomic revision of Eocene Antarctic penguins based on tarsometatarsal morphology. *Polish Polar Research,* 23: 5–46.

Rohlf, F.J. (2005) tpsDig, digitize landmarks and outlines, version 2.04. Department of Ecology and Evolution, State University of New York at Stony Brook.

Simpson, G.G. (1946) Fossil penguins. *Bulletin of the America Museum of Natural History,* 87: 1–99.

Slack, K.E., Jones, C.M., Ando, T., Harrison, G.L., Fordyce, R.E., Arnason, U. and Penny, D. (2006) Early penguin fossils, plus mitochondrial genomes, calibrate avian evolution. *Molecular Biology and Evolution,* 23: 1144–1155.

Walsh, S.A., MacLeod, N.M. and O'Neill, M. (in press) Analysis of spheniscid humerus and tarsometatarsus morphological variability using DAISY automated image recognition. Papers arising from the 6th International Meeting of the Society of Avian Paleontology and Evolution, Quillan, 2004. Oryctos.

Walsh, S.A. and Suárez, M. (2006) New penguin remains from the Pliocene of northern Chile. *Historical Biology,* 18: 115–126.

Zusi, R.L. (1975) An interpretation of skull structure in penguins. In *The Biology of Penguins* (ed B. Stonehouse), Macmillan, London, pp. 59–84.

14 A New Semi-Automatic Morphometric Protocol for Conodonts and a Preliminary Taxonomic Application

David Jones and Mark Purnell

CONTENTS

Introduction 240
The Species Concept and *Ozarkodina excavata* 240
Materials and Method 242
  Materials 242
  Data Acquisition: Theoretical Considerations 243
  Data Acquisition: Empirical Protocol 244
    Length 246
    Interprocess Angle 247
    Cusp-Base Width 247
    Denticle Packing 247
    Denticle Number 247
  Analysis of Error 247
Analysis 248
  Morphological Variation in the *O. excavata* Hypodigm 248
    Results 248
  Temporal Variation in the *O. excavata* Hypodigm 250
    Results 250
  Spatial Variation in the *O. excavata* Hypodigm 252
    Results 252
Discussion 253
  Temporal Variation in the *O. excavata* Hypodigm 253
  Spatial Variation in the *O. excavata* Hypodigm 255
  Biological Interpretation of Morphological Variation in the *O. excavata* Hypodigm 255
Conclusions 255
Acknowledgements 256
References 256

INTRODUCTION

Conodonts are a large, extinct clade of stem gnathostomes, possessing a skeleton composed of phosphatic tooth-like elements that formed an oropharyngeal feeding apparatus (Aldridge and Purnell, 1996, Donoghue et al., 1998). The conodont fossil record consists dominantly of disarticulated elements; the huge abundance of these elements throughout the 300 million year span of the clade's existence has made them invaluable tools for establishing and constraining relative ages in the geological record. The exceptionally high quality of the conodont fossil record also offers an unparalleled opportunity to study evolutionary rates, patterns and processes within a vertebrate group. Additionally, by virtue of their phylogenetic position, conodonts have a key role to play in elucidating the sequence of character acquisition near the base of the vertebrate clade.

Exploiting the potential of the conodont fossil record obviously requires a stable taxonomic foundation. However, conodont taxonomy is frequently problematic. As with many palaeontological studies, species boundaries must be delineated based solely on partial skeletal material, which often displays extensive and complex morphological variation.

This would seem an ideal problem for the application of morphometric analysis. Yet no comprehensive morphometric work has been conducted on the clade. Conodont morphometric studies have been published previously, but individually have had rather narrow aims. They range from simple biometric studies to more sophisticated outline analyses. Significantly, they have had a wide variety of goals, demonstrating the utility and flexibility of morphometrics. These goals include analysis of size distributions (e.g. Jeppsson, 1976), examining ontogeny and survivorship (Tolmacheva and Löfgren, 2000, Tolmacheva and Purnell, 2002), testing hypotheses of feeding mechanisms (Purnell, 1993, 1994), identifying biostratigraphically useful morphologies (Barnett, 1972; Murphy and Cebecioglu, 1984; Murphy and Springer, 1989), uncovering ontogenetic patterns (Murphy and Cebecioglu, 1986), taxonomy and species recognition (Croll et al., 1982; Klapper and Foster, 1986, 1993; Sloan, 2000; Girard et al., 2004b), and investigating evolutionary trends (Barnett, 1971; Murphy and Cebecioglu, 1987; Renaud and Girard, 1999; Girard et al., 2004a; Roopnarine et al., 2004).

Nevertheless, the potential of morphometric analysis of conodonts that is suggested by these studies has yet to be fully realized. Here we present a brief review of the hitherto unacknowledged difficulties inherent in morphometric analyses of conodonts, provide a new standardized morphometric protocol with wide cross-taxon applicability and demonstrate its utility by presenting an example of its application to analysis of morphological variation in a conodont species within a taxonomic context.

THE SPECIES CONCEPT AND *OZARKODINA EXCAVATA*

Our focus in this chapter is the conodont species *Ozarkodina excavata* (Branson and Mehl, 1933). This species has a global distribution (Jeppsson, 1974) and a stratigraphic range extending from at least the mid-Silurian to the Early Devonian (Murphy and Cebecioglu, 1986), perhaps originating far earlier (Jeppsson, 1974; Cooper 1975, 1976; Aldridge and Mabillard, 1985). Most authorities currently consider *O. excavata* to be a single species displaying a high degree of continuous morphological variation, which appears not to vary systematically through time. However, the degree of morphological variation that can be included within *O. excavata* is uncertain (Jeppsson, 1974). This problem is difficult to address using traditional methods of qualitative observation: the relative morphological simplicity of the elements within the *O. excavata* skeleton and the complex yet subtle variation they display has led to considerable subjectivity and inconsistency in determining the taxonomic boundaries of the species (Jeppsson, 1974). These uncertainties make *O. excavata* an ideal choice of species for a preliminary application of our morphometric protocols. Our goal is to attempt to test quantitatively the hypothesis that the *O. excavata* hypodigm represents a single species (Our use of the term 'hypodigm' follows Mayr et al., 1953: p. 237, "A hypodigm is all the available material of a species.").

Testing the morphological boundaries of a species begs the question of what a species is and, in order to justify our approach to the specific problem of *O. excavata,* some theoretical consideration of species concepts are required. Readers primarily interested in the methodological aspects of our study, however, may wish to move on to the 'Materials and Method' section.

We frame our hypotheses and interpret our results within the context of a general species concept chosen *a priori*, as recommended by Wiens (2004). Full discussion of the continuing debate over the various merits of different species concepts is beyond the scope of this chapter, but one fact seems unequivocal: despite implicit suggestions to the contrary by many authors (e.g. see contributions to Wheeler and Meier, 2000), no single species concept is universally applicable. Rather than selecting a particular concept – and, inevitably, its associated conceptual baggage – we adopt a pragmatic approach, delineating species as follows.

Extinct and most extant sexually reproducing species are operationally identified through morphological features, or phenetic clusters in a quantitative sense (Sokal and Crovello, 1970). Yet such morphospecies are often implicitly, if not explicitly, considered as proxies for biological species. This is because biological species are composed of reproductively isolated populations (Mayr, 1942, 1969) and the sharing of morphological characters is taken to indicate a shared, common gene pool. In this theoretical definition, biological species are deemed significant because this genetic coherence means that they approach closest to real entities or individuals (Mishler and Donoghue, 1982; Baum, 1998), in comparison with somewhat arbitrary supraspecific taxa.

Unfortunately, identification of biological species is not simply a question of discerning morphological differences; the use of morphology alone is frequently insufficient because often there is not an exact correspondence between morphological distinctiveness and reproductive isolation. Numerous instances of morphologically indistinguishable sibling species are now known in a range of animal groups (see Knowlton, 1993, for a review of marine examples), and intraspecific differences can exceed those between species (e.g. Bell et al., 2002). Polymorphisms such as ecophenotypy can also produce a range of morphologies within one species (e.g. Peijnenburg and Pierrot-Bults, 2004). Moreover, delineating biological species is problematic because the biological species concept is ahistorical and emphasizes intrinsic reproductive isolation mechanisms for species maintenance. Reproductive isolation is obviously impossible to test for in fossil populations, and the inapplicability of species concepts based around potential interbreeding is clear where populations are separated in time, perhaps by millions of years. Of course, in the case of fossils, determining where the boundaries between potential species may lie must be based on recognition of morphological discontinuities. But evaluating the biological significance of these discontinuities requires them to be interpreted within their spatial, temporal and ecological contexts. Only then can the likelihood that distinct morphologies represent reproductively isolated biological species be assessed.

Spatial information can be incorporated by considering the geographic distribution of morphologically distinct fossil populations; extrinsic spatial separation can prevent gene flow between populations, creating the potential for phenotypic differentiation. Moreover, such vicariance is easier to demonstrate in fossil populations than the intrinsic reproductive barriers required by the biological species concept. Environments will also vary across a species' geographic range, and the contrasting selection pressures that result will favour genetic and morphological divergence (the former enhancing reproductive isolation, the latter producing visible change), potentially reflecting the evolution of new species.

Initially, temporal information need only consist of whether a fossil population differs significantly in its morphology from those stratigraphically above or below, where all are initially considered to be the same species. Induction of these patterns explicitly as ancestor–descendent relationships through time is not necessary (but may be undertaken). Nevertheless, these patterns can be assessed to determine whether they would be most sensibly interpreted as the evolving sequence of populations forming a single species or, for example, as an anagenetic pathway, where one species evolves gradually into another. Other factors (e.g. sampling density, stratigraphic completeness) will heavily influence any decision as to which of these alternative evolutionary-taxonomic scenarios is determined to be most probable.

Ecological and biological interpretations, including functional morphology, are also required to better assess the taxonomic significance of observed differences – aiding, for example, in the identification of confounding intraspecific variation caused by ontogenetic change or ecophenotypy. We also emphasize a population-based rather than strictly type-based approach to taxonomy. This is complementary to the application of quantitative analysis involving a large number of specimens and provides a clearer picture of the variation within the population by better constraining non-taxonomic aspects of variation. If the morphological differences between a given population and the type specimens are statistically significant to a standardized level, we feel this is justification for assigning those morphotypes to different species.

Based on the foregoing discussion, we have formulated two levels of hypotheses. The initial null hypothesis is that the morphological variation within the *O. excavata* hypodigm is continuously distributed, supporting the general consensus that the hypodigm is a single morphospecies. The alternative hypothesis is that morphological variation within the hypodigm displays discrete or semidiscrete clustering. If the null hypothesis is falsified and multiple morphological clusters are detected, then, if the morphological clusters correspond to temporally and/or spatially discrete populations and characters that define them have biological significance, the most parsimonious interpretation is that these populations represent separate species.

MATERIALS AND METHOD

MATERIALS

In this study of *O. excavata,* we have chosen to focus on elements that are thought to have occupied the P_1 position in the skeleton. This is partly because P_1 elements are regarded as taxonomically useful (Sweet, 1988), but particularly because this element has a comparatively large number of continuously varying features for measurement and is thus amenable to morphometric analysis. Consideration of multiple quantitative characters is advantageous in a taxonomic context (c.f. Murphy and Cebecioglu, 1986), as it avoids the problem of dividing a sample into different phenetic clusters depending on which particular variable is being considered. This approach also means that correlations between characters are incorporated into the analysis; omission of such correlations can produce misleading results (Willig et al., 1986; Knowlton, 1993).

In order to capture as much of the potential variation as possible, the samples analysed included P_1 elements of *O. excavata* from most of its spatiotemporal range (see Table 14.1). Assignment of these elements to *O. excavata* was based on published opinions and the active input of conodont workers with experience in this taxon. The full data set is provided in Appendix 5. Original sample sizes ranged between 9 and 44 elements. Inequality in sample size can mask differences in variance between samples, so samples were reduced through random subsampling to produce sample sizes of between 10 and 20 (except for the poorly preserved American topotype material).

Including such a 'global' sample of the hypodigm produces an empirical morphospace for *O. excavata* within which individual samples lie. The use of empirical morphospaces has been criticized because of their potential instability with changing sample number and size; however, theoretical morphospaces, although more stable, are also problematic. For example, Villier and Eble (2004) have noted that theoretical morphospaces are dependent on *a priori* models of which variables (and usually a small number) best describe aspects of form, and this may result in unsatisfactory descriptions of object form. To test the robustness of the empirical morphospace with changing sample sizes, we also conducted an analysis of the original, non-subsampled dataset, following McClain et al. (2004). Between the original (n = 536) and subsampled (n = 454) sets, eigenvalues were within 0.02 variance units of each other, variance partitioning between components was identical and variable loadings were all similar. This supports the interpretation that the empirical hyperspace generated by this analysis does approach the stability of a theoretical morphospace.

TABLE 14.1
Sample Information Showing Locality, Age and Sample Size for Samples of *O. excavata* P_1 Elements Subjected to PCA

Locality	Age	N
Lithium, Missouri, USA	mid Ludlow	9
Broken River, Queensland, Australia	Silurian-Devonian boundary	11
Broken River, Queensland, Australia	mid Ludlow	20
East Mount Cellon, Carnic Alps, Austria	mid Lower Ludlow	17
Muslovka Quarry, Barrandian region, Bohemia	Upper Pídolí	12
Netherton, Worcestershire, England	Ludlow, Ludfořdian	13
Ludlow, Shropshire, England	Ludlow, Ludfořdian	20
Hepworth, Ontario, Canada	Wenlock	20
Pusku, Estonia	Llandovery, late Rhuddanian	20
Gerete 2, Gotland, Sweden	Ludlow, Gorstian	20
Alsvik 7, Gotland, Sweden	Ludlow, Gorstian	19
Alsvik 4, Gotland, Sweden	Ludlow, Gorstian	20
Lilla Hallvards 3, Gotland, Sweden	Ludlow, Gorstian	20
Lukse 1, Gotland, Sweden	Ludlow, Gorstian	20
Smiss 2, Gotland, Sweden	Ludlow, Gorstian	17
Snoder 1, Gotland, Sweden	Ludlow, Gorstian	20
Smissarvestrand, Gotland, Sweden	Ludlow, Gorstian	20
Bodbacke 3, Gotland, Sweden	Ludlow, Gorstian	20
Urgude, Gotland, Sweden	Ludlow, Gorstian	20
Sigdarve 1, Gotland, Sweden	Wenlock, Homerian	20
Sudervik 2, Gotland, Sweden	Wenlock, Homerian	16
Svarvare 3, Gotland, Sweden	Wenlock, Homerian	20
Svarvare 1, Gotland, Sweden	Wenlock, Homerian	20
Östergårde 2, Gotland, Sweden	Wenlock, Sheinwoodian	20
Östergårde 1, Gotland, Sweden	Wenlock, Sheinwoodian	20
Total		469

Note: Swedish samples are ordered according to relative age at a series of intervals through the Ludlow and Wenlock Series.

Data Acquisition: Theoretical Considerations

The effective incorporation of biological homology is generally considered key to the power of landmark-based morphometrics, as outlined, for example, by Bookstein (1991). Biological homology is identified through morphological and topological similarity of structures shared with a common ancestor, and thus justifies comparability among these structures in different individuals (see Purnell et al., 2000, for a discussion of homology in the conodont skeleton; Smith, 1988). Unfortunately, biological homology is difficult to incorporate when analysing features *within* conodont elements, owing to their growth pattern (Donoghue, 1998; Donoghue et al., 2007, in press). Element growth is indeterminate, so the number of potential landmarks varies widely, even between P_1 elements of a similar size and, presumably, ontogenetic stage. Furthermore, element growth is accretionary, with the element increasing in size through apposition of apatite lamellae, making empirical identification of homologous landmarks difficult.

Consequently, in the majority of euconodont elements, only two types of homologous landmarks can be relocated consistently. These define biologically homologous structures preserved in the basal cavity: its apex and its distal extremities. Both types of landmarks represent developmentally significant locations. The apex of the basal cavity marks the point of initiation of element growth; the distal extremities of the cavity are the points of incremental addition of crown and basal body tissue along the axis of growth (i.e. the distal tips of the enamel–dentine interface; Sansom, 1996). Moreover, these landmarks occur only along the lower surface of the element crown (the crown being all that is normally preserved in the vast majority of conodonts). Purely landmark-based approaches are thus unable to capture a comprehensive picture of conodont element morphology.

Although biological homology cannot easily be incorporated into measurements of conodont elements, morphological and topological equivalence between measures can be maintained using other approaches. Several outline techniques, including some that respect biological homology, have previously been used to analyse conodonts, but detailed discussion is beyond the scope of this chapter. Traditional multivariate techniques (the application of multivariate analyses to simple biometric variables such as distances, angles, ratios, etc.; Marcus, 1988) can also be utilized, but have several drawbacks in comparison with landmark-based approaches. For example, they do not incorporate biological homology as effectively as landmark-based methods (Bookstein, 1991), so measured variables must have sufficient topological equivalence to ensure biological comparability from element to element. Furthermore, in the absence of a landmark-based framework (e.g. truss analysis), these traditional approaches tend to sample forms in an unsystematic, arbitrary way and thus cannot be used to recover the original form (Strauss and Bookstein, 1982).

However, traditional multivariate techniques do have major strengths. First, they are applicable to incomplete skeletal elements, thus maximizing sample sizes (an important advantage when dealing with fossil material). Second, they can be easily applied by non-experts in morphometrics, which is an obvious benefit. Third, and more specific to this work, because the number of traditional measurements theoretically obtainable from *O. excavata* is limited by the morphological simplicity of its elements, we have not engaged in *a priori* selection of characters to be measured based on which are deemed to be taxonomically useful (unlike many previous studies); we include every measurement that it is theoretically valid to make.

Data Acquisition: Empirical Protocol

Conodont elements were extracted from the host rock using standard acid dissolution techniques (see Stone, 1987, for a review). The elements were placed in 10-well black-field slides, one element to a well. This ensured that each element could be easily relocated. Elements were photographed on the slide. The measurements taken were insensitive to orientation within the x–y plane of the slide. Orientation of this plane with respect to the z-axis did, however, have an effect on the measurements and so was kept as constant as possible through use of a universal stage, which allows independent tilting of the x–y plane. This stage is more sophisticated than Barnett's (1970) design and allows finer control of specimen orientation.

Images were acquired using a Qimaging Evolution Micropublisher 3 color digital CCD camera mounted on a Leica Wild M8 light microscope. Magnification was fixed such that a 2048 × 1536 pixel image captured a field of view approximately 7 × 3 mm in size, the maximum able to accommodate all elements at constant magnification. This apparatus is faster, cheaper and easier to use than scanning electron microscopy (SEM) imaging. Although image quality decreases marginally when photographing smaller elements, this can be overcome in principle by increasing the magnification for these elements, but time constraints prevented this. Specimens were illuminated with both a ring source, to avoid shadows, and directed fibre optic lighting, to maximize incident light. Polarizing filters were required to eliminate the obscuring glare of reflected light. Images were captured as tif files and measurements were acquired using the Media Cybernetics ImagePro Plus® (version 4.5) software on a desktop PC.

Despite the clarity of the resulting images, some enhancement (as defined by Bengtson, 2000) was required before measurements could be made. This was kept to a minimum, to avoid introducing visual artefacts into the images. A HiGauss filter was applied within ImagePro Plus, which proved superior to the unsharp mask generally used for increasing image sharpness (Bengtson, 2000). Slight adjustment of image brightness and contrast was also occasionally required, particularly on darker coloured elements (CAI 4-5). Wherever image clarity following enhancement was poor enough to introduce uncertainty into the measures, the original element was rechecked.

Taking measurements from digital images onscreen ensured greater accuracy and precision than that obtainable using conventional ocular graticule measures. Captured images of a stage micrometer were used to calibrate within the ImagePro Plus software. A data collection protocol that combined manual and semi-automated procedures was adopted. Working speed was under 3 minutes per element, from image capture to data entry into a spreadsheet.

Since there are potentially several methods for the measurement of some element features, we outline our protocols below to justify our particular approach. Element fragments lacking a cusp were not considered, to avoid measuring different parts of the same element. Biological anatomical notation is used throughout (Purnell et al., 2000). Figure 14.1 illustrates this notation together with terms specific to this work. All measurements were obtained from images acquired for the purpose; illustrations from publications were not used (c.f. Sloan, 2000) to avoid potential bias resulting from uncertainty over details of preparation, orientation, etc. All raw measurement data are provided in Appendix 5.

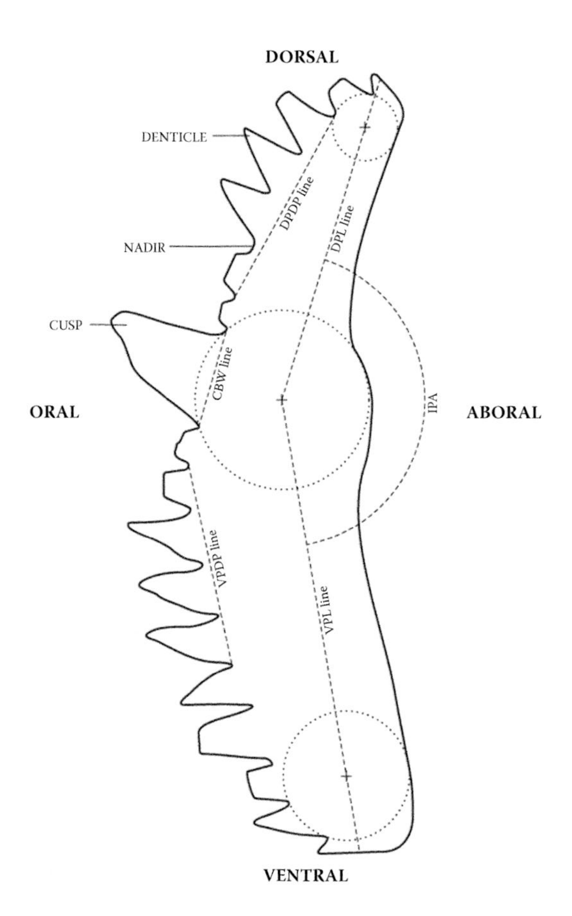

FIGURE 14.1 *O. excavata* P_1 element in lateral view, showing anatomical notation used in this work. Dashed lines represent the measures, as discussed and outlined in Table 14.2. Anchored circles are dotted. See Table 14.2 for key to abbreviations.

TABLE 14.2
Summary of Morphometric Variables Measured in *O. excavata* P_1 Elements

Measurement name	Abbreviation	Measurement description
Ventral process length Dorsal process length	VPL DPL	Linear distance from the cusp anchored point to the distal process terminus, measured along a line passing through the anchored point of the penultimate denticle
Total length of both processes	TL	Sum of dorsal and ventral processes lengths
Interprocess angle	IPA	Angle between dorsal and ventral process length lines
Cusp base width	CBW	Linear distance between interspace nadirs immediately adjacent to cusp
Ventral process denticle packing Dorsal process denticle packing	VPDP DPDP	Calculated using the linear distance from interspace nadir proximal but one from cusp, along the base of four denticles distally
Ratio of cusp base width: mean denticle base width for ventral process	CBW:VPDW	Ratio of cusp base width to mean ventral denticle width along denticle packing line
Ratio of cusp base width: mean denticle base width for dorsal process	CBW:DPDW	Ratio of cusp base width to mean dorsal denticle width along denticle packing line
Ventral denticle number	VDN	Enumeration of denticles on the ventral process
Dorsal denticle number	DDN	Enumeration of denticles on the dorsal process

Note: Variables are illustrated in Figure 14.1.

Length

Length measures were acquired using a technique based on symmetric or median axis analysis. First used for representing outlines by Blum (1973) and later elaborated on by Straney (1988), this technique internalizes the outline of a two-dimensional shape using symmetrical points. These are the centres of circles that contact the margin of a shape tangentially at two or more points. A line drawn through these points forms the symmetric axis of that shape, representing a description of the object's outline. It is perhaps a justification from precedence that Bookstein (1991) and Velhagen and Roth (1997) used the technique in the analysis of jaw shape. We used a variant of the median axis method in this investigation, with the circumference of the circles defined by two interdenticle nadirs and the tangent point to the aboral margin, where a nadir is defined as the point of contact between the free tips of adjacent denticles or between denticle tip and cusp (see Figure 14.1). These circles provide a consistent means of obtaining an effective and intuitive median of upper and lower margins from which to garner data.

This modified technique is fast, easy and accurate. It is important to emphasize that these were not triple point circles *sensu stricto,* since they frequently extended outside the element margin; to distinguish them, they are referred to here as anchored circles. Each process of an element was measured separately, and from this three other measures are automatically obtained: total length as the sum of the process, interprocess angle (see following discussion) and the total element length from dorsal to ventral tip (available but not used in this work).

The precision of the anchored circle technique in comparison with using linear measures constrained directly by the landmark points was tested empirically by repeat measurements where the positions of the anchoring points were varied to mimic the situation in which the placement of the points is unclear on an image. Even when two of the anchoring points for a circle were uncertain, using anchored circles still produced more precise measurements than using a simple line. Thus, the use of anchored circles also reduces measurement error.

Interprocess Angle

Because the anchored circles are affected by both the oral and aboral margins of an element, the difference in height between proximal and distal parts of a process, often observed in carminate elements (classic *O. excavata* P_1 element morphology), will concomitantly drag the circle centre up or down, resulting in different interprocess angles than would be obtained from direct measurement of the aboral margin. This effect should be borne in mind when interpreting the results, as they will not always be directly comparable to interprocesses angles measured along the aboral margin, as used by some previous workers (e.g. Barnett, 1971).

Moreover, in P_1 elements where the aboral margin is curved, as is often the case in *O. excavata,* our approach avoids the ambiguity of placing straight lines along a curved margin to measure the arching of the element. Obviously, the anchored circle technique does not capture information regarding the angular relationship between the oral and aboral margins of a process, but this shape information can be obtained using outline analysis (Jones, 2006). Within the framework of this study, however, interprocess angle is being used to analyse the processes, for which it does provide an appropriate measurement (see Figure 14.1).

Cusp-Base Width

Appositional growth of the element crown may lead to incorporation into the cusp of the denticles adjacent to it. This will vary with ontogeny, and the vertical shifting of the nadirs between the cusp and adjacent denticles that results is expected to render of the measurement cusp base width relatively noisy (see Figure 14.1). This should be taken into consideration when interpreting results.

Denticle Packing

Denticle packing was used as a morphometric variable by Croll et al. (1982), and also for *O. excavata* by Murphy and Cebecioglu (1986), as a measure of denticle number per unit length along a process. The measurement protocol was designed to be applicable to all the elements within the *O. excavata* skeleton. Four denticles were measured, starting at the second nadir from the cusp (see Figure 14.1).

Denticle Number

Enumeration of denticles is occasionally problematic owing to element growth pattern (Donoghue, 1998). Initially, incipient denticles are produced that, with further growth, develop sufficiently to be counted. Such nascent denticles have previously been coded as one-half (Tolmacheva and Löfgren, 2000), but this begs the question of what counts as half a denticle. We use only integer counts here. The problem will only become acute with smaller elements, where a difference of a single denticle can represent a major proportion of the total number. This is more an interpretive caveat than a limitation on the usefulness of the measure.

Analysis of Error

The omission of discussions of error has beset many previous morphometric studies of conodonts. Barnett (1971) made no mention of it, but subsequently (Barnett, 1972) controlled for error to some extent, stating vaguely that two or three operators measured some specimens several times. Error in Barnett (1972) was quoted as ±1° or 2° for angular measures and ±0.01 mm for linear measures. Croll et al. (1982) went no further than simply asserting that their technique was operator independent. Klapper and Foster (1986) constrained error by redigitizing 15 specimens; however, it seems this was conducted once only, and not by different operators. Tolmacheva and Purnell (2002)

TABLE 14.3
Statistics from Repeat Measures Made within ImagePro Plus to Assess Error

	Calibration error	Orientation error (linear)	Orientation error (angular)	Measurement error (linear)	Measurement error (angular)
Replications	50	12	12	12	12
mean	0.842 μm/pixel	0.24 mm	127.5°	0.117 mm	147.7°
Standard deviation	0.008 μm/pixel	0.001	0.477	0.005	0.761
error	±0.03 μm/pixel	±0.004 mm	±1.8°	±0.005 mm	±2.2°

Notes: Error values are based on the difference between mean and the upper/lower value of the range, whichever produces the larger value. Linear orientation error was assessed using the M element of the *O. excavata* skeleton, since this element is the most three-dimensionally curved and so its measurements are most prone to error of this kind. Angular orientation error was assessed using the S_0 element of the *O. excavata* skeleton. This element also is three-dimensionally curved but, unlike the M element, possesses an angular measure. Measurement error was assessed through measurement on an image of an *O. excavata* P_1 element. An image of a small element of lower resolution was selected so as to maximize the potential error and obtain a "pessimistic" estimate.

quoted error at less than 10 μm for their linear measures. No consideration of error was given in any other studies of conodont morphometrics.

Three potential sources of significant operator error were identified in this study: systematic error in calibration within ImagePro Plus was assessed through repeat calibrations of stage micrometer images; errors in linear and angular measures arising from inconsistencies in orientation of the *x–y* plane were assessed by multiple re-acquisition of images for an element, with the stage adjusted afresh each time; pure measurement inconsistencies were assessed by repeat measurements of one linear and one angular measure on one image. For each of these analyses of potential errors, repeat measures were obtained by one operator, with measurements taken days apart to minimize bias from recall. Results are shown in Table 14.3. Analysis of interoperator error (repeat measurement obtained by multiple workers) is in preparation.

ANALYSIS

Morphological Variation in the *O. excavata* Hypodigm

To test the hypothesis that variation within the *O. excavata* hypodigm is continuously distributed, we subjected our full data-set (all variables, 25 samples considered to belong to the *O. excavata* hypodigm; see Table 14.1) to a principal components analysis (PCA). Principal components analysis is a standard technique for reducing dimensionality in multivariate data and is also useful for visualization and exploration of data structure. The samples were not subdivided *a priori* by locality or age, and PCA makes no assumptions that clusters are present within the data. A correlation matrix was used for the PCA because of the different units and scales of the variables. The total length variable was excluded owing to the strength of its correlation with several other variables; this allows it to be used as an independent variable for investigating potential ontogenetic patterns. Eight per cent of the data were missing; this was handled through within-group mean replacement, which reduces the variation within each sample.

Results

The eigenvalues and variance partitioning for the extracted principle components are shown in Table 14.4. Table 14.5 records the loadings of variables on each component. The first three components accounted for 77 per cent of the total variation within the hypodigm, so only these

TABLE 14.4
Eigenvalues and Percentage Variance Explained for the First Three Principal Components of the PCA Conducted on the Global Sample of *O. excavata* P_1 Elements

Component	Eigenvalue	Percent of variance	Cumulative percentage
1	3.593	35.9	35.9
2	2.179	21.8	57.7
3	1.947	19.5	77.2

TABLE 14.5
Variable Loadings on the First Three Principal Component Axes of the PCA Conducted on the Global Sample of *O. excavata* P_1 Elements

	Variable loadings		
Variable	PC-1	PC-2	PC-3
VPL	0.786379	0.078479	0.480474
DPL	**0.925808**	**–0.17235**	–0.01935
IPA	–0.36243	–0.12626	0.575112
CBW	0.615933	0.616282	**–0.28534**
VPDP	–0.60424	0.098606	0.657262
DPDP	**–0.67723**	0.324077	0.206139
CBW:VPDW	0.102695	0.889932	0.224584
CBW:DPDW	–0.0875	**0.899198**	–0.20582
VDN	0.433473	0.109357	**0.832007**
DDN	0.759578	–0.14293	0.209709

Note: Figures in bold are the values for the variables with the heaviest loading on that component. See Table 14.1 for key to abbreviations.

components are considered further. The variance is, however, rather evenly distributed among these components. This probably reflects the large range of conflicting variation that has been observed qualitatively within the *O. excavata* hypodigm, such that this variation cannot be partitioned into one dominant component by the PCA. The morphological variables that weigh most heavily on the component axes, as identified by the analysis, are shown in bold in Table 14.5.

The first principal component (PC-1) primarily represents a contrast between dorsal process length (DPL) and dorsal process denticle packing (DPDP). Elements with high scores on PC-1 have long dorsal processes and wide dorsal process denticle bases; elements with low scores have short dorsal processes and narrow dorsal process denticle bases. Typically, PC-1 is associated predominantly with size differences, such as the increase in dorsal process length.

The second principal component (PC-2) is a contrast between ratio of cusp base width to dorsal process denticle width (CBW:DPDW) and dorsal process length (DPL). Elements with high scores on PC-2 have wide cusp bases relative to the average basal width of dorsal denticles and shorter dorsal processes; elements with low scores have narrow cusp bases relative to the average basal width of dorsal process denticles and have longer dorsal processes.

The third principal component (PC-3) is a contrast between ventral process denticle number (VPDN) and cusp base width (CBW). Elements with high scores on PC-3 have large numbers of denticles on their ventral process and narrow cusp bases; elements with low scores have small numbers of denticles on their ventral process and wide cusp bases.

In order to determine whether there were significant morphological differences between the samples – indicating discontinuities within the range of variation encompassed by the *O. excavata* hypodigm – we conducted a multivariate analysis of variance (MANOVA) on the principal component scores. This produced significant results (Wilk's $\lambda = 0.08$, $F = 22.8$, $p < 0.001$), pointing towards rejection of the null hypothesis of continuous morphological variation.

Unfortunately, although the large sample number necessitated the use of a MANOVA to test for discontinuities within the global range of variation, the data-set showed significant deviation from a multinormal distribution (Mardia multivariate skewness and kurtosis test, $p < 0.001$) and heterogeneous variance (Box's M test, $p < 0.001$), both of which violate the assumptions of a MANOVA. Consequently, the results of the analysis of the global data-set must be treated with caution. Nevertheless, if we accept with this caveat that the null hypothesis can be rejected, further testing of the nature of the discontinuities in the data is worthwhile. Do significant morphological discontinuities within the hypodigm correspond to populations of *O. excavata* that are separated in space and time?

In order to address this question, we conducted separate investigations into the spatial and temporal variation in *O. excavata*. For each sample, eigenscores from the PCA were plotted to facilitate visual examination of the data, and canonical variates analysis (CVA) was used to optimize any clustering. The smaller sample number required for the spatial and temporal analyses allowed a non-parametric MANOVA (NPMANOVA) to be conducted on the data using a Bray–Curtis distance measure, following the procedures described by Anderson (2001).

Temporal Variation in the *O. excavata* Hypodigm

Variation through time was investigated through analysis of a stratigraphic sequence of 12 samples derived from a small area (approximately 10 × 50 km) in the south-west of the island of Gotland, Sweden, for which excellent material is available in the collections of Lennart Jeppsson at Lund University.

Results

Figure 14.2 illustrates how the morphology of the elements from the 12 Swedish samples varies through time. The specimens are ordinated based on eigenscores on the first two principal component axes that bound the morphospace of the hypodigm. As points of reference, the American topotype elements are also plotted on the same axes; interestingly, they lie at the centre of the morphospace. Clusters are evident in the Swedish data, varying through time both in position within the morphospace and in volume of morphospace occupied. The NPMANOVA demonstrates that morphological differences between the Swedish samples are indeed significant ($F = 18.76$, $p < 0.001$). Previous conodont studies have considered such a result sufficient to accept the presence of multiple species within a sample (e.g. Girard et al., 2004b). In actuality, this result provides limited information in a taxonomic context, since it does not indicate *which* samples differ significantly from which. However, pair-wise comparisons between all samples to establish the pattern of significant differences are unsatisfactory because each comparison will discriminate sample pairs based upon different variables, depending on the pattern of variation within those samples.

Furthermore, it is not differences between *all* pairs of samples that are important; rather, it is differences between temporally sequential populations. Therefore, to gain an indication of the significance of the differences in morphology between each sample, a global CVA of the full data

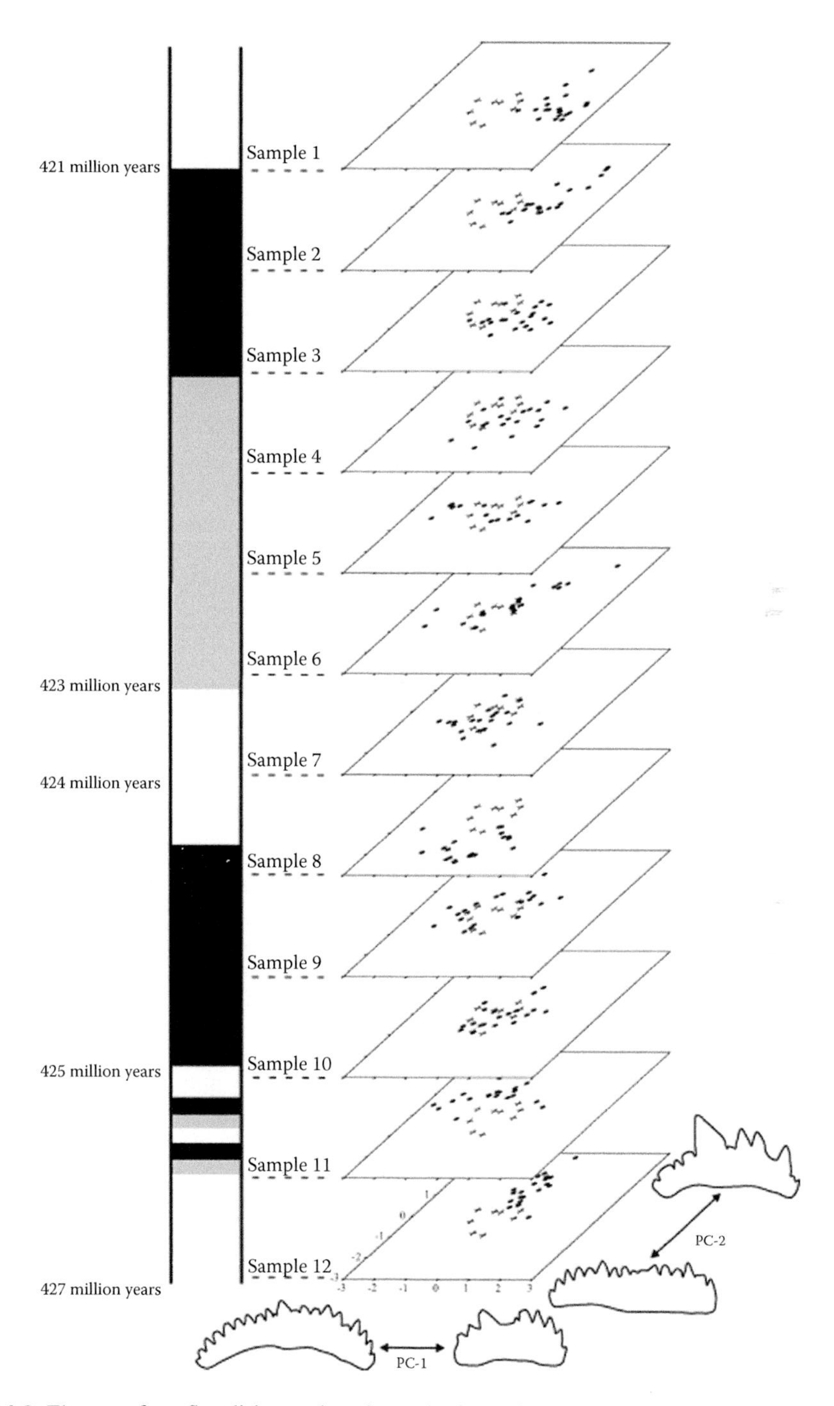

FIGURE 14.2 Elements from Swedish samples taken at horizons through a stratigraphic succession, ordinated against the first two principal component axes, based on eigenscores. The axes are standardized, defining the empirical morphospace of the entire hypodigm. The topotype material (shown as crosses) is plotted for reference. Elements representing the morphological endmembers of the component axes are illustrated. Grey and white sections of column represent Primo and Secundo episodes, respectively; black sections represent events separating these episodes (see text). Approximate age dates are given to indicate sampling density. (Stratigraphic data from Aldridge et al., 1993 and Jeppsson and Aldridge, 2000.)

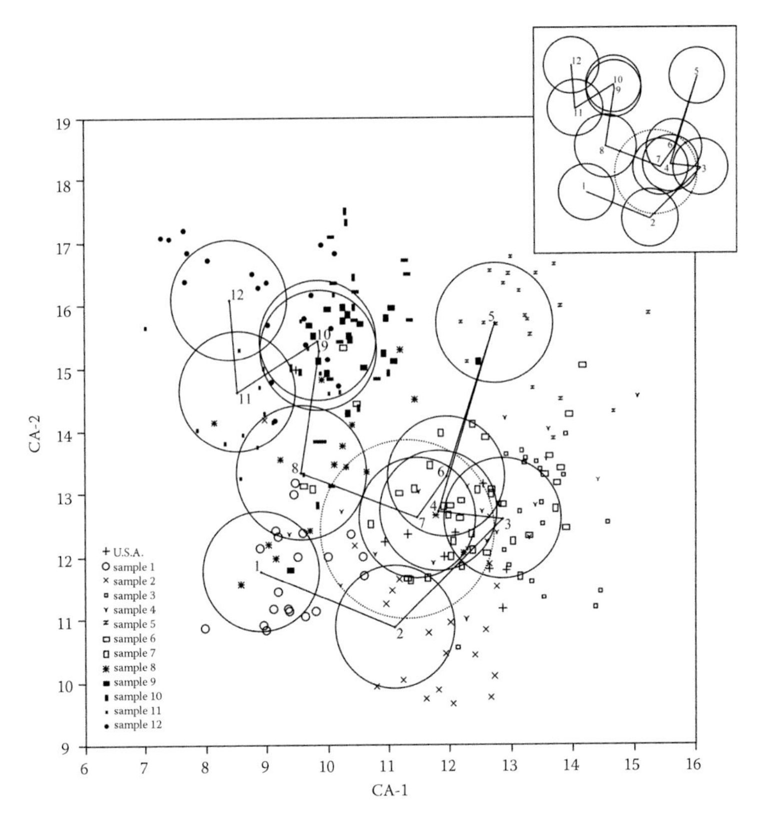

FIGURE 14.3 Ordination of *O. excavata* P_1 elements on the first two canonical axes, based on scores from the global canonical variates analysis. The 12 Swedish samples and topotype material are plotted. Circles represent 95 per cent confidence intervals; that of the topotype material is dotted. The line links samples in stratigraphic order, starting with the oldest: sample 12. Inset shows plot with data points removed for clarity.

was undertaken. The Swedish elements and topotype material are ordinated in Figure 14.3, based on the scores from the global CVA, on canonical axes one and two. Ninety-five per cent confidence ellipses for the samples are shown. Although not strictly equivalent, these confidence ellipses can provide an approximate surrogate for statistical tests: if the ellipses do not overlap, then the samples are significantly different at $p = 0.05$ (Simpson et al., 1960). The line connecting the distribution centres in Figure 14.3 links the samples in temporal sequence, starting with the oldest sample, 12. Confidence intervals for the oldest populations (12–6) overlap, forming a single morphological continuum through time. However, sample 5 differs significantly from samples 4 and 6. Samples 4 and 3 are not significantly different, and the two youngest populations (2–1) differ significantly from sample 3 and each other. Only samples 1 and 5 are morphologically discontinuous with the topotype material.

Spatial Variation in the *O. excavata* Hypodigm

Variation through space was investigated through analysis of four spatially separated samples (Austria, Bohemia, the American topotype material and sample 1 from Sweden). The samples were coeval to within a million years (see Boucot, 1958; Walliser, 1964; and Walmsley et al., 1974, for age data).

Results

Figure 14.4 shows the elements from the four samples ordinated on the first two principal components, scaled as in the previous plots. Some degree of clustering is apparent. The NPMANOVA

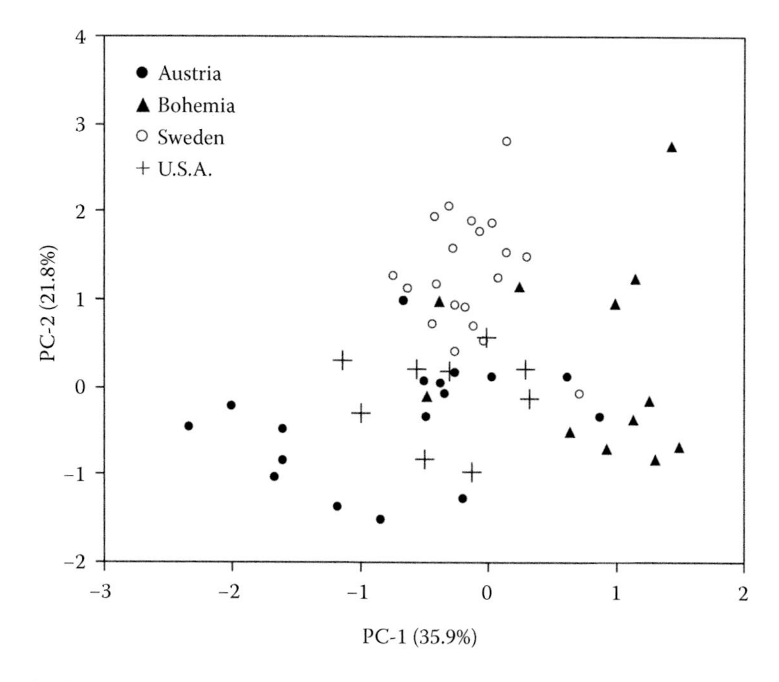

FIGURE 14.4 Ordination of *O. excavata* P_1 elements from four approximately coeval samples (421–422 million years ago) on the first two principal component axes that bound the entire hypodigm, based on eigenscores.

demonstrates that the morphological differences among the four samples are indeed significant ($F = 24.66$, $p < 0.001$). Figure 14.5 shows the elements plotted based on the scores from the global CVA, on the first two canonical axes. Ninety-five per cent confidence ellipses are shown. Confidence intervals for the Austrian, Bohemian and topotype samples all overlap, indicating these populations do not differ significantly in morphology. The Swedish elements are significantly different from all three samples.

DISCUSSION

Temporal Variation in the *O. excavata* Hypodigm

The tests conducted on the Swedish samples demonstrated that statistically significant morphological differences exist between samples of different ages. Examination of 95 per cent confidence intervals provides an indication of the pattern of these differences. For example, populations 12 and 6 differ significantly in morphology; however, the confidence intervals of the stratigraphically intervening populations show overlap, so each population is not significantly different from the populations temporally above and below it (see Figure 14.3). Thus, there is a morphological continuity through time amongst the oldest seven samples. The five younger samples also display significant morphological differences, and there is less morphological continuity between them. The topotype material overlaps or is morphologically continuous with all samples, save 1 and 5.

The significant morphological discontinuities are not correlated with the major environmental changes occurring during the Silurian, which involved switches in oceanic circulation that affected the degree of nutrient up-welling and planktonic abundance (Primo and Secundo Episodes, Aldridge et al., 1993; Jeppsson, 1990). For example, three samples from one Primo Episode (samples 4, 5

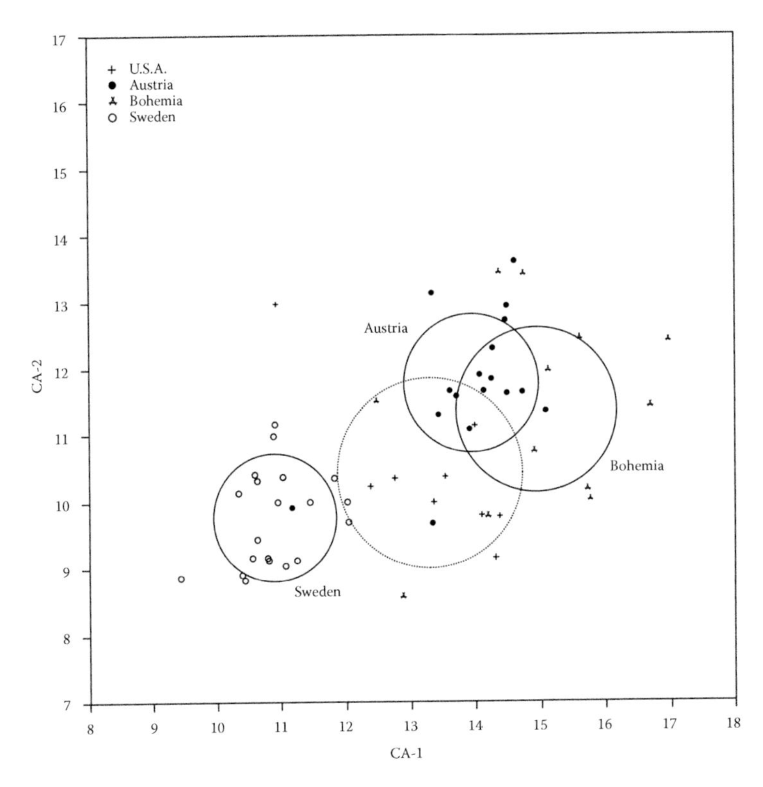

FIGURE 14.5 Ordination of *O. excavata* P_1 elements from four approximately coeval samples (421–422 million years old) on the first two canonical axes, based on scores from the global canonical variates analysis. Circles represent 95 per cent confidence intervals; that of the topotype material is dotted.

and 6 from the Sproge Primo Episode) differ significantly from one another (NPMANOVA, F = 20.85, $p < 0.001$; see also Figure 14.3), suggesting that populations do not exhibit consistent morphological changes in response to such broad-scale changes in environment. More parochial effects may be overriding the general environmental influences and producing the differences, raising the possibility of local ecophenotypic responses. This could be tested by examining the immediate lithological and biotic context of each sample to provide detailed environmental information. If a given morphology is consistently associated with particular local environment across temporal boundaries, ecophenotypy is likely.

One potential difficulty in interpreting the results arises from the nature of the sampling. No sample locality provided a sequence covering the complete temporal range of *O. excavata,* so although the Swedish populations inhabited the same region (an area of approximately 50 × 10 km of present-day south-west Gotland) through time, each is from a different locality. Consequently, the apparent changes through time may represent local spatial differences. For example, morphology may differ between localities and yet be relatively stable at each, so sampling at different localities through time will produce a false appearance of temporal changes. This possibility was evaluated by analysing three coeval samples from different localities on Gotland, each separated by 5–10 km. An NPMANOVA showed no significant differences among the three samples (F = 1.839, $p > 0.1$). Of course, sampling of multiple populations from different localities at every time horizon is necessary to eliminate the confounding effect of spatial variation completely, but this result suggests that the morphological differences observed through time are not reflecting geographic variation.

If spatial and ecophenotypic variation is unlikely, then the differences between the samples probably represent genuine morphological changes through time. Those younger samples whose confidence intervals overlap the topotype material probably do belong to *O. excavata*; conversely, populations 1 and 5 do not appear to be part of *O. excavata*. The non-directional and continuous nature of the morphological change in the sequence of older populations (12–6) would suggest stasis within a single species, in accordance with our null hypothesis. However, the morphological continuity is also compatible with anagenetic speciation, perhaps also reflected in the significance of the morphological differences between sample 12 and sample 6. A higher sampling density is required to better test these alternatives against a random walk (Bookstein, 1987).

Spatial Variation in the *O. excavata* Hypodigm

The tests conducted on the four coeval samples demonstrated that statistically significant morphological differences exist between spatially separated populations. Examination of 95 per cent confidence intervals indicated that a population from Sweden differed significantly from coeval samples in Austria and Bohemia, and also from the topotype material. During the Silurian, some of these populations were separated by major landmasses and wide oceans (Scotese, 2001). These would probably present significant barriers to reproductive coherence between populations. Moreover, these populations would be exposed to different environments and selection pressures. Such differences would be expected to promote genetic and morphological divergence; indeed, significant morphological differences are present between the samples, including that of the topotype material. Again, more detailed study of local conditions will be required to rule out ecophenotypic explanations for the differing morphologies.

Biological Interpretation of Morphological Variation in the *O. excavata* Hypodigm

Understanding the biological context of the observed differences is also critical to test for ecophenotypic or ontogenetic causes of the morphological variation demonstrated earlier to exist in spatiotemporally discrete samples of *O. excavata*. It may also provide additional information regarding the processes that could potentially produce speciation.

The variables that load most heavily on the first principal component relate to the size and shape of the dorsal process. In conodont taxa for which function has been investigated rigorously, this is the food-handling process of the P_1 element (Donoghue and Purnell, 1999). Increasing the length and denticulation of the process has the obvious potential advantage of increasing surface area for shearing of food particles, along with possible benefits from the shape alteration that change in denticulation produces. Thus, the morphological variation may reflect differences in diet between the different populations. Such resource partitioning in extant populations is known to lead to divergent natural selection and morphological differentiation (see Dayan and Simberloff, 2005, for a review with examples). Variations in diet can be tested for using other lines of evidence – for example, examining whether element microwear differs between populations (Purnell, 1995), indicating that different food items are being processed. Also, the different morphologies could be modelled (Evans and Sanson, 2003) to investigate how observed changes in element structure might affect function, helping to constrain the efficiency with which elements of different configurations could process various hypothesized food types. The latter approach may also provide indications of how other characters of the element contribute to food processing or element articulation, particularly those that load heavily on the other component axes.

CONCLUSIONS

In this chapter we have reviewed the generally unacknowledged difficulties in morphometric analysis of conodonts and have presented a new, standardized, semi-automated and widely applicable morphometric protocol that addresses these difficulties. We have provided an example of its application by testing the hypothesis that the taxonomically problematic conodont species *O. excavata* is monospecific. This has demonstrated the efficiency of the protocol in constraining the nature of the morphological variation within the *O. excavata* hypodigm. We have determined that significant morphological discontinuities are present between *O. excavata* populations that are separated in time and space. Moreover, many of these populations have been shown to differ significantly in their morphology from the topotype material of *O. excavata*. Analysis of these data produced by the protocol has also revealed which morphological characters in *O. excavata* best characterize these discontinuities. This permits interpretation of the variation in a biological context, which suggests that these differences are most probably associated with variations in diet.

Biological and spatiotemporal interpretation of the morphological discontinuities suggests that there may be multiple species present in the *O. excavata* hypodigm. We have outlined further independent tests that could more rigorously test these hypotheses, but they are beyond the scope of the work presented here. If the morphotypes identified here actually represent multiple species, then the high discriminatory power of our protocols to distinguish between them suggests that these techniques have potential to provide a generalized tool for conodont species identification that is methodologically standardized and can produce repeatable and reproducible results. Hopefully, the methods and results presented in this chapter will also catalyze more comprehensive morphometric analysis of conodonts using these protocols. We believe such engagement with morphometrics is a crucial first step if a fully automated protocol is to be developed and used effectively for conodont taxonomy, a methodological advance that would contribute substantially towards realizing the potential of the rich fossil record of conodonts for testing evolutionary hypotheses.

ACKNOWLEDGEMENTS

We thank Lennart Jeppsonn, Lund University and Peep Männik, Institute of Geology, Tallinn, for free access to their collections of conodont elements, and Richard J. Aldridge for his valued comments on the manuscript. The funding of NERC (Studentship NER/S/A/2002/10486 to DOJ and Advanced Fellowship NER/J/S/2002/00673 to MAP) is gratefully acknowledged.

REFERENCES

Aldridge, R.J., Jeppsson, L. and Dorning, K.J. (1993) Early Silurian oceanic episodes and events. *Geology*, 150: 501–513.

Aldridge, R.J. and Mabillard, J.E. (1985) Microfossil distribution across the base of the Wenlock Series in the type area. *Palaeontology*, 28: 89–100.

Aldridge, R.J. and Purnell, M.A. (1996) The conodont controversies. *Trends in Ecology & Evolution*, 11: 463–468.

Anderson, M.J. (2001) A new method for non-parametric multivariate analysis of variance. *Austral Ecology*, 26: 32–46.

Barnett, S.G. (1970) A new stage for orientating microfossils. *Journal of Paleontology*, 44: 1133–1144.

Barnett, S.G. (1971) Biometric determination of the evolution of *Spathognathodus remscheidensis*: a method for precise intrabasinal time correlations in the Northern Appalachians. *Journal of Paleontology*, 45: 274–300.

Barnett, S.G. (1972) The evolution of *Spathognathodus remscheidensis* in New York, New Jersey, Nevada, and Czechoslovakia. *Journal of Paleontology*, 46: 900–917.

Baum, D.A. (1998) Individuality and existence of species through time. *Systematic Biology*, 47: 641–653.

Bell, A.S., Sommerville, C. and Gibson, D.I. (2002) Multivariate analyses of morphometrical features from *Apatemon gracilis* (Rudolphi, 1819) Szidat, 1928 and *A. annuligerum* (v. Nordmann, 1832) (Digenea: Strigeidae) metacercariae. *Systematic Parasitology*, 51: 121–133.

Bengtson, S. (2000) Teasing fossils out of shales with cameras and computers. *Palaeontologia Electronica*, 3, article 4, http://palaeo-electronica.org/2000_1/fossils/issue1_00.htm.

Blum, H. (1973) Biological shape and visual science (Part 1). *Journal of Theoretical Biology,* 38: 205–287.

Bookstein, F.L. (1987) Random walks and the existence of evolutionary rates. *Paleobiology,* 13: 446–464.

Bookstein, F.L. (1991) *Morphometric Tools for Landmark Data: Geometry and Biology,* Cambridge University Press, Cambridge, 435 pp.

Boucot, A.J. (1958) Age of the Bainbridge Limestone. *Journal of Paleontology,* 32: 1029–1030.

Branson, E.B. and Mehl, M.G. (1933) Conodonts from the Bainbridge (Silurian) of Missouri. *University of Missouri Studies,* 8: 39–53.

Cooper, B.J. (1975) Multi-element conodonts from the Brassfield Limestone (Silurian) of southern Ohio. *Journal of Paleontology,* 49: 984–1008.

Cooper, B.J. (1976) Multi-element conodonts from the St-Clair Limestone, Silurian of southern Illinois, USA. *Journal of Paleontology,* 50: 205–217.

Croll, V.M., Aldridge, R.J. and Harvey, P.K. (1982) Computer applications in conodont taxonomy: characterization of blade elements. Miscellaneous paper. *Geological Society of London,* 14: 237–246.

Dayan, T. and Simberloff, D. (2005) Ecological and community-wide character displacement: the next generation. *Ecology Letters,* 8: 875–894.

Donoghue, P.C.J. (1998) Growth and patterning in the conodont skeleton. *Philosophical Transactions of the Royal Society of London, Series B,* 353: 633–666.

Donoghue, P.C.J. and Purnell, M.A. (1999) Mammal-like occlusion in conodonts. *Paleobiology,* 25: 58–74.

Donoghue, P.C.J., Purnell, M.A. and Aldridge, R.J. (1998) Conodont anatomy, chordate phylogeny and vertebrate classification. *Lethaia,* 31: 211–219.

Donoghue, P.C.J., Purnell, M.A., Aldridge, R.J. and Zhang, S. (in press) The interrelationships of complex conodonts (Vertebrata). *Systematic Palaeontology.*

Evans, A.R. and Sanson, G.D. (2003) The tooth of perfection: functional and spatial constraints on mammalian tooth shape. *Biological Journal of the Linnean Society,* 78: 173–191.

Girard, C., Renaud, S. and Korn, D. (2004a) Step-wise morphological trends in fluctuating environments: evidence in the Late Devonian conodont genus *Palmatolepis. Geobios,* 37: 404–415.

Girard, C., Renaud, S. and Sérayet, A. (2004b) Morphological variation of *Palmatolepis* Devonian conodont: species versus genera. *Comptes Rendus Palevol,* 3: 1–8.

Jeppsson, L. (1974) Aspects of Late Silurian conodonts. *Fossils and Strata,* 6: 1–54.

Jeppsson, L. (1976) Autecology of Late Silurian conodonts. In *Conodont Paleoecology* (ed C.R. Barnes), Geological Society of America Special Paper 15, Geological Society of America, Boulder, CO, pp. 105–118.

Jeppsson, L. (1990) An oceanic model for lithological and faunal changes tested on the Silurian record. *Journal of the Geological Society,* 147: 663–674.

Jeppsson, L. and Aldridge, R.J. (2000) Ludlow (Late Silurian) oceanic episodes and events. *Journal of the Geological Society,* 157: 1137–1145.

Jones, D. (2006) Morphometric analysis of taxonomy, evolution, antecology and homology in ozarkodinid conodonts. PhD thesis, University of Leicester, Leicester, UK.

Klapper, G. and Foster, C.T. (1986) Quantification of outlines in Frasnian (Upper Devonian) platform conodonts. *Canadian Journal of Earth Sciences,* 23: 1214–1222.

Klapper, G. and Foster, C.T. (1993) Shape analysis of Frasnian species of the Devonian condont genus *Palmatolepis. Supplement to Journal of Paleontology,* 67: 1–35.

Knowlton, N. (1993) Sibling species in the sea. *Annual Review of Ecology and Systematics,* 24: 189–216.

Marcus, L.F. (1988) Traditional morphometrics. In *Michigan Morphometrics Workshop* (eds F.J. Rohlf and F.L. Bookstein), University of Michigan Museum of Zoology, University of Michigan, Ann Arbor, pp. 179–200.

Mayr, E.E. (1942) *Systematics and the Origin of Species,* Columbia University Press, New York, 334 pp.

Mayr, E.E. (1969) *Principles of Systematic Zoology,* McGraw–Hill, New York, 428 pp.

Mayr, E.E., Limsley, E.G. and ??, R.L. (1953) *Methods and Principles of Systematic Zoology,* McGraw-Hill, New York.

McClain, C.R., Johnson, N.A. and Rex, M.A. (2004) Morphological disparity as a biodiversity metric in lower bathyal and abyssal gastropod assemblages. *Evolution,* 58: 338–348.

Mishler, B.D. and Donoghue, M.J. (1982) Species concepts: a case for pluralism. *Systematic Zoology,* 31: 491–503.

Murphy, M.A. and Cebecioglu, M.K. (1984) The *Icriodus steinachensis* and *I. claudia* lineages (Devonian conodonts). *Journal of Paleontology,* 58: 1399–1411.

Murphy, M.A. and Cebecioglu, M.K. (1986) Statistical study of *Ozarkodina excavata* (Branson and Mehl) and *O. tuma* Murphy and Matti (Lower Devonian, delta Zone, conodonts, Nevada). *Journal of Paleontology,* 60: 865–869.

Murphy, M.A. and Cebecioglu, M.K. (1987) Morphometric study of the genus *Ancyrodelloides* (Lower Devonian, Conodonts), Central Nevada. *Journal of Paleontology,* 61: 583–594.

Murphy, M.A. and Springer, K.B. (1989) Morphometric study of the platform elements of *Amydrotaxis praejohnsoni* n. sp. (Lower Devonian, Conodonts, Nevada). *Journal of Paleontology,* 63: 349–355.

Peijnenburg, K. and Pierrot-Bults, A.C. (2004) Quantitative morphological variation in *Sagitta setosa* Muller, 1847 (Chaetognatha) and two closely related taxa. *Contributions to Zoology,* 73: 305–315.

Purnell, M.A. (1993) Feeding mechanisms in conodonts and the function of the earliest vertebrate hard tissues. *Geology,* 21: 375–377.

Purnell, M.A. (1994) Skeletal ontogeny and feeding mechanisms in conodonts. *Lethaia,* 27: 129–138.

Purnell, M.A. (1995) Microwear on conodont elements and macrophagy in the first vertebrates. *Nature,* 374: 798–800.

Purnell, M.A., Donoghue, P.C.J. and Aldridge, R.J. (2000) Orientation and anatomical notation in conodonts. *Journal of Paleontology,* 74: 113–122.

Renaud, S. and Girard, C. (1999) Strategies of survival during extreme environmental perturbations: evolution of conodonts in response to the Kellwasser crisis (Upper Devonian). *Palaeogeography Palaeoclimatology Palaeoecology,* 146: 19–32.

Roopnarine, P.D., Murphy, M.A. and Buening, N. (2005) Microevolutionary dynamics of the Early Devonian conodont *Wurmiella* from the Great Basin of Nevada. *Palaeontologia Electronica,* 8, http://palaeo-electronica.org/2005_2/dynamics/issue2_05.htm.

Sansom, I.J. (1996) *Pseudooneotodus*: a histological study of an Ordovician to Devonian vertebrate lineage. *Zoological Journal of the Linnean Society,* 118: 47–57.

Scotese, C.R. (2001) *Altas of Earth History,* volume 1, Paleogeography, PALEOMAP Project. http://www.scotese.com/.

Simpson, G.G., Roe, A. and Lewontin, R. (1960) *Quantitative Zoology,* Harcourt Brace, New York, 440 pp.

Sloan, T.R. (2000) Results of a new outline-based method for the differentiation of conodont taxa. In *Second Australian Conodont Symposium* (AUSCOS II) (eds R. Mawson and J.A. Taylor). Courier Forschung-institut Senckenberg, Orange, Australia, pp. 389–404.

Smith, G.R. 1988. Homology in morphometrics and phylogenetics. In *Michigan Morphometrics Workshop* (eds F.J. Rohlf and F.L. Bookstein). University of Michigan Museum of Zoology, University of Michigan, Ann Arbor, pp. 325–338.

Sokal, R.R. and Crovello, T.J. (1970) The biological species concept: a critical review. *American Naturalist,* 104: 127–153.

Stone, J. (1987) Review of investigative techniques used in the study of conodonts. In *Conodonts: Investigative Techniques and Applications* (ed R.L. Austin). Ellis Horwood Ltd, Chichester, pp. 17–34.

Straney, D.O. (1988) Median axis methods in morphometrics. In *Michigan Morphometrics Workshop* (eds F.J. Rohlf and F.L. Bookstein), University of Michigan Museum of Zoology, University of Michigan, Ann Arbor, pp. 179–200.

Strauss, R.E. and Bookstein, F.L. (1982) The truss: body form reconstructions in morphometrics. *Systematic Zoology,* 31: 113–135.

Sweet, W.C. (1988) *The Conodonta: Morphology, Taxonomy, Paleoecology, and Evolutionary History of a Long-Extinct Animal Phylum,* Clarendon Press, Oxford, 212 pp.

Tolmacheva, T. and Löfgren, A. (2000) Morphology and paleogeography of the Ordovician conodont *Paracordylodus gracilis* Lindström, 1955. *Journal of Paleontology,* 74: 1114–1121.

Tolmacheva, T.Y. and Purnell, M.A. (2002) Apparatus composition, growth, and survivorship of the lower Ordovician conodont *Paracordylodus gracilis* Lindström, 1955. *Palaeontology,* 44: 209–228.

Velhagen, W.A. and Roth, V.L. (1997) Scaling of the mandible in squirrels. *Journal of Morphology,* 232: 107–132.

Villier, L. and Eble, G.J. (2004) Assessing the robustness of disparity estimates: the impact of morphometric scheme, temporal scale, and taxonomic level in spatangoid echinoids. *Paleobiology,* 30: 652–665.

Walmsley, V.G., Aldridge, R.J. and Austin, R.L. (1974) Brachiopod and conodont faunas from the Silurian and lower Devonian of Bohemia. *Geologica et Palaeontologica,* 8: 39–47.

Wheeler, Q.D. and Meier, R. (2000) *Species Concepts and Phylogenetic Theory: A Debate,* Columbia University Press, New York, 256 pp.

Wiens, J.J. (2004) What is speciation and how should we study it? *American Naturalist,* 163: 914–923.

Willig, M.R., Owen, R.D. and Colbert, R.L. (1986) Assessment of morphometric variation in natural populations – The inadequacy of the univariate approach. *Systematic Zoology,* 35(2): 195–203.

15 Decision Trees: A Machine-Learning Method for Characterizing Morphological Patterns Resulting from Ecological Adaptation

Manuel Mendoza

CONTENTS

Introduction 261
Method 262
Results and Discussion 266
Identification of the Morphological Patterns 266
Morphological Characterization of Extant Omnivorous Ungulates 266
Morphological Characterization of the Selective Feeders 267
Morphological Characterization of Browsers and Mixed Feeders 268
Morphological Characterization of High-Level Browsers 270
Simultaneous Characterization of the Feeding Guilds 270
Assessment of the Methodology 270
Characterization of the Feeding Adaptations of *Prosthennops xiphodonticus* 273
Geometry of the Morphospace and Principle of Actualism 274
Summary 275
Acknowledgements 275
References 276

INTRODUCTION

Comparative studies of a broad spectrum of living mammal species have provided many morphological correlates of ecological categories. Thus, certain morphological features enable some degree of discrimination among various mammalian ecological guilds, making inferences about the adaptations of extinct species possible. For example, the crown length of the cheek teeth (hypsodonty) has been used often to distinguish grass-eating grazers from leaf-eating browsers (Janis, 1988). Several other morphological features (e.g. skull length, muzzle width) are also correlated with grazing or browsing feeding strategies (Janis 1990, 1995). However, complete discrimination among such categories is not achieved by any single measurement or observation. That is, the range of values for any morphological variable in any individual group, whether using raw variables or variables that have been size adjusted, may exhibit considerable overlap with that of the other feeding groups. As a consequence, inferences about the adaptations of past species often cannot be precise.

The main reason for the existence of this overlap in variable ranges is the fact that any morphological adaptation involves a complex pattern of covariation among many skeletal characters. Thus, the manner in which any morphological trait is modified for a given adaptation depends on the nature of modifications occurring in the rest of the skeletal morphology. Indeed, ecological adaptations may be better characterized by complex morphological patterns that correspond to different regions of a multidimensional theoretical morphospace than defined by single variables. Such multidimensional regions cannot be represented directly within a bidimensional surface, but subspaces optimized for the task of group characterization can be recognized with the help of multivariate statistical techniques.

Identification-based problems like this have been commonly dealt with using principal components analysis (PCA) or discriminant analysis (DA). Interestingly, machine learning is also a process concerned with uncovering patterns, associations, and statistically significant structures in data, though it employs a different approach. Principal components analysis, for example, describes the diversity of shapes, simplifying descriptions of variation among individuals. Similar to PCA, Bayesian networks identify hidden regularities in data without using direct information about class membership (Friedman and Koller, 2003). In essence, the first principal component corresponds to the longest axis (linear combination of the variables) of the cloud of samples in the morphospace defined by the variables, and each new component corresponds to the longest axis orthogonal to the previous ones. Bayesian networks identify minimum sets of variables among which a conditional dependence exists. The analysis of the conditional distributions allows identifying regions with a special concentration of samples (Friedman and Koller, 2003).

Similar to the discriminant analysis, decision trees (DTs) are designed to identify patterns characterizng a specified number of disjoint groups, using direct information about the membership of the samples (Quinlan, 1985). While DA generates equations consisting of linear combinations of the variables, decision trees provide branching trees with decisions at each branch point.

In this study DA and DT will be compared in the context of identifying morphological patterns in the jaws of ungulates characterizing a number of feeding groups. The goal of these analyses is to develop tools for the automated recognition of the feeding guild (e.g. grazers, browsers) of any ungulate, extant or extinct. As an example of application, these tools will be used to recognize the feeding adaptations of *Prosthennops xiphodonticus*, a peccary (tayassuid) from the middle Miocene in North America.

METHOD

Decision trees represent a type of machine learning whereby computer systems acquire knowledge inductively from the input of a number of samples. The product of this learning is a piece of procedural knowledge that can assign a hitherto unidentified object to one of a specified number of disjoint classes. Knowledge is explicitly represented as a branching tree with decisions at each branch point, rather than being implicit in algorithms (as in the case with DA). In this study, the initial input is morphological information from the jaw of extant species of ungulates, plus information about their feeding guild membership. All this information is used by the DT technique to characterize each guild according to the morphology of the jaw of its species.

Supplying taxonomical information instead of ecological, DTs could also be used to characterize species, subspecies or even populations or higher taxa morphologically (e.g. genera, family). An interesting advantage of this methodology is that qualitative and quantitative variables can be used simultaneously.

The trees generated contain a great amount of information (knowledge) about the relationship between the variables and the categories. In this study, this knowledge is about the relationship between morphology and adaptation. Thus, the system learns and 'explains', at the same time. This

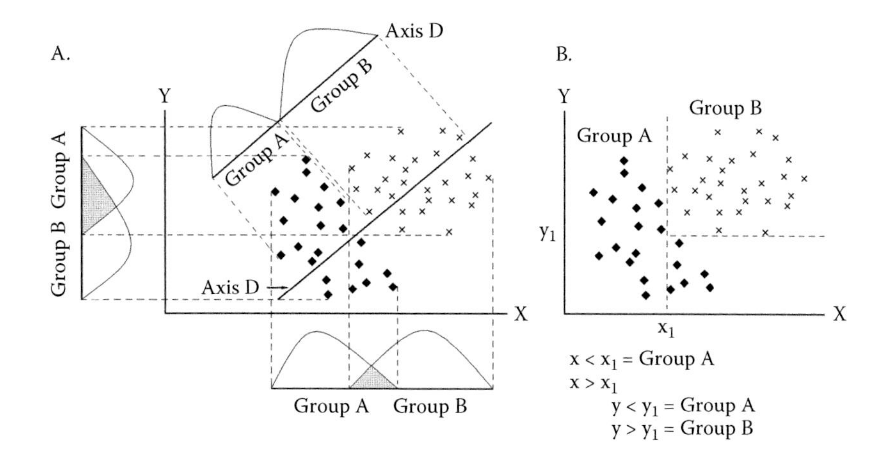

FIGURE 15.1 Characterization of two hypothetical taxa using discriminant analysis (A) and decision trees (B). Although both variables in each group show a considerable overlap, both methodologies allow the identification of the limits of each group in the morphospace defined by the variables.

is an important difference with regard to other techniques of machine learning, such as the neural networks, which work as a 'black box'.

Discriminant analysis uses a different approach. It describes the diversity of morphologies simplifying descriptions of differences between groups, generating equations and linear combination of the variables. The variables involved in the discriminant functions are those that contribute to minimize the ratio of between-groups to within-groups variance (Davis, 1986). From a geometric point of view, stepwise discriminant analysis generates new axes in the space defined by those variables involved in the pattern that characterize each group, in the direction in which the specified classes are farthest apart (Figure 15.1a). A DT directly assesses the region occupied by each group in that morphospace.

Using DA, a good discrimination between the groups compared (e.g. between grazers and browsers) is often too hastily interpreted as evidence that the discriminant functions captured the patterns characterizing each group. However, there is a probability of obtaining a good discrimination merely by chance. The possibility that correlations are not merely obtained by chance is usually tested with the statistical significance of the result, usually estimated using the Mahalanobis distance between the group centroids and the value of Wilk's lambda (Davis, 1986). However, as with other statistical techniques, DA assumes the statistical independence of the samples. Species do not represent independent samples of a statistical distribution as a consequence of their phylogenetic inter-relationships (Felsenstein, 1985; Harvey and Pagel, 1991). This is a variant of the classic type-1 error in statistics, where the apparent relationship is due to the operation of an extraneous variable (phylogeny in this case) on the observed variables. Thus, simple statistical analysis cannot be performed in studies of this nature, though this has been the most common approach.

There are several methodologies whereby continuous variable data can be corrected for the effects of phylogenetic patterning (e.g. phylogenetic autocorrelation; see Gittleman and Kot, 1990). However, in removing the effect of the phylogeny, some important information about morphology and adaptation can be lost. For this reason, instead of removing the effect of the phylogeny, a non-statistical approach is used in this study, and no descriptive statistics are so presented. By using a non-statistical approach, the goals of DA can be realized in a way that allows the recognition of phylogenetic patterning, thus allowing data analysts to compensate for its effect (see Mendoza and Palmqvist, in press).

The probability of obtaining a good discrimination by chance increases with the number of variables involved in the discriminant function, so by minimizing the number of variables this probability is also minimized. This can be achieved by increasing the level of significance required for their incorporation within the discriminant functions. On the other hand, the effects of phylogeny can be minimized as well, maximizing the taxonomic evenness of the sample. This can be achieved by weighting the species of those taxonomic groups (e.g. families) under-represented in the sample, with respect to those that are better represented (this weighting is one of the possibilities that the program used, SPSS 13.0, provides to perform DA). In statistical analyses, this weighting would violate the rules of phylogenetic independence, but not using this non-statistical approach.

Weighting the species of the under-represented taxonomic groups, the analyses can be performed with an equal contribution from all of the groups. In addition, this can be made keeping constant the total weight of the samples. This allows us to assess more realistically the significance of the variables in the discrimination. In order to adjust the evenness at the family level, for example, the species from each family have to be weighted according to the following equation:

$$WF_x = MNS/NSF_x \tag{15.1}$$

where WF_x is the weighting factor of the species from family x, MNS is the mean number of species per family (total number of species divided by the number of families), and NSF_x is the number of species belonging to family x.

In this way, the final weight of each family in the analysis is the same, and the weight of the total number of samples does not change. Using a similar set of species and morphological measurements to infer the body mass of the extinct mammal species, Mendoza et al. (2006) showed that increasing the taxonomic evenness in this way also increased the predictive capacity of the resulting algorithms over 20 test species that were not used to obtain the functions.

As mentioned earlier, the DT technique is based in the division of the multidimensional morphospace into boxes containing (to the extent possible) species of each defined group (Figure 15.1B). In this way, the region of the morphospace occupied by the group of species belonging to the same ecological guild is outlined. These divisions are represented explicitly as branching trees with decisions defining each branch point (see Quinlan, 1985, and Michie et al., 1994, for more detailed information about how the branch points are obtained). These trees allow new individuals to be assigned to a specified number of groups (e.g. feeding guilds, Figure 15.1B).

Although one of the main goals of this study is to compare the DT technique with the DA performing recognition of feeding guilds, no discriminant functions or decision trees are directly compared. Instead, a comparison is made between the patterns identified by both techniques as characteristic of known feeding guilds. The partial representation of these patterns can be used as tools for guild recognition (Mendoza and Palmqvist, 2006a, b; in review). This alternative procedure is less automated than the direct application of trees and functions, but may lead to the understanding of the relationship between the morphology of the jaw of the ungulates and their feeding adaptations. In fact, the partial representation of these patterns provides the exact localization of the samples, and so contains much more information than the trees and functions from which they were inferred.

For identifying the patterns, a simple DA was not performed, but instead was used as a technique of data mining of knowledge discovery (Cios et al., 1998). Discriminant analysis exploits the complementary information contained in morphological variables that would otherwise be dismissed because their contribution is only in combination with others. In addition, it rules out variables for which discriminant information is already present in the result. Thus, under this protocol some variables may not be included in the discriminant function, in spite of being involved in the morphological characterization of a group. Since our focus is not to obtain a function that allows a good discrimination per se, but to understand as much as possible about the pattern characterizing the groups analysed, we repeated the analysis, excluding some of the variables important in the previous analysis. In this way, more than one function involving different combinations of variables was obtained. These results were then analysed for changes in the significance values of the variables throughout the stepwise process to identify the patterns (Mendoza and Palmqvist, 2006a; in review).

Because there are different ways to divide a multidimensional morphospace into boxes containing samples of each defined group, different trees can be generated for each discrimination. Similar to DA, the probability of obtaining a good discrimination by chance increases with the number of variables involved in the tree solution. Thus, a number of decision trees are generated for each discrimination in order to select the ones that have the lowest number of branches and misclassifies the minimum number of species.

As a final step of the data mining performed with both methodologies, the patterns identified (partial patterns indeed) are represented graphically. These representations contain much more information than the trees and functions from which they were inferred.

Adaptive boosting is a different approach in the use of DTs, proposed by Freund and Schapire (1996). The notion is to generate many DTs, rather than just one. Each tree pays special attention to those cases misclassified by the other trees. The process continues for a predetermined number of times, generating the same number of trees, each one misclassifying a number of samples. When a new sample (e.g. an extinct species) is analysed, each tree predicts a category (e.g. its guild membership). The number of predictions for each category is counted to determine the final classification. In this fashion, a result is obtained that takes into account many different features (e.g. morphological traits) related with the difference between the categories. This approach is not used here for identifying patterns, but to supply additional information about the feeding characterization of *P. xiphodonticus* based in a wider range of morphological information.

The computer system was trained on 138 extant species of ungulates to perform the morphological characterization of five feeding guilds characteristic from forested habitats: omnivores (six species vs. 132 herbivores), selective feeders (13 species vs. 45 non-selective brachydont species), mixed feeders from forested habitats (21 species vs. 24 browsers), browsers (24 species vs. 21 mixed feeders) and high-level browsers (6 species vs. 37 other brachydont species, including 24 browsers and 13 selective feeders). Jaw morphology was captured using six size-adjusted variables. Size is an important indicator of adaptation, but using mixed size–shape variables is more difficult than using only shape information.

In order to obtain these variables, six out of seven measurements depicted in Figure 15.2 (all of them less the lower molar row length LRML) were divided by LMRL – one of the craniodental variables best correlated with body mass in extant ungulates (Janis, 1990, 1995) scaled isometrically, with r^2 values of 0.94. The seventh variable is the hypsodonty index, estimated as crown height of the third lower molar divided by the width of the same tooth. It was not size transformed using LMRL because it is already size independent.

The feeding adaptations of *P. xiphodonticus*, a tayassuid that lived during the middle Miocene in North America, were analysed as an example of application of this methodology. The only specimen available from this species was a jaw fragment from Railway Quarry A (UNSM 70078, Valentine Formation, middle Miocene of Nebraska, ~11.5 Mya).

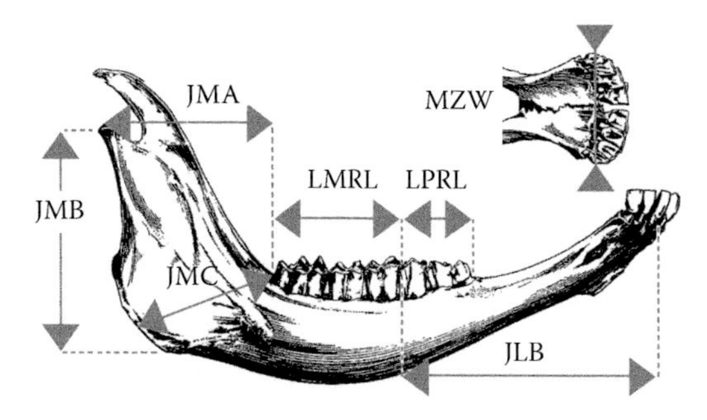

FIGURE 15.2 Measurements used in this investigation, in addition to the hypsodonty index (HI = unworn third lower molar height/width).

RESULTS AND DISCUSSION

Identification of the Morphological Patterns

Morphological Characterization of Extant Omnivorous Ungulates

No discriminant functions obtained allowed for a full discrimination between omnivores and herbivores among the entire range of extant ungulates. However, changes in the significance values of the variables throughout the stepwise procedure of discriminant analysis (see Mendoza and Palmqvist, 2007a) revealed a simple pattern that includes only two variables (HI and JLB) provides an almost perfect discrimination (Figure 15.3A). In the morphospace defined by these two variables, the six species classified as omnivores occupy the lower left vertex (true omnivores, black symbols).

There are some herbivores near this region of the morphospace. Some of these do not belong to the Order Artiodactyla (which includes the suines plus most of the herbivorous ungulates) and include some taxa with a low hypsodonty index that belong to the orders Perissodactyla (i.e. tapirs and certain rhinos) and Hyracoidea (i.e. tree hyraxes). Most non-equid perissodactyls and hyracoids have relatively short lower jaws in comparison with most artiodactyls (i.e. a low value of JLB; see Janis, 1990, 1995). Thus, their placement in this segment of the morphospace likely reflects a phylogenetic pattern rather than a similarity of functional morphology.

However, there are also a number of artiodactyls traditionally classified as herbivores that, while they cannot be considered as true omnivores, take a very broad spectrum of food items in their diet, many of them even including some meat, eggs, carrion or crustaceans. One of them, the Reeves's muntjac (*Muntiacus reevesi*), is even reported to be able to hunt small mammals. These taxa were classified as herbivores to perform the analyses, but identified as semi-omnivores in Figure 15.3. They include, in addition to the Reeves's muntjac, the water chevrotain (*Hyemoschus aquaticus*), the Indian muntjac (*Muntiacus muntjak*), the blue duiker (*Cephalophus monticola*), the mouse-deers (*Tragulus napu, T. javanicus* and *T. memmina*), the giant forest hog (*Hylochoerus*

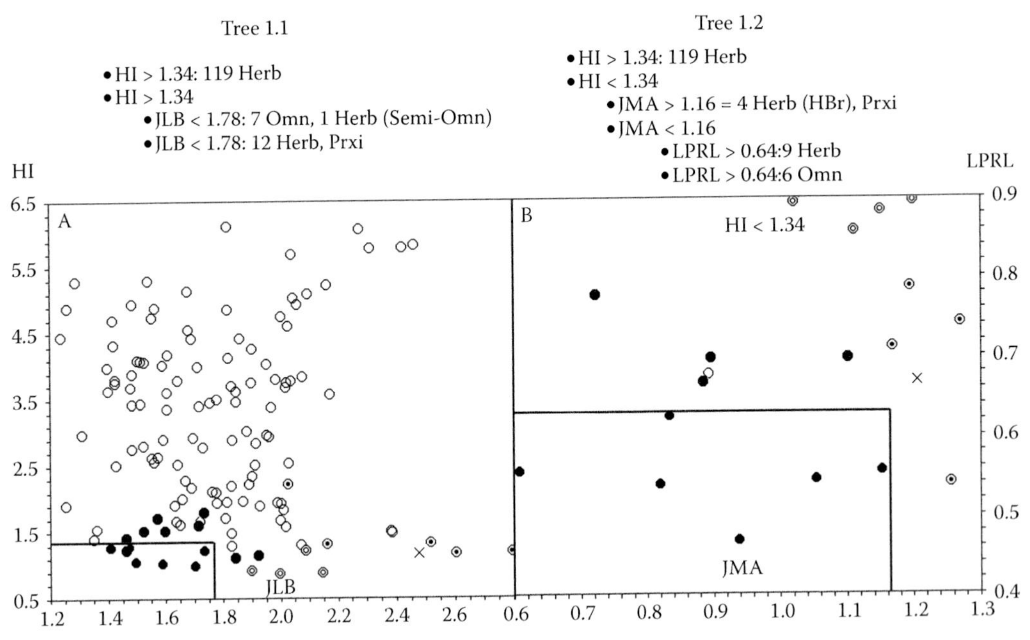

FIGURE 15.3 Trees 1.1 and 1.2 and distribution of ungulates in two partial representations of the pattern characterizing the omnivores, revealed by both trees. Herb: herbivores (open circles); Omn: omnivores (black circles); Prxi: *P. xiphodonticus* (cross); semi-omnivores (grey circles); high-level browsers (dotted circles); rings: tapirs.

meinertzhageni) and the duiker (*Cephalophus dorsalis*). The four tapir species are also situated next to the omnivores (Figure 15.3, ring symbols). As mentioned previously, non-equid perissodactyls could occupy this region because of phylogenetic autocorrelation rather than a similarity of feeding adaptation. However, although tapirs can be considered generalized browsers, the diet of the whole family includes fruits, leaves, stems, sprouts, small branches, grasses, aquatic plants, tree bark, aquatic organisms, cane, melon, cocoa, rice and corn from plantations (Jackson et al., 1977; Dubost 1984; Walther, 1990; d'Huart, 1993; Nowak, 1999; Wenninger and Shipley, 2000; Huffman, 2005; Myers et al., 2005).

The DT technique also revealed the pattern HI-JLB (tree 1.1, Figure 15.3A) and another one involving three that also distinguishes omnivorous ungulates from herbivorous ones (tree 1.2, Figure 15.3B). Only those species in which the HI value is lower than 1.34 are depicted in Figure 15.3B. This value corresponds to the first division of tree 1.2 (Figure 15.3; see first branch). There are only 13 species sharing these low HI values with the six true omnivores. Four are tapirs (Figure 15.3B, ring symbols), another four are semi-omnivores (grey symbols) and the other four are species adapted to browse from high levels of the canopy (identified as high-level browsers, dotted circles). The four semi-omnivores share HI < 1.34 with the true omnivores, in addition to similarly low values of JMA (relative posterior jaw length), but higher values of LPRL (relative lower premolar row length). The four high-level browsers differ from true and semi-omnivores in having higher values of JMA.

According to these results, truly omnivorous ungulates are characterized as being different from herbivorous ungulates by a very low hypsodonty index (HI) in combination with a relatively short anterior and posterior jaw (JLB and JMA, respectively), as well as a relatively short lower premolar row (LPRL). Thus, four variables, at least, are involved in the morphological patterns characterizing the omnivores. In the four-dimensional morphospace defined by these variables, all the species would be contained in a polyhedron with perpendicular edges. Omnivores would occupy a small region in one of its 16 vertices and other ungulate species with a broader diet (i.e. semi-omnivores) would be placed around them.

The fact that semi-omnivorous species show a morphological pattern similar to the one captured by two independent trees as characteristic of the omnivores (1.1 and 1.2) supports these results, and hence the suitability of the DT technique for identifying morphological patterns.

It is worth mentioning that, through use of this technique, not only was it possible to identify the more complex pattern captured by tree 1.2, but it was also possible to understand how the morphology of the omnivores differed from the rest of the ungulate species. There are discriminant functions involving three or four variables that allow for a full discrimination between two groups, but the inspection of their coefficients and variables has not led to a direct understanding of the relationship between morphology and adaptation.

Morphological Characterization of the Selective Feeders

These species mainly feed on fruits and other non-fibrous soft material. A discriminant analysis between this group and the other herbivores (browsers and mixed feeders from forested habitats) identifies a simple pattern that characterizes the selective feeder ungulates – that is, a low hypsodonty index (HI) – in combination with a relatively short posterior portion of the jaw (JMA, Figure 15.4A). The same combination (HI-JMA) appears in most of the decision trees generated. Using this combination, tree 2.1 misclassifies four species, but in combination with MZW (relative muzzle width), tree 2.2 misclassifies only two of these four species (Figure 15.4B) and shows that selective feeders are characterized as different from the rest of the brachydont species not only by a low hypsodonty combined with a relatively short posterior portion of the jaw (JMA), but also by a relatively narrow muzzle (MZW). In Figure 15.4B, only species where HI < 1.95 were depicted. As in the case of tree 1.2, this value corresponds to the first division of tree 2.2 (Figure 15.4, first branch).

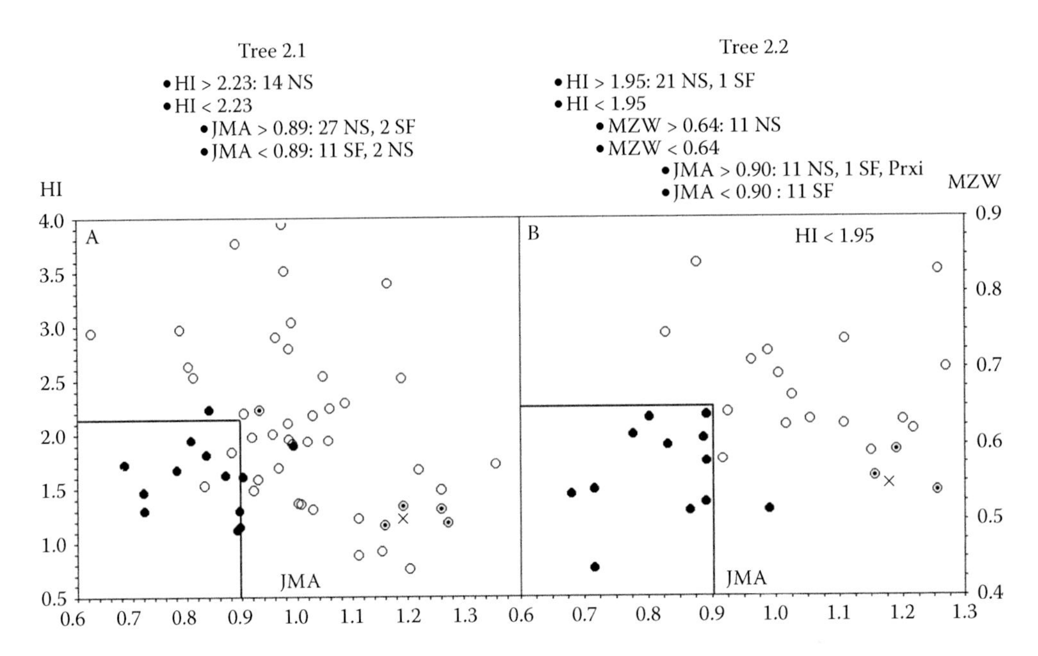

FIGURE 15.4 Trees 2.1 and 2.2 and distribution of ungulates in two partial representations of the pattern characterizing the selective feeders, revealed by a discriminant analysis and trees 2.1(A) and tree 2.2(B). SF: selective feeders (black circles); NS: non-selective feeders (open circles); high-level browsers (dotted circles); Prxi: *P. xiphodonticus* (cross).

According to these results, HI, JMA and MZW, at least, are involved in the morphological patterns characterizing the species adapted to feed mainly on fruits and other non-fibrous soft material. In the three-dimensional, perpendicular-edges polyhedron containing all the species, selective feeders occupy a small region around one of its eight vertices.

Morphological Characterization of Browsers and Mixed Feeders

Browsers (Br) feed predominantly on dicotyledonous plants, whereas mixed feeders (MF) feed on both grass and dicotyledonous plants, depending on the availability. The best function obtained to discriminate between both groups involves only two variables, JMC (relative maximum width of the mandibular angle) and MZW (relative muzzle width), but their combined representation does not support a good discrimination.

In contrast, several trees involving only three variables (four branches), misclassify very few species. These trees always group the mixed feeders in the same branch, while the browsers are often distributed in different branches. This shows that there is not a single pattern characterizing the whole group of browsers with regard to the mixed feeders. In spite of all browser species sharing a similar diet, this broad group is made up of different, more narrowly defined morphological groups. The inspection of tree 3.1 (Figure 15.5) reveals that, in addition to the typical browsers from forested habitats (made up by 13 species in the data-base), there is a small group of five species that dwell in more open and dry non-forested habitats, and another five species adapted to feed from high levels of the canopy (the aforementioned high-level browsers).

The results obtained using DTs are presented in Figure 15.5. The detailed analysis of this figure supplies a great amount of information about the morphological adaptations of mixed feeders and the three ecological groups of browsers. Figure 15.5 (tree 3.1) shows that mixed feeders (fourth branch, black circles) mainly differ from those browsers dwelling in non-forested habitats (third branch, grey circles) in having a wider muzzle (MZW). However, both groups share a short mandibular angle (JMB, plot a), a short lower premolar row (LPRL, plots b, d), a short anterior

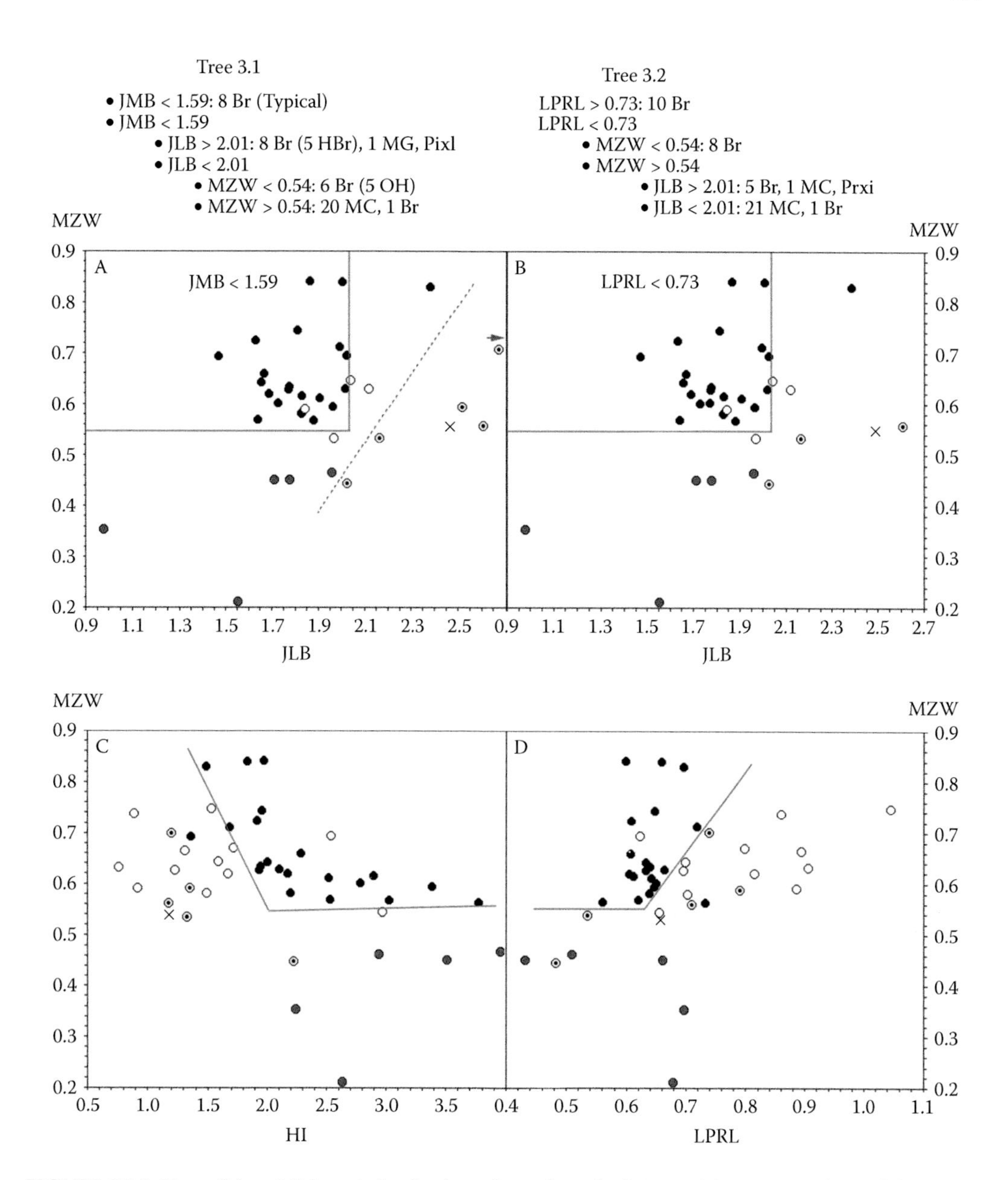

FIGURE 15.5 Trees 3.1 and 3.2, and distribution of ungulates in four partial representations of the pattern characterizing mixed feeders from closed habitats with regard to three types of browsers (see text), revealed by trees 3.1 and 3.2 and another two trees not represented. Br: typical browsers (open circles); MC: mixed feeders (black circles); HBr: high-level browsers (dotted circles); OH: browsers from non-forested habitats (grey circles); Prxi: *P. xiphodonticus* (cross).

jaw (JLB, plots a, b) and higher values of hypsodonty (HI, plot c). In fact, both groups are so similar that trees involving fewer variables often misclassify this group of browsers as mixed feeders.

From high-level browsers (second branch, dotted circles), mixed feeders mainly differ in the shorter anterior jaw (JLB), although this difference is less marked in those species with a narrower muzzle (MZW, Figures 15.5A and 15.5B). A broken line outlines the region occupied by high-level browsers in the subspace defined by these two variables, in combination with JMB. Both groups share a shorter mandibular angle (JMB, Figure 15.5A). Finally, mixed feeders are very different from the typical browsers (first branch, empty circles) because these show a deeper mandibular angle (JMB, Figure 15.5A), are less hypsodont (HI, Figure 15.5C) and have longer lower premolar rows (LPRL, Figures 15.5B and 15.5D).

In short, five variables, at least, are involved in the morphological patterns characterizing mixed feeders regard to these three ecological types of browsers. In the five-dimensional, perpendicular-edges polyhedron containing all these species, mixed feeders would occupy a region around one of its 32 vertices, and each kind of browser would also occupy more or less well-defined regions.

Morphological Characterization of High-Level Browsers

In the former section, tree 3.1 was generated to characterize morphologically browsers and mixed feeders with regard to each other. However, it revealed a simple pattern involving three variables (i.e. JMB, JLB, MZW), allowing for a full discrimination between high-level browsers and both other types of browsers and mixed feeders from forested habitats (see Figure 15.5A, dotted line). The new analyses described in this section were performed specifically to characterize the group of ungulates adapted to feed from high levels of the canopy, with regard to other browsers and selective feeders.

Both the DA and the DTs revealed three different partial patterns related to the relative length of the anterior portion of the jaw (JLB) that make a full discrimination between both groups possible (Figure 15.6). JLB alone does not support a full discrimination, but combined with the width of the jaw at the level of the muzzle (MZW, Figure 15.6A), the maximum width of the mandibular angle of the jaw (JMC, Figure 15.6B) or the lower premolar tooth row length (LPRL, Figure 15.6C), the discrimination is complete. Thus, the anterior jaw is longer in the high-level browsers than in other types of browser species, but this feature is less pronounced in those species having a narrower muzzle (MZW, as revealed by tree 3.1), a wider mandibular angle (JMC, with which JLB seems to be correlated; Figure 15.6B) or a shorter lower premolar row (LPRL). The posterior jaw (JMA) is also especially longer in high-level browsers, except in the dibitag (*Ammodorcas clarkei*), and both JLB and JMA (anterior and posterior portions of the jaw) are correlated (Figure 15.6D). According to the results obtained with both methodologies, one simple morphological trait characterizes the jaw of the high-level browsers. This is elongation, which seems an obvious adaptation to browse from trees and bushes above its body height.

Simultaneous Characterization of the Feeding Guilds

Six disjoint groups were used in these analyses because browsers from non-forested habitats were specified as a feeding guild different from typical and high-level browsers. Trees 5.1 and 5.2 (Figure 15.7) were selected. Both have nine branches, involve six variables (all less JD, which is not involved by any pattern) and misclassify only six species. Some complementary information about the pattern characterizing each feeding guild can also be obtained from these trees. However, they are mainly proposed here as tools for a completely automated recognition of the feeding guild in ungulates from the morphology of the jaw. In fact, the program used (See5, vers. 2.01), allows the automated assignment of any new sample (e.g. fossil species) to one of the groups. Using the adaptive boosting approach, the program allows the generation of up to 100 trees and determination of the percentages that classify the unknown sample into each group.

ASSESSMENT OF THE METHODOLOGY

The patterns proposed here, especially those identified using DT, show a very high capacity of discrimination. However, the high percentage of species correctly reclassified by these patterns cannot be considered, at least in principle, as a good estimate of their predictive accuracy for new species (e.g. extinct ones) because the samples discriminated are the same ones that were used for identifying the patterns. However, if one or more of the species that are well reclassified by one

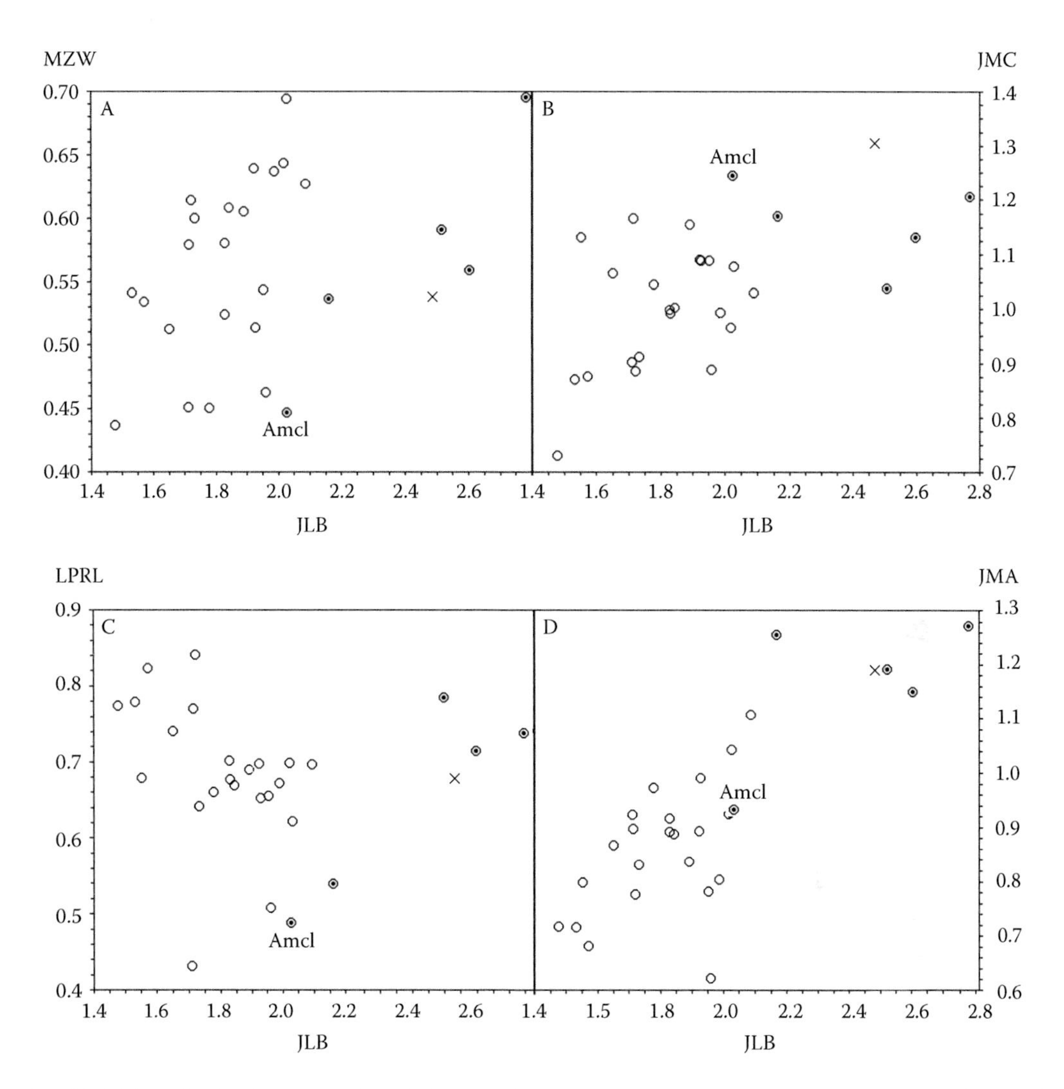

FIGURE 15.6 Distribution of ungulates in four partial representations of the pattern characterizing high-level browsers. Open circles: non-high-level browsers and selective feeders; dotted circles: high-level browsers; cross: *P. xiphodonticus*; Amcl: *Ammodorcas clarkei*.

of these patterns are excluded before performing the same analysis, and the same pattern is obtained, then it can be accepted that the methodology shows a predictive capacity.

In order to assess the ability of this method for generating tools with predictive capacity, different test analyses were conducted for each discrimination process, excluding one or more species each time. Although the coefficients of the functions obtained changed slightly and sometimes the values of bifurcation of the trees as well, the variables involved and the patterns revealed by them are almost always the same as the ones obtained before by excluding the test species. Only when the test species was one of the few placed in a border zone between the guilds was its characterization uncertain. This could be considered as a limitation of the methodology or may reflect a limited capacity of the shape variables for capturing completely relevant aspects of the morphology of the specimens (in terms of autecological adaptations).

However, it could also be that the morphological adaptation of such species was really intermediate, as in the case of the semi-omnivores (grey circles, Figure 15.4). Regardless, most of the species are not situated very close to the limits between the guilds. Thus, we can accept that most

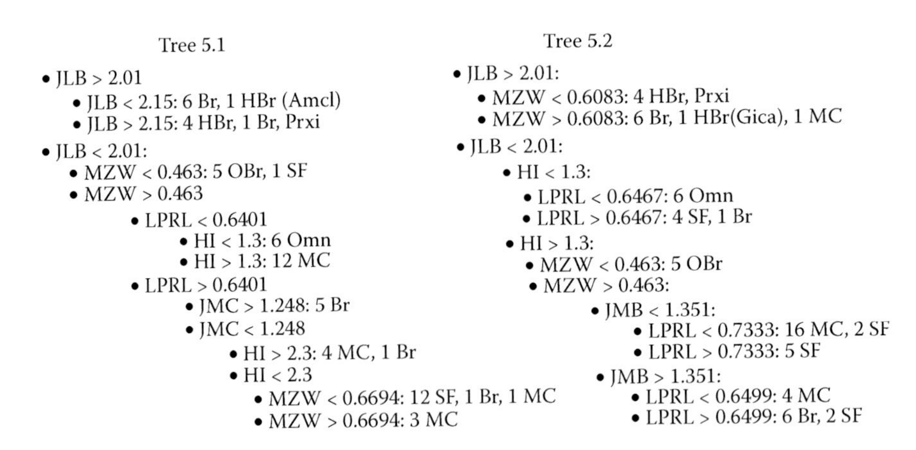

FIGURE 15.7 Trees 5.1 and 5.2 for the automated recognition of the feeding adaptations of ungulates. Br: typical browsers; HBr: high-level browsers; Obr: browsers from non-forested habitats; Omn: omnivores; SF: selective feeders; MC: mixed feeders from forested habitats.

are well characterized. The high percentage of species correctly reclassified by these patterns can be considered as an estimate of their predictive capacity over new species (e.g. extinct species). The robustness of the patterns identified here is a consequence of its simplicity and of the fact that they involve only very few and very significant variables.

For example, in the case of the discrimination between herbivores and omnivores using DT, 15 randomly selected herbivores were excluded at the same time, and an identical HI-JLB pattern was obtained. Visual inspection of Figure 15.3 reveals that the weight of a few herbivorous species is not enough to change the high significance of this morphological pattern. Even when the suiform species (including the six omnivores and the three herbivores), which were over-weighted, were excluded (one of them each time), the same HI-JLB pattern was obtained. The only exception was the Chacoan peccary (*Catagonus wagneri*). When this species was previously excluded, a different morphological pattern was obtained, but this pattern also classified the Chacoan peccary correctly, placing it close to other omnivores.

The rest of the discriminations always resulted in the same patterns when different species (one each time) were previously excluded. Only in the case of the characterization of the high-level browsers were the results different. More specifically, when the dibitag (*Ammodorcas clarkei*, Amcl; Figure 15.6) was excluded, the patterns identified were different. Since both JLB and JMA alone allow a complete discrimination when the dibitag was excluded, trees including any of them are obtained, inducing one to conclude that the relative length of the anterior or posterior portion of the jaw is enough to determine whether an ungulate species is or was a high-level browser. In fact, since only one sample is involved, when JLB is combined with any other variable, there is a not inconsiderable probability that the dibitag becomes correctly reclassified only by chance.

Using discriminant analysis, when the dibitag was excluded, the pattern depicted in Figure 15.6D was obtained. As this figure shows, this pattern also misclassifies the dibitag. However, when the dibitag is excluded but high-level browsers are characterized regard to the rest of feeding guilds (all together), the same pattern as tree 3.1 is obtained. This pattern characterizes high-level browsers completely and classifies the dibitag correctly. This could mean that the wider the taxonomic and ecological diversity of the sample is with regard to which a group is discriminated, the higher is the predictive capacity of the pattern identified.

According to these results, we could conclude that the DT technique is superior to the non-statistical approach of discriminant analysis, especially because it makes an easy interpretation of the patterns possible. However, other investigations (Mendoza and Palmquist, in review) have revealed that some important patterns can only be identified with the help of discriminant analysis.

Thus, both methodologies are useful for identifying morphological patterns underlying different ecological guilds (or other biological entities), and should be considered as complementary.

CHARACTERIZATION OF THE FEEDING ADAPTATIONS OF *PROSTHENNOPS XIPHODONTICUS*

There are three extant species of peccaries in the family Tayassuidae, also known as the pigs of the New World. All are omnivorous. However, among the more diverse Old World suid family, the Suidae (true pigs), while most species are omnivorous, some are herbivorous (e.g. the warthog, *Phacochoerus aethiopicus*). Thus, it is interesting to investigate whether any extinct peccary, such as *P. xiphodonticus,* was omnivorous, as with the extant peccaries, or whether it had a different diet, such as a greater degree of herbivory. *P. xiphodonticus* shares with the suiform omnivores a very low hypsodonty (HI), but both the anterior and posterior jaw, as well as the lower premolar row, are too long for being omnivorous or even semi-omnivorous (see Figure 15.3). Using the adaptive boosting algorithm, 7 out of 10 trees generated classify *P. xiphodonticus* as herbivore.

Prosthennops xiphodonticus was extremely brachydont (very low hypsodonty index), so it is possible to rule out the hypothesis that it was a grazer, or even a mixed feeder in open habitats (see Janis, 1988). Within the categories of brachydont herbivorous ungulates, it could be, at first, selective feeder, browser or mixed feeder from closed habitats.

Prosthennops xiphodonticus shares with selective feeders both a low hypsodonty and a narrow muzzle, but its posterior portion of the jaw is too long to be specialized in feeding on fruits and other non-fibrous soft material (see Figure 15.4). In addition, using the adaptive boosting algorithm, *P. xiphodonticus* is also classified as a non-selective feeder by 9 out of 10 trees generated.

Prosthennops xiphodonticus shares with the mixed feeders a shorter mandibular angle (JMB, plots a, b), but it differs in the longer anterior jaw (JLB), showing, in addition, a relationship with the width of the muzzle characteristic of the high-level browsers (Figure 15.5A). In fact, tree 3.1 places it in the second branch, in which most of the species are high-level browsers (Figure 15.5). Moreover, many of the partial representations of the patterns characterizing different feeding guilds place to *P. xiphodonticus* very close to the high-level browsers (see Figures 15.3A, 15.3B, 15.4A and 15.4B), and it shows very clearly the typical elongated jaw characterizing this group of species adapted to browse from high levels of the canopy (Figure 15.6).

When the six feeding guilds were characterized simultaneously, trees 5.1 and 5.2 were selected (Figure 15.7). Both have nine branches, misclassify only six species, and classify to *P. xiphodonticus* as a high-level browser (Figure 15.7). Using the boosting algorithm, 90 out of 100 trees generated classify this extinct peccary as a high-level browser.

As we saw in the characterization of the high-level browsers, the length of the anterior jaw plays a very important role. Thus, in order to test if this variable alone determines the characterization of *P. xiphodonticus* as a high-level browser, new analysis was performed excluding JLB. Out of 100 trees, 64 classified *P. xiphodonticus* as a high-level browser. Excluding JLB and HI, 77 out of 100 trees also classified it in the same feeding guild and, excluding the width of the muzzle (MZW) alone, 86 out of 100 trees classified it in the same way.

According to the results of this full set of analyses, *P. xiphodonticus* was most probably a high-level browser. Taking into account the feeding behaviour and morphology of the present-day species of tayassuids (Figure 15.8A), the notion of *P. xiphodonticus* being a high-level browser is a very unexpected result. However, as Figure 15.8 shows, living peccaries are not representative of the whole group, if extinct species are also taken into account. *P. xiphodonticus* may have looked more like the younger late Pleistocene *Mylohyus* sp. shown in Figure 15.8B, which does appear to be longer jawed and longer legged than extant peccaries. Even in that case, *P. xiphodonticus* lacked the long legs and long neck typical of extant high-level browsers, and also had the type of bunodont molar crown occlusal pattern typical of an omnivorous diet (Wright, 1998). This might mean that

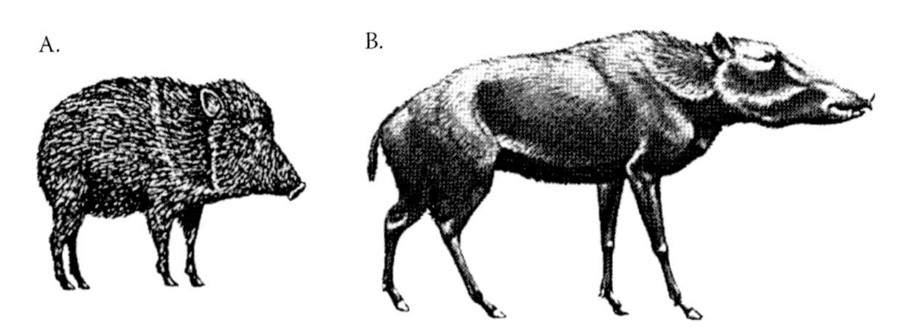

FIGURE 15.8 A: extant peccary (Chacoan peccary, *Catagonus wagneri*); B: rendering of an extinct late Pleistocene species (*Mylohyus* sp.).

it habitually selected foliage from a level above its own body height, not necessarily that it fed from a high level of the canopy, as do extant high-level browsers. It thus seems that *P. xiphodonticus* does not have a precise ecological equivalent among extant ungulates.

GEOMETRY OF THE MORPHOSPACE AND PRINCIPLE OF ACTUALISM

Each individual of a given species can be represented as a point in a multidimensional theoretical morphospace defined by the shape variables of the jaw. Thus, each species occupies a small region (like a small cloud) included in a larger region occupied by the species that share the same feeding adaptations (e.g. grazers). Each species can also be represented as a point, the result of averaging the variables of the individuals of this species. There are different combinations of these variables best suited for each ecological adaptation, which correspond to different morphological solutions to deal with the same (or similar) biomechanical problems (e.g. eating grass with a high silicophytolith content).

Each living species corresponds to one of these combinations, but they are just some examples of how the jaw of an ungulate can be adapted to its feeding behaviour. The real pattern that morphologically characterizes each feeding adaptation corresponds to the whole region of the theoretical morphospace that would occupy, at least in theory, all the imaginable ungulate species from the same feeding guild. If we knew all the possible morphologies that may represent each adaptation (the whole region), we could determine with accuracy the paleoecology of any extinct species. However, we only have some examples of morphological adaptations depicted by the living species, which occupy only a part of the whole region. Inevitably, we can only identify a partial pattern.

Subsets of species, such as taxonomic groups, will occupy defined regions inside the zone of the morphospace inhabited by the species that show the same ecological adaptation. Thus, as the number of species and the taxonomic range of the sample are wider, the patterns identified will be closer to the real ones.

From this context arises the problem of the capacity that methodologies based on living species have for inferring the adaptations of past ones (principle of actualism). The patterns that can be identified with the living species do not always capture the way in which certain extinct species were adapted to display a given feeding behaviour. The closer the patterns identified are to the real ones, the higher the probability is that they will capture the morphological solution of any past species. In the best of cases, the pattern identified is as close as possible to the real one, according to the information available, but this depends on the performance of the analytical method used to identify it. For example, though the pattern identified here for the high-level browsers fits the real one, analysis performed without the dibitag would result in a partial pattern based only on JLB and JMA that would not include the way that the dibitag is adapted to browse from high levels of the canopy. This mistake would not be a consequence of a limitation of the methodology, but rather of the information available.

Finally, both discriminant functions and trees involving many variables with regard to the number of samples fit the region of the morphospace to small subgroups of very similar samples, thus capturing the particular traits of these small groups, instead of identifying wide regions of the morphospace that correspond to the general traits of the group. These functions and trees often allow a very good discrimination but, as expected, show very low predictive capacity. This is because it is not probable that it includes any new combination of the variables or points in the morphospace.

SUMMARY

Suiform omnivorous ungulates are characterized as being different from herbivorous ungulates by an extremely low hypsodonty combined with a short anterior and posterior jaw, as well as a short lower premolar row. Other non-suiform species with a broader diet, which could even be considered as semi-omnivores, are morphologically very similar. *P. xiphodonticus* shares with them a very low hypsodonty (HI), but both the anterior and posterior jaw, as well as the lower premolar row, are too long for being omnivorous or even semi-omnivorous.

Selective feeders are characterized as different from the rest of the brachydont species by a very low hypsodonty combined with both a short posterior portion of the jaw and a narrow muzzle. *Prosthennops xiphodonticus* shares with them both a low hypsodonty and a narrow muzzle, but its posterior portion of the jaw is too long to be specialized in feeding on fruits and other non-fibrous soft material.

There is not a single pattern characterizing the whole group of browsers with regard to the mixed feeders because browsers are made up by different relatively defined morphological groups: typical browsers from forested habitats, species dwelling in more open and dry non-forested habitats and species adapted to browse from high levels of the canopy. Mixed feeders are very different from the typical browsers because these show a deeper mandibular angle, are less hypsodont and have longer lower premolar rows. However, with those browsers from more open habitats, mixed feeders share a mandibular angle, lower premolar row and anterior jaw shorter than in the other groups, as well as higher values of hypsodonty. Both groups only differ in the mixed feeders having a wider muzzle. From high-level browsers, mixed feeders mainly differ in the shorter anterior jaw, although this difference is less marked in those species with a narrower muzzle. *Prosthennops xiphodonticus* shares with the mixed feeders a shorter mandibular angle, but it differs in the longer anterior jaw, showing in addition a relationship with the width of the muzzle characteristic of the high-level browsers.

High-level browsers are mainly different from other types of browsers and selective feeders in having an elongated jaw. *Prosthennops xiphodonticus* shows this morphological trait very clearly. All these patterns show a high robustness, so their high capacity of discrimination can be considered as a good estimate of their predictive accuracy over new species, such as the extinct ones.

Finally, *P. xiphodonticus* shows very clearly all the morphological patterns identified here as characteristic of the high-level browsers but, like the extant peccaries, it lacked the long legs and long neck typical of extant high-level browsers, and also had the type of bunodont molar crown occlusal pattern typical of an omnivorous diet. Thus, it could be that it habitually selected foliage from a level above its own body height, so it would not have a precise ecological equivalent among extant ungulates.

ACKNOWLEDGEMENTS

The author has been funded by a postdoctoral grant from the Spanish CICYT and the Fulbright Visiting Scholar Program.

REFERENCES

Cios, K., Pedryez, W. and Swinianski, R.W. (1998) *Data Mining Methods in Knowledge Discovery,* Kluwer, Boston/Dordrecht/London, 491 pp.

Davis, J. (1986) *Statistics and Data Analysis in Geology,* 2nd edition, John Wiley & Sons, New York, 646 pp.

d'Huart, J. (1993) The forest hog (*Hylochoerus meinertzhageni*). In *Pigs, Peccaries, and Hippos* (ed P. Oliver), Gland, Switzerland, IUCN, pp. 84–92.

Dubost, G. (1984) Comparison of the diets of frugivorous forest ruminants of Gabon. *Journal of Mammalogy,* 65: 298–316.

Felsenstein, J. (1985) Phylogenies and the comparative method. *American Naturalist,* 125: 1–15.

Freund, Y. and Schapire, R.E. (1996) Experiments with a new boosting algorithm. In *Machine Learning: Proceedings of the Thirteenth International Conference,* Morgan Kauffman, San Francisco, pp. 148–156.

Friedman, N. and Koller, D. (2003) Being Bayesian about Bayesian network structure: a Bayesian approach to structure discovery in Bayesian networks. *Machine Learning,* 50: 95–125.

Gittleman, J.L. and Kot, M. (1990) Adaptation: statistics and a null model for estimating phylogenetic effects. *Systematic Zoology,* 39: 227–241.

Harvey, P.H. and Pagel, M.D. (1991) *The Comparative Method in Evolutionary Biology,* Oxford University Press, Oxford, 239 pp.

Huffman, B. (2005) The ultimate ungulate page, http://www.ultimateungulate.com.

Jackson, J., Chapman, D. and Dansie, O. (1977) A note on the food of Muntjac deer (*Muntiacus reevesi*). *Journal of Zoology,* 183: 546–548.

Janis, C.M. (1988) An estimation of tooth volume and hypsodonty indices in ungulate mammals, and the correlation of these factors with dietary preference. In *Teeth Revisited* (eds D.E. Russell, J.P. Santoro and D. Sigogneau), Mémoires du Muséum national d'Histoire naturalle du Paris (série C) 53, Paris, pp. 367–387.

Janis, C.M. (1990) Correlation of cranial and dental variables with dietary preferences: a comparison of macropodoid and ungulate mammals. *Memoirs of the Queensland Museum,* 28: 349–366.

Janis, C.M. (1995) Correlations between craniodental morphology and feeding behaviour in ungulates: reciprocal illumination between living and fossil taxa. In *Functional Morphology in Vertebrate Paleontology* (ed J.J. Thomason), Cambridge University Press, Cambridge, pp. 79–98.

Larose, D.T. (2004) *Discovering Knowledge in Data: an Introduction to Data Mining.* John Wiley and Sons, New York.

Mendoza, M., Janis, C.M. and Palmqvist, P. (2002) Characterizing complex craniodental patterns related to feeding behaviour in ungulates: a multivariate approach. *Journal of Zoology,* London, 256: 223–246.

Mendoza, M., Janis, C.M. and Palmqvist, P. (2006). Estimating the body mass of extinct ungulates: a study on the use of multiple regression. *Journal of Zoology,* London, 270(1): 90–101.

Mendoza, M. and Palmqvist, P. (2006a) Characterizing adaptive morphological patterns related to diet in Bovidae (Mammalia, Artiodactyla). *Acta Zoologica Sinica,* 52(6): 988–1008.

Mendoza, M. and Palmqvist, P. (2006b) Characterizing adaptive morphological patterns related to habitat use and body mass in bovidae (mammalia, artiodactyla). *Acta Zoologica Sinica,* 52(6): 971–987.

Mendoza, M. and Palmqvist, P. (in review) Hyprodonty: an adaptation for grass consuming or foraging in open habitats. *Journal of Zoology,* London.

Michie, D., Spiegelhalter, D.J. and Taylor, C.C. (1994) *Machine Learning, Neural and Statistical Classification,* Prentice Hall, Englewood Cliffs, NJ, 289 pp.

Myers, P., Espinosa, R., Parr, C.S., Jones, T., Hammond, G.S. and Dewey, T.A. (2005) The animal diversity web, http://www.animaldiversity.org.

Nowak, R.M. (1999) *Walker's Mammals of the World,* 6th edition, The Johns Hopkins University Press, Baltimore, MD, 280 pp.

Quinlan, J.R. (1985) Introduction to decision trees. *Machine Learning,* 1: 81–106.

Walther, F.R. (1990) Duikers and dwarf antelopes. In *Grzimek's Encyclopedia of Mammals* (ed S.P. Parker), McGraw–Hill, New York, pp. 325–343.

Wenninger, P. and Shipley, L. (2000) Harvesting, rumination, digestion, and passage of fruit and leaf diets by a small ruminant, the blue duiker. *Oecologia,* 123: 466–474.

Wright, D.B. (1998) Tayassuidae. In *Evolution of Tertiary Mammals of North America* (eds C.M. Janis, K.M. Scott and L.L. Jacobs), Volume 1, Cambridge University Press, Cambridge, p. 389.

16 Data Integration and Multifactorial Analyses: The Yeasts and the BioloMICS Software as a Case Study

Robert Vincent

CONTENTS

Introduction 277
Comparison Methods 279
BioloMICS Software as a Solution to Store and Analyse Biological Data 281
Creating Self-Designed Data-Bases 282
Managing Any Biological Data 283
Analyzing Any Biological Data 283
Conclusions 286
Acknowledgements 286
References 286

INTRODUCTION

Yeasts are microscopic organisms that show very few distinguishable morphological features. Early taxonomic studies were carried out using a mixture of morphological, sexual and fermentation, or assimilation features (Hansen, 1888, Beijerinck, 1889; Hansen, 1891, 1898, 1902). Later, more assimilation and growth abilities were included in standard testing panels used for identification and classification (Lodder, 1934; Wickerham and Burton, 1948; Wickerham, 1951; Lodder and Kreger-Van Rij, 1952; Lodder, 1970; Kreger-Van Rij, 1984). Chemical and biochemical properties also play a role, but nowadays the most prominent characters are molecular. While one-dimensional electrophoresis has been used extensively and is still an inexpensive and interesting methodology – especially at the strain and specimen level – sequencing is the tool of choice.

Kurtzman (1993), Fell (1995) and Fell et al. (2000), as well as many others, were pioneers and sequenced the D1/D2 region of the 26S ribosomal DNA of all type strains of known and recognized yeast species. Later, the internal transcribed spacers of rDNA were introduced to reinforce the quality of the classifications. Recently, Kurtzman and Robnett (2003) published a partial classification of Ascomycetes based on eight different DNA regions. Rokas et al. (2003) and Kuramae et al. (2006) have studied how many genes or regions need to be used in order to obtain reliable phylogenies. For the first, 30 genes would be 'sufficient' while the latter suggest that the number of genes necessary should be a function of the groups and taxonomic levels studied (species, genus, families, etc.).

When including 4850 orthologous coding genes (KOGS) of all eukaryote genomes (38) published at the time of the study, the optimum number of regions to be used for phylogenetic analyses

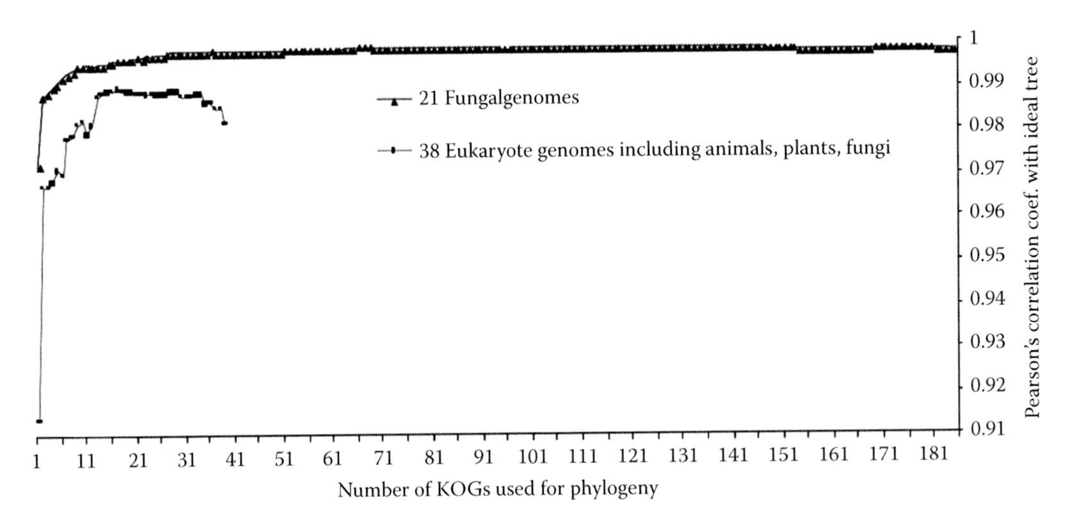

FIGURE 16.1 Evolution of the correlation (Pearson's coef.) or the overall fitting with a complete genome (ideal) phyologeny when the number of included KOGs is increasing. Two different groups of organisms are displayed. The first group is composed of fungi only (triangular symbols) and represents a less heterogeneous group. The second comprises animals, plants and fungi (square symbols).

should be 17 (Figure 16.1). When considering the fungi only, 68 genes is the optimum number (Figure 16.1). A higher degree of fitting with the overall phylogenetic content will be reached in more homogeneous groups. Even with two or three regions sequenced, the fitting in small groups (21 fungi in our example) will be higher than the optimum obtained in heterogeneous, larger and more divergent groups (38 eukaryotes in our example).

As can be seen, the number and the types of characters that must be used to classify accurately – and so obtain correct identifications – is changing with time and is a function of the taxonomic group under study. There is no doubt that new methods (e.g. DNA microarrays and others) will be introduced in the near future. While the number and the diversity of characters examined is increasing, the number of strains isolated and deposited in major culture collections is also rising dramatically. To give an idea of the rate of increase, in 1945 the CBS yeast collection contained 791 strains and at the same moment 134 species of yeasts were described (Figure 16.2). Today, the collection holds almost 7000 strains, around 1100 species of yeasts have been described and many more are expected to be published in the near future.

The CBS collection is the largest and the most complete yeast biodiversity collection in the world. But as can be seen from Figure 16.3, the geographic origin of available isolates is far from uniform. This suggests that only a small portion of the potential diversity has been sampled. Accordingly, what we know today is only a fraction of what exists. This confirms the predictions of Hawksworth (2001) and of Rosa and Peter (2006) that suggest that we know only a small percentage (5 and 1%, respectively) of fungal biodiversity.

While working at the species level has several advantages in terms of communication and makes sense in terms of evolution, the circumscription of yeast species is not always a trivial endeavour. Many strains do not show any sexual properties and application of the biological species concept remains, therefore, rather subjective. Many yeast species are known from a single or few isolates, making it more difficult to understand the whole potential and properties of those species. Even with well-established species (i.e. with a large number of well-described isolates), discrepancies within groups or subgroups are often recognized. The *Saccharomyces* genus is a very nice example of such problems (Kurtzman and Robnett, 2003).

Many researchers are more interested in strains and their particular potential and original combination of characteristics than in species. Therefore, existing yeast species data-bases are far

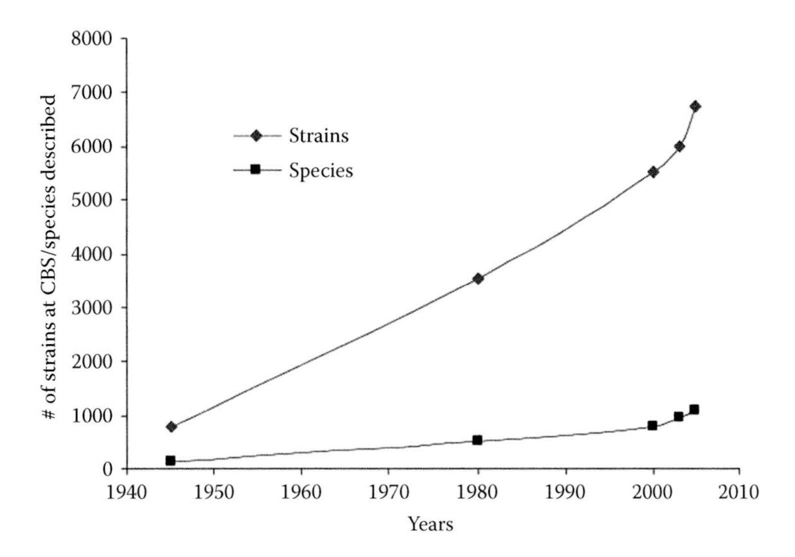

FIGURE 16.2 Evolution of number of species described and of the strains deposited at the CBS yeasts collection (Utrecht, the Netherlands) since 1945.

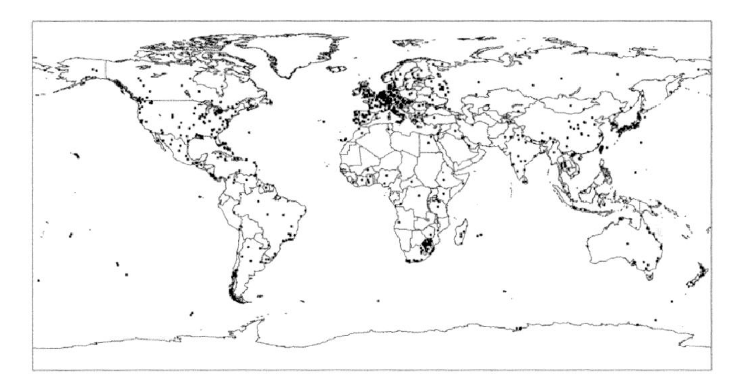

FIGURE 16.3 Geographic distribution of the strains available from the CBS yeasts collection (Utrecht, the Netherlands). Dots represent the original location of the strains.

from providing a satisfactory solution. To address this concern, it is necessary to record all the data available at different taxonomic levels, including the strain level. This has serious implications as the system would not have a fixed or partly fixed numbers of records or characters. Such a convention disqualifies some interesting identification methods – for example, artificial neural networks (ANN), which require training of the system on a stable or fixed number of reference operational taxonomic units (OTUs). Of course, some ANN variants incorporate some dynamic learning properties, but the learning process can still take too long to accommodate very large data-bases with data that are constantly changing. Knowing the types and the amount of data to be used for the classification of this continuously increasing taxonomic group, it becomes more obvious that flexible data-bases and adaptable algorithms for comparisons are needed.

COMPARISON METHODS

In the early stages, but not exclusively, and as in most other taxonomic groups, identification was based on dichotomous keys. Kreger-van Rij (1984), Kurtzman and Fell (1998) and Barnett et al. (2000a) presented keys based on physiological tests only or physiological tests together with morphological and sexual characters, which lead directly to the species names. Some keys include only a selected set of species that were considered the most likely to occur in a given situation. An isolate that does not belong to this set may be misidentified or be unidentifiable with such a key. On the other hand, a key that includes all yeast species would be very long and require many tests to be used (see Barnett et al., 2000a, where the key to all 704 species involved more than 100 tests). If not all the results required by a key are available, the identification could be completed until they are done. Moreover, if an erroneous or an unexpected result has been recorded for one of the tests, then either an incorrect identification would be made or the organism found to be unidentifiable. In non-computerized keys, the order of the questions is important since the tests or the observations have to be done sequentially.

Computerized keys are usually multiple-entry keys (MEKs) that allow the user to ask any question in any order. Overall, MEKs are superior to printed or dichotomous keys. Some of the problems highlighted previously for predefined printed dichotomous keys can be avoided. Using MEKs, identifications may be quick and easy, but a single mistake or difference in the observed results may lead to errors. Therefore, identification keys should be used with care. However, MEK are very useful, for example, to search for a set of properties in a given data-base.

The wider use of computers has led some systematists to employ other methods of identifying or classifying species. Probabilistic methods based on the use of Bayes's theorem for the computation of a probabilistic coefficient have gained a lot of favour since they are easy to implement and quick when calculated on slow computers (Willcox and Lapage, 1975). Barnett et al. (2000b) proposed a Bayesian implementation that has been used for a long time by many zymologists. Identifications obtained with this technique are usually clear-cut and users tend to like it, even if it can easily be misleading.

However, probabilistic methods can be applied only to discrete ordered monotonous characters. Continuous or unordered, non-monotonous characters can only be analysed after transformation or reduction. This can lead to many additional problems that are beyond the scope of the present chapter (but see Grzymala-Busse, 2002, for an overview on data reduction and discretization problems). The method used to create probabilistic profiles for species is another major issue since the number and the diversity of strains taken into account can have a profound impact on the basic probabilities and, therefore, on the resulting identifications.

The diversity of data to be analyzed, including the requirements of the users of our data-bases, and the aforementioned considerations suggest that the best solution for comparing two OTUs, species or strains, should be based on the computation of similarity coefficients. Many such coefficients have been described in the literature (e.g. Sneath and Sokal, 1973; Legendre and Legendre, 1998) and new ones can easily be implemented in response to the properties and requirements of any existing data types. It is easy to understand that sizes, colours, discrete characters, electrophoretic profiles, DNA sequences, time series or latitude–longitude positioning cannot be handled using the same methods. The properties of the different types of data, the way they are obtained, their reliability and their importance must all be considered when selecting a given method for the computation of a similarity coefficient. The context of a comparison is also crucial. Identification is not classification and a query for a set of properties to select the best candidates for an industrial process is a different process that requires different algorithmic approaches.

Methods based on similarity do not necessitate data transformations (often a source of information loss) and can be adapted to a wide range of problems and objectives. Weighting of characters can be done in accordance with their relative importance (see Neff, 1986, for a review on character weighting). However, it remains difficult to attribute the right value using objective criteria. Similarity or distance matrices can be summarized by available clustering techniques (Sneath and Sokal, 1971; Jain and Dubes, 1988; Kaufman and Rousseeuw, 1990; Legendre and Legendre, 1998).

Some disadvantages of similarity methods must also be considered. Identification reports may be more difficult to analyse. Similarity coefficients based on a given set of data may vary greatly as a function of the methods used for the comparisons. The following trivial example – a comparison of two size ranges using three algorithms that present different properties – illustrates this problem.

The sizes of the cells of a strain range from 4 to 7 µm. This strain has to be compared to a species that has cells ranging from 3 to 9 µm. A first option could be to consider that sizes of the strain are within the range (3 < 4 and 7 < 9) of the given species and the similarity coefficient is then equal to 100.0 per cent. A second solution could be based on overlaps of ranges and the coefficient would then be equal to (7 – 4)/(9 – 3) = 50.0 per cent. Another method based on comparison of medians gives a distance coefficient: [(3 + 9)/2] – [(4 + 7)/2] = 6 – 5.5 = 0.5. The latter value can be standardized by dividing the obtained distance by the largest median (0.5/7 = 0.71). The equivalent similarity is 1 – 0.71 = 92.9 per cent. As can be seen with this simple example, selection of algorithms is crucial.

The complexity of some algorithms, like sequence alignments, may lead to heavy computational loads and the time to obtain results may become an issue when very large numbers of records and characteristics have to be compared. Missing data and non-equivalent sets of data can be an issue as well. For example, if 6 criteria out of 18 are identical, then a similarity of 30 per cent is obtained. The same coefficient will be obtained when 600 criteria out of 1800 are identical. In the latter case, more data are included and the set of characteristics employed is obviously larger and hence different.

In identification procedures, pairwise similarity comparisons of an unknown against a series of OTUs results in the production of ordered lists where the best matching OTUs appear first. Interpretation of such lists is usually rather straightforward. Some problems can be encountered when several candidates have similar or identical similarity coefficients. This can be alleviated by showing the characters that differ and letting the users or experts decide the best or most likely option.

In (phenetic) classification procedures, similarity comparisons between the members of a given group give symmetrical or non-symmetrical similarity or distance matrices that can be difficult to interpret or to summarize, especially when the number of OTUs is large. In order to group items, many clustering methods are available. This is not the place to discuss those methods in detail, but both divisive and agglomerative clustering are used in the BioloMICS software solution (see later discussion). There are also other ordination methods such as principal coordinate analysis (PCoA). For small groups of OTUs, one of the numerous agglomerative clustering methods implemented in BioloMICS would be used (UPGMA, UPGMC, WPGMA, WPGMC, single and complete linkages, Ward, Lance and Williams flexible method, and neighbour joining).

For large groups of OTUs, it is well known that agglomerative clustering methods fail to provide phenetic trees that well represent the underlying distance or similarity matrix. The cophenetic coefficient of correlation between the matrix and the tree decreases as the number of OTUs increases (Rohlf, 1998; Robert, 2003). Divisive or ordination methods can more adequately define larger uniform groups, but are slow and computer intensive; all types of characters cannot be used, and no specific, user-friendly graphical representations can be provided with such methods. One idea would be to allow for the combination of several methods (MEK, then divisive clustering or ordination, then agglomerative clustering) in sequences in order to benefit from their advantages and to minimize their disadvantages. This allows more accurate grouping and greater flexibility. See Sneath and Sokal (1973) and Legendre and Legendre (1998) for more details on these methods and their advantages and disadvantages.

BIOLOMICS SOFTWARE AS A SOLUTION TO STORE AND ANALYSE BIOLOGICAL DATA

In 1990, while starting a biodiversity study of the yeast flora in wet and dry forests of central Africa, it became obvious that identification using conventional methods was barely possible. The standard physiological tests were tedious, slow and difficult to interpret unambiguously, and the identification of isolates using the classical dichotomic keys was very difficult. As a consequence, it was decided to first develop software to replace the existing keys (Robert et al., 1994). Soon after, a miniaturized system using 96-well microplates to perform physiological tests was created (Robert et al., 1997). This was a success, but the number of characteristics that needed to be introduced into the system was growing continuously with the addition of morphological and molecular features (among others). To endow the system with improved scalability, it was necessary to develop new and adapted algorithms, along with tools to first retrieve the data, then to manage and to analyze them.

This revised software, now called BioloMICS, was first developed as an MS-Windows client/server multi-user application (Robert and Szoke, 2003). It has since been developed further and can be used for retrieval, management, and analysis (e.g. identification, classification and statistical summary) of any biological material or research experiments. In 2000, another software package called BioloMICSWeb (Robert, 2000) was introduced to complement the first version and allow online Internet publication of data prepared with BioloMICS. This package also permits real polyphasic identifications of yeasts to be performed online against the species or strain data-bases (see http://www.cbs.knaw.nl/Yeast.htm; 6800 strains, 901 species, ±350 characters; Robert et al., 2004), *Penicillium* (see http://www.cbs.knaw.nl/Penicillium.htm; 58 species, ±60 characters; Samson and Frisvad, 2004) or *Phaeoacremonium* (see http://www.cbs.knaw.nl/Phaeoacremonium.htm; 22 species, ±40 characters; Mostert et al., 2005).

CREATING SELF-DESIGNED DATA-BASES

Most researchers have no time to spend on the creation of data-bases and the needed data-base-creation software. BioloMICS was therefore developed as a system that allows any user to create his or her own custom data-bases without any prior knowledge of data-basing or programming. Users choose the types of fields they need to store their biological data and define a few parameters related to them, such as the algorithms that should be used when performing comparisons (i.e. identifications, classifications), the tolerance (for fuzzy comparisons), the weighting, etc. The following field types can now be included:

- continuous data, range (min, low percentile, high percentile, max) (e.g. the size of the cells such as (5) 5.5–7 (8);
- continuous data, single value (e.g. a pH value of 9);
- discrete data, single state (e.g. presence or absence of a property);
- discrete data, multiple states (e.g. colour of the colony [not red, not green, white, cream, not brown, not black, etc.]);
- 96- or 384-well microplate, continuous data (absorbance values) (e.g. microplate containing user-defined tests, data stored as continuous values);
- 96- or 384-well microplate, discrete data (–, +, weak, etc.) (e.g. self-designed microplate containing user-defined tests, data; stored as discrete values);
- free array data, continuous data (e.g. array of any size of positive continuous data);
- free array data, discrete data (e.g. array of any size of discrete data);
- molecular weights from electrophoresis analysis (e.g. RAPD, RFLP, PCR, AFLP);
- time series and chromatograms;
- DNA sequences;
- protein sequences;
- text, administrative field;
- date field;
- bibliographic information field (e.g. bibliographic references such as articles, books, etc.);
- link to external files, including URL, pictures and others files; and

- geographical (lat, long.) coordinates stored at the record level and an unlimited number of coordinates stored and viewed on maps within the BioloMICS software; a precision factor can also be included.

Access to the data-base can be restricted (read, write, delete) by user (each user has a specific login and password allowing differential access to data-bases) and per field within a data-base. Some users can see all fields, whereas others can only view and use a portion. Data-base access is defined by the data-base owner/administrator. Any changes to any records and fields are recorded and a history of all the changes is kept. This allows tracking all the modifications made by any given user and it permits re-introduction of a previously stored value if needed.

Fields/characters can be added, modified (e.g. add/delete states for a given character, change the title of a character) or deleted at any time by the owner/administrator of the database. Fields can be re-ordered, re-organized and grouped by the administrator.

Managing Any Biological Data

The BioloMICS Software also supports the basic management of the created data-bases, such as the addition, modification or deletion of records and tables. Data can also be imported from, and exported to, a variety of formats (text, tab delimited, html, XML, MS-Excel or MS-Word, etc.).

Records can be linked with the taxonomic data-base at levels ranging from kingdom to species, subspecies, variety or form. All relevant taxonomic information (e.g. synonymy, basionym, anamorph–teleomorph connections for fungi, type strain, bibliographic information) is available. When the classification or nomenclature is changed, records are automatically updated. Bibliographic data-bases can be created and managed from within the software. These can be queried and linked to external data-bases such as PubMed, for example. Pictures can be displayed and measurements of the sizes of structures (e.g. cells, ascospores, colony diameters) can be performed on any electronic images, which may be stored in 13 different formats. Basic descriptive statistics and information (number of observations, mean, minimum, maximum, median, percentiles, standard deviation and variance) can be obtained and be stored in fields.

The software allows the automated reading and importation (through the RS-232 interface) of data from 96-well microplates with five different brands of microplate readers. Pictures of electrophoretic gels can be fully analyzed and molecular weights retrieved and properly stored. The virtual editor allows visual comparison of several profiles and their compilation to obtain consensus and/or agglomerated profiles.

The geographic-data viewer allows maps (shape and MrSid files) to be displayed. Records can be plotted to create distribution maps that can be exported to the clipboard or to Google Earth. A very large data-base of world locations has also been compiled and is supplied with the software. This contains the exact coordinates (latitudes–longitudes) of more than 5.5 million locations from all over the world, as well as altitudes and monthly precipitation and temperature data from more than 10,000 meteorological stations worldwide. Additional features are also available but will not be described here. Basic statistics on all type of fields like frequencies, data distribution, mean, minima, maxima, standard deviation and variance are provided on request.

Analyzing Any Biological Data

As discussed before, the analysis of stored data is of primary importance and should be possible using a variety of tools in response to a large panel of users having different goals in mind. Therefore, both basic and advanced MEK searching tools have been developed to allow querying any type of field using the best possible method. For example, a field containing size data (e.g. from 5 to 10) could be searched in up to five different ways, while DNA sequences can be aligned using four different algorithms (e.g. Blastn; see Altschul et al., 1990). This functionality allows users to select

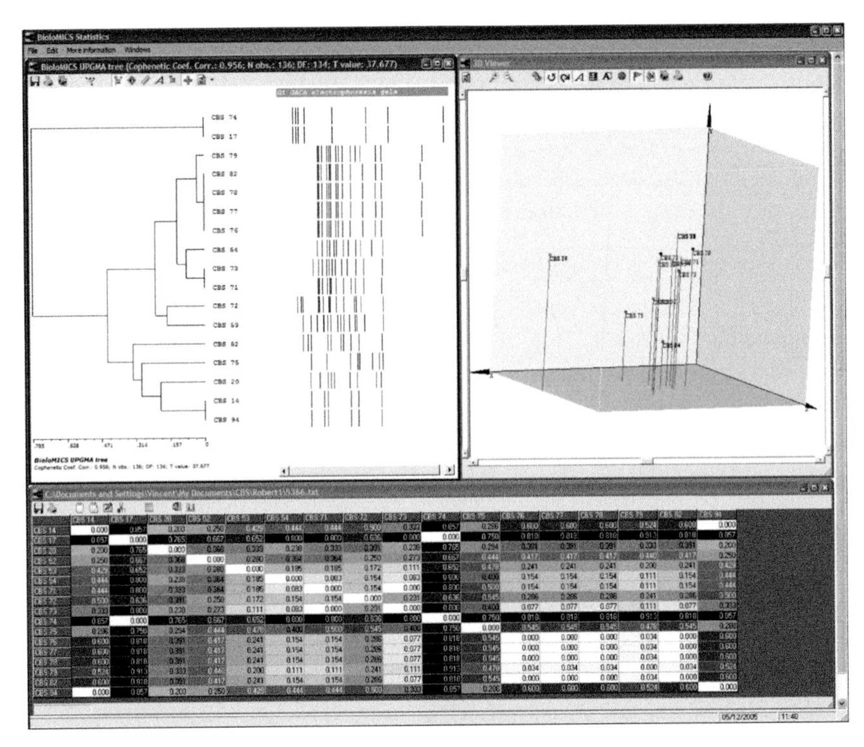

FIGURE 16.4 Results of a similarity-based comparison producing a distance square matrix (lower window) on which two different clustering methods were applied. The top-left window represents a UPGMA tree where one additional criterion was printed (bands of one-dimensional electrophoresis profiles). The top-right window is a three-dimensional representation of the OTUs presented on the distance matrix below that have been reordered using the principal coordinate analysis ordination method.

records with a given set of properties or to identify an unknown organism at the species level, for example. Advanced and complex queries containing questions separated by AND, OR or NOT and using brackets for their grouping can be performed.

Polyphasic identification and classification modules are available and permit a wide range of comparisons (already discussed). Several two- or three-dimensional graphical displays are available to represent the results of the combination of similarity- or correlation-based comparison methods and clustering techniques (see Figure 16.4 for an example). Comparison methods can be used singly or in combination to take the best of the different techniques. For example, one can first perform a preliminary selection of records based on an MEK query, compare or identify an unknown against the selected set of records, apply a divisive clustering method on the best matching records and then finally analyse by agglomerative clustering some of the groups obtained in the previous steps and draw a dendrogram or a three-dimensional display, as seen in Figure 16.4.

Another type of available comparison is called 'functional analysis'. This is a method allowing the grouping of objects/records and of their characters/fields at the same time using agglomerative clustering methods. Objects are compared with one another on the basis of a selection of characters. The obtained distance matrix is later represented as a phenotypic tree using one of the selected agglomerative clustering methods. The characters are first normalized and reduced (note that DNA or protein sequences cannot be used in such a method), then correlated with one another to obtain a correlation matrix that can be displayed in a tree-like representation using one of the available agglomerative clustering methods.

The result is displayed as a double tree (Figure 16.5). The first is a vertical tree showing the relationships between the objects/records. The second tree is horizontal and shows the groups of

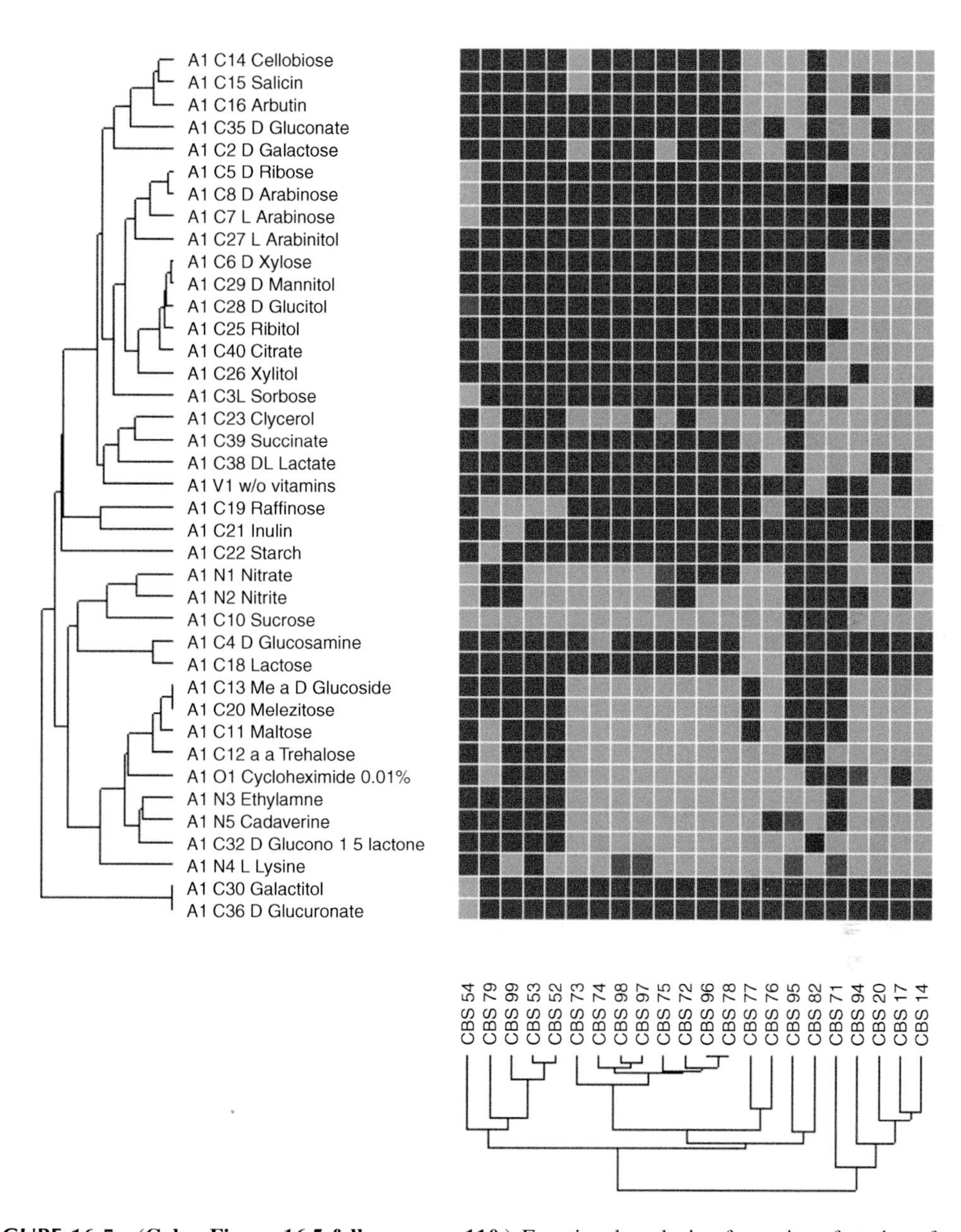

FIGURE 16.5 (Color Figure 16.5 follows page 110.) Functional analysis of a series of strains of yeast (UPGMA left-vertical tree) based on a set of physiological features (UPGMA top-horizontal tree). Normalized and reduced states of the above characters are displayed in light grey (negative result or absence of activity), in dark grey (positive result or presence of activity), in medium grey (unknown result) or in an intermediate shade for intermediate results. [Editor's note: medium and intermediate greys are not apparent in this image. Please see color figure for positions of these cells.]

characters/fields that are positively – or negatively, depending on the display method used ('positive' or 'negative' clustering) – correlated. Between the two trees a coloured 'heat map' shows the states of the characters/fields for the different objects/records. With this method, it is easier to isolate characters or groups of characters that are associated with some groups of objects. It is also possible to infer possible relations between different types of criteria – for example, the pathogenicity and a given physiological feature or the activity of a gene on a microarray.

CONCLUSIONS

Modern taxonomists do more than just discover and describe species. They record data related to their specimens or strains as well and publish them directly on the Internet. It is now possible to build data-bases of experiments, strains and species and to use a diversity of tools and algorithms to compare OTU. The reference systems based on biological, phylogenetic and/or phenetic species concepts can be used together with additional reference systems at other levels, such as the strains, for example. In addition, many taxonomically uninformative or non-relevant criteria can be introduced in data-bases dedicated to researchers working in industrial or clinical settings. This enables a wider use of such data-bases and will hopefully attract more attention to the work of the new taxonomists developing such resource centres.

Such data-rich centres require continuous software improvements in order to incorporate the best data retrieval, management and analysis tools. Bioinformatics is therefore at the core of the business of systematics and should be encouraged. Also, it is crucial that scientists and technicians get some form of training in bioinformatics since the tools that they are or will be using are becoming more powerful but at the same time more complex. Many biologists are hermetic to mathematical and computer science and some claim that one does not need to understand the basics behind the software being used daily. This assertion is not true anymore. The poor understanding of the algorithms hidden in software can and will lead scientists to draw erroneous conclusions. Software providers have to compile help files that are explicit enough and understandable by users who have no acquaintance with mathematics. These users also have to make the needed effort and should be trained properly. They also have to develop stronger critical thinking towards bioinformatics, not only to improve their research, but also to help bioinformaticians further develop and correct their applications. Bioinformatics is not a science of recipes and users should therefore use the methods that best fit with their needs instead of blindly applying methods used in previously published papers.

Bioinformaticians also have to develop new algorithms and tools to respond to today's biological questions. I strongly believe that polyphasic or multifactorial or holistic approaches have to be favored over single-sided studies. The complexity of holistic studies should not be underestimated and it will be quite challenging to mix wide arrays of criteria and to analyze them in sensible ways. The BioloMICS software solution is certainly a good step in this direction, but much remains to be done, especially with the always increasing flow and diversity of data. Polyphasic analyses mean different specialists will necessarily need to become involved in software developments. It implies that pure programmers will have to collaborate closely with taxonomists, phylogeneticists, ecologists, molecular biologists, physiologists, mathematicians, statisticians and many others. In itself, this is a real challenge, but bioinformatics is the catalyst that could allow this high 'potential chemistry' to produce unprecedented results. Bioinformatics is one of the most important factors in current biological science and should be accounted as such by both group leaders and funding agencies.

ACKNOWLEDGEMENTS

Many other tools are available with the BioloMICS software that are not mentioned in this chapter. See http://www.bio-aware.com for more information.

REFERENCES

Altschul, S.F., Gish, W., Miller, W., Myers, E.W. and Lipman, D.J. (1990) Basic local alignment search tool. *Journal of Molecular Biology,* 215: 403–410.

Barnett, J.A., Payne, R.W. and Yarrow, D. (2000a) *The Yeasts: Characteristics and Identification,* 3rd edition, Cambridge University Press, Cambridge, 1150 pp.

Barnett, J.A., Payne, R.W. and Yarrow, D. (2000b) Yeasts Identification PC Program. Version 5, Barnett, Norwich, UK.

Beijerinck, W.M. (1889) L'auxanographie, ou la méthode de l'hydrodiffusion dans la gelatine appliquée aux recherches biologiques. *Archives Néerlandaises des Sciences Exactes et Naturelles,* 23: 367–372.

Fell, J.W. (1995) rDNA targeted oligonucleotide primers for the identification of pathogenic yeasts in a polymerase chain reaction. *Journal of Industrial Microbiology,* 14: 475–477.

Fell, J.W., Boekhout, T., Fonseca, A., Scorzetti, G. and Statzell-Tallman, A. (2000) Biodiversity and systematics of basidiomycetous yeasts as determined by large-subunit rDNA D1/D2 domain sequence analysis. *International Journal of Systematic and Evolutionary Microbiology,* 50: 1351–1371.

Grzymala-Busse, J.W. (2002) Data reduction: discretization of numerical attributes. In *Handbook of Data Mining and Knowledge Discovery* (eds W. Klösgen, J.M. Zytkow and J. Zyt), Oxford University Press, Oxford, pp. 218–225.

Hansen, E.C. (1888) Recherches sur la physiologie et la morphologie des ferments alcooliques. VII. Action des ferments alcooliques sur les diverses espèces de sucre. *Comptes Rendus des Travaux du Laboratoire Carlsberg,* 2: 143–192.

Hansen, E.C. (1891) Recherches sur la physiologie et la morphologie des ferments alcooliques. *Comptes Rendus des Travaux du Laboratoire Carlsberg,* 3: 44–66.

Hansen, E.C. (1898) Recherches sur la physiologie et la morphologie des ferments alcooliques. IX. Sur la vitalité des ferments alcooliques et leur variation dans les milieux nutritifs et a l'état sec. *Comptes Rendus des Travaux du Laboratoire Carlsberg,* 4: 93–121.

Hansen, E.C. (1902) Recherches comparatives sur les conditions de la croissance végétative et le développement des organes de reproduction des levures et des moisissures de la fermentation alcoolique. *Comptes Rendus des Travaux du Laboratoire Carlsberg,* 5: 68–107.

Hawksworth, D.L. (2001) The magnitude of fungal diversity: the 1.5 million species estimate revisited. *Mycological Research,* 105: 1422–1432.

Jain, A.K. and Dubes, R.C. (1988) *Algorithms for Clustering Data,* Prentice Hall, Englewood Cliffs, NJ, 320 pp.

Kaufman, L. and Rousseeuw, P.J. (1990) *Finding Groups in Data. An Introduction to Cluster Analysis,* Wiley, New York, 368 pp.

Kreger-Van Rij, N.J.W. (1984) *The Yeasts: A Taxonomic Study,* 3rd edition, Elsevier Scientific, Amsterdam, 1076 pp.

Kuramae, E.E., Robert, V., Snel, B. and Boekhout, T. (in press) Phylogenomics reveal a robust fungal tree of life. *FEMS Yeast Research.*

Kurtzman, C.P. (1993) Systematics of the ascomycetous yeasts assessed from ribosomal RNA sequence divergence. *Antonie van Leeuwenhoek,* 63: 165–174.

Kurtzman, C.P. and Fell, J.W. (1998) *The Yeasts: A Taxonomic Study,* 4th edition, Elsevier, Amsterdam, 1076 pp.

Kurtzman, C.P. and Robnett, C.J. (2003) Phylogenetic relationships among yeasts of the '*Saccharomyces* complex' determined from multigene sequence analyses. *FEMS Yeast Research,* 4: 417–432.

Legendre, P. and Legendre, L. (1998) *Numerical Ecology,* Elsevier, Amsterdam, 870 pp.

Lodder, J. (1934) Die anaskosporogenen Hefen, I. Halfte. Verhandelingen der Koninklijke Akademie van Wetenschappen te Amsterdam. *Afdeeling Natuurkunde* Sectie II, 32: 1–256.

Lodder, J. (1970) *The Yeasts: A Taxonomic Study,* 2nd edition. North-Holland, Amsterdam, 1, 385 pp.

Lodder, J. and Kreger-van Rij, N.J.W. (1952) *The Yeasts: A Taxonomic Study,* North-Holland, Amsterdam, 713 pp.

Mostert, L., Groenewald, J.Z., Summerbell, R.C., Robert V., Sutton, D.A., Padhye, A.A. and Crous, P.W. (2005) Species of *Phaeoacremonium* associated with human infections and environmental reservoirs in infected woody plants. *Journal of Clinical Microbiology,* 43: 1752–1767.

Neff, N.A. (1986) A rational basis for a priori character weighting. *Systematic Zoology,* 35: 110–123.

Robert, V. (2000) BioloMICSWeb software for online publication and polyphasic identification of biological data. Version 1, BioAware, Hannut, Belgium.

Robert, V. (2003) Data management and bioinformatics. In *Yeasts and Food* (eds T. Boekhout and V. Robert), Behr's Verlag, Hamburg, Germany, pp. 139–170.

Robert, V., de Bien, J.-E. and Hennebert, G.L. (1994) ALLEV, a new program for computer-assisted identification of yeasts. *Taxon,* 43: 433–439.

Robert, V., Epping, W., Boekhout, T., Smith, M.Th., Poot, G. and Stalpers, J. (2004) *CBS Yeasts Database,* Centraalbureau voor Schimmelcultures, Utrecht, the Netherlands.

Robert, V., Evrard, P. and Hennebert, G.L. (1997) BCCM/Allev 2.00 an automated system for the identification of yeasts. *Mycotaxon,* 64: 455–463.

Robert, V. and Szoke, S. (2003) BioloMICS, Biological Manager for Identification, Classification and Statistics. Version 6.2, BioAware, Hannut, Belgium.

Rohlf, F.J. (1998) NTSYSpc 2.00 software. Applied Biostatistics Inc., New York.

Rokas, A., Williams, B.L., King, N. and Carroll, S.B. (2003) Genome-scale approaches to resolving incongruence in molecular phylogenies. *Nature,* 425: 798–804.

Rosa, C.A. and Peter, G. (2006) *Biodiversity and Ecophysiology of Yeasts,* Springer–Verlag, Berlin, 580 pp.

Samson, R.A. and Frisvad, J.C. (2004) *Penicillium* subgenus *Penicillium*: new taxonomic schemes, mycotoxins and other extrolites. *Studies in Mycology,* 49: 1–257.

Sneath, P.H.A. and Sokal, R.R. (1973) *Numerical Taxonomy,* W.H. Freeman, San Francisco, 573 pp.

Wickerham, L.J. (1951) Taxonomy of yeasts, United States Department of Agriculture, Technical Bulletin no. 1029, Washington, 56 pp.

Wickerham, L.J. and Burton, K.A. (1948) Carbon assimilation tests for the classification of yeasts. *Journal of Bacteriology,* 56: 363–371.

Willcox, W.R. and Lapage, S.P. (1975) Methods used in a program for computer-aided identification of bacteria. In *Biological Identification with Computers* (ed R.J. Pankhurst), Academic Press, London, pp. 103–119.

17 Automatic Measurement of Honeybee Wings

Adam Tofilski

CONTENTS

Introduction 289
Wing Mounting 290
Image Acquisition 290
Wing Measurements..291
Image Analysis 291
    Detection of the Wing Outline 291
    Detection of Junctions 293
Reliability and Precision of the Automatic Measurements..293
    Material and Methods 294
    Results 294
    Discussion 294
Wing-Based Species Discrimination 295
Future Prospects...296
Acknowledgements 296
References 296

INTRODUCTION

Honeybee wings are one of the most frequently measured insect body parts. These measurements are most often used to distinguish between honeybee subspecies (Ruttner, 1988a). This discrimination is required in honeybee breeding in order to preserve the morphological characteristics of selection lines (Ruttner, 1988b), with breeding programs often being related to conserving honeybee biodiversity (Rortais et al., 2004). Discrimination between subspecies is essential for monitoring and controlling Africanized honeybees on the American continents (Strauss and Houck, 1994). There is also a growing number of studies using honeybee wing measurements, not only in biogeography but also genetics (Brückner, 1976), development (Smith et al., 1997) and ecology (Higginson and Barnard, 2004).

The process of measuring of honeybee wings is tedious and time consuming. In subspecies discrimination studies, the measurement of a single individual is not enough to obtain 100 per cent accuracy (Daly and Balling, 1978). Usually, more than 10 workers from one colony are used. In biogeography studies, it is recommended that measurement of 10 workers from five to six colonies per apiary be taken (Radloff et al., 2003). Also, the number of wing measurements per individual is large, usually ranging from 15 (Ruttner, 1988a) to 38 (Smith et al., 1997).

WING MOUNTING

Before a wing can be measured it needs to be mounted. Mounting allows the wing to be flattened and positioned on a plane perpendicular to the optic axis of the measuring instrument or image acquisition system. The mounted wing can also be easily labelled and stored.

Specimens are usually mounted on microscopic slides or glass photographic slide frames. Wings mounted on microscopic slides are fixed in place by Canada balsam (Adsavakulchai et al., 1999; Weeks et al., 1999b), Euparal (Batra, 1988; Smith et al., 1997) or transparent tape (Yu et al., 1992; Kokko et al., 1996). Using mounting substances is tedious and their optical properties deteriorate with time. Mounting with the use of transparent plastic tape is quick; however, the optical quality of images obtained from such preparations is markedly reduced.

The optimal solution seems to be mounting the wings in photographic glass frames. These do not reduce optical properties because the wing is kept in place by the glass and no mounting medium is required. Moreover, mounting is quick and easy and remounting is possible if the wing needs to be cleaned of dust or dirt. Wings mounted this way are ready for scanning, projecting with a slide projector or photographing under a microscope. If numerous samples of worker front wings are taken from a single colony, they can be mounted in one slide frame. On the other hand, if right and left wings are required (e.g. in fluctuating asymmetry studies), all the wings of a single individual can be mounted in one slide frame. The position of the wings in the frame can provide information as to whether it is the left or right wing. A slight problem is that the smaller hind wings, which are thinner than the front wings, tend to slide down because the glass plates are pushed away by the front wings. This problem can be solved by using a small amount of water-soluble glue. This should be applied at the wing base, which provides little information for the automatic identification of species.

IMAGE ACQUISITION

In most contemporary studies, a wing image is acquired before measurements are made. In earlier studies, PAL or NTSC video cameras were mainly used to obtain the images. These required a frame grabber that converted an analog video signal into a digital image. Resolutions obtained by PAL or NTSC video cameras can be relatively low (often less than 0.5 Mpixel). This can be improved by the use of digital still cameras, which produce images of up to 12 Mpixels. The cameras can be connected to various microscopes or lenses to obtain an image of the appropriate magnification and quality. They are versatile and can take wing images even in the case of live insects and museum specimens, when wings cannot be detached from the insect's body (Houle et al., 2003). Another advantage of these cameras is fast image acquisition, which can be important in the case of live insects anaesthetized for only a short time.

When the wing can be detached from an insect's body and mounted in a photographic slide frame, scanners might be preferred. These are easily accessible and available in a wide range of prices – from inexpensive flatbed scanners with a slide adapter to more expensive high-resolution photographic slide scanners. There are attachments to some of the scanners allowing the automatic scanning of multiple frames, making image acquisition faster and requiring less human intervention. The resolution of a photographic slide scanner can be up to 4000 dpi (equivalent of about 22 Mpixels when a whole slide frame is scanned), which is more than most digital cameras produce. When scanners are used, scaling is not required because the resolution of the image (in dpi) is known. Images obtained by scanning have a uniform background, which makes wing detection and analysis easier (Weeks and Gaston, 1997). Producing a uniform background with a microscope is much more difficult and requires a precisely aligned light source.

WING MEASUREMENTS

The first large-scale measurements of honeybee wings were undertaken by Alpatov (1929). At that time microscopes equipped with ocular micrometers were used. DuPraw (1964) used a microscope-slide projector to obtain a wing image on a piece of paper, which he pricked with a pin to mark the vein junctions. He then measured angles and distances in the traditional way with a ruler and protractor.

Daly et al. (1982) combined the projector with a digitizer tablet. In doing this they were able to indicate the landmarks that were automatically recorded by the computer. This procedure increased the speed of measurements, their precision and their repeatability, and eliminated errors related to data input or calculations.

A more sophisticated system was described by Batra (1988). In this system wing images were obtained through the use of a microscope equipped with a camera. This system required substantial user intervention, but allowed the automatic detection of 16 vein junctions. It was based on specific and expensive hardware, which reduced its accessibility.

In most recent studies wing images acquired by a camera or scanner are transferred to the computer and analysed using data acquisition software. Either specialized honeybee wing measurement software (e.g. Beemorph; see Talbot, 2002) or general-purpose data acquisition software (e.g. tpsDig; see Rohlf, 2005) can be used. Usually, a computer mouse is used to indicate the vein junctions on a computer screen and the data acquisition software provides the coordinates, which can be used to calculate distances and angles.

IMAGE ANALYSIS

Despite several attempts to automate honeybee wing measurements, the methods currently in use are, to a large extent, manual. This is surprising because image analysis allows the automatic detection of vein junctions on which the measurements are based (Batra, 1988; Steinhage et al., 1997). One of the computer programs specialized in the image analysis of insect wings is *DrawWing* (Tofilski, 2004). It is particularly suitable for automatically measuring wings because it is able to recognize and position wings in an image. *DrawWing* software is able to recognize all vein junctions of the honeybee forewing, which allows for fully automatic measurements. The image analysis of honeybee wings is performed in two main steps: the first step is the detection of the wing outline and the second step is the detection of the wing venation.

DETECTION OF THE WING OUTLINE

In a fully automated measurement system, wings need to be detected in an image without user intervention. The process of wing detection is based on the image histogram, which represents the number of pixels at each greyscale value within the image. This works very well if the wings are darker and presented against a relatively uniform light background (Figure 17.1), which is always the case if the image has been obtained by a scanner.

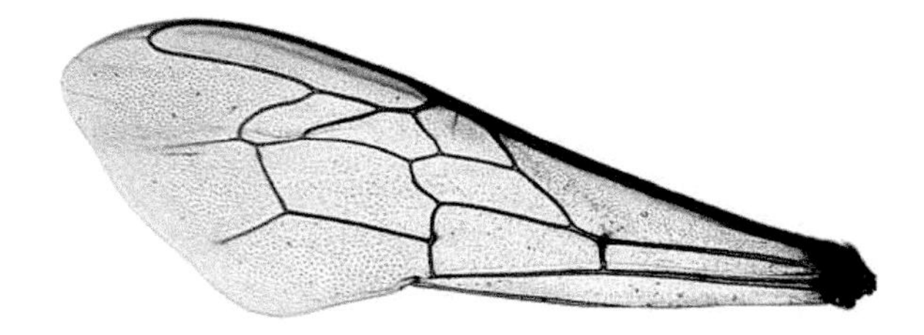

FIGURE 17.1 The front wing of a worker honeybee.

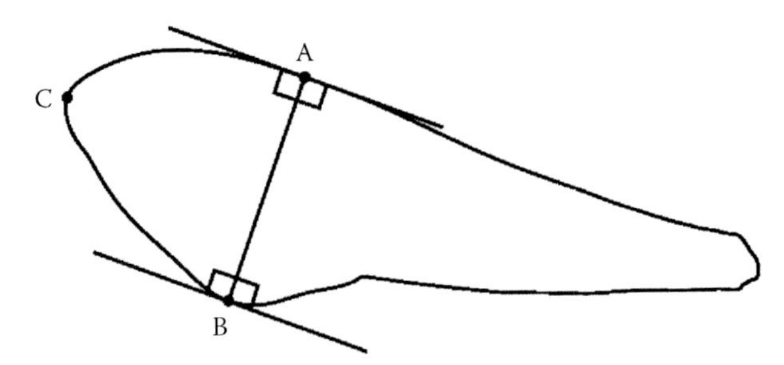

FIGURE 17.2 The wing outline with three characteristic points marked: A: anterior point, B: posterior point and C: apex point. Tangents to the outline at the anterior and posterior points are perpendicular to the line crossing the points. The anterior and posterior points demarcate the wing width, which can be used for wing positioning and scaling. There are two extrema of the outline on opposite sides of the wing width. The apex point is the extremum at which the outline curvature is smaller.

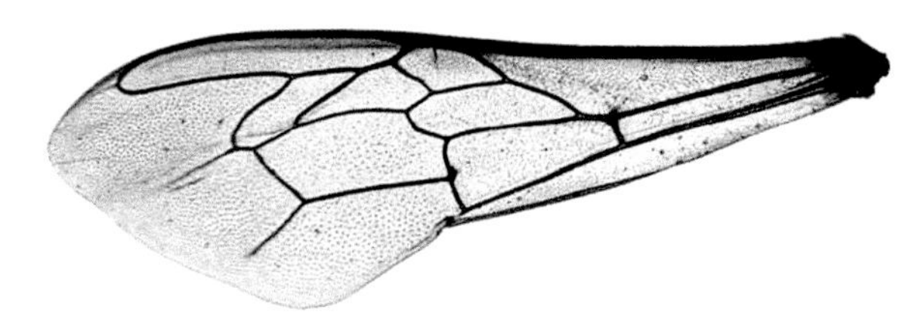

FIGURE 17.3 The standard wing image produced by *DrawWing* from the image depicted in Figure 17.2.

If the background is uniform, the image histogram is markedly bimodal. The first mode corresponds to wing pixels and the second mode corresponds to the background. The colour corresponding to the minimum between the maxima is used as a threshold for conversion of the greyscale image to black and white. At this stage wings (and some other dark artefacts) are represented as black objects. Outlines of all the objects are detected by contour tracing (Rohlf, 1990) and their size and shape are analysed to decide whether or not they are potential wings. Only the potential wings are subjected to further processing.

In order to remove artefacts (e.g. hairs or dust particles) from the wing edge, the outline is smoothed. The smoothing is achieved by contour dilation equal to earlier erosion. The smoothing algorithm removes only the thin protrusions of the outline and minimally affects the wing shape.

On the honeybee wing outline there are three characteristic points: anterior, posterior and apex (Figure 17.2). The anterior and posterior points demarcate the wing width. The tangents to the wing outline at those two points are parallel to each other and perpendicular to the line crossing the points. The anterior and posterior points are automatically detected by examining outline points in pairs until they meet the aforementioned criteria. There are two extrema of the outline on opposite sides of the wing width. The apex point is the extremum at which the outline curvature is smaller. Finally, the wing is rotated in software and cropped from the original image. The resulting image is called a 'standard wing image' (Figure 17.3).

In order to avoid negative coordinates, the origin of the Cartesian coordinate system is defined by the bottom and left extreme of the wing outline. Traditionally, right wings are depicted in the drawings and images (Mason, 1986). However, if the right wing is presented in the Cartesian coordinates system, zero on the x-axis has to be assigned to the wing base. It is better to avoid this because the wing base is not very well defined. When the wing is detached from the body, it can break off in different places. Moreover, the wing base can be invisible when wing images are acquired from live specimens. Therefore, in a standard wing image, left wings are depicted. Then the zero on the x-axis is defined by the wing apex, which is always visible. Right wings are flipped

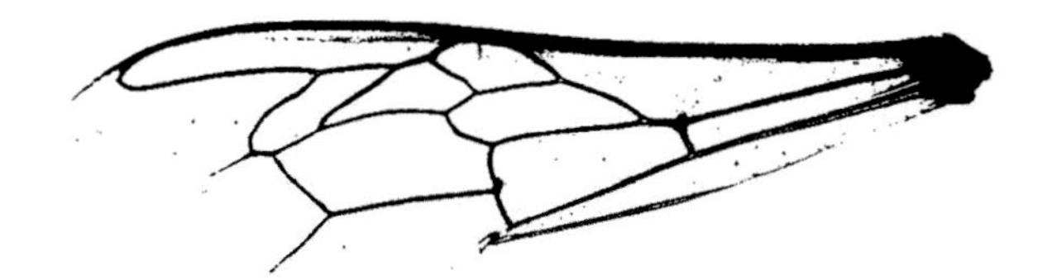

FIGURE 17.4 The honeybee front wing after extraction of venation outline.

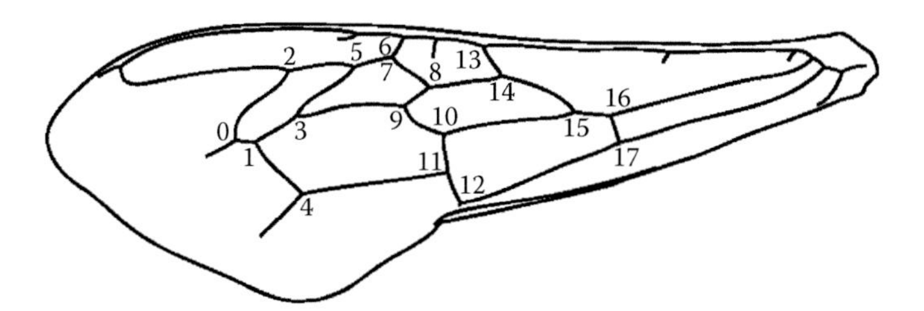

FIGURE 17.5 The wing diagram depicting only the wing outline and the venation skeleton.

horizontally before they are presented in the standard wing image. Multiple standard wing images are produced if the input image contains more than one wing.

Detection of Junctions

The veins are darker than the membranous parts of the wing, but the intensity of their colour varies. This makes it difficult to choose the optimal threshold for extracting the venation outline. In this case, the method used to detect the wing outline is not always satisfactory and a more complex method needs to be used. A gradient for the image (Shen and Castan, 1992) can be calculated and the gradient's maxima found. The maxima at which the gradient is relatively high correspond to the veins' edges. The median of the maxima can then be used as the threshold. Pixels darker than the threshold are considered 'veins' and their outline (Figure 17.4) is found by contour tracing (Rohlf, 1990). The vein outline is reduced to its skeleton using a thinning algorithm (Rohlf, 1990). The skeleton pixels with three of four neighbours are vein junctions. The junctions are compared with the expected junctions from a typical honeybee wing. Junctions that fit those expected are reported to the user. In order to avoid misidentification, both coordinates of the junctions and tangents at which veins approach the junction are used in the comparison. Apart from the list of vein junctions, a wing diagram is produced (Figure 17.5). It depicts the wing in the same position as the standard wing image but only the wing outline and skeleton of venation are visible.

RELIABILITY AND PRECISION OF THE AUTOMATIC MEASUREMENTS

As a rule, differences between wings of individuals belonging to the same species are small. For example, venation of two honeybee subspecies – *Apis mellifera carnica* and *A. m. ligustica* – differs mainly by lengths of a cubital vein, which are 0.56 and 0.59 mm, respectively (Nazzi, 1992). This difference is small in comparison with precision of the venation measurements: 0.023 mm for the ocular micrometer and 0.007 mm for the tablet (Daly et al., 1982). The measurements of the wing length and width proved to be even less precise (Dedej and Nazzi, 1994). Therefore, the precision is one of the most important characteristics of any automated wing-measurement method. Other important factors are reliability and the speed of measurement.

Material and Methods

In order to test reliability of the *DrawWing* software, measurements from 300 honeybee workers were used. Their wings were torn off and mounted in glass photographic frames (Rowi 260). In every frame all wings of one worker were mounted (two front wings and two hind wings). Images of the wings were obtained using a Nikon Coolscan 5000 ED scanner equipped with SF-210 slide feeder. The original resolution of the images was 4000 dpi. However, this was converted to 2400 dpi before analysis. The images were analysed by *DrawWing* software and the number of correctly recognized front wings was counted. Correctly recognized wings were checked if all vein junctions were detected.

In order to compare precision and speed of the automatic and manual wing measurement, 40 front wings of honeybee workers were used. Three images were produced for every wing and the positions of 18 landmarks determined. All landmarks were vein junctions. These landmarks were either manually pointed with a computer mouse using tpsDig2 software (Rohlf, 2005), or determined automatically by *DrawWing*. Manual measurements were made by three different inexperienced persons. Procrustes superimposition was used to align the junctions found on three images of the same wing. The precision was measured as the mean distance between the three junctions and their centroid (Arnqvist and Martensson, 1998). The results are reported as mean ± standard deviation.

Results

Out of 600 front wings in the 300 images, 552 wings (92.0%) were correctly detected and saved as standard wing images. In the standard wing images, out of 9936 expected vein junctions, 9894 (99.6%) were detected. The automatic measurements proved to be significantly more precise than the manual measurements (Mann–Whitney test: $Z = -27.38$, $N_1 = 720$, $N_2 = 2160$, $p < 0.001$; Table 17.1). The mean precision of the automatic and manual measurements was 5.20 ± 4.44 and 9.37 ± 5.22 μm, respectively. In the 2400-dpi images, those values correspond to 0.49 ± 0.42 and 0.88 ± 0.49 pixels, respectively. There were significant differences between precision of measurements performed by three different manual data collectors (Kruskal–Wallis test: $H = 283$, df = 2, $N = 2160$, $p < 0.001$; Table 17.1). The automatic measurements were also much faster than the manual measurements (Mann–Whitney test: $Z = -9.47$, $N_1 = 40$, $N_2 = 120$, $p < 0.001$). The time required to analyse one wing automatically and manually was 6.4 ± 0.83 and 68.8 ± 18.40 s, respectively.

Discussion

The detection rate of the front wings of honeybee workers was satisfactory, but it can be improved further. The main reason for the detection failures was damages of the wing edge, which led to incorrect detection of wing width. This problem can be solved to some degree by changing the algorithm of the wing width search; however, if the wing edge is damaged at the anterior or posterior point (Figure 17.2), correct detection of the wing width is impossible. In a few cases, there were also problems with artefacts near the wing base arising from tearing the wings off the thorax. This problem could be avoided by cutting the wings off with scissors. The detection of vein junctions was much better than the detection of wings – close to 100 per cent. The problems with the vein junctions' detections were mainly caused by unusual shapes of the venation.

These results show clearly that the automatic measurements are more precise and much faster than the manual measurements. This test used only one scanner, but it is probable that differences between scanners will be much smaller than the differences between persons. This is important when results from different studies are compared. The precision of automatic measurements is close to the limits set by resolution of the images. In order to increase the precision further, images of higher resolution need to be used, but this requires changes to the software, which is optimized for 2400-dpi images.

TABLE 17.1
Precision[a] of Automatic and Manual Measurements of Honeybee Wing Venation

Junction no.	Automatic	Manual			
		A	B	C	Mean
0	4.31 ± 3.68	6.94 ± 3.54	8.26 ± 4.59	10.57 ± 4.57	8.59 ± 4.23
1	4.70 ± 4.14	6.30 ± 3.57	9.54 ± 4.66	10.49 ± 5.79	8.78 ± 4.68
2	5.69 ± 3.30	7.42 ± 3.80	9.06 ± 5.25	10.83 ± 5.87	9.10 ± 4.98
3	3.70 ± 2.69	6.90 ± 3.71	8.51 ± 4.53	10.56 ± 5.54	8.66 ± 4.59
4	5.94 ± 3.78	7.67 ± 3.59	9.85 ± 5.08	12.19 ± 5.68	9.91 ± 4.78
5	4.08 ± 2.59	6.75 ± 3.88	8.87 ± 5.05	10.50 ± 5.54	8.71 ± 4.83
6	5.77 ± 3.58	9.19 ± 5.71	15.90 ± 9.17	12.32 ± 6.37	12.47 ± 7.08
7	3.83 ± 2.34	6.92 ± 4.16	8.83 ± 4.94	9.89 ± 5.56	8.55 ± 4.89
8	4.07 ± 3.04	6.42 ± 4.00	7.64 ± 3.57	10.55 ± 5.18	8.20 ± 4.25
9	4.01 ± 2.34	7.04 ± 6.46	8.36 ± 7.37	12.10 ± 6.54	9.17 ± 6.79
10	4.67 ± 3.26	7.14 ± 3.84	8.44 ± 4.46	11.10 ± 5.94	8.89 ± 4.75
11	5.11 ± 4.30	7.85 ± 4.03	8.46 ± 4.91	10.80 ± 5.65	9.03 ± 4.87
12	5.94 ± 4.33	6.30 ± 3.26	8.38 ± 5.08	10.60 ± 5.22	8.43 ± 4.52
13	11.21 ± 13.28	12.26 ± 7.08	19.54 ± 11.08	14.76 ± 8.19	15.52 ± 8.78
14	5.33 ± 5.66	8.64 ± 5.84	8.28 ± 4.41	10.60 ± 5.68	9.17 ± 5.31
15	5.11 ± 7.38	6.94 ± 3.83	7.83 ± 4.23	10.12 ± 6.17	8.30 ± 4.74
16	5.81 ± 7.64	6.85 ± 4.24	9.02 ± 5.01	9.76 ± 5.18	8.54 ± 4.81
17	4.28 ± 2.60	7.57 ± 4.08	8.92 ± 5.38	9.40 ± 5.57	8.63 ± 5.01
Mean	5.20 ± 4.44	7.50 ± 4.37	9.65 ± 5.49	10.95 ± 5.79	9.37 ± 5.22

Note: The manual measurements were performed by three persons (A, B and C). The veins' junctions are numbered according to Figure 17.5.

[a] In microns; mean ± SD.

WING-BASED SPECIES DISCRIMINATION

Wings seem to be the best morphological structure for the automatic identification of insects. They differ much more between species than within species (Matias et al., 2001; Baylac et al., 2003). Even within species the variation is large enough to distinguish subspecies (Daly and Balling, 1978; Ruttner, 1988a). There is no problem with the positioning of wings in three dimensions because they are flat and can be considered as two dimensional (Weeks and Gaston, 1997). High-quality wing images can be obtained using relatively inexpensive scanners, which allows the process to be semi-automated.

When wings of different species are compared there is a need for scaling and rotating them to a standard position (Lane, 1981; Albrecht and Kaila, 1997). The wing outline can be used for the positioning because, unlike venation, its shape is similar in all species of insects. The outline is always rounded at the apex and narrower at the wing base, which can be used for wing detection and positioning. The width of such outline can be chosen as a baseline and the two-point registration (Bookstein, 1991) can be used to produce a standard wing image and wing diagram by a combination of scaling, translating and rotating. Ideally, the baseline should be positioned along the longest wing diameter and across the wing centroid (Zelditch et al., 2004). However, this is difficult to achieve because one side (the wing base) of the longest diameter is not well defined and can be invisible.

Both the standard wing image and the wing diagram can be important elements of an automatic species discrimination system. In comparison with unprocessed images, these are a better form of storage and medium of exchange for information about wings. They contain only relevant information, which reduces both storage space and retrieval time. The standard wing image can be used by pixel-based species identification systems that, instead of extracting characteristic points from the image, use the intensity of all pixels (Weeks et al., 1997; Roth et al., 1999a; MacLeod et al., in press). In those systems, the position of a wing in the image is essential for the correct discrimination of species. The manual positioning of wings has proven to be time consuming and imprecise (Weeks et al., 1999a, 1999b).

FUTURE PROSPECTS

The system described here has been tested on honeybee wings; however, it should work with other species of insects as well. The detection of a wing outline is based on properties common to the wings of all insects; therefore, *DrawWing* should generate the standard wing image, even for butterfly wings. If wings are membranous with clearly visible veins, a wing diagram can also be produced. Extraction of the standard wing image was also tested on three species of wasps and the results were satisfactory (unpublished data). The earlier version of this software was already able to produce wing diagram of various species (Tofilski, 2004). Only the detection of junctions in the wing of a particular species requires a list of expected junction coordinates for this species.

It has been shown that insects' wings can be used to discriminate between species (Weeks and Gaston, 1997). Methods based on wing outline (Rohlf and Archie, 1984), vein junctions (Schröder et al., 2002) and all image pixels (Weeks et al., 1997; Roth et al., 1999a) have proven successful. Various statistical methods of species discrimination have been described as well (Strauss and Houck, 1994; Roth et al., 1999b). However, there are relatively few tools for efficient data acquisition and wing visualization. *DrawWing* proved to be successful in this area. It will be developed further to become a general-purpose tool for the processing of insect wing images.

ACKNOWLEDGEMENTS

I thank Norman MacLeod for helpful comments on earlier versions of this paper. This work was supported by EC Marie Curie European Reintegration Grant MERG-6-CT-2005-517576 and MEiN grant 2P06Z01328.

REFERENCES

Adsavakulchai, S., Baimai, V., Prachyabrued, W., Grote, P.J. and Lertlum, S. (1999) Morphometric study for identification of the *Bactrocera dorsalis* complex (Diptera: Tephritidae) using wing image analysis. *Biotropica,* 13: 37–48.

Albrecht, A. and Kaila, L. (1997) Variation of wing venation in Elachistidae (Lepidoptera: Gelechioidea): Methodology and implications to systematics. *Systematic Entomology,* 22: 185–198.

Alpatov, W.W. (1929) Biometrical studies on variation and races of the honey bee (*Apis mellifera* L.). *Quarterly Review of Biology,* 4: 1–58.

Arnqvist, G. and Martensson, T. (1998) Measurement error in geometric morphometrics: empirical strategies to asses and reduce its impact on measures of shape. *Acta Zoologica Academiae Scientiarum Hungaricae,* 44: 73–96.

Batra, S.W.T. (1988) Automatic image analysis for rapid identification of Africanized honey bees. In *Africanized Honey Bees and Bee Mites* (ed G.R. Needham), Ellis Horwood, Chichester, UK, pp. 260–263.

Baylac, M., Villemant, C. and Simbolotti, G. (2003) Combining geometric morphometrics with pattern recognition for the investigation of species complexes. *Biological Journal of the Linnean Society,* 80: 89–98.

Bookstein, F.L. (1991) *Morphometric Tools for Landmark Data: Geometry and Biology.* Cambridge University Press, Cambridge, 435 pp.

Brückner, D. (1976) The influence of genetic variability on wing symmetry in honeybees (*Apis mellifera*). *Evolution,* 30: 100–108.
Daly, H.V. and Balling, S.S. (1978) Identification of Africanized honeybees in the Western Hemisphere by discriminant analysis. *Journal of the Kansas Entomological Society,* 51: 857–869.
Daly, H.V., Hoelmer, K., Norman, P. and Allen, T. (1982) Computer-assisted measurement and identification of honey bees (Hymenoptera: Apidae). *Annals of the Entomological Society of America,* 75: 591–594.
Dedej, S. and Nazzi, F. (1994) Two distances of forewing venation as estimates of wing size. *Journal of Apicultural Research,* 33: 59–61.
DuPraw, E.J. (1964) Non-Linnean taxonomy. *Nature,* 202: 849–852.
Higginson, A.D. and Barnard, C.J. (2004) Accumulating wing damage affects foraging decisions in honeybees (*Apis mellifera* L.). *Ecological Entomology,* 29: 52–59.
Houle, D., Mezey, J., Galpern, P. and Carter, A. (2003) Automated measurement of *Drosophila* wings. *BMC Evolutionary Biology,* 3: 1–13.
Kokko, E.G., Floate, K.D., Colwell, D.D. and Lee, B. (1996) Measurement of fluctuating asymmetry in insect wings using image analysis. *Annals of the Entomological Society of America,* 89: 398–404.
Lane, R.P. (1981) A quantitative analysis of wing pattern in the *Culicoides pulicaris* species group (Diptera: Ceratopogonidae). *Zoological Journal of the Linnean Society,* 72: 21–41.
MacLeod, N., O'Neill, M.A. and Walsh, S.A. (in press) A comparison between morphometric and artificial neural-net approaches to the automated species-recognition problem in systematics. In *Biodiversity Database: From Cottage Industry to Industrial Network* (eds G. Curry and C. Humphries), Taylor & Francis, London.
Mason, W.R.M. (1986) Standard drawing conventions and definitions for venational and other features of wings of hymenoptera. *Proceedings of the Entomological Society of Washington,* 88: 1–7.
Matias, A., de la Riva, J.X., Torrez, M. and Dujardin, J.P. (2001) *Rhodnius robustus* in Bolivia identified by its wings. *Memorias do Instituto Oswaldo Cruz,* 96: 947–950.
Nazzi, F. (1992) Morphometric analysis of honey bees from an area of racial hybridization in northeastern Italy. *Apidologie,* 23: 89–96.
Radloff, S.E., Hepburn, R. and Bangay, L.J. (2003) Quantitative analysis of intracolonial and intercolonial morphometric variance in honeybees, *Apis mellifera* and *Apis cerana. Apidologie,* 34: 339–351.
Rohlf, F.J. (1990) An overview of image processing and analysis techniques for morphometrics. In *Proceedings of the Michigan Morphometrics Workshop* (eds F.J. Rohlf and F.L. Bookstein), The University of Michigan Museum of Zoology, Ann Arbor, pp. 37–60.
Rohlf, F.J. (2005) tpsDig, digitize landmarks and outlines, version 2.04. Department of Ecology and Evolution, State University of New York at Stony Brook.
Rohlf, F.J. and Archie, J.W. (1984) A comparison of Fourier methods for the description of wing shape in mosquitoes (Diptera: Culicidae). *Systematic Zoology,* 33: 302–317.
Rortais, A., Arnold, G., Baylac, M. and Garnery, L. (2004) Analysis of the genetic diversity of honeybees in France and establishment of bees conservatories. In *First European Conference of Apidology,* Udine, Italy, p. 48.
Roth, V., Pogoda, A., Steinhage, V. and Schröder, S. (1999a) Integrating feature-based and pixel-based classification for the automated identification of solitary bees. In *Jahrestagung der Deutschen Gesellschaft für Mustererkennung,* 21, pp. 120–129.
Roth, V., Steinhage, V., Schröder, S. and Cremers, A. B. (1999b) Pattern recognition combining de-noising and linear discriminant analysis within a real world application. *8th International Conference on Computer Analysis of Images and Patterns,* Ljubljana, pp. 251–266.
Ruttner, F. (1988a) *Biogeography and Taxonomy of Honeybees,* Springer, Berlin, 284 pp.
Ruttner, F. (1988b) *Breeding Techniques and Selection for Breeding of the Honeybee,* BIBBA, London.
Schröder, S., Wittmann, D., Drescher, W., Roth, V., Steinhage, V. and Cremers, A.B. (2002) The new key to bees: automated identification by image analysis of wings. In *Pollinating Bees – The Conservation Link between Agriculture and Nature* (eds P. Kevan and V.L. Imperatiz-Fonseca), Ministry of Environment, Brasilia, pp. 209–216.
Shen, J. and Castan, S. (1992) An optimal linear operator for step edge detection. *CVGIP: Graphical Model and Image Processing,* 54: 112–133.
Smith, D.R., Crespi, B.J. and Bookstein, F.L. (1997) Fluctuating asymmetry in the honey bee, *Apis mellifera*: effects of ploidy and hybridization. *Journal of Evolutionary Biology,* 10: 551–574.

Steinhage, V., Kastenholz, B., Schröder, S. and Drescher, W. (1997) A hierarchical approach to classify solitary bees based on image analysis. In *DAGM-Symposium* (eds E. Paulus and F.M. Wahl), Springer, Berlin, pp. 419–426.

Strauss, R.E. and Houck, M.A. (1994) Identification of Africanized honeybees via non-linear multilayer perceptrons. *Proceedings of the IEEE International Conference on Neural Networks,* 5: 3261–3264.

Talbot, R. (2002) Beemorph, version 1.0.7. http://www.hockerley.plus.com/.

Tofilski, A. (2004) *DrawWing*, a program for numerical description of insect wings. *Journal of Insect Science,* 4: 1–5.

Weeks, P.J.D. and Gaston, K.J. (1997) Image analysis, neural networks, and the taxonomic impediment to biodiversity studies. *Biodiversity and Conservation,* 6: 263–274.

Weeks, P.J.D., Gauld, I.D., Gaston, K.J. and O'Neill, M.A. (1997) Automating the identification of insects: a new solution to an old problem. *Bulletin of Entomological Research,* 87: 203–211.

Weeks, P.J.D., O'Neill, M.A., Gaston, K.J. and Gauld, I.D. (1999a) Automating insect identification: exploring the limitation of a prototype system. *Journal of Applied Entomology,* 123: 1–8.

Weeks, P.J.D., O'Neill, M.A., Gaston, K.J. and Gauld, I.D. (1999b) Species-identification of wasps using principal component associative memories. *Image and Vision Computing,* 17: 861–866.

Yu, D.S., Kokko, E.G., Barron, J.R., Schaalje, G.B. and Gowen, B.E. (1992) Identification of ichneumonid wasps using image analysis of wings. *Systematic Entomology,* 17: 389–395.

Zelditch, M.L., Swiderski, D.L., Sheets, H.D. and Fink, W.L. (2004) *Geometric Morphometrics for Biologists: A Primer,* Elsevier/Academic Press, London, 443 pp.

18 Good Performers Know Their Audience! Identification and Characterization of Pitch Contours in Infant- and Foreigner-Directed Speech

Monja A. Knoll, Stig A. Walsh, Norman MacLeod, Mark A. O'Neill and Maria Uther

CONTENTS

Introduction 299
Analysis of Pitch Contour Shape 301
Methods 301
Qualitative Analysis 302
DAISY 302
Eigenshape Analysis (EA) 303
Results 303
Qualitative Analysis 303
DAISY 304
Eigenshape Analysis 305
Discussion 306
Function of IDS Pitch Contours 306
Evaluation of Novel Approaches 307
Conclusions 308
Acknowledgements 309
References 309

INTRODUCTION

Most readers browsing through the content pages of this volume might well be surprised to find a chapter that concerns itself with pitch contour analysis and infant-directed speech (IDS). The reader could be excused for feeling the urge to turn quickly to the next chapter. However, in the following pages we will attempt to show that techniques used for shape characterization also have a place in psychological research. In fact, as with research into automated taxonomic identification, pattern recognition is of considerable interest in a psychological context. In the case of human speech, the potential to detect and characterize discrete acoustic patterns represents the key to a storehouse of information that can shed light on the origin, diversity, and mechanics of language. This research direction has much in common with the identification of different animal sounds in the natural environment, but here we are particularly concerned with the diversity within our own species. Herein we investigate the characterization of sound waveforms that are common to speech from the same speakers, but vary with changing audiences.

Spoken language is arguably the single most important human ability, yet the apparent ease with which language is acquired during early development belies its complexity. It is therefore not surprising that the way language is transferred, and subsequently acquired, has been the subject of considerable debate and research (e.g. Werker and Tees, 1984; Kuhl, 2000; Burnham et al., 2002). It should not be taken for granted that all meaning is restricted to semantics; acoustic features of the voice of the speaker can carry as much meaning as the words themselves. Most research into these acoustic features has concentrated on prelingual infants, as they constitute the most obvious group as a focus of investigation into speech production and acquisition. It is well established that adults tend to talk differently in interactions with infants compared to conversation with other adults (e.g. Fernald and Kuhl, 1987; Werker et al., 1994; Burnham et al., 2002). This seems to be the case for speech directed to infants by adults of both genders, and also by infants' older siblings (Jacobson et al., 1983).

The acoustic and phonetic characteristics of this IDS are well recognized, and seem to be universal (e.g. Grieser and Kuhl, 1988; Kuhl et al., 1997; Kuhl, 2000). It has been found that IDS is characteristically higher in pitch than adult-directed speech (ADS), and contains exaggerated pitch contours, shorter utterances, longer pauses and hyperarticulation of vowels (Stern et al., 1983; Fernald and Simon 1984; Andruski and Kuhl, 1996; Kuhl et al., 1997; Trainor et al., 2000; Burnham et al., 2002).

Researchers now believe that acoustic modifications in IDS probably have a role in language transference, but that they might also have emotional or attention-gaining functions (e.g. Fernald and Simon, 1984; Karzon, 1985; Grieser and Kuhl, 1988; Kuhl, 1994; Kitamura and Burnham, 2003). Despite detailed investigation of some of these IDS aspects, the importance and function of several other acoustic attributes (e.g. exaggerated pitch contours and intensity) remain controversial. Previous research has investigated these roles by comparing IDS with non-emotional ADS, mainly in imaginary interactions with the help of scenarios (Trainor et al., 2000). However, it could be argued that ADS as a comparator to IDS is not sufficient as ADS lacks both the emotional and linguistic requirements of IDS. Recent research has attempted to counter these limitations by comparing IDS with emotional ADS in imaginary scenarios (Trainor et al., 2000), emotional pet-directed speech (PDS) (Burnham et al., 2002) and foreigner-directed speech (FDS) representing a linguistic comparison group (Knoll and Uther, 2004).

For the present study, we were interested in determining the linguistic role of IDS, and for this reason we used a linguistic comparison group (foreigners). Foreigner-directed speech should present a particularly valid linguistic comparison group to IDS for a variety of reasons. Firstly, foreigners displaying language and pronunciation problems are likely to be perceived as being in need of linguistically modified speech input. Yet, they are unlikely to engender a high emotional response in the speaker, particularly if they are strangers. Secondly, the adult status of this group also makes them directly comparable with an ADS control group, which should itself evoke neither a linguistic nor an emotional response. As such, we would expect that acoustic features common to both IDS and FDS might provide some important information about the linguistic function of IDS.

Earlier studies partially investigated similarities between IDS and FDS in both tonal (Mandarin Chinese) (Papousek and Hwang, 1991) and non-tonal (English) (Biersack et al., 2005) languages. The pitch range (an aspect of the pitch contour) was found to increase in FDS but not in tonal language IDS, whereas the opposite was apparent in the non-tonal variant. Because pitch is used to convey linguistic meaning in tonal languages, these findings suggest pitch contours might have a linguistic function. However, these studies investigated only the pitch range. The extent to which pitch contour *shape* in those conditions varies remains unclear. Consequently, our primary aim was to investigate the function of pitch contour shape modifications in IDS compared to FDS (as a linguistic group) and ADS (as a non-linguistic and non-emotional control group) in natural interactions. If the main role of exaggerated pitch contours is language transference, exaggerated contours should occur in interactions with infants and with foreigners. Conversely, if the exaggerated pitch contours have an emotional function, they would be expected to occur only in IDS.

Analysis of Pitch Contour Shape

A major impediment to pitch contour analysis is that pitch contour shape has traditionally been analysed qualitatively using human raters (e.g. Fernald and Simon, 1984). There have recently been several attempts to find reliable quantitative methods, but no single approach has emerged as either simple in its execution or capable of being applied to a wide variety of speech types. For instance, Trainor et al. (2000) used measurements of pitch values at prespecified points within a contrived stock phrase. This approach was unable to provide a sensitive characterization of pitch contours. By contrast, Moore et al. (1994) successfully applied mathematical modelling to the problem. Their approach would potentially allow a wide variety of speech types to be addressed, but the complexity of their approach seems to have prevented its wide acceptance within psychology.

Another approach by Tian and Nurminen (2004), based on principal component analysis (PCA) of syllable pitch values (eigenpitch), is a far less complex technique. Although this method remains to be tested in comparative speech conditions, it offers the potential to characterize pitch contours in a generalized context. However, a preprocessing stage involving careful scaling of contours would be required to avoid inclusion of other non-shape-related variables (e.g., contour length, referential pitch) without excessive abstraction of the contour shape. At present, the eigenpitch approach is not suitable for accurate characterization of pitch contour shape.

Clearly, if the role of pitch contours is to be explored adequately, a method must be found that can readily be applied to a variety of experimental conditions, while remaining straightforward enough for general (non-specialist) use. This approach must be as reliable, consistent and free of subjectivity as possible, while providing an easily usable method of characterizing the shape of the pitch curves. Two approaches currently exist that might satisfy most, if not all, of these requirements. The first of these is eigenshape analysis (EA), which uses the x–y coordinates of the curve to build a PCA-like ordination of the actual shape variation of pitch contours in series of conditions (MacLeod, 1999). The second approach involves unsupervised artificial neural net (uANN) analysis of contour curves in the form of images. For this, the digital automated identification system (DAISY) at The Natural History Museum (London) was used. Our secondary aim was to evaluate the utility of these approaches for pitch contour analysis by comparing them to the more traditional approach of using qualitative analysis.

METHODS

A pre-existing data-set of infant-, foreigner- and adult-directed speech samples in WAV format formed the basis of the data-set (Knoll and Uther, 2004). Pitch contours were randomly selected from these groups and then modified to a standardized format that could be used for the DAISY, EA and qualitative conditions. The extracted pitch contours comprised the specific target words 'sheep', 'shark' and 'shoe', which were chosen because they had been found to be prosodically highlighted in IDS and FDS in a previous study (Knoll and Uther, 2004). A total of 167 pitch contours (IDS = 60, ADS = 53, FDS = 54) were extracted from these words.

Pitch contour extraction was achieved using the freely available speech analysis program, Praat 4.1.19 (Boersma and Wernicke, 2004). The pitch range for the extraction was set at a floor of 100 Hz and a ceiling of 600 Hz (recommended setting in Praat 4.1.19 for female voices). It should be noted that although the whole word was used for the extraction process, only visible pitch contours were extracted (which often do not encompass whole words). The 'smooth' function was then used to smooth the contours (at 10-Hz bandwidth). This is a standard procedure to obtain

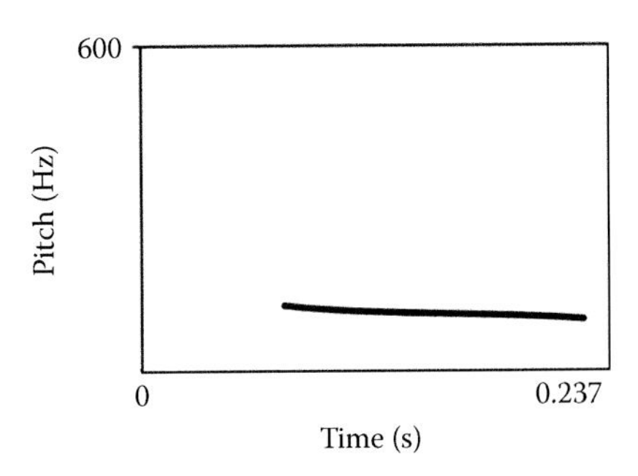

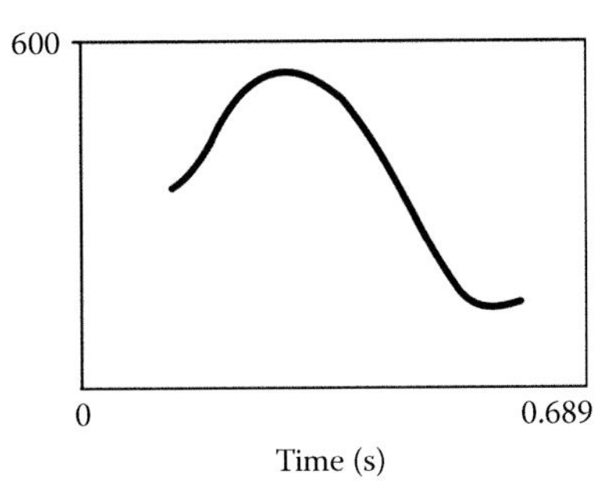

FIGURE 18.1 Comparison of pitch contours in ADS (left) and IDS (right). Note that the duration is very different, but that the actual pitch contours are drawn in the same size. Using the set window of 0.700 avoids this distortion as a durational pitch contour of 0.237 would be drawn relative to the 0.700 seconds window.

more homogeneous contours and was also used by Moore et al. (2004). Finally, the contours were drawn using the Praat 4.1.19 'draw' function (set at pitch range of 100–600 Hz, duration of 0.7 seconds).

Ensuring standardization of sample duration was problematic, as previous research (e.g. Fernald and Simon, 1984) has shown that words in IDS are generally of a longer duration than words in ADS. Using the Praat 4.1.19 'drawing' function with a time range of 0.00 to 0.00 [all] was not viable, as it would have led to a distortion of the smaller contours in comparison with the larger contours (Figure 18.1). The pitch contours were therefore drawn within a set window of 0.7 seconds (chosen on the basis of the largest contour), thus representing the true duration of the pitch contours in relation to each other. However, the durational length of the drawn pitch contours exhibited wide variation, which would have made comparison difficult. To counter this problem, the pitch contours were exported into CorelDraw 11.0 and standardized for duration without distorting the shape. The width of the lines was increased to 8-point thickness and then the images were re-exported as a standardized 500 × 500 pixel grid TIFF format at 72 dpi in CorelPaint 11.0. Widening of the line thickness was required to enhance the pixel contrast pattern for the DAISY analysis.

Qualitative Analysis

For the qualitative analysis we recruited five raters who were required to rate the 167 pitch contour images using a predesigned rating scheme. It would have been problematic to average the ratings for each shape, as the ratings were based on categories rather than on nominal data, such as that obtained, for instance, on a Likert-scale. In order to deal with this problem, each rater's rating for each image was taken into account, resulting in five data points per image (e.g. 60 IDS images × 5 raters = 300 data points).

The rating scheme consisted of graphic representation of five different shapes with one option for undecided. The shapes consisted of (1) bell, (2) complex, (3) falling, (4) rising and (5) level shapes (Figure 18.2). These shapes were chosen on the basis of previous work (Fernald and Simon, 1984). Raters were provided with two trial shapes before the main procedure commenced, and the shapes were presented in counterbalanced order to avoid anomalies caused by rater fatigue. The raters were not aware that the shapes constituted three different speech conditions. In order to determine reliability, intraclass correlation for inter-rater (reliability coefficient α = 0.97) and intrarater (reliability coefficient α = 0.96) reliability was carried out.

DAISY

A detailed discussion of the concept and mechanics of DAISY is given in O'Neill (this volume). The present study follows the standard procedure used to build DAISY training sets as described

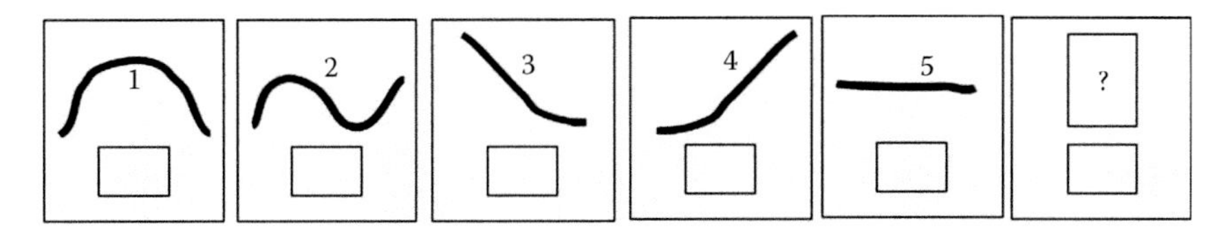

FIGURE 18.2 These are the shapes that were presented to the raters in the rating scheme. See main text for definition of each shape. Note that numbers were not included in original rating scheme.

by Walsh et al. (this volume). Once built, the consistency of the training set was tested using the JACKTOOL cross-validation algorithm included with the DAISY system. Pass results are provided at two main levels. This pass measure is given as a coordination number between 3 and 10, which indicates that the cross-validated image is positioned within a cluster of *n* nearest neighbours of the same class. Images that have less than three neighbours of the same class may still be identified if they are positioned close to the perimeter of a class cluster, in which case a Sill-level identification is given. Identifications that fall below these levels are classed as fails for the purposes of this study.

Eigenshape Analysis (EA)

The difficulty in locating consistent geometric landmarks along these pitch contours meant that standard eigenshape analysis was used rather than the landmark-registered extended eigenshape approach (MacLeod, 1999). Thirty-seven paired *x–y* coordinates were collected using tpsDIG 2.04 (Rohlf, 2005). For accuracy on the thickened line, nodes were positioned along the upper edge of the pitch, and started at the extreme left of the curve. Interpolation at the 99 per cent tolerance level resulted in 18 nodes for the three-class data-set. When the classes were analysed separately, this same value was calculated for IDS, whereas ADS and FDS required only three. Pitch curves for each class were also analysed separately to determine the mean shape for each category. Eigenscores for the complete 167 data-set were analysed using canonical variates analysis (CVA) implemented through SPSS 13.0 software.

RESULTS

Qualitative Analysis

Of the original six categories, the 'undecided option' was not chosen by any of the raters, indicating that the raters believed they were able to categorize each of the 167 images into the provided five pitch contour categories (Table 18.1). Rated pitch contour category and type of speech recipient group variables were found to be associated, and not independent of each other ($\lambda^2 = 544.038$, df $= 8$, $p < 0.001$). Cramer's V produced a value of 0.571, indicating a strong relationship between the two variables ($p < 0.001$). Goodman and Kruskal's lambda was also calculated for type of speech recipient group ($\lambda = 0.338$, $p < 0.001$) and rated pitch contour category ($\lambda = 0.314$, $p < 0.001$), the result of which showed that both variables (contour shapes and speech groups) were equally and significantly predictive of each other. Over 60 per cent of the IDS pitch contours were characterized as bell contours, followed by 16 per cent level contours (Table 18.1).

In both ADS and FDS this result was reversed. Here, the majority of pitch contours were characterized as level contours (ADS = 81.1%; FDS = 78.9%), with less than 2 per cent being characterized as bell contours in both conditions. Interestingly, none of the ADS or FDS pitch contours were characterized as complex contours, whereas 12 per cent of the IDS pitch contours fell into this category. With regards to the falling contours, more ADS pitch contours were categorized as falling contours (15.5%) than FDS (13.3%) and IDS (3.3%), although the difference between ADS and FDS is minimal (2.2%). In the category for rising contours, the highest frequency

TABLE 18.1
Distribution of Ratings for Each of the Five Contours across IDS, FDS and ADS

Speech groups	Shape contours Bell	Complex	Rising	Falling	Level
IDS	191	36	14	10	49
% within group	63.7	12	4.7	3.3	16.3
% within shape	96.5	100	37.8	11.5	10.3
ADS	3	0	6	41	215
% within group	1.1	0	2.3	15.5	81.1
% within shape	1.5	0	16.2	47.1	45.1
FDS	4	0	17	36	213
% within group	1.5	0	6.3	13.3	78.9
% within shape	2	0	45.9	41.4	44.7

Note: Percentage distribution for each group per shape and each shape per group are also displayed.

was achieved by the FDS contours (6.3%) followed by IDS contours (4.7%), with ADS contours achieving the lowest frequency (2.3%). The conclusion that can be drawn from these results is that IDS pitch contours are clearly distinct from ADS and FDS pitch contours, and were mainly rated as consisting of 'exaggerated' contours (bell shapes and complex contours). In contrast, both ADS and FDS seem to consist mainly of level contours with no indication of exaggeration, and it was not possible to separate these two speech categories reliably with the qualitative analysis.

DAISY

Each of the classes (IDS, ADS and FDS) was recognized by DAISY as a discrete group, but the higher number of nearest neighbours in IDS (mean pass = 8) indicates a tighter clustering within the IDS data-set compared with the adult conditions (mean pass = 4). Note that although ADS and FDS share the same mean pass coordination rate of 4, this is an indication of clustering consistency within each class, rather than an indication of a similar clustering between these two classes. ADS and FDS images are more homogeneously clustered within their class shape space. The individual results for each class revealed that IDS achieved the highest pass rate of 80 per cent (Table 18.2), indicating that the IDS images exhibited an inherent similarity to each other.

In contrast, in ADS, 45 per cent of the ADS images were recognized as belonging to the same (ADS) class, with 55 per cent rejected. This result is, to a certain degree, repeated in FDS, where

TABLE 18.2
Confusion Matrix for Pass/Fail Rates for IDS, ADS and FDS

Groups classified as	IDS	ADS	FDS
IDS	80%[a]	6%	9%
ADS	5%	45%[a]	51%
FDS	15%	49%	40%[a]

[a] Correct classification of image into its group.

40 per cent of the images were recognized as belonging to the FDS class, with 60 per cent rejected. This association between the type of speech recipient and pass and fail rates was found to be significant ($\lambda^2 = 21.616$, df = 2, $p < 0.001$). Pass rates of below 50 per cent suggest the presence of discrete groupings, but indicate the interdigitation of class clusters; the IDS cluster was clearly better separated from ADS and FDS than the two adult classes were from each other.

DAISY provides a list of what class each failed image was mistaken for as part of a partition function measure. The percentage of failed images in each class provides a measure of which classes interdigitate with which. Fail rates indicate that most of the IDS failed images were mistaken for FDS (15%), whilst both FDS (51%) and ADS (49%) were mistaken for each other (see Table 18.2). These results indicate that the majority of IDS pitch contours represent a more or less coherent group despite showing greater variability within the IDS space. In contrast, both ADS and FDS pitch contours were difficult to separate due to interpenetration of the group clouds for each class, despite exhibiting less variability within their respective classes. The most important finding is that all three classes (IDS, ADS and FDS) were recognized by the system as belonging to separate groups, although with weaker groupings in the adult conditions than in IDS.

Eigenshape Analysis

Almost 70 per cent of the shape variation in all three conditions was attributable to the first eigenshape axis. As such, all outlines exhibited a fundamental similarity with each other. Separate analysis of each of the three groups demonstrated that most of the variation on the second eigenshape axis is indeed due to IDS. ADS and FDS pitch contour variation is due entirely to the first eigenshape axis, therefore indicating that these two groups are similar in that they possess very little shape variation. In contrast, IDS variation is due to both the first and the second eigenshape axes and, as such, indicates wider variation across the IDS shape space. The IDS mean eigenshape is characterized by a more exaggerated curve than either ADS or FDS, and could be characterized as a bell curve. However, the top of the bell curve is flattened due to shape variation within the IDS pitch contours. The most interesting finding is that the mean shapes of ADS and FDS are almost identical (Figure 18.3).

A standard CVA was performed with speech recipient groups as category variable, and the scores on the 16 axes derived from EA as the predictor variables. Univariate ANOVAs revealed that the three groups differed significantly on the eigenscores of axis 1 ($F_{(2, 164)} = 66.397, p < 0.001$) and axis 10 ($F_{(2, 164)} = 4.448, p < 0.013$). Two discriminant functions were calculated. The values of the first of these functions were significantly different for IDS, ADS and FDS ($\lambda^2 = 130.228$, df = 32, $p < 0.001$, Wilk's $\lambda = 0.453$), whereas the second function was found not to be significantly different. The correlation between predictor variables and the discriminant functions suggested that both axis 1 (on discriminant function 1) and axis 10 (on function 2) would be the best predictors for membership of future pitch contours. Overall, the discriminant functions successfully predicted

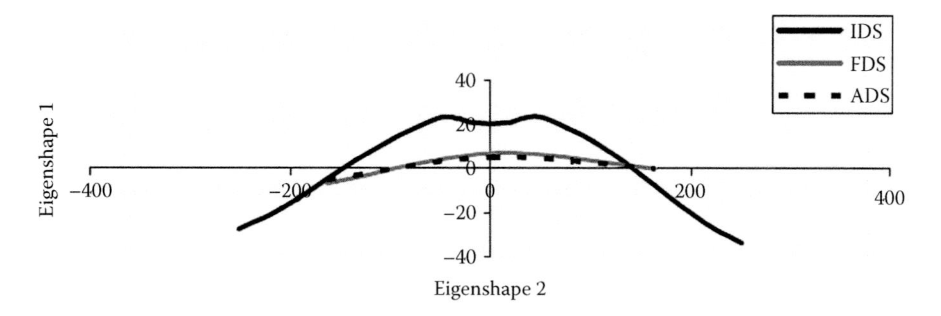

FIGURE 18.3 Comparison of mean shapes for IDS, ADS and FDS plotted on eigenshape 1 versus eigenshape 2.

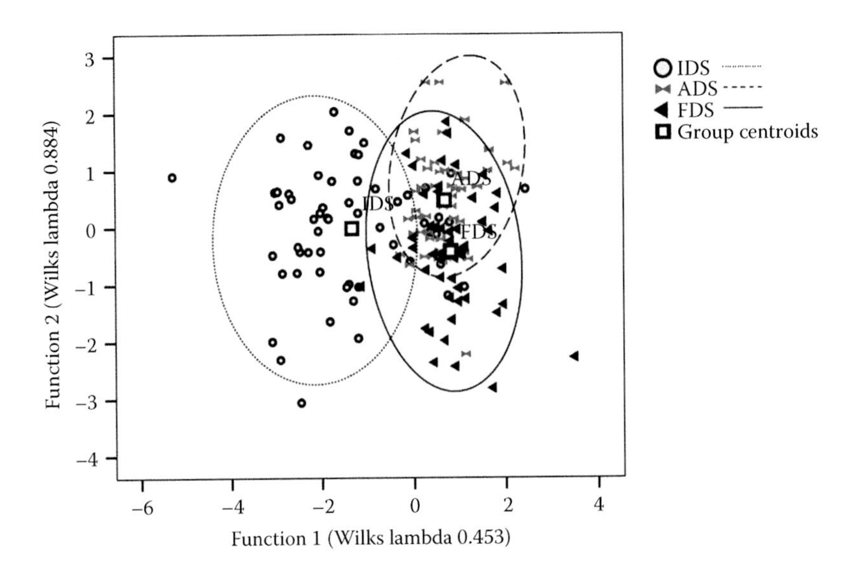

FIGURE 18.4 Combined CVA plot for IDS, ADS and FDS. Ellipses provide more effective presentation of clustering of the groups.

outcome for 71 per cent of all cases, with accurate predictions for IDS of 75 per cent, for ADS of 77 per cent and for FDS of 59 per cent. Figure 18.4 shows plotting of each of the pitch contours for IDS, ADS and FDS on functions 1 and 2. IDS is distinct from both ADS and FDS. FDS and ADS share more of their variance, but still represent discrete groupings (confirmed by the position of the group centroids).

From this EA it is clear that IDS is a variable group that is nevertheless distinct from ADS/FDS. These adult conditions exhibit very little shape variability, but do provide notable characteristics that separate one from the other. The mean ADS and FDS eigenshapes confirm this similarity, indicating that the separation relates to a shape 'tendency' rather than to a strong shape characterization, as is the case with the mean IDS bell contour.

DISCUSSION

Function of IDS Pitch Contours

This study supports the findings of previous research (e.g. Fernald and Simon, 1984, Fernald and Mazzie, 1991) that pitch contours in IDS are not only very distinctive, but also are noticeably different from those of speech directed to adults. The pitch contours of FDS were found to be more similar to ADS, which is also consistent with earlier findings (Biersack et al., 2005). Contrary to research comparing tonal IDS and FDS (Papousek and Hwang, 1991), we found no evidence of linguistic exaggeration of pitch contours in non-tonal FDS. The most likely reason for this difference is that the present study centred on a non-tonal language (English), and was therefore not directly comparable with Papousek and Hwang's (1991) findings, which focused on Mandarin Chinese. In such a language, an over-emphasis on the pitch changes would be expected when talking to foreigners. Conversely, pitch changes in English are not linguistically relevant and do not require emphasis.

The qualitative analysis found that 76 per cent of IDS contours consisted of bell and complex shapes. The majority of ADS and FDS contours were characterized as level shapes, with almost no occurrence of bell or complex contours. Level contours, therefore, seem to be characteristic of ADS and FDS, whereas bell and complex contours seem to be indicative of IDS, at least within the words analysed in this study. Although the two algorithmic approaches evaluated here do not provide named classes for the contour shapes (a fundamental requirement for the raters), the results of those techniques are in close agreement with the qualitative analysis.

An association between these exaggerated IDS shapes and particular emotional and turn-taking interactions has already been noted, and it was suggested that they were responsible for the melodic quality of IDS (Trainor et al., 2000). Our findings are consistent with this viewpoint. They also naturally lead to the conclusion that, in non-tonal languages, the speaker talking to an infant primarily uses pitch contour exaggeration to convey emotion or gain attention rather than as a tool for language transference. However, the results of the qualitative analysis also indicate a tendency for rising contours to form part of the characteristic shape space of FDS compared with ADS. This slight but tantalizing finding might provide information as to how the two algorithmic approaches were able to separate ADS from FDS. Rising contours might be associated with questioning, possibly in the context of comprehension. The reliability and implications of this observation obviously require further investigation.

Because we investigated the importance of the linguistic IDS component by comparing it with a linguistic condition (FDS), our results are not useful for discussion of the relative independence of the emotional and attentional components. Using an emotional speech recipient group such as partner- or pet-directed speech (e.g. Burnham et al., 2002) could potentially be more informative about the interdependence of these two components.

Evaluation of Novel Approaches

The major advantages of qualitative analysis are that it provides an intuitive insight into the data and that it can easily be applied, since in most psychology departments there is normally a readily available source of potential raters. However, it also relies on preselection of shapes that are assumed to represent fundamental groups within the data-set. This is unavoidable because, without restricting the field of presented shapes, the raters would have had difficulty distinguishing contours that grade between well-defined end member shapes. This difficulty was commented on by most of the raters, who reported that this was a task that ostensibly seemed straightforward, but in practice proved quite the opposite. Considering the apparent difficulty of the task, it is surprising that none of the raters selected the 'undecided' option from the rater sheets. Perhaps a belief in their own discriminatory abilities prevented them from admitting that they could not make a decision with conviction, although the inter- and intrarater reliability suggests that this is unlikely. It seems more probable that the raters possessed an ability to recognize these broad shape categories that allowed them to get close to a core shape characterization, but were unable to detect the subtle shape changes that might have occurred between these examples.

In theory, any ANN that is capable of assessing digital images should be able to provide a similar analysis of a given data-set to human raters, but the reduction in subjectivity should result in a more reliable characterization of shapes within a given category. For instance, DAISY fails gracefully when presented with marginal images, whereas a human rater may be tempted to make a potentially inaccurate snap decision. Moreover, DAISY provides consistent, yet objective, classifications based on reliable, well-defined algorithmic parameters. For instance, a weakly characterized bell shape among a series of well-defined bell shapes might be characterized by a human rater as a level shape, while the same shape might also be characterized as a bell shape among a series of flat lines. Although counterbalancing should reduce this context effect, counterbalancing is nevertheless a random procedure that might not alleviate the problem. By comparison, DAISY is not affected by the human need to classify objects on the basis of context. Neither is DAISY affected by rater fatigue or differences between individuals. One of the major advantages of DAISY is that it provides the user with ready-made, easily interpreted statistical output that requires little statistical grounding on the part of the user. As such, DAISY is readily adaptable across a wide variety of disciplines (including psychology). DAISY is also fast and comparatively easy to operate. It does not depend on the recruitment of human raters, and thus can potentially avoid preselection biases.

However, the preprocessing of image preparation suitable for DAISY input is time consuming and requires expert knowledge of graphic file manipulation. This is important for this kind of analysis because DAISY requires files that contain an adequate sample of the signal classes for processing. In this study, the thickness of the line of the pitch contours had to be magnified in order to provide the minimum required input needed for signal processing. We suspect that the thickened pitch contour line may still not include a strong enough signal. An improvement of this would consist of amplification of the signal by generating a pattern of greyscale pixel brightness values below the pitch contour, based on the pitch *y*-coordinate value at each step along the pitch curve. The need to standardize the images for comparability between approaches meant that this was not possible here, although future work might involve such a procedure.

DAISY could also have dealt with the raw spectrogram output from Praat, and normally this kind of patterned image is the basic input for the system. However, this would have meant that other prosodic modifications (e.g. mean pitch, duration, intensity and formant) would also have been represented in the image files. While this may have been desirable in some circumstances, for this analysis it would have been impossible to conclude that any groupings supported by DAISY are solely the result of pitch contour shape. As such, DAISY image preprocessing is the basis of an effective analysis and requires careful consideration. It is worth noting here that the DAISY algorithm can theoretically be applied to sound samples, and the potential of the system in speech analysis remains to be fully tested.

The eigenshape approach clearly offers an objective, relatively fast, independent and statistically viable way of analysing curves and outlines. Although some minor knowledge of geometry is required to carry out EA, this should generally not present an obstacle within psychological research. Time-intensive image preparation is unnecessary for EA, and in fact coordinates could be captured directly from the spectrogram if necessary. The major advantage of EA is the fact that its output provides scores that can be used for further multivariate analysis (CVA), which supplies the researcher with a visual representation of the groupings. Furthermore, unlike the two other methods, the visual presentation of the mean shape of each group provides a quick approximation of the shapes of the figures. Using EA, the region-specific shape variation noted between ADS and FDS contours could also be analysed using a partial least squares approach (Rohlf and Corti, 2000). However, because this study aimed to evaluate these techniques for psychological research by analysing a simple three-class case study, we have chosen to pursue this when we have new comparative data from other speech recipient conditions.

CONCLUSIONS

In summary, our findings suggest that the function of the exaggerated pitch contours in IDS is emotional–attentional rather than a device to highlight important linguistic features. Because we investigated the importance of the linguistic component of IDS, our results are not useful for discussion of the relative dependence of the emotional and attentional components. Using an emotional speech recipient group such as partner- or pet-directed speech could potentially be informative about the nature and importance of the emotional component, and future research should concentrate on these groups. We suggest that the algorithmic approaches evaluated here present considerable potential for future pitch contour analysis and offer an alternative to the subjectivity of teams of human raters. However, because the results of each of the evaluated approaches were consistent with each other, we note that human raters may not be as unreliable as might be imagined. The potential of uANN technology in speech analysis is of considerable interest, and we suspect that data obtained from research into acoustic characterization of speech directed to specific groups will feed back into such areas as synthetic speech production and recognition.

ACKNOWLEDGEMENTS

We thank Professor Alan Costall for reading through an earlier manuscript. We would also like to thank all mothers and their infants, the confederates and the raters that took part in this study. This research was supported by a student bursary to MAK from the Economic and Social Research Council (ESRC).

REFERENCES

Andruski, J.E. and Kuhl, P.E. (1996) The acoustic structure of vowels in mothers' speech to infants and adults. In *Proceedings of the Fourth International Conference on Spoken Language Processing,* Philadelphia, PA, pp. 1545–1548.

Biersack, S., Kempe, V. and Knapton, L. (2005) Fine-tuning speech registers: a comparison of the prosodic features of child-directed and foreigner-directed-speech. *Proceedings of the 9th European Conference on Speech Communication and Technology,* Lisbon.

Boersma, P. and Werninke, D. (2004) *praat: Doing phonetics by computer* (Version 4.1.19) [Computer program]. Retrieved June 2004 from http://www.praat.org/.

Burnham, D., Kitamura, C. and Vollmer-Conna, U. (2002) What's new pussycat? On talking to babies and animals. *Science,* 296: 1095.

Fernald, A. and Kuhl, P. (1987) Acoustic determinants of infant preference for motherese speech. *Infant Behavior and Development,* 10: 279–293.

Fernald, A. and Mazzie, C. (1991) Prosody and focus in speech to infants and adults. *Developmental Psychology,* 27(2): 209–221.

Fernald, A. and Simon, T. (1984) Expanded intonation contours in mother's speech to newborns. *Developmental Psychology,* 20: 104–113.

Grieser, D.L. and Kuhl, P. (1988) Maternal speech to infants in a tonal language: support for universal prosodic features in motherese. *Developmental Psychology,* 24(1): 14–20.

Jacobson, J.L., Boersma, D.C., Fields, R.B. and Olson, K.L. (1983) Paralinguistic features of adult speech to infants and small children. *Child Development,* 54: 436–442.

Karzon, R.G. (1985) Discrimination of polysyllabic sequences by one- to four-month-old infants. *Journal of Experimental Child Psychology,* 39: 326–342.

Kitamura, C. and Burnham, D. (2003) Pitch and communicative intent in mother's speech: adjustments for age and sex in the first year. *Infancy,* 4(1): 85–110.

Knoll, M.A. and Uther, M. (2004) "Motherese" and "Chin-ese": evidence of acoustic changes in speech directed at infants and foreigners. *Journal of the Acoustical Society of America,* 116(4): 2522.

Kuhl, P.K. (1994) Learning and representation in speech and language. *Current Opinion in Neurobiology,* 4: 812–822.

Kuhl, P.K. (2000) Language, mind, and brain: experience alters perception. In *The new cognitive neuroscience,* 2nd edition (ed M.S. Gazzaniga), MIT Press, Cambridge, MA, pp. 99–115.

Kuhl, P.K., Andruski, J.E., Chistovich, I.A., Chistovich, L.A., Kozhevnikova, E.V., Ryskina, V.L., Stolyarova, E.I., Sundberg, U. and Lacerda, F. (1997) Cross-language analysis of phonetic units in language addressed to infants. *Science,* 277: 684–686.

MacLeod, N. (1999) Generalizing and extending the eigenshape method of shape space visualization and analysis. *Paleobiology,* 25(1): 107–138.

Moore, C.A., Cohn, J.F. and Katz, G.S. (1994) Quantitative description and differentiation of fundamental frequency contours. *Computer Speech and Language,* 8: 385–404.

Papousek, M. and Hwang, S-F.C. (1991) Tone and intonation in Mandarin babytalk to presyllabic infants: comparison with registers of adult conversation and foreign language instruction. *Applied Psycholinguistics,* 12: 481–504.

Rohlf, F.J. (2005) TpsDig, digitise landmarks and outlines (Version 2.0) [Computer program]. Department of Ecology and Evolution, State Univesity of New York at Stony Brook. Retrieved June 2005 from http://life.bio.sunysb.edu/morph/.

Rohlf, F.J. and Corti, M. (2000) Use of two-block partial least-squares to study covariationin shape. *Systematic Biology,* 49(4): 740–753.

Stern, D.N., Spieker, S., Barnett, R.K. and MacKain, K. (1983) The prosody of maternal speech: infant age and context related changes. *Journal of Child Language,* 10: 1–15.

Tian, J. and Nurminen, J. (2004) On analysis of eigenpitch in Mandarin Chinese. In *International Symposium on Chinese Spoken Language Processing,* ICSLP, Jeju Island, S. Korea, pp. 89–92.

Trainor, L.J., Austin, C.M. and Desjardins, R.N. (2000) Is infant-directed speech prosody a result of the vocal expression of emotion? *Psychological Science,* 11(3): 188–195.

Walsh, S.A., MacLeod, N.M. and O'Neill, M. (in press) Analysis of spheniscid humerus and tarsometatarsus morphological variability using DAISY automated image recognition. Oryctos.

Werker, J.F., Pegg, J.E. and McLeod, P.J. (1994) A cross-language investigation of infant preferences for infant-directed communication. *Infant Behavior and Development,* 17: 323–333.

Werker, J.F. and Tees, R.C. (1984) Cross-language speech perception: evidence for perceptual reorganization during the first year of life. *Infant Behavior and Development,* 7: 49–63.

Appendix 1

Appendix
Cynipid-p Images

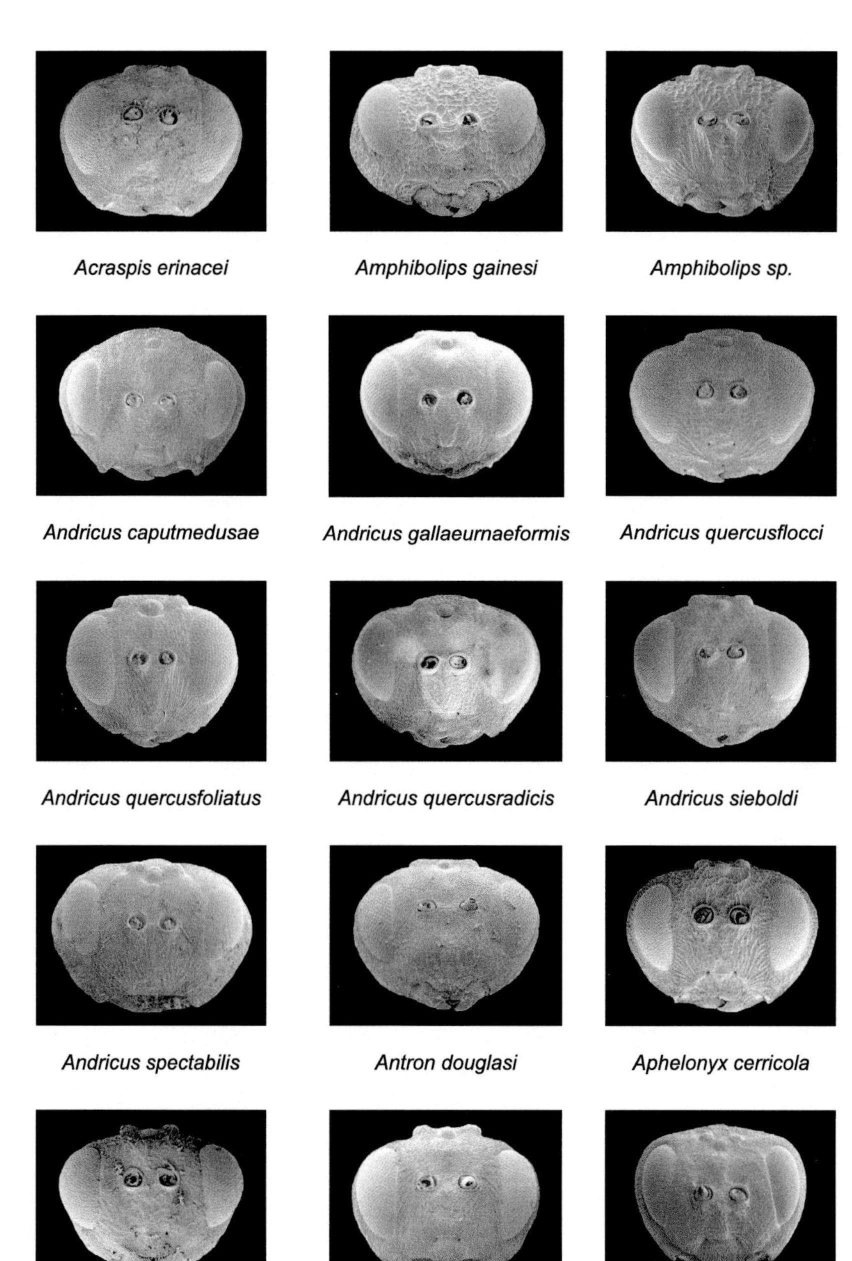

Acraspis erinacei · *Amphibolips gainesi* · *Amphibolips sp.*

Andricus caputmedusae · *Andricus gallaeurnaeformis* · *Andricus quercusflocci*

Andricus quercusfoliatus · *Andricus quercusradicis* · *Andricus sieboldi*

Andricus spectabilis · *Antron douglasi* · *Aphelonyx cerricola*

Atrusca emergens · *Besbicus conspicuus* · *Biorhiza pallida*

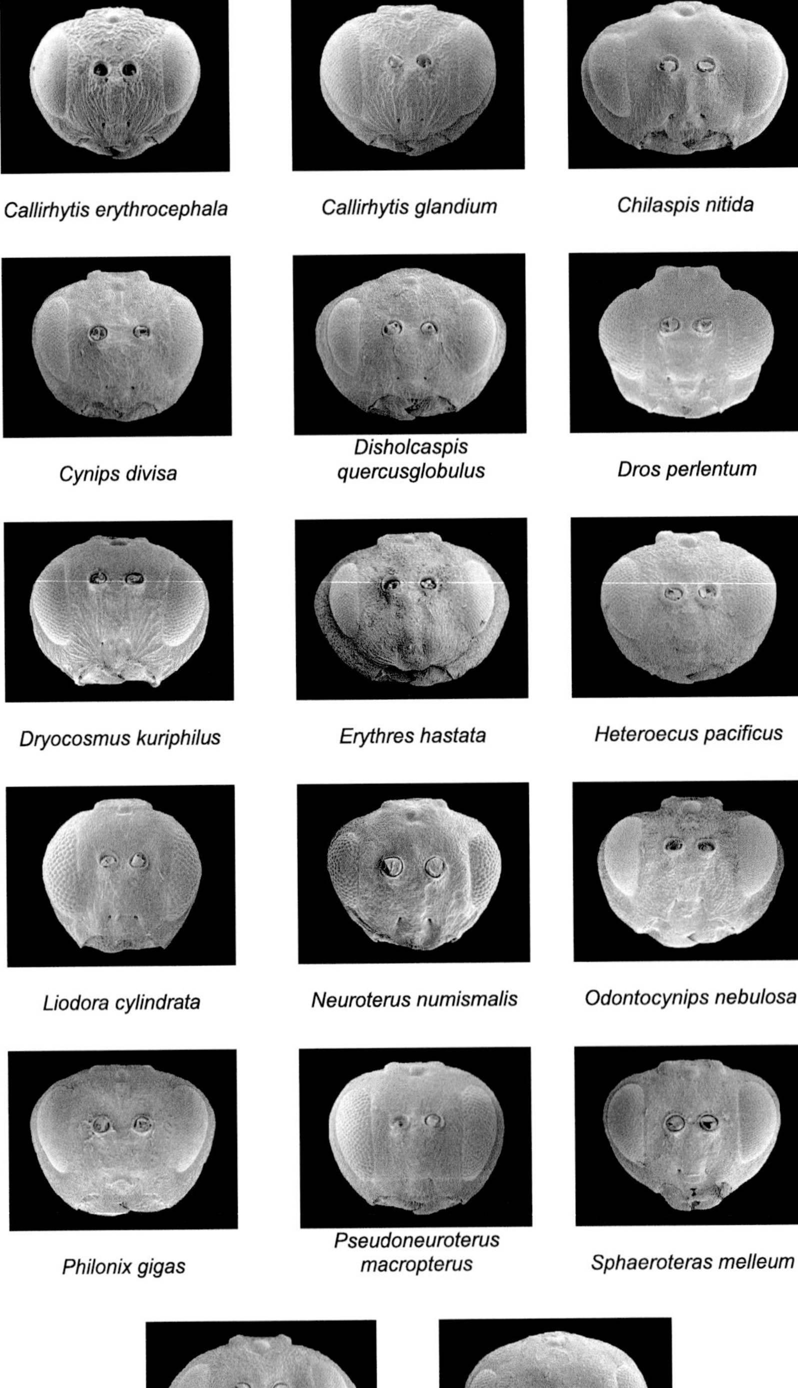

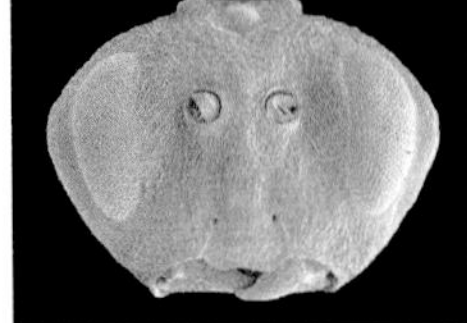
Trichagalma serratae

Xanthoteras quercusforticorne

Appendix 2

Cynipid-s Images

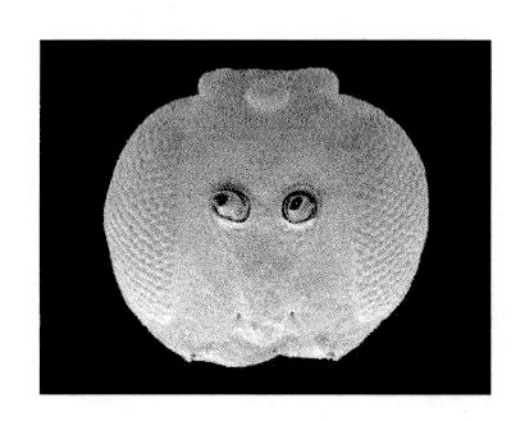

Acraspis erinacei

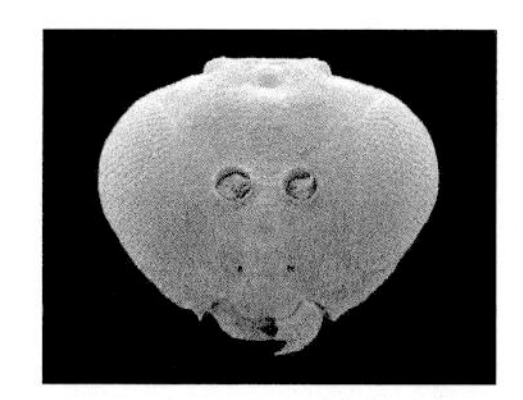

Andricus curvator

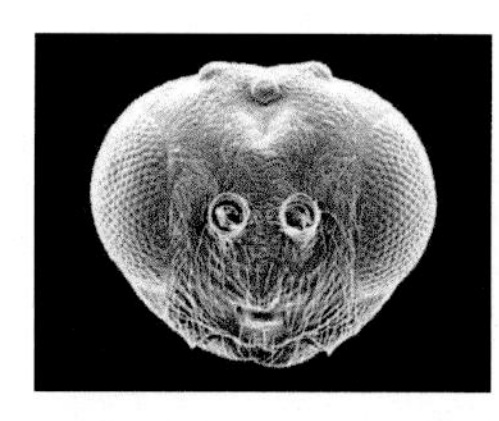

Andricus gallaeurnaeformis

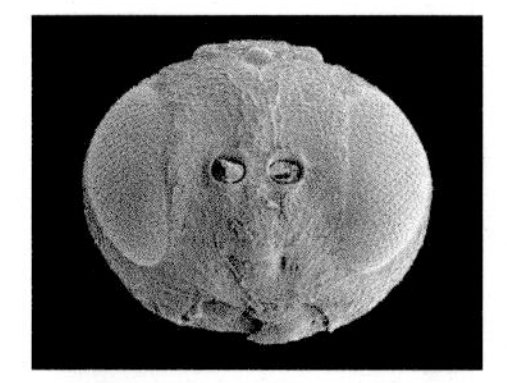

Andricus grossulariae

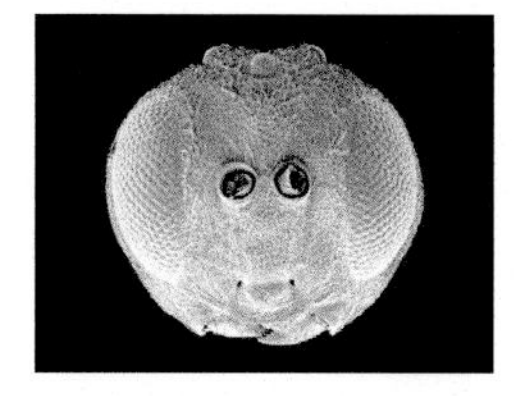

Andricus kollari

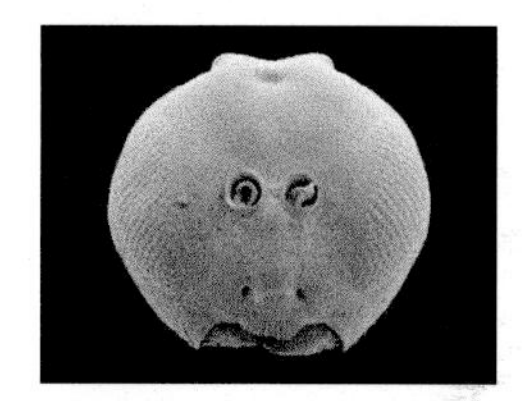

Andricus quercusramuli

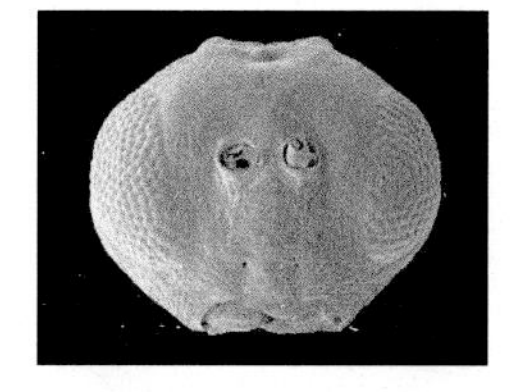

Andricus sieboldi

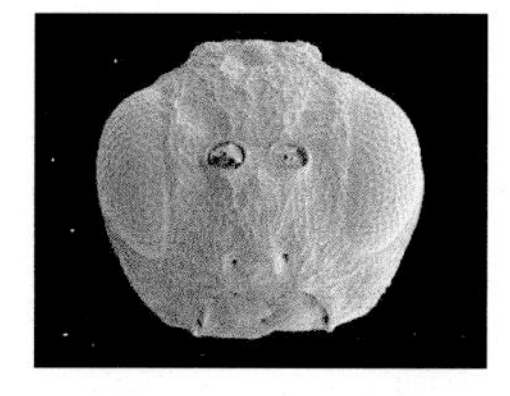

Andricus solitarius

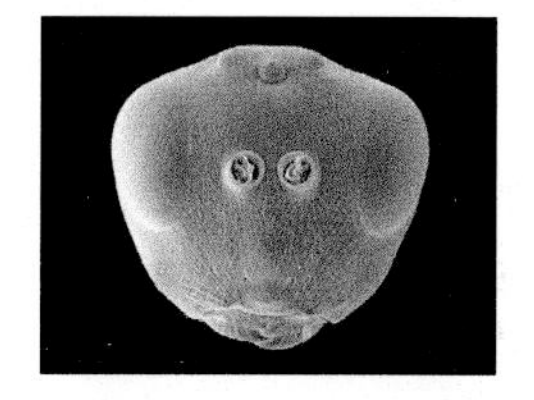

Aulacidea martae

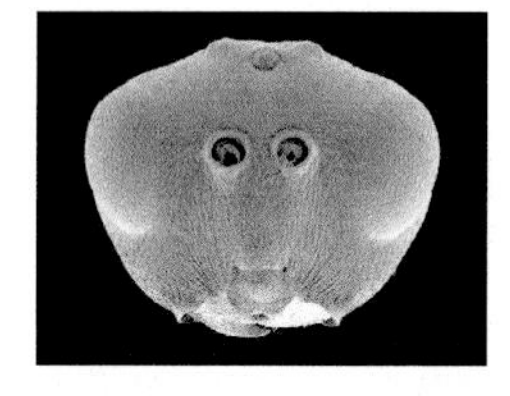

Aylax papaveris

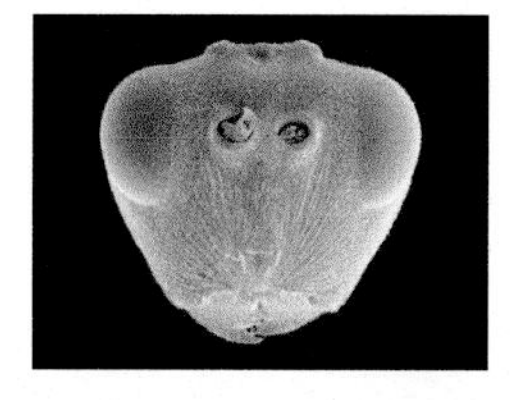

Barbotinia oraniensis

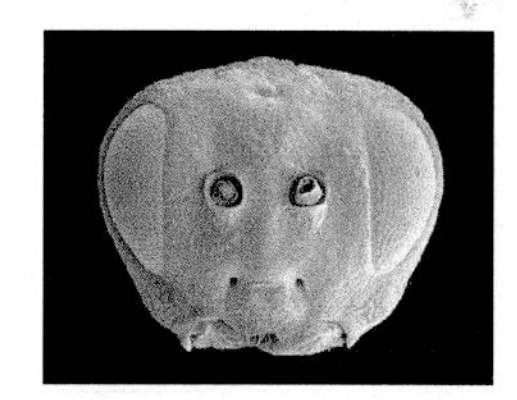

Belizinella gibbera

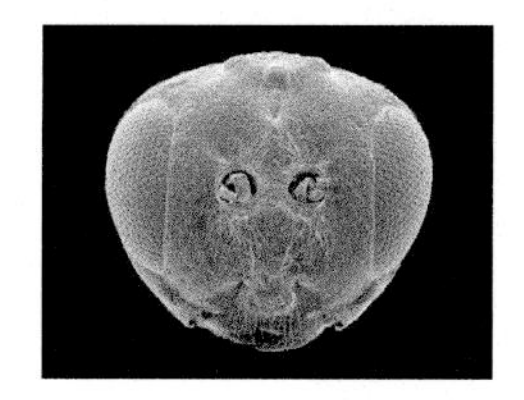

Biorhiza pallida

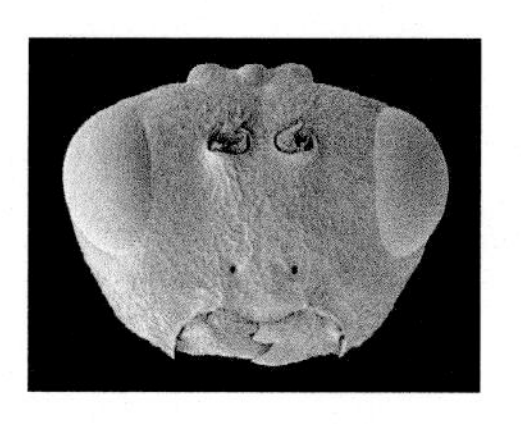

Diplolepis triforma

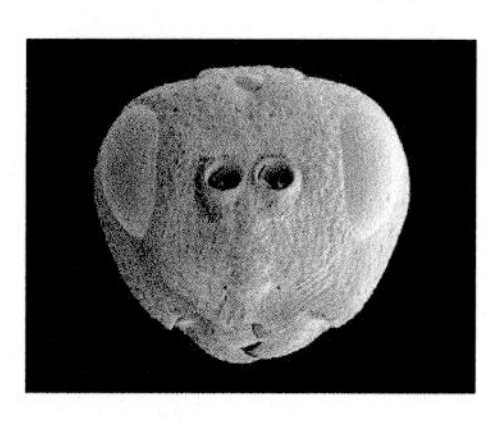

Eumayria floridana

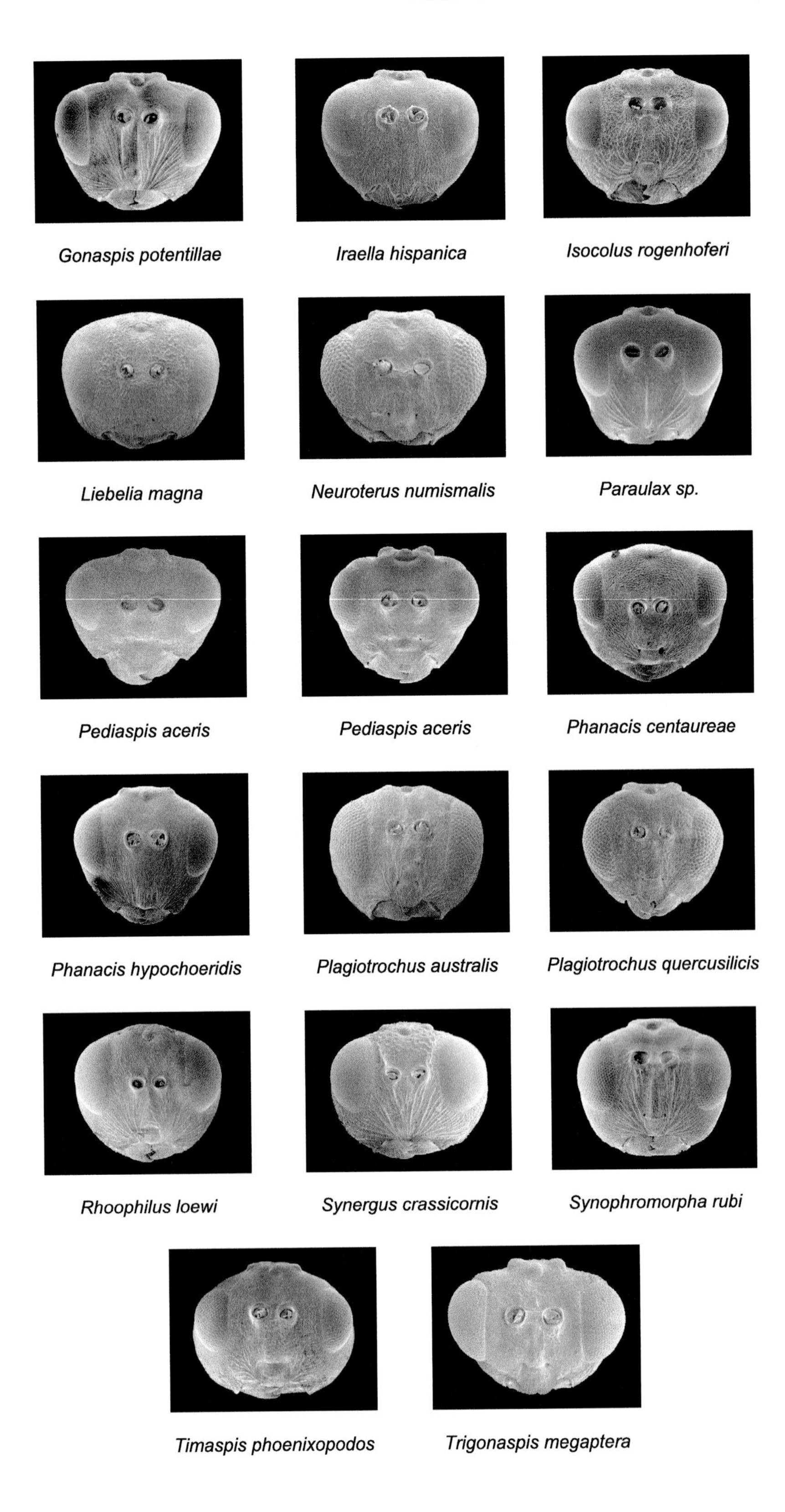
Gonaspis potentillae
Iraella hispanica
Isocolus rogenhoferi
Liebelia magna
Neuroterus numismalis
Paraulax sp.
Pediaspis aceris
Pediaspis aceris
Phanacis centaureae
Phanacis hypochoeridis
Plagiotrochus australis
Plagiotrochus quercusilicis
Rhoophilus loewi
Synergus crassicornis
Synophromorpha rubi
Timaspis phoenixopodos
Trigonaspis megaptera

Appendix 3

Figitid Images

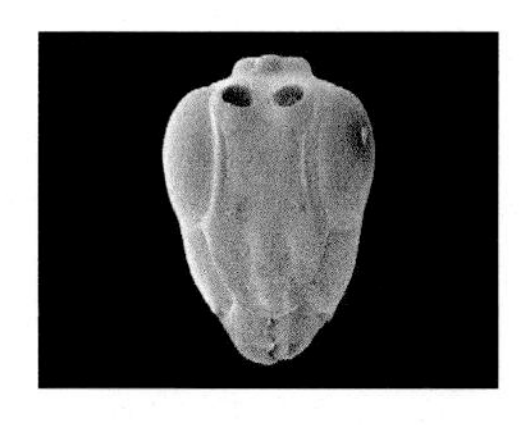

Acantheucoela sp.

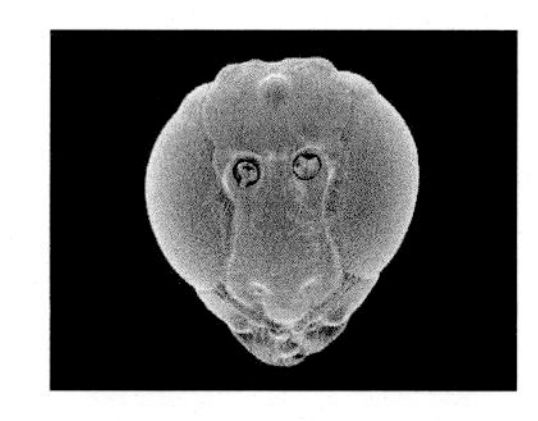

Aegeseucoela flavotincta

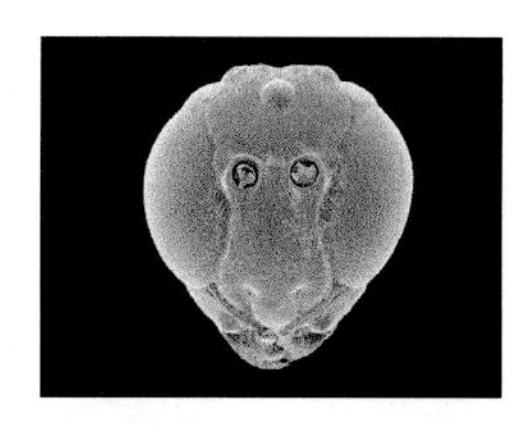

Aegeseucoela flavotincta

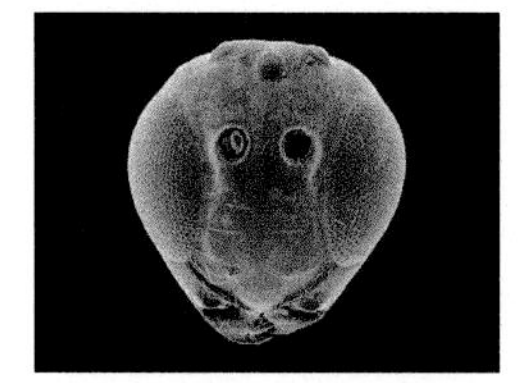

Agrostocynips clavatus

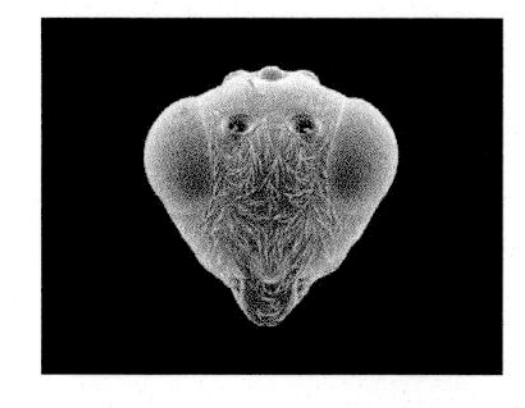

Anacharis eucharioides

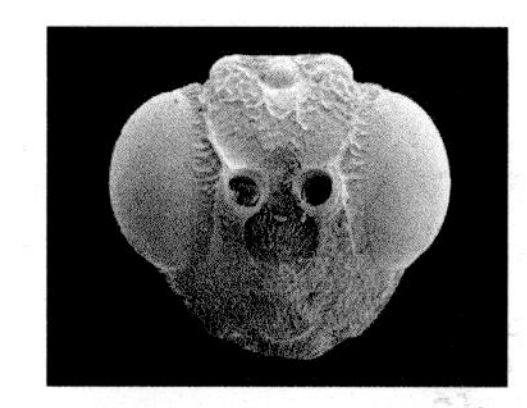

Aspicera scutellata

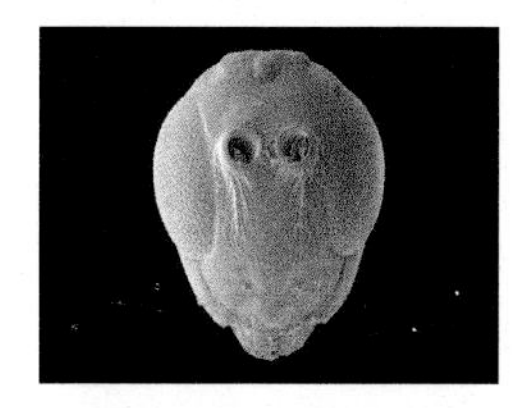

Chrestosema erythropum

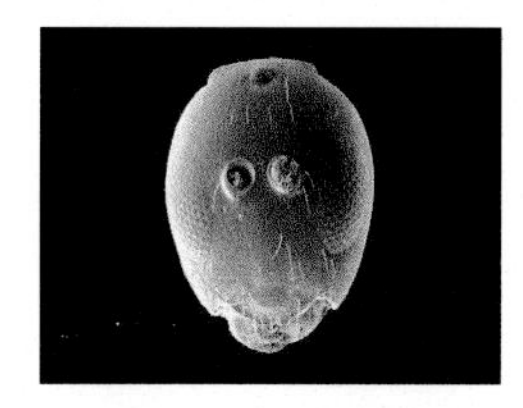

Cothonaspis longula

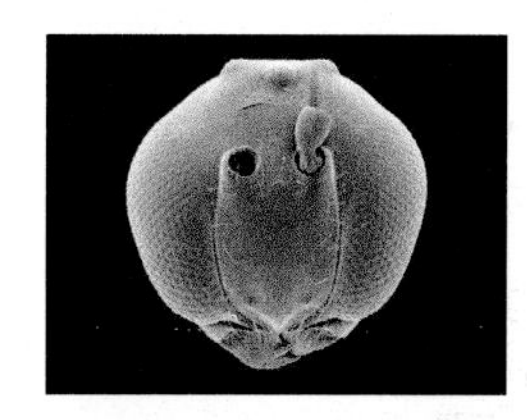

Dicerataspis grenadensis

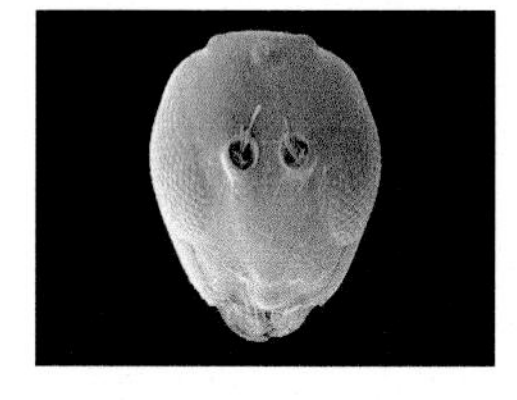

Didyctium nigriclava

Dieucoila sp.

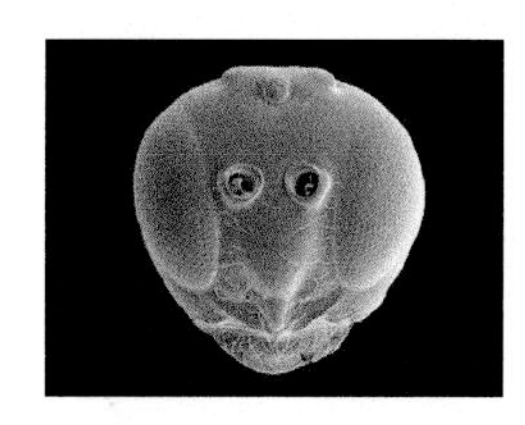

Disorygma depile

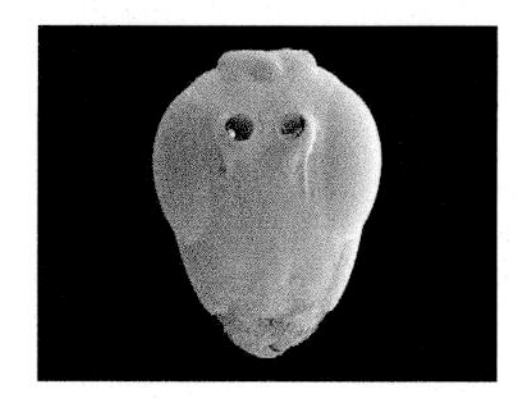

Epicoela sp.

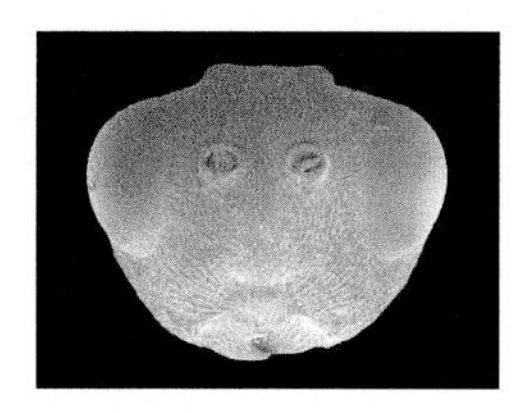

Euceroptres montanus

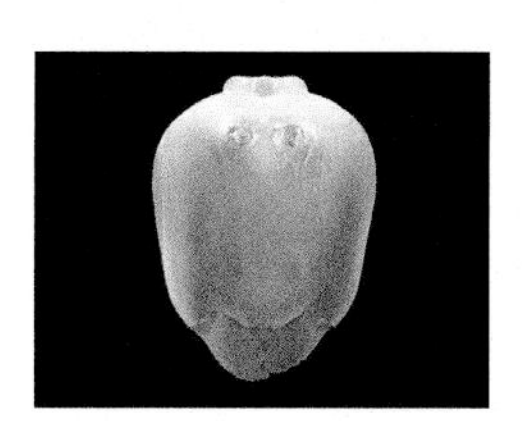

Eucoila crassinerva

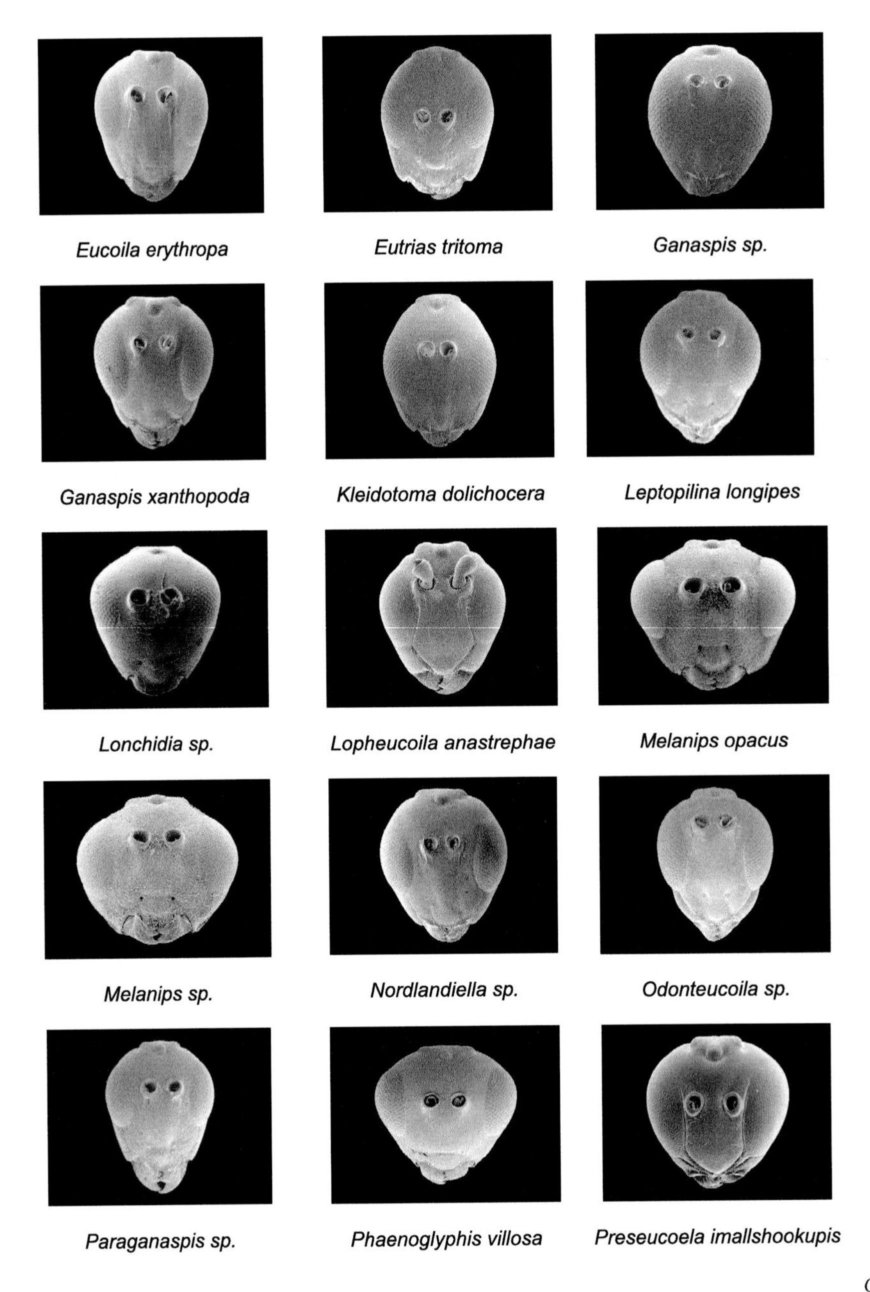
Eucoila erythropa
Eutrias tritoma
Ganaspis sp.
Ganaspis xanthopoda
Kleidotoma dolichocera
Leptopilina longipes
Lonchidia sp.
Lopheucoila anastrephae
Melanips opacus
Melanips sp.
Nordlandiella sp.
Odonteucoila sp.
Paraganaspis sp.
Phaenoglyphis villosa
Preseucoela imallshookupis

Continued.

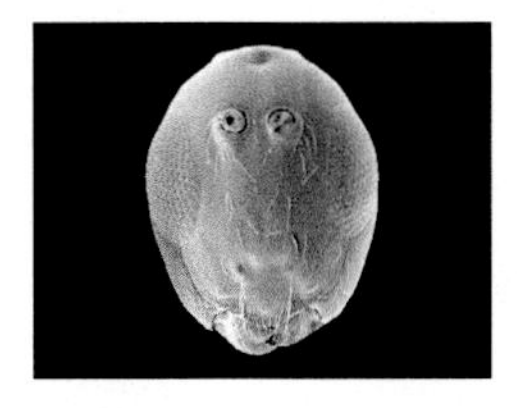

Rhoptromeris heptoma

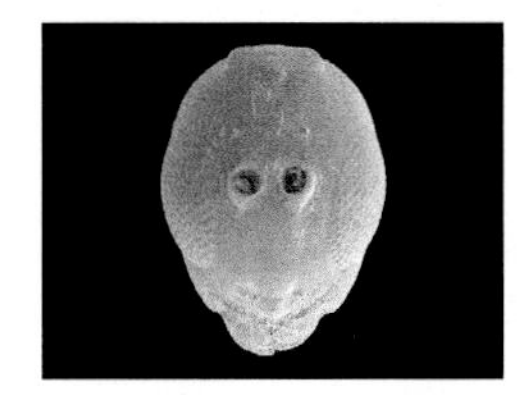

Triplasta sp.

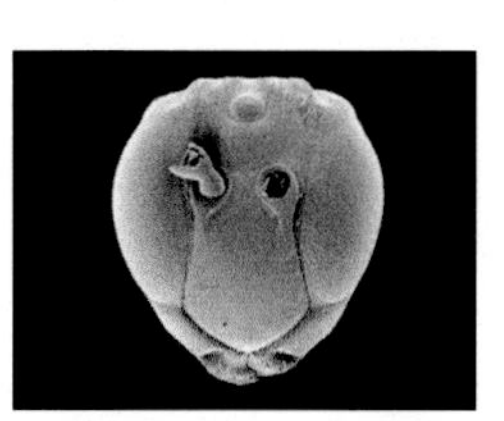

Tropieucoila nr. rufipes

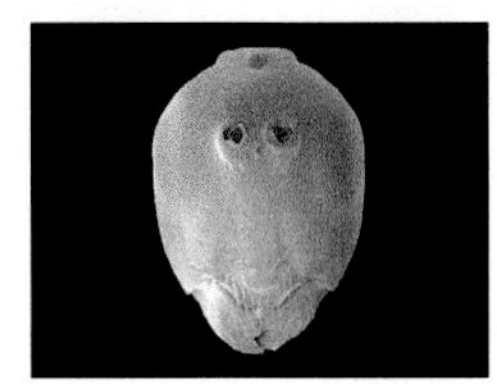

Trybliographa rapae

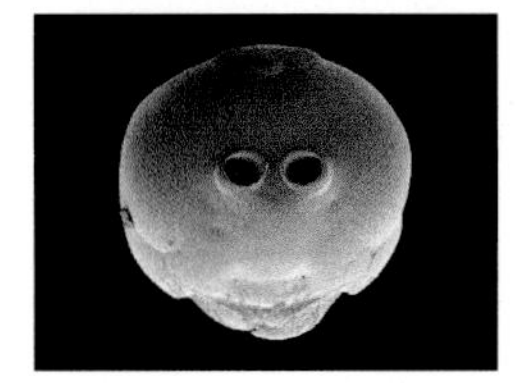

Trybliographa sp.

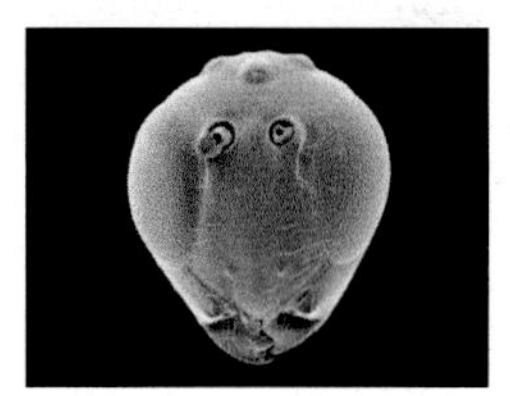

Zaeucoila sp.

Appendix 4

COMPARATIVE MATERIAL

Aptenodytes patagonicus. 1846.4.15.33; 1846.4.15.31; 1846.4.15.32; S/1972.1.24; S/1952.1.28; S/1952.1.29

Aptenodytes forsteri. 1846.4.15.26; 1846.4.15.27; 1846.4.15.28; 1850.9.7.2; 1998.55.2; S/1972.1.25

Pygoscelis papua. 1860.12.19.5; 1884.3.26.1; 1895.7.4.1; 1900.8.17.1; S/1952.3.135; 1846.4.15.29

Pygoscelis adeliae. 1849.10.2.2; 1910.11.5.1; 1846.4.15.34; 1850.9.7.1; S/1952.1.31; S/1952.1.32; S/1952.1.36; S/1965.10.1; 1966.4.2

Pygoscelis antarctica. S/1973.66.6; S/1966.4.1

Eudyptes crestatus. 1898.7.1.12; 1898.7.1.13; 1852.1.17.92; 1869.2.24.6; S/1956.14.1; S/1952.1.39; S/1952.1.136; 1998.12.6; S/1964.14.2

Eudyptula minor. S/2002.2.1; S/1966.51.1; 1896.2.16.38; S/1952.1.41

Spheniscus demersus. S/1998.23.2; 1905.7.23.1; 1898.7.1.8; 1898.7.1.9; S/1952.3.144; S/1998.48.24

Spheniscus humboldti. S/2000.7.1; S/1961.15.1; S/1952.1.42

Spheniscus magellanicus. 1891.7.20.133; S/1952.1.43; S/2001.45.1; S/1972.1.27; 1869.2.27.7.

Appendix 5

Raw data for all *O. excavata* P_1 elements used in this study (sub-sampled set). Each block of values represents a different sample of elements. Element side: 0 = dextral, 1 = sinistral. See Table 14.2 for key to measurement abbreviations. Values in italics are missing values replaced with mean.

ELEMENT	SIDE	APL	PPL	IPA	CBW	PDP	ADP	ACW:DW	PCW:DW	AND	PDN
P-1-2	0	0.306	0.213	148.9	0.072	*20.0*	18.4	1.333	*1.410*	6	*6*
P-1-6	0	0.452	*0.293*	*146.3*	0.085	*20.0*	20.3	1.723	*1.410*	9	*6*
P-1-8	1	0.392	0.285	142.9	*0.083*	25.3	18.5	*1.588*	*1.410*	7	7
P-1-10	0	0.512	*0.293*	146.3	0.098	*20.0*	19.8	1.947	*1.410*	10	*6*
P-1-5	1	0.418	0.319	145.2	0.062	20.3	19.8	1.227	1.258	8	6
P-1-7	0	0.320	0.195	155.7	0.082	*20.0*	23.3	1.903	*1.410*	7	*6*
P-1-9	1	0.647	*0.293*	*146.3*	0.092	*20.0*	14.5	1.329	*1.410*	9	*6*
P-4-78d	1	0.332	0.392	147.5	0.073	16.1	14.2	1.041	1.180	5	6
P-4-82e	0	0.528	0.351	137.9	0.094	14.6	17.3	1.629	1.375	8	5
P-1-a	1	0.610	0.460	164.2	0.092	14.0	16.7	1.541	1.292	9	7
P-1-b	1	0.478	0.329	172.0	0.071	18.0	20.2	1.427	1.272	9	7
P-1-c	1	0.611	0.429	155.2	0.120	15.0	15.0	1.806	1.806	10	7
P-2-a	1	0.552	0.395	159.9	0.080	19.0	20.7	1.658	1.522	11	8
P-2-b	0	0.536	0.315	164.2	0.081	17.8	20.2	1.636	1.442	10	5
P-2-c	0	0.510	0.332	157.6	0.063	15.7	19.2	1.203	0.983	9	5
P-2-d	1	0.567	0.380	158.8	0.076	16.9	18.7	1.421	1.284	10	6
P-2-e	0	0.501	0.277	163.4	0.100	*16.8*	17.0	1.700	*1.403*	7	4
P-3-a	1	0.349	0.261	168.7	0.073	19.0	18.2	1.333	1.391	7	5
P-3-b	0	0.441	0.244	166.4	0.087	*16.8*	20.3	1.773	*1.403*	9	4
P-4-a	0	0.484	0.311	148.8	0.102	16.0	13.8	1.409	1.633	7	5
P-2-a	0	0.729	0.381	165.5	0.127	13.0	14.1	1.796	1.656	10	6
P-2-b	0	0.768	0.389	161.4	0.157	*16.1*	8.9	1.399	*1.831*	7	4
P-3-a	0	0.746	0.417	164.8	0.127	13.9	14.7	1.873	1.771	9	5
P-3-c	0	0.776	0.318	165.2	0.079	16.3	19.1	1.503	1.283	13	6
P-3-f	1	0.751	0.505	149.5	0.152	12.2	13.8	2.103	1.859	10	7
P-4-a	0	0.756	0.341	157.5	0.095	19.2	16.3	1.550	1.826	14	6
P-4-b	1	0.625	0.375	159.2	0.134	15.6	14.6	1.953	2.087	9	6
P-4-c	1	0.608	0.333	158.0	0.114	17.2	11.5	1.309	1.958	7	5
P-5-a	1	0.655	0.418	166.8	0.124	13.2	10.4	1.291	1.639	8	5
P-5-b	1	0.617	0.376	169.5	0.140	15.4	15.4	2.163	2.163	8	5
P-6-a	1	0.663	0.397	163.6	0.147	13.5	13.0	1.916	1.989	8	6
P-6-b	0	0.544	0.343	171.1	0.099	16.1	14.6	1.452	1.601	8	5
P-7-a	1	0.653	0.451	166.6	0.113	14.1	14.6	1.644	1.588	9	6
P-7-c	0	0.605	0.390	162.5	0.118	14.6	12.3	1.446	1.716	8	5
P-7-d	0	0.660	0.308	162.0	0.100	17.6	18.8	1.874	1.754	11	6
P-8-a	1	0.635	0.441	161.7	0.123	16.2	13.9	1.704	1.986	10	6
P-8-b	1	0.518	0.312	158.7	0.109	17.1	16.5	1.795	1.861	8	5
P-8-c	0	0.547	0.329	174.4	0.113	14.8	15.6	1.760	1.670	8	5
P-9-a	1	0.542	0.371	158.9	0.116	20.0	15.0	1.747	2.329	9	6
P-9-b	0	0.536	0.257	162.7	0.090	21.9	23.2	2.078	1.962	10	5
P-1-e	0	0.532	0.310	170.9	0.087	18.3	19.4	1.683	1.587	9	5
P-2-d	1	0.609	0.381	162.8	0.081	23.2	18.3	1.473	1.868	11	6
P-3-b	0	0.658	0.383	156.7	0.096	17.1	16.1	1.542	1.638	10	6
P-4-a	0	0.422	0.316	161.8	0.086	18.3	18.5	1.584	1.567	8	5
P-5-c	1	0.630	0.460	168.5	0.088	18.4	17.6	1.553	1.624	11	8
P-6-b	1	0.479	0.310	166.5	0.090	18.3	18.0	1.628	1.655	8	6
P-7-c	1	0.465	0.345	162.1	0.079	21.1	18.9	1.491	1.665	9	6
P-14-1b	0	0.312	0.203	164.3	0.063	24.5	18.7	1.173	1.536	6	5
P-16-b	0	0.492	0.345	164.1	0.065	18.3	17.8	1.149	1.181	9	6
P-17-a	0	0.447	0.330	160.0	0.074	20.0	20.1	1.484	1.477	8	6
P-18-1b	0	0.296	0.216	169.5	0.060	23.3	18.9	1.128	1.390	6	5
P-20-1a	1	0.233	0.216	170.8	0.064	24.8	22.5	1.442	1.590	5	5
P-22-3c	0	0.266	0.231	172.2	0.057	23.3	19.2	1.101	1.336	6	5
P-27-c	0	0.350	0.273	146.8	0.098	19.8	16.1	1.573	1.934	6	5
P-45-1a	1	0.164	0.151	150.5	0.048	*19.9*	22.7	1.091	*1.640*	5	3
P-61-1a	0	0.342	0.268	170.7	0.057	20.2	18.8	1.076	1.156	7	5

P-61-2f	1	0.377	0.289	167.2	0.060	17.7	16.1	0.971	1.068	6	5
P-1-b	0	0.804	0.451	169.8	0.094	15.6	12.2	1.149	1.469	12	6
P-2-d	0	0.690	0.380	168.4	0.084	17.9	17.1	1.431	1.498	13	8
P-4-c	0	0.818	0.477	160.1	0.082	16.5	20.9	1.713	1.353	15	8
P-5-c	0	0.785	0.369	146.5	0.146	17.2	16.5	2.410	2.512	12	7
P-8-c	1	0.847	0.393	154.9	0.082	21.0	20.0	1.631	1.712	14	9
P-9-d	1	0.591	0.402	162.5	0.084	16.6	19.0	1.595	1.394	12	7
P-10-c	1	0.686	0.454	165.9	0.092	15.9	18.1	1.665	1.463	11	8
P-12-b	1	0.740	0.384	164.3	0.112	15.5	20.1	2.256	1.739	14	6
P-15-e	1	0.417	0.294	159.4	0.086	18.0	15.9	1.367	1.547	7	5
P-18-c	0	0.753	0.365	158.3	0.124	16.0	16.1	1.991	1.979	13	6
P-19-b	1	0.637	0.328	166.9	0.104	17.8	22.5	2.337	1.849	12	6
P-20-a	1	0.490	0.373	160.7	0.084	26.2	21.9	1.843	2.205	10	7
P-1-1	0	0.440	0.323	145.4	0.094	17.8	14.1	1.325	1.673	7	6
P-1-10	0	0.464	0.297	156.1	0.096	19.4	14.7	1.414	1.866	7	5
P-1-2	1	0.296	0.206	170.6	0.087	*18.5*	20.6	1.798	*1.857*	5	3
P-1-3	1	0.302	0.186	156.7	0.078	*18.5*	21.8	1.708	*1.857*	7	4
P-1-4	1	0.512	0.312	166.5	0.103	19.9	18.1	1.866	2.052	10	5
P-1-5	1	0.433	0.290	144.2	0.104	17.5	16.4	1.712	*1.857*	8	4
P-1-6	0	0.430	0.337	147.8	0.093	16.8	17.3	1.611	1.564	9	5
P-2-6	0	0.363	0.153	170.2	0.074	*18.5*	16.6	1.235	*1.857*	7	4
P-4-10	0	0.345	0.211	161.3	0.091	*18.5*	18.4	1.672	*1.857*	7	3
P-4-2	0	0.242	0.150	151.2	0.078	*18.5*	25.0	1.951	*1.857*	5	3
P-4-9	1	0.490	0.284	153.6	0.109	19.5	17.1	1.868	2.130	9	6
P-6-5	1	0.502	0.254	164.7	0.089	*18.5*	16.3	1.444	*1.857*	8	4
P-8-2	1	0.211	0.175	154.8	0.066	*18.5*	22.1	1.452	*1.857*	4	4
P-1-10	0	0.588	0.360	157.3	0.123	14.7	16.0	1.970	1.810	9	6
P-1-7	1	0.623	0.447	157.8	0.104	15.0	14.0	1.450	1.555	9	7
P-1-8	1	0.682	0.462	153.0	0.130	14.0	13.4	1.740	1.820	9	6
P-2-2	1	0.453	0.299	165.8	0.111	20.0	14.3	1.590	2.230	7	5
P-2-6	1	0.286	0.215	169.6	0.077	*15.5*	21.6	1.670	*1.853*	5	3
P-3-4	1	0.570	0.353	157.4	0.116	16.6	11.6	1.340	1.920	7	5
P-3-8	1	0.495	0.275	166.3	0.122	*15.5*	13.6	1.650	*1.853*	7	3
P-4-2	0	0.499	0.324	166.7	0.097	*15.5*	16.3	1.570	*1.853*	9	4
P-4-5	0	0.311	0.221	161.3	0.089	*15.5*	18.8	1.660	*1.853*	5	3
P-5-5	1	0.249	0.147	169.2	0.088	*15.5*	24.0	2.110	*1.853*	4	3
P-6-5	1	0.594	0.397	163.5	0.138	14.2	13.9	1.920	1.960	9	6
P-6-7	0	0.645	0.358	162.9	0.146	19.6	12.4	1.810	2.860	8	5
P-6-9	1	0.685	0.490	144.4	0.135	12.8	13.3	1.800	1.730	10	7
P-7-1	1	0.623	0.386	155.9	0.107	14.9	15.2	1.620	1.590	9	5
P-7-10	0	0.633	0.431	158.2	0.134	13.4	9.3	1.250	1.800	6	5
P-7-2	0	0.703	0.335	155.8	0.144	*15.5*	9.4	1.360	*1.853*	8	4
P-7-4	0	0.453	0.396	161.7	0.112	14.2	13.6	1.520	1.590	7	5
P-7-5	0	0.608	0.362	168.1	0.149	*15.5*	13.0	1.940	*1.853*	9	5
P-7-6	0	0.658	0.383	165.7	0.084	16.3	*14.8*	*1.649*	1.370	7	4
P-7-8	1	0.303	0.203	163.3	0.083	*15.5*	16.6	1.370	*1.853*	6	4
P-1-4	1	0.394	0.356	154.1	0.104	16.2	12.5	1.295	1.678	6	6
P-1-1	1	0.157	0.232	156.9	0.065	26.7	*13.0*	*1.453*	1.748	4	5
P-2-4	1	0.277	0.336	164.4	0.098	15.8	*13.0*	*1.453*	1.544	4	5
P-2-9	1	0.172	0.258	158.5	0.085	22.0	*13.0*	*1.453*	1.881	2	6
P-3-4	1	0.232	0.315	141.9	0.099	21.0	*13.0*	*1.453*	2.074	4	7
P-3-5	1	0.409	0.363	154.1	0.092	17.1	17.3	1.586	1.567	6	7
P1-2226	1	0.422	0.396	157.2	0.094	15.6	15.5	1.455	1.464	7	7
P1-2259	1	0.386	0.412	166.7	0.137	13.5	12.3	1.691	1.856	5	5
P1-2384	1	0.485	0.441	140.4	0.099	13.8	10.8	1.065	1.361	5	6
P2-2299	1	0.369	0.398	140.2	0.113	17.1	16.2	1.825	1.927	6	7
P2-2410	1	0.126	0.213	150.4	0.053	31.4	*13.0*	*1.453*	1.659	2	5
P2-2458	1	0.574	0.486	139.4	0.130	13.6	12.7	1.648	1.764	8	6
P1-2305	?	0.468	0.439	165.0	0.129	13.3	11.2	1.448	1.720	5	5
P1-2407	?	0.473	0.386	161.0	0.117	*17.6*	11.1	1.298	*1.725*	5	4
P2-2366	0	0.198	0.252	148.3	0.080	21.0	*13.0*	*1.453*	1.682	3	5
P1-105	?	0.655	0.535	135.0	0.120	17.5	12.3	1.472	2.094	*6*	*6*
P1-26	?	0.386	0.398	155.9	0.081	*17.6*	14.5	1.177	*1.725*	*6*	*6*

P1-43	?	0.500	0.413	141.2	0.114	14.9	13.2	1.499	1.692	*6*	*6*
P1-128	?	0.722	0.492	130.8	0.130	13.3	9.7	1.264	1.733	6	8
P1-173	?	0.621	0.431	147.8	0.093	10.9	11.4	1.064	1.017	*6*	*6*
P1-O-1	1	0.527	0.456	139.0	0.102	13.7	12.1	1.228	1.391	7	6
P1-O-15	1	0.332	0.350	139.4	0.102	15.6	*13.5*	*1.180*	1.594	4	4
P1-O-14	1	0.304	0.262	136.0	0.087	*14.1*	14.6	1.269	*1.354*	5	5
P1-O-11	1	0.387	0.344	132.2	0.091	14.7	11.6	1.053	1.334	6	5
P1-O-10	1	0.362	0.463	138.4	0.112	12.4	12.0	1.347	1.392	6	5
P1-O-21	1	0.240	0.287	140.4	0.079	*14.1*	16.9	1.329	*1.354*	5	4
P1-O-2	1	0.412	0.446	131.6	0.118	12.0	13.3	1.570	1.417	6	5
P1-O-19	1	0.263	0.256	143.4	0.075	*14.1*	18.9	1.421	*1.354*	5	3
P1-O-18	1	0.318	0.324	136.4	0.097	14.7	14.2	1.380	1.428	5	4
P1-O-17	1	0.388	0.303	137.4	0.091	14.3	12.3	1.122	1.305	6	4
P1-O-22	1	0.302	0.256	136.3	0.071	*14.1*	14.9	1.064	*1.354*	5	4
P1-O-24	1	0.356	0.297	132.3	0.068	16.0	14.0	0.950	1.086	5	4
P1-O-23	1	0.267	0.223	137.3	0.065	*14.1*	17.0	1.102	*1.354*	5	4
P1-O-28	1	0.338	0.413	134.4	0.108	13.8	12.8	1.384	1.492	6	4
P1-O-3	1	0.410	0.356	127.6	0.104	14.4	11.4	1.181	1.491	4	6
P1-O-4	1	0.430	0.423	135.8	0.093	12.6	11.5	1.075	1.178	5	5
P1-O-5	1	0.370	0.409	145.2	0.094	13.2	12.5	1.172	1.237	6	5
P1-O-7	1	0.416	0.402	134.3	0.086	12.3	10.8	0.932	1.062	5	5
P1-O-6	1	0.334	0.379	135.8	0.103	*14.1*	11.7	1.201	*1.354*	5	4
P1-O-9	1	0.367	0.265	140.8	0.098	17.4	*13.5*	*1.180*	1.700	4	5
P-R-1	0	0.531	0.296	144.3	0.117	*17.1*	13.1	1.534	*2.008*	8	4
P-R-13	0	0.454	0.269	146.9	0.127	18.3	13.8	1.751	2.322	6	4
P-R-11	0	0.423	0.282	147.9	0.120	*17.1*	15.0	1.795	*2.008*	7	4
P-R-20	0	0.347	0.251	151.6	0.099	*17.1*	*14.2*	*1.636*	*2.008*	7	4
P-R-2	1	0.485	0.340	149.7	0.117	17.4	16.7	1.954	2.036	9	6
P-R-19	1	0.448	0.409	137.9	0.101	13.6	13.5	1.358	1.368	8	6
P-R-17	1	0.434	0.330	151.5	0.128	17.0	12.5	1.601	2.178	7	5
P-R-21	1	0.433	0.321	146.5	0.156	15.6	13.2	2.054	2.427	6	4
P-R-22	0	0.421	0.276	155.7	0.108	19.8	14.2	1.530	2.133	7	5
P-R-26	0	0.438	0.300	144.7	0.138	*17.1*	13.0	1.797	*2.008*	7	4
P-R-25	1	0.406	0.299	147.3	0.108	16.3	13.2	1.430	1.765	6	5
P-R-24	1	0.397	0.305	152.8	0.101	17.5	13.0	1.310	1.764	6	5
P-R-28	1	0.372	0.260	148.8	0.107	*17.1*	14.8	1.587	*2.008*	6	3
P-R-8	1	0.452	0.281	155.6	0.095	*17.1*	15.3	1.457	*2.008*	8	4
P-R-7	1	0.415	0.254	149.8	0.136	*17.1*	14.0	1.911	*2.008*	6	5
P-R-6	1	0.391	0.332	150.7	0.107	18.1	14.2	1.520	1.937	7	5
P-R-5	0	0.420	0.345	152.4	0.122	18.1	15.4	1.881	2.211	7	5
P-R-3	0	0.404	0.269	152.1	0.121	19.7	14.5	1.748	2.375	6	5
P-R-15	1	0.354	0.287	155.6	0.114	12.9	*14.2*	1.468	*2.008*	7	4
P-R-4	1	0.428	0.330	149.7	0.101	16.6	15.3	1.550	1.682	8	5
P-A-10	1	0.209	0.137	151.6	0.056	*23.2*	22.9	1.291	*1.979*	5	3
P-A-10	0	0.460	0.250	138.9	0.092	*23.2*	18.9	1.733	*1.979*	9	5
P-A-11	0	0.251	0.167	167.2	0.072	*23.2*	23.1	1.666	*1.979*	6	3
P-A-12	0	0.180	0.219	157.7	0.073	*23.2*	*19.8*	*1.642*	*1.979*	4	5
P-A-16	0	0.405	0.263	144.8	0.091	21.5	19.1	1.733	1.950	9	6
P-A-18	0	0.433	*0.277*	*144.4*	0.087	*23.2*	18.1	1.569	*1.979*	7	6
P-A-1	0	0.384	0.208	142.5	0.087	*23.2*	19.1	1.666	*1.979*	8	4
P-A-23	?	0.134	0.105	168.7	0.047	*23.2*	*19.8*	*1.642*	*1.979*	4	3
P-A-22	0	0.407	*0.277*	*144.4*	0.080	*23.2*	21.1	1.693	*1.979*	9	4
P-A-25	0	0.345	0.200	153.1	0.101	*23.2*	21.7	2.182	*1.979*	7	3
P-A-2	0	0.223	0.112	144.9	*0.083*	*23.2*	25.5	*1.642*	*1.979*	6	4
P-A-29	1	0.424	*0.277*	*144.4*	0.106	18.5	19.2	2.030	1.956	9	4
P-A-28	1	0.379	0.505	132.3	0.093	19.3	15.6	1.453	1.797	7	9
P-A-27	?	0.408	0.230	145.7	0.083	24.2	21.8	1.816	2.015	9	5
P-A-9	1	0.288	0.189	147.7	*0.083*	25.5	*19.8*	*1.642*	*1.979*	5	7
P-A-6	1	0.410	0.272	148.3	0.098	*23.2*	19.4	1.902	*1.979*	9	5
P-A-5	0	0.381	0.248	151.6	0.081	*23.2*	22.0	1.778	*1.979*	8	4
P-A-4	0	0.305	0.216	146.0	0.064	22.2	22.9	1.459	1.415	6	5
P-A-9	0	0.574	0.302	120.2	0.092	18.6	15.9	1.469	1.718	10	6
P-Q-10	0	0.405	0.273	158.6	0.063	21.4	20.2	1.273	1.349	9	5

P-Q-14	1	0.472	0.303	162.2	0.070	17.5	19.0	1.334	1.229	9	5
P-Q-16	0	0.519	0.331	141.6	0.100	17.8	19.5	1.952	1.782	11	5
P-Q-19	0	0.436	0.255	158.3	0.064	23.4	19.4	1.233	1.488	9	5
P-Q-18	0	0.578	0.358	147.0	0.095	15.9	19.4	1.833	1.503	12	7
P-Q-22	1	0.406	0.243	162.8	0.080	21.7	25.0	1.994	1.731	10	5
P-Q-26	0	0.482	0.277	156.1	0.081	20.8	18.8	1.519	1.681	10	5
P-Q-25	1	0.682	0.386	159.8	0.092	15.9	19.6	1.811	1.469	12	6
P-Q-24	1	0.532	0.339	167.8	0.109	16.5	20.5	2.229	1.794	10	6
P-Q-23	0	0.595	0.305	162.1	0.105	20.4	19.9	2.093	2.146	12	7
P-Q-28	1	0.523	0.380	164.9	0.077	18.5	20.4	1.569	1.423	11	6
P-Q-32	1	0.571	0.344	169.1	0.093	17.2	17.6	1.638	1.600	10	5
P-Q-31	0	0.697	0.406	152.3	0.097	15.3	20.1	1.947	1.482	12	8
P-Q-30	0	0.694	0.338	153.5	0.082	*19.2*	19.1	1.567	*1.783*	14	4
P-Q-2	0	0.651	0.342	155.5	0.084	15.2	21.2	1.771	1.270	13	6
P-Q-33	0	0.463	0.265	151.9	0.070	*19.2*	22.8	1.607	*1.783*	10	4
P-Q-8	1	0.422	0.301	169.0	0.074	20.6	19.2	1.414	1.517	9	5
P-Q-5	0	0.517	0.283	163.5	0.074	20.1	18.5	1.378	1.497	9	6
P-Q-35	0	0.541	0.303	159.6	0.085	17.0	19.1	1.625	1.446	10	5
P-Q-9	1	0.522	0.266	165.9	0.074	20.5	18.5	1.360	1.507	9	6
P-E-11	0	0.599	0.355	145.1	0.097	18.6	15.4	1.500	1.812	10	7
P-E-14	1	0.501	0.335	160.1	0.099	24.7	18.3	1.808	2.440	9	6
P-E-16	1	0.475	0.356	156.7	0.070	19.8	16.5	1.148	1.378	8	6
P-E-17	0	0.466	0.282	162.2	0.079	23.9	17.4	1.366	1.877	9	6
P-E-18	1	0.562	0.437	141.3	0.147	*21.0*	11.8	1.737	*1.863*	8	4
P-E-19	1	0.378	0.283	158.2	0.084	23.9	17.5	1.466	1.877	7	5
P-E-1	0	0.430	0.233	157.2	0.084	*21.0*	17.0	1.431	*1.863*	8	3
P-E-24	1	0.328	0.227	174.7	0.083	*21.0*	17.9	1.487	*1.863*	6	4
P-E-25	1	0.122	0.138	150.9	0.053	35.4	*17.1*	*1.518*	1.865	3	5
P-E-28	0	0.302	0.204	166.4	0.086	*21.0*	19.2	1.644	*1.863*	6	3
P-E-29	1	0.327	0.290	162.1	0.074	*21.0*	17.8	1.322	*1.863*	6	4
P-E-32	1	0.444	0.275	166.3	0.086	37.1	21.0	1.814	3.204	9	5
P-E-33	0	0.536	0.280	147.3	0.092	22.8	16.7	1.530	2.089	10	5
P-E-36	0	0.429	0.255	151.1	0.082	20.4	14.8	1.221	1.683	7	5
P-E-37	0	0.390	0.493	137.6	0.161	17.0	15.8	2.545	2.738	7	7
P-E-38	1	0.321	0.218	172.7	0.084	*21.0*	18.3	1.529	*1.863*	6	4
P-E-45	1	0.494	0.315	155.2	0.090	18.3	19.8	1.777	1.643	10	5
P-E-4	1	0.463	0.316	155.1	0.074	17.8	18.0	1.330	1.316	8	6
P-E-9	0	0.542	0.385	151.8	0.090	17.9	15.7	1.418	1.617	10	6
P-E-40	1	0.651	*0.317*	*155.6*	0.115	*21.0*	17.1	1.961	*1.863*	11	*5*
P-L-15	1	0.681	0.391	179.6	0.088	19.4	20.0	1.758	1.706	13	7
P-L-14	1	0.314	0.190	166.7	0.050	*19.6*	19.9	0.997	*1.801*	6	3
P-L-11	0	0.578	0.312	161.4	0.095	19.0	19.4	1.840	1.802	11	6
P-L-1	0	0.552	0.314	157.0	0.112	20.1	19.9	2.221	2.244	10	5
P-L-8	0	0.514	0.352	141.0	0.102	15.1	17.4	1.779	1.544	8	5
P-L-7	1	0.676	0.390	170.5	0.100	19.7	18.3	1.824	1.964	12	8
P-L-4	0	0.440	*0.304*	*159.5*	0.118	14.9	15.5	1.827	1.756	7	*5*
P-L-3	0	0.241	0.159	162.2	0.069	*19.6*	21.3	1.470	*1.801*	5	3
P-L-9	0	0.499	0.258	167.8	0.070	21.1	21.0	1.471	1.478	10	6
P-L-10	0	0.468	0.237	169.5	0.091	22.0	19.4	1.760	1.996	8	5
P-L-15	0	0.350	0.231	155.5	0.070	24.0	19.2	1.351	1.689	7	5
P-L-14	1	0.314	0.255	162.8	0.065	*19.6*	17.7	1.144	*1.801*	7	4
P-L-20	1	0.342	0.207	170.3	0.074	*19.6*	20.4	1.513	*1.801*	7	4
P-L-19	0	0.421	0.368	142.2	0.111	15.0	14.8	1.649	1.672	7	5
P-L-27	1	0.441	0.286	170.6	0.097	20.3	19.3	1.868	1.965	9	4
P-L-25	1	0.301	0.245	165.4	0.072	22.7	20.3	1.453	1.625	7	4
P-L-29	1	0.419	0.287	169.8	0.070	18.5	18.3	1.284	1.299	8	5
P-L-2	0	0.264	0.247	139.7	0.068	22.3	*18.9*	*1.666*	1.519	4	5
P-L-8	0	0.438	0.283	163.7	0.068	20.3	20.4	1.392	1.385	9	6
P-L-9	1	0.693	0.423	162.8	0.112	16.7	17.9	2.010	1.876	14	7
P-3-10	0	0.534	0.366	159.4	*0.061*	18.0	21.5	*1.056*	*0.937*	11	7
P-3-12	0	0.550	0.345	163.5	0.064	19.0	20.6	1.313	1.209	10	6
P-3-13	1	0.586	0.501	158.9	0.049	13.0	17.9	0.872	0.633	10	7
P-3-14	1	0.701	0.415	164.9	0.056	14.6	18.3	1.023	0.816	11	6
P-3-15	0	0.658	0.511	154.0	0.062	14.7	15.4	0.950	0.907	11	8

P-3-16	0	0.618	0.451	167.4	0.068	14.8	15.8	1.081	1.013	10	7
P-3-17	1	0.655	0.435	176.2	0.054	16.1	16.7	0.896	0.864	10	7
P-3-18	1	0.560	0.388	159.9	0.057	15.4	19.7	1.125	0.879	11	7
P-3-19	0	0.641	0.368	170.7	0.062	13.9	17.7	1.091	0.857	11	5
P-3-1	0	0.614	0.417	162.4	0.063	13.4	17.4	1.105	0.851	10	6
P-3-20	0	0.652	0.413	156.4	0.074	12.9	16.1	1.196	0.958	11	7
P-3-3	1	0.583	0.502	158.6	0.049	13.0	17.7	0.876	0.643	11	6
P-3-5	0	0.621	0.399	168.8	0.064	13.8	17.8	1.132	0.877	10	6
P-3-6	0	0.576	0.416	169.0	0.054	16.3	14.7	0.797	0.884	8	7
P-3-7	0	0.510	0.281	173.2	0.054	24.6	21.1	1.148	1.338	9	5
P-3-8	1	0.642	0.370	161.8	0.073	15.4	17.1	1.255	1.130	9	6
P-3-9	0	0.684	0.449	161.1	0.074	15.4	14.2	1.045	1.133	11	7
P-B-12	1	0.402	0.312	171.0	0.054	18.9	19.8	1.071	1.022	9	5
P-B-13	1	0.440	0.338	153.0	0.051	16.8	18.7	0.961	0.863	9	7
P-B-15	1	0.439	0.366	161.6	0.055	15.1	17.6	0.961	0.825	8	6
P-B-21	1	0.288	0.208	170.7	0.058	*16.6*	22.2	1.288	*1.077*	6	3
P-B-1	0	0.665	0.470	167.7	0.083	12.3	17.8	1.470	1.016	10	7
P-B-22	0	0.643	0.449	166.6	0.059	16.6	18.6	1.101	0.982	12	5
P-B-24	0	0.546	0.382	169.5	0.062	16.2	19.4	1.202	1.003	10	7
P-B-26	1	0.556	0.327	160.6	0.075	15.1	18.5	1.379	1.126	9	5
P-B-27	0	0.304	0.228	165.2	0.055	*16.6*	20.1	1.109	*1.077*	7	4
P-B-28	1	0.427	0.281	154.4	0.074	19.7	19.4	1.437	1.459	8	5
P-B-29	0	0.384	0.282	166.1	0.048	18.8	20.3	0.971	0.900	8	5
P-B-2	1	0.738	0.384	167.1	0.077	12.7	17.0	1.305	0.975	11	5
P-B-3	0	0.651	0.586	158.5	0.067	17.3	18.4	1.235	1.161	11	7
P-B-5	1	0.566	0.407	166.5	0.065	17.0	20.4	1.328	1.107	11	7
P-B-6	0	0.637	0.290	163.5	0.058	15.8	18.1	1.055	0.921	10	5
P-B-7	0	0.577	0.272	166.3	0.076	19.2	19.2	1.456	1.456	10	5
P-B-8	0	0.546	0.324	173.5	0.059	18.4	22.7	1.336	1.083	11	6
P-B-9	1	0.473	0.334	159.1	0.082	19.6	17.9	1.460	1.599	9	6
P-B-3	1	0.699	0.528	166.9	0.056	14.3	17.7	0.996	0.805	10	9
P-B-4	0	0.596	0.419	157.6	*0.344*	15.0	17.0	*1.217*	*1.077*	10	7
P-C-11	1	0.715	0.348	168.0	0.076	18.6	19.4	1.477	1.416	11	6
P-C-12	1	0.554	0.496	159.5	0.074	15.8	19.4	1.441	1.174	11	8
P-C-15	1	0.435	0.364	161.9	0.050	14.6	17.6	0.879	0.729	8	5
P-C-18	1	0.503	0.374	172.6	0.055	14.5	18.1	0.987	0.791	8	5
P-C-19	1	0.550	0.457	161.3	0.075	15.7	20.1	1.507	1.177	10	7
P-C-1	1	0.663	0.513	147.0	0.063	16.5	17.4	1.104	1.047	12	9
P-C-22	1	0.491	0.395	164.4	0.050	15.1	17.1	0.857	0.757	9	6
P-C-26	1	0.538	0.387	168.0	0.086	15.4	16.4	1.413	1.327	9	6
P-C-2	0	0.705	0.450	165.6	0.079	14.7	15.5	1.230	1.167	9	7
P-C-30	1	0.470	0.406	164.6	0.048	17.3	17.5	0.844	0.834	8	7
P-C-32	0	0.439	0.300	167.4	0.057	20.7	19.2	1.101	1.187	9	6
P-C-33	1	0.437	0.312	172.7	0.059	16.8	16.2	0.952	0.987	7	5
P-C-34	0	0.437	0.289	166.2	0.054	18.9	18.1	0.968	1.011	8	5
P-C-3	1	0.539	0.415	167.8	0.060	14.0	16.6	0.998	0.842	9	6
P-C-41	0	0.210	0.150	179.6	0.061	*17.0*	25.1	1.627	*1.172*	5	4
P-C-42	1	0.273	0.261	174.2	0.050	18.8	17.2	0.861	0.941	5	5
P-C-5	1	0.548	0.413	165.1	0.059	13.9	16.5	0.977	0.823	9	6
P-C-6	0	0.602	0.477	168.5	0.074	16.0	20.0	1.486	1.189	10	8
P-C-7	0	0.654	0.410	172.7	0.057	17.1	19.2	1.096	0.976	12	6
P-C-9	1	0.536	0.472	159.1	0.053	17.6	18.7	1.000	0.941	11	7
P-H-10	1	0.296	0.367	143.9	0.080	16.9	15.2	1.221	1.357	6	6
P-H-11	0	0.402	0.563	129.9	0.129	15.1	14.3	1.844	1.947	9	6
P-H-17	0	0.274	0.200	153.6	0.061	*17.2*	17.3	1.061	*1.674*	5	3
P-H-16	0	0.331	0.247	153.8	0.071	21.4	17.0	1.211	1.574	6	5
P-H-26	0	0.295	0.385	150.8	0.081	20.8	17.0	1.378	1.686	6	7
P-H-23	1	0.446	0.289	163.8	0.065	17.3	16.8	1.099	1.131	7	5
P-H-22	0	0.408	0.348	149.1	0.092	16.1	13.6	1.254	1.484	7	6
P-H-27	0	0.464	0.375	143.4	0.101	20.3	15.0	1.517	2.053	8	6
P-H-32	1	0.306	0.231	162.9	0.066	*17.2*	23.4	1.546	*1.674*	6	4
P-H-30	0	0.356	0.275	157.1	0.076	*17.2*	15.9	1.211	*1.674*	7	4
P-H-29	0	0.298	0.195	160.0	0.052	23.3	23.7	1.233	1.212	7	5
P-H-34	1	0.178	0.173	167.9	0.076	*17.2*	15.4	1.176	*1.674*	4	3

P-H-39	0	0.431	0.330	146.5	0.097	17.7	17.6	1.707	1.717	7	6
P-H-38	0	0.546	0.354	157.1	0.077	15.4	14.3	1.107	1.192	8	5
P-H-37	1	0.490	0.359	153.0	0.081	17.8	16.7	1.360	1.449	10	6
P-H-35	1	0.394	0.358	154.0	0.089	16.2	15.1	1.339	1.436	7	5
P-H-42	0	0.453	0.278	157.4	0.080	18.0	14.1	1.127	1.438	7	5
P-H-41	0	0.410	0.294	150.1	0.076	17.2	15.7	1.187	0.300	7	5
P-H-9	1	0.372	0.380	155.2	0.078	*17.2*	15.1	1.179	*1.674*	6	4
P-H-45	1	0.378	0.501	124.5	0.104	17.1	12.0	1.250	1.781	5	7
P-O-15	1	0.428	0.324	159.7	0.085	*19.0*	18.6	1.583	*1.500*	8	4
P-O-14	1	0.290	0.235	164.6	0.063	*19.0*	21.2	1.333	*1.500*	7	4
P-O-13	0	0.400	0.338	147.6	0.102	16.4	15.3	1.556	1.667	6	6
P-O-12	0	0.224	0.197	157.2	0.058	*19.0*	22.1	1.272	*1.500*	5	4
P-O-16	0	0.434	0.332	158.2	0.070	18.0	18.1	1.261	1.254	9	6
P-O-17	1	0.474	0.305	156.1	0.077	18.9	18.5	1.417	1.447	10	6
P-O-1	0	0.615	0.291	151.1	0.082	18.8	19.3	1.575	1.534	10	6
P-O-19	1	0.327	0.220	158.9	0.064	*19.0*	16.2	1.031	*1.500*	6	4
P-O-18	0	0.323	0.228	157.9	0.063	*19.0*	19.5	1.233	*1.500*	7	4
P-O-27	0	0.438	0.270	162.9	0.065	18.3	20.9	1.358	1.189	9	5
P-O-25	1	0.498	0.311	175.6	0.078	18.2	20.9	1.634	1.423	9	5
P-O-29	1	0.293	0.364	154.9	0.064	17.4	15.3	0.975	1.109	5	6
P-O-2	1	0.497	0.308	165.8	0.080	19.5	20.0	1.598	1.558	10	6
P-O-8	1	0.554	0.531	135.9	0.094	12.0	15.1	1.420	1.128	10	7
P-O-6	0	0.611	0.375	154.2	0.080	15.9	19.6	1.576	1.279	12	6
P-O-5	0	0.527	0.325	155.6	0.066	20.9	19.5	1.286	1.379	10	7
P-O-4	0	0.518	0.291	161.5	0.080	20.4	20.4	1.623	1.623	11	6
P-O-3	0	0.665	0.409	160.3	0.065	16.3	17.9	1.170	1.066	12	7
P-O-30	0	0.418	0.268	153.8	0.092	21.2	17.1	1.571	1.948	8	5
P-O-9	0	0.435	0.238	148.7	0.069	20.0	19.3	1.325	1.374	8	5
P-J-10	0	0.551	0.312	154.4	0.094	15.7	17.5	1.637	1.469	10	5
P-J-13	0	0.551	0.381	153.2	0.097	*16.4*	16.4	1.593	*1.591*	11	4
P-J-11	0	0.407	0.255	160.8	0.066	20.3	21.2	1.409	1.349	9	5
P-J-14	0	0.458	0.258	163.0	0.080	18.8	18.8	1.501	1.501	8	5
P-J-15	0	0.510	0.343	151.9	0.091	17.5	22.0	2.005	1.595	10	6
P-J-1	1	0.618	0.473	153.8	0.117	16.1	19.0	2.230	1.890	10	7
P-J-19	1	0.442	0.375	155.4	*0.092*	17.6	19.9	*1.688*	*1.591*	10	6
P-J-18	0	0.402	0.370	155.5	*0.092*	17.7	17.7	*1.688*	*1.591*	9	6
P-J-17	0	0.409	0.261	157.2	0.077	*16.4*	19.6	1.504	*1.591*	9	4
P-J-16	0	0.457	0.262	149.3	0.071	*16.4*	19.0	1.352	*1.591*	9	4
P-J-1	1	0.631	0.458	156.4	0.113	16.1	19.5	2.195	1.812	9	7
P-J-20	1	0.517	*0.338*	*155.3*	0.099	15.6	18.5	1.823	1.537	9	5
P-J-24	0	0.408	0.306	162.0	0.072	17.7	18.2	1.315	1.279	8	6
P-J-23	1	0.163	0.141	177.9	0.058	*16.4*	*18.5*	*1.688*	*1.591*	4	2
P-J-22	1	0.220	0.141	162.4	0.049	*16.4*	21.3	1.053	*1.591*	5	3
P-J-3	0	0.676	0.329	148.4	0.119	15.9	16.8	1.998	1.891	10	6
P-J-8	1	0.499	0.343	160.7	0.109	18.1	18.6	2.022	1.968	10	5
P-J-7	0	0.608	0.315	145.7	0.087	15.4	19.6	1.713	1.346	11	5
P-J-6	0	0.603	*0.338*	*155.3*	0.135	13.3	12.7	1.714	1.795	7	5
P-J-5	0	0.817	0.453	138.4	0.142	15.1	16.9	2.393	2.139	12	8
P-P-10	0	0.197	0.174	160.1	0.051	*17.4*	22.5	1.156	*1.767*	5	3
P-P-11	0	0.199	0.152	164.7	0.055	*17.4*	*14.3*	*1.432*	*1.767*	4	3
P-P-15	1	0.482	0.394	153.4	0.111	14.2	13.9	1.543	1.576	8	5
P-P-16	0	0.153	0.192	162.9	0.061	27.3	14.3	1.432	1.655	3	5
P-P-18	0	0.594	0.436	143.3	0.095	13.5	12.7	1.202	1.280	9	6
P-P-19	1	0.473	0.386	147.9	0.095	14.3	14.1	1.342	1.362	8	6
P-P-1	1	0.597	0.464	165.0	0.111	14.9	15.0	1.671	1.662	9	7
P-P-25	0	0.159	0.129	159.0	0.056	*17.4*	*14.3*	*1.432*	*1.767*	3	3
P-P-26	1	0.342	0.449	134.3	0.096	18.2	14.0	1.349	1.752	5	8
P-P-32	1	0.349	0.312	156.6	0.084	17.9	15.3	1.276	1.498	7	5
P-P-33	0	0.528	0.482	136.0	0.116	13.3	12.7	1.476	1.536	9	7
P-P-37	0	0.383	0.236	158.8	0.083	20.0	19.1	1.576	1.648	8	5
P-P-38	1	0.378	0.376	159.7	0.087	15.6	13.4	1.173	1.364	6	5
P-P-35	1	0.547	0.369	158.6	0.091	14.8	15.5	1.404	1.346	8	6
P-P-3	0	0.416	0.323	146.9	0.083	15.9	14.6	1.216	1.330	7	6
P-P-6	0	0.214	0.191	162.0	0.060	*17.4*	*14.3*	*1.432*	*1.767*	4	3

P-F-12	1	0.316	0.305	159.7	0.085	19.3	14.6	1.248	1.650	5	5
P-F-13	1	0.244	0.237	165.1	0.066	*16.5*	19.0	1.263	*1.660*	5	8
P-F-15	1	0.532	*0.446*	*158.3*	0.102	16.0	13.8	1.404	1.627	7	7
P-F-17	1	0.352	0.383	168.2	0.082	16.7	14.3	1.168	1.364	6	6
P-F-18	0	0.370	0.294	159.6	0.097	19.3	14.5	1.399	1.863	6	5
P-F-19	1	0.648	0.572	157.1	0.095	13.2	14.0	1.328	1.252	10	7
P-F-20	0	0.520	0.388	144.2	0.098	14.7	13.4	1.307	1.434	7	6
P-F-21	0	0.479	0.332	165.5	0.112	34.3	12.8	1.440	3.859	6	5
P-F-22	0	0.593	0.481	153.5	0.080	14.2	12.9	1.026	1.130	9	7
P-F-1	0	0.781	0.555	159.4	0.125	12.7	10.9	1.364	1.590	9	6
P-F-24	0	0.575	*0.446*	*158.3*	0.083	*16.5*	12.7	1.054	*1.660*	8	7
P-F-26	1	0.439	0.466	165.4	0.094	14.7	14.2	1.341	1.388	6	6
P-F-27	1	0.329	0.304	159.2	0.061	16.7	14.4	0.879	1.019	5	5
P-F-28	0	0.452	*0.446*	*158.3*	0.074	*16.5*	13.4	0.996	*1.660*	7	7
P-F-2	0	0.623	0.457	158.8	0.103	15.6	14.1	1.455	1.609	9	7
P-F-30	0	0.315	*0.446*	*158.3*	0.080	*16.5*	15.9	1.278	*1.660*	6	7
P-F-5	1	0.380	0.398	171.5	0.082	16.6	16.4	1.351	1.367	7	6
P-F-6	1	0.588	0.561	162.6	0.110	12.3	12.1	1.327	1.349	8	7
P-F-7	1	0.491	0.411	166.0	0.067	13.7	13.6	0.912	0.919	7	6
P-F-9	1	0.507	0.481	158.1	0.114	15.5	13.1	1.497	1.771	9	7
P-I-13	1	0.469	0.368	171.4	0.098	14.3	12.4	1.214	1.399	6	6
P-I-16	1	0.528	0.543	159.9	0.098	15.1	14.1	1.376	1.473	8	7
P-I-15	1	0.467	0.485	157.5	0.111	15.3	13.9	1.543	1.699	8	7
P-I-17	0	0.465	0.414	153.1	0.075	14.3	13.1	0.983	1.073	7	6
P-I-18	1	0.548	0.459	157.0	0.094	14.9	12.0	1.132	1.406	8	7
P-I-20	0	0.494	0.460	154.5	0.105	15.5	13.5	1.412	1.621	8	7
P-I-1	1	0.537	0.505	153.2	0.109	15.6	13.2	1.442	1.705	8	8
P-I-19	0	0.448	0.399	151.3	0.089	15.0	12.6	1.124	1.338	6	6
P-I-25	1	0.359	0.286	168.0	0.074	*15.5*	14.5	1.080	*1.571*	6	4
P-I-28	1	0.621	0.598	154.1	0.090	13.6	12.5	1.121	1.219	8	7
P-I-27	0	*0.497*	0.495	*152.3*	0.087	13.0	14.1	1.223	1.127	8	7
P-I-26	0	0.602	0.484	147.9	0.085	15.9	12.3	1.050	1.357	8	8
P-I-2	1	0.606	0.575	166.5	0.113	11.6	12.5	1.415	1.313	8	7
P-I-25	0	0.429	0.308	159.0	0.081	16.9	12.8	1.036	1.368	6	5
P-I-30	1	0.405	0.368	166.5	0.085	15.7	14.9	1.270	1.338	7	6
P-I-8	1	0.409	0.343	168.9	0.077	20.0	18.7	1.433	1.533	8	8
P-I-5	1	0.435	0.442	160.4	0.096	15.5	15.1	1.447	1.485	7	7
P-I-4	0	0.404	0.382	144.8	0.090	18.1	14.2	1.281	1.633	7	4
P-I-3	0	0.726	0.559	146.4	0.101	11.6	13.2	1.334	1.173	10	7
P-I-31	0	0.584	0.455	154.4	0.099	14.8	14.5	1.431	1.460	9	7
P-T-10	1	0.637	0.513	142.4	0.158	13.5	11.1	1.751	2.134	8	7
P-T-11	0	0.493	0.460	163.7	0.128	16.1	13.4	1.717	2.057	7	7
P-T-16	0	0.655	0.518	129.2	0.105	15.0	10.8	1.135	1.567	8	7
P-T-13	1	0.565	0.503	146.6	0.106	14.8	12.9	1.363	1.557	8	8
P-T-18	1	0.456	0.467	163.8	0.108	14.7	12.1	1.300	1.589	7	6
P-T-19	1	0.437	0.474	162.8	0.096	12.8	13.7	1.312	1.226	7	6
P-T-23	1	0.565	0.480	143.7	0.108	12.9	10.6	1.143	1.389	7	6
P-T-22	0	0.516	0.465	131.7	0.113	11.7	9.7	1.093	1.318	6	5
P-T-20	1	0.497	0.422	156.1	0.106	15.5	13.6	1.431	1.640	7	7
P-T-1	0	0.399	0.412	154.9	0.094	15.5	13.0	1.229	1.464	6	6
P-T-24	1	0.545	0.477	136.4	0.120	15.4	12.7	1.530	1.852	8	8
P-T-2	0	0.464	0.414	146.1	0.098	16.2	11.5	1.128	1.588	6	7
P-T-28	0	0.567	*0.495*	*143.6*	0.099	14.3	13.5	1.346	1.422	9	7
P-T-26	0	0.709	*0.495*	*143.6*	0.148	13.7	12.2	1.804	2.036	10	7
P-T-30	1	0.700	*0.495*	*143.6*	0.112	*14.4*	11.9	1.328	*1.670*	9	7
P-T-31	1	0.480	0.620	132.4	0.131	14.5	11.2	1.475	1.903	6	9
P-T-32	0	0.499	0.369	138.3	0.107	14.0	10.7	1.149	1.495	6	5
P-T-7	0	0.658	0.515	145.6	0.159	13.4	11.8	1.881	2.138	8	7
P-T-6	0	0.415	0.350	148.5	0.095	13.9	12.7	1.202	1.310	5	5
P-T-3	0	0.485	*0.495*	*143.6*	0.105	15.0	11.7	1.223	1.568	6	7
P-U-16	0	0.563	0.509	150.4	0.122	12.2	15.5	1.893	1.487	8	8
P-U-14	0	0.578	0.490	137.6	0.102	12.1	15.9	1.621	1.232	8	8
P-U-12	0	0.653	0.590	141.8	0.103	12.9	11.6	1.192	1.323	9	8

P-U-11	1	0.601	*0.549*	*147.8*	0.117	12.4	11.4	1.334	1.453	7	*8*
P-U-23	1	0.483	*0.549*	*147.8*	0.103	*0.1*	13.0	1.338	*1.599*	7	*8*
P-U-21	0	0.575	*0.549*	*147.8*	0.131	*0.1*	12.1	1.581	*1.599*	8	*8*
P-U-1	1	0.494	0.539	*147.8*	0.140	13.7	16.2	2.268	1.920	8	8
P-U-24	1	0.540	*0.549*	*147.8*	0.114	*0.1*	11.6	1.317	*1.599*	7	*8*
P-U-25	1	0.404	0.462	160.3	0.097	15.7	14.8	1.431	1.513	7	7
P-U-2	1	0.607	0.646	143.1	0.097	14.6	13.4	1.303	1.418	10	10
P-U-29	0	0.589	*0.549*	*147.8*	0.122	*0.1*	13.5	1.649	*1.599*	*8*	9
P-U-28	0	0.458	*0.549*	*147.8*	0.103	*0.1*	14.9	1.531	*1.599*	7	*8*
P-U-27	0	0.570	*0.549*	*147.8*	0.106	*0.1*	12.5	1.328	*1.599*	7	*8*
P-U-26	0	0.607	0.493	150.2	0.104	12.9	10.5	1.098	1.346	8	7
P-U-30	1	0.484	0.519	160.0	0.119	13.9	13.2	1.576	1.653	8	6
P-U-8	0	0.471	0.550	156.8	0.117	13.7	11.8	1.378	1.604	7	7
P-U-6	0	0.589	0.586	139.9	0.106	14.3	11.6	1.229	1.508	7	9
P-U-4	1	0.514	0.529	167.6	0.109	14.3	13.0	1.420	1.568	7	7
P-U-3	1	0.577	0.501	151.5	0.120	11.5	10.5	1.253	1.373	7	6
P-U-9	1	0.505	0.531	151.6	0.131	12.9	14.4	1.889	1.691	7	6

Subject Index

A

Active sensor(s) [see Sensor(s): active]
Adanson, Michel ix
Adaptation(s) 262
Adaptive resonance theory [see Neural network(s): adaptive resonance theory]
Adaptivity 48
Adult-directed speech (ADS) 300-308
Agriculture and Food Research Council 102
Agricultural management [see Management: agricultural]
Agglomerative clustering [see Clustering: agglomerative]
Aitchison, John xi
Aleksander, Igor 104
Algorithmic Approaches to the Identification Problem in Systematics symposium 4
Amateurs 20
Amino acid 17
Ampliture probability density function(s) 89
Ancestor(s): human 20
Anderson, T.W. xi
Animals 278
Antenna base 170, 172, 173
Anthropology: physical 69
Anthropometrics 70
Aphid(s) 92
Applied taxonomy [see Taxonomy: applied]
Arachnids 132
Arthropods 102, 194
Artificial
  intelligence 116, 154
  neural networks (ANN) [see Neural network(s): artificial (ANN)]
Autecology 92
Autocorrelation 89
Automatic
  Bee Identification System (ABIS) 5, 102, 116-128, 148, 183
  Diatom Identification and Classification (ADIAC) 36, 41
Automated
  systems [see Systems: automated]
  taxon identification (ATI) 83,93
  taxonomy [see Taxonomy: automated]

B

Backpropagation 54, 107, 162, 209
cascade correlation 138
Bacteria 133
Bagging 201
Barnard, M.M. xi
Basswoord(s) 209
Bayes' Theorem [see Theorem: Bayes']
Bayesian
  functions [see Functions: Bayesian]
  logic 26
  matting [see Matting: Bayesian]
Bee(s) x, 116-128
  bumblebee(s) 108
  honeybee(s) 5, 92, 289
  solitary 92
Beetle larvae [see Larvae: beetle]
Bending energy 75
Berlese funnel extraction 195
Biodiversity 10, 194
  assessment 84, 153
  crisis 21
  insect 192
Biogeography 289
Bioinformatics 18, 286
Biology
  developmental 20
  evolutionary 17
BioloMICS 281-286
Biometric(s)
  multivariate 1
  systems [see Systems: biometric]
Biomonitoring 192
Bio-NET–INTERNATIONAL Group for Computer-aided Taxonomy 84
Biosphere 19
Biostratigraphy 240
Birds 21
Blackith, R.E. 70
Biodiversity: soil 190
Bootstrap [see Statistics: bootstrap]
Botanists 207
Brachial cells [see Cell(s): brachial]
Brachiopod(s) x, xi
Braconid wasps [see Wasp(s): braconid]
Bray-Curtis similarity [see Similarity: Bray-Curtis]
Browser(s) [see Feeding: browser(s)]
Bumblebees [see Bees: bumblebee(s)]

C

Camera(s)
  digital 290
  video (NTSC) 290
  video (PAL) 290
Caminacules 33
Campbell, Norman A. x
Cane toad(s) 91

Canonical variates analysis (CVA) 156-157, 160, 161, 167, 169, 171, 172, 176, 177, 182, 229, 230, 250, 303, 305
Carbon dioxide (CO_2) 193
Cell(s)
    brachial 119
    cubital 120,121
    discoidal 119
    radial 121
Character(s) 212
    coding 218
    continuous 210
    distributions 14
    transformation(s) 16
Charismatic groups [see Group(s): charismatic]
Chemical pollution [see Pollution: chemical]
China ix
Clade(s) 16, 21
Cladistic(s) 84
    analyses 18
Cladogram(s) 153
Classification(s) 10, 13, 48, 51, 62, 195
    first-past-the-post (FPTP) 104,108
    nearest neighbour 103
    normalized vector difference (NVD) 104, 169
    phenetic 2, 281
    two-dimensional 196
Classifier(s)
    continuous *n*-tuple 138, 169
    scalable 92
Cluster
    analysis 90
    threshold 65,66
Clustering 48, 51, 62, 280
    agglomerative 284
    set 202
Clypeus 170, 173
Coefficient(s)
    linear predictive 89
Colour images [see Image(s): colour]
Collections 15, 19
Comb filters [see Filter(s): comb]
Comparative
    ethology [see Ethology: comparative]
    morphology [see Morphology: comparative]
Compound eye 170, 172, 173
Computational anatomy 69
Computed homology [see Homology/homologies: computed]
Computer-aided
    identification systems [see System(s): computer-aided identification]
    keys [see Key(s): computer-aided]
    taxonomy [see Taxonomy: computer-aided]
Computer-assisted tomography (CT) 12, 87, 111
Computer
    scientists 190
    vision 154
Confidence 49
Confusion matrix 37
Conodont(s) 5, 240-256
    Taxonomy [see Taxonomy: conodont]
Conservation monitoring [see Monitoring: conservation]
Continuous characters [see Character(s): continuous]
Contour tracing 293
Converter
    analogue-to-digital 86
Copepod(s) 31
Cornell University 19
Correspondence analysis 126
Cramer's V 303
Cross-correction 89
Cubital cells [see Cell(s): cubital]
Curse of dimensionality 181
Curve(s) 69,70
Cyanobacteria 93
Cyberinfratructure 13, 18, 21

D

DAISY (see Digital automated identification system [DAISY])
    BUILDTOOL 227
    JACKTOOL 227, 303
    nearest neighbour classifier (NNC DAISY, DAISY II) 104,105-109
Darwin initiative 103
Data mining 264
Daubechies 4 wavelet function [see Function(s): Daubechies 4 wavelet]
de Buffon, Georges Louis Leclerc ix
Decapod larvae [see Larvae: decapod]
Decision
    theory 90
    trees (DT) 201, 262, 263, 265, 268-269, 270
        boosted 203
Delphi protocol 38
DELTA key generation software 207, 212
Descriptive taxonomy [see Taxonomy: descriptive]
Descriptor(s)
    bagged local region 203
    vector [see Vector(s): descriptor]
Detector(s)
    region 197
Development 289
Developmental Biology [see Biology: developmental]
Diatom(s) 92
Dibitag 270, 274
Dichotomous keys [see Key(s): dichotomous]
Digital
    automated identification system (DAISY) 5, 101-112, 148, 169, 171, 174-180, 183, 226-236, 301, 304-305
    camera(s) [see Camera(s): digital]
Dinoflagellate Identification by Artificial Neural Network (DiCANN) 36
Dinoflagellate(s) 36, 92
Dinosaur(s) 20, 21, 48-67
Discoidal cells [see Cell(s) discoidal]

Discriminant analysis
linear (LDA) 0, 119, 120, 122, 123, 156-157, 160, 161, 181, 263, 264
non-linear kernel (NKDA) 122,123,124
Distance(s)
Euclidean 59,63,64,65,66
Mahalanobis 167, 263
matrix [see Matrix: distance]
Procrustes 71, 167, 168, 171, 174, 176, 177
Diversity: taxonomic 14
DNA 1, 13, 33, 154, 277, 278, 280, 283
barcoding 20, 153
sequences 18
taxonomy [see Taxonomy: DNA-based]
Domesticated cow(s) 91
DrawWing 291, 294
Duiker 267
blue 266
Dynamic neural network(s) [see Neural network(s): dynamic]

E

E-publication 22
Echo size 89
Ecological
theory of perception 26
guilds [see Guild(s): ecology]
monitoring [see Monitoring: ecological]
Ecology 289
Ecophenotypy 241, 242, 254, 255
Ecosystem
complexity 10
science 20
Edge elements (edgels) 118,120
Eigenpitch 301
Eigenshape analysis (EA) 5, 301, 303, 305-306, 308
extended 5, 228
standard 5
Eigenvalues 120
Electronic monographs [see Monographs: electronic]
Emotional pet-directed speech (PDS) 300
Empiricism 18
Encoding 50
Endo, Kazuyoshi xi
Endoparasitoids: koinobiont 165
Entomologists 190
Epigyum 134, 135, 148
Epistomal sulcus 170, 173
Ethology
comparative 20
Euclidean distance(s) [see Distance(s): Euclidean]
Eutrophication 192
Evolution
macro-12
micro-12
Evolutionary
biology [see Biology: evolutionary]
history 10
trends 240
Expectation-maximization (EM) algorithm 200
Expert systems [see System(s): expert]
Extended eigenshape analysis [see Eigenshape analysis: extended]

F

Fallow deer 91
False
feature hypotheses 119
killer whale 91
Fast Fourier Transform (FFT) (see Transform: Fast Fourier [FFT])
Fault tolerance 49
Feature extraction 190
Feeding
browser(s) 261, 268
grazer(s) 261
high-level browser(s) 265
mixed 265, 268, 269, 275
omnivorous 265
selective 265
strategies 261
Filter(s)
comb 87
Gabor 139
Gaussian 122
First-past-the-post (FPTP) metric [see Metric: first-past-the-post (FPTP)]
Fisher, Ronald A. xi
Focus 58-59, 63-64,66
Foraminifera
planktonic 109
Foreigner-directed speech (FDS) 300-308
Forest management [see Management: forest]
Forested habitats [see Habitats: forested]
Fossils 17
Fourier shape coefficients [see Shape coefficients: Fourier]
Function(s)
Bayesian 90
Daubechies 4 wavelet 138, 142, 144
Gabor wavelet 138, 142
radial basis 90
squashing 209
Fungi 14, 20, 278

G

Gabor
Filter(s) [see Filter(s): Gabor]
masks [see Mask(s): Gabor]
wavelet function [see Function(s): Gabor wavelet]
Gaston, Kevin 101,102,103
Gauld, Ian 101,102,103
Gaussian
Filters [see Filter(s) Gaussian]
kernel 117
mixture model (GMM) 200, 202

Geometry
 Minikowski 74
GenBank 4
Gene(s) 18
 orthologous coding 277
Generalization 48
Genetics 289
Genomics 20
Genotype 17
Genus template [see Template: genus]
Geometry
 Minikowski 74
Geometric morphometrics [see Morphometrics: geometric]
Giant forest hog 266
Global Biodiversity Information Facility (GBIF) 19, 20
Google Earth 283
Grazer(s) [see Feeding: grazer(s)]
Groups
 charismatic 132
Growing neural gas with utility (GNG-U) 61
Guild(s)
 ecological 261
Gunnerson's prairie dogs 91

H

Habitat(s)
 forested 265
Harmful Algal Bloom (HAB) Buoy 37, 38, 41
Hennig, Willi 10, 12
Hidden layer [see Layer(s): hidden]
High-level browser(s) [see Feeding: high-level browser(s)]
Holomorphy 10
Homology/homologies 11, 16, 18, 78, 243, 244
 computed 70,78
Honeybee(s) [see Bees: honeybee(s)]
Human Genome Project 1, 16
Humerus 225
Hyperdiverse taxa [see Taxa: hyperdiverse]
Hypertext keys [see Key(s): hypertext]
Hypsodonty Index (see Index: hypsodonty)

I

Ichnumonid wasps [see Wasp(s): ichnumonid]
Identification 13
 multifactorial 286
 polyphasic 284
 single-sided 286
Image(s)
 histogram 89, 291
 segmentation 26
 colour 26, 116
 matting [see Matting: image]
 monochomatic 26, 116
 spin 197
Integrated taxonomy [see Taxonomy: integrated]
Index
 Hypsodonty 265, 267, 269, 275
Indian muntjac 266
Infant-directed speech (IDS) 299, 301-308
Infants
 prelingual 300
Information technologies 9
Input
 dimension 52
 layer [see Layer(s): input]
 vector [see Vector(s): input]
Insect(s) 14, 20, 102, 116, 132, 190, 289
Instituto Nacional de Biodiversidad (INBio) 132
Internet 4, 93
Interpolation theory 73
Intrinsic
 random field (IRF) 70,73,77,78,79
 warp(s) [see Warp(s): intrinsic (IW)]

J

Jackknife [see Statistics: jackknife]
Janzen, Daniel 1, 102, 103
Java servlets [see Servlets: Java]
Jurassic 21

K

Kaesler, Roger 3
Kendall
 shape manifold 171, 174
 David 70
Kendall's tau 104
Kent, John 69,73
Key(s) 279
 artificial neural network 209
 computer-aided 84
 dichotomous 132, 279
 hypertext 132
 multiple entry (MEK) 280, 281, 283, 284
 taxonomic 207

L

Landmark point(s) 69, 70, 161, 226
Larvae
 beetle 85
 decapod 31
 parasitic 133
 stonefly 190-204
Laser vibrometer 85
Lateral ocellus [see Ocellus: lateral]
Layer(s)
 hidden 56, 208
 input 208
 output 54-55, 208
Learning rate 54, 58, 63
Learning machine 262
Learning
 supervised 51, 162
 unsupervised 51, 90, 162

Linden(s) 209
Linear distance measurements [see Measurement(s): linear distance]
Linearity 103
Link-ageing parameter 62,66,67
Linnaeus 20, 32
Linné, Carl von ix (see Linnaeus)

M

Machine learning [see Learning: machine]
Macro-evolution [see Evolution: macro-]
Mahalanobis distance(s) [see Distance(s): Mahalanobis]
Malar area 170, 173
Mammals 21
Management
  agricultural 192
  forest 192
MANOVA
  non-parametric 250
Mardia, Kanti 69,73
Mardia-Dryden
  distribution 71
  models 71
Marine zooplankton [see Zooplankton: marine]
Marmot(s) 169
Mask(s)
  Gabor 139
Matheron, Georges x, 73
Matrix
  distance 280
  similarity 280
Matting
  Bayesian 196
  image 195
Maximum likelihood 169
Measurement(s)
  linear distance 226
Mechanical engineers 190
Medial ocellus [see Ocellus: medial]
Megafauna 194
Metadata 92
Metric
  coordination 170
  first-past-the-post (FPTP) 170
  Procrustes distance 69, 174
  SILL 170
Micro-evolution [see Evolution: micro-]
Micro-organisms 133, 194
Minikowski geometry (see Geometry: Minikowski)
Mixed feeding [see Feeding: mixed]
Molecular diagnostics 13
Moments 89
Momentum 213
Monitoring
  conservation 84
  ecological 84
  water quality 190
Monochromatic image(s) [see Image(s): monochromatic]
Monographs 12, 15
  electronic 15
Monophyletic groups 14
Monophyly 11
Montage 195
Monte Carlo simulations 158
MorphBank 12, 165
MorphoBank 12, 165
MorphoBox 124
Morphological divergence 255
Morphologists 18
Morphology
  comparative 17
Morphometric synthesis 70, 155
Morphometrics 5
  geometric x, 2, 69, 155
Morphospace(s)
  theoretical 242, 262
Moths
  sphingid 110
Mouse deer 266
Multifactorial identification [see Identification: multifactorial]
Multilayer(ed) perceptron (MLP) [see Neural network(s): multilayer(ed) perceptron (MLP)]
Multiple entry keys (MEK) [see Key(s): multiple entry (MEK)]
Museums 14

N

n-tuple classifier [see Classifiers: continuous *n*-tuple]
National
  Biodiversity Institute, Costa Rica (INBIO) 103
  Environmental Research Council (NERC) 16
  Science Foundation (NSF) 19
Nearest neighbour classification [see Classification: nearest neighbour]
Neighbourhood 59
Neighbourhood size 58, 59
Network(s)
  Bayesian 262
Neural network(s) x, 5, 160, 182
  plastic self-organizing map (PSOM) 5, 61-67, 164
  self-organizing map (SOM) 51, 57-60, 67, 90, 92, 104, 162, 163, 167, 182
  adaptive resonance theory 90
  artificial (ANN) 47, 90, 133, 134, 137, 138, 147, 208, 212, 215, 226-236, 279, 307
  multilayer(ed) perceptron (MLP) 5, 51-57, 58, 61, 67, 90, 107, 162, 208-210
  time-delay 90
  unsupervised, artificial (uANN) 308
  dynamic 47, 61-67
  fully connected 52
  static 48, 51-67
  traditional 47, 51-67
Neuron activation 52
Neuron(s) 51-67, 138, 141, 163
  output 141

Node-building parameter 62, 66
Non-parametric MANOVA [see MANOVA: non-parametric]
Nonlinearity 48
Normalized vector difference (NVD) [see Classification: normalized vector difference (NVD)]
Numerical taxonomy [see Taxonomy: numerical]
Nutrient cycling 194
Nyquist frequency 85-86

O

Object-class recognition [see Recognition: object-class]
Ocellus
  lateral 170, 179
  medial 172, 179
Omnivore(s) [see Feeding: omnivorous]
Ontogeny 17, 240, 242, 255
Ontology 116
Operational taxonomic unit(s) (OUT) 280, 281
Operator(s): interest 197
Outline(s) 161, 226, 244, 292, 295
Output
  layer [see Layer(s): output]
  vector [see Vector(s): output]
Over-fitting 49
Overpopulation x
Overtraining 209

P

P_1 element(s) 242, 253, 255
Palaeontology 3, 17
Paleocene
  Early 225
Parasitic
  larvae [see Larvae: parasitic]
  wasps [see Wasp(s): parasitic]
Parataxonomist(s) 83, 132
Partial
  least squares (PLS) analysis 308
  warp(s) [see Warp(s): partial]
Passive sensor(s) [see Sensor(s): passive]
Pattern recognition 90, 93, 154, 167, 190, 203
Peccary
  Chacoan 272, 274
Peer review 3
Penguins 225-236
Pentland, Alex (Sandy) 102
Personal digital assistant (PDA) 111
Phenetics [see Classification: phenetic]
Phenotype 17
Phylogenetic
  analysis 10, 226
  inference 153
  patterning 263
Phylogenetics 17, 22
Phylogeny/phylogenies 15, 17, 18
Phytoplankton 133
Pipesnake(s) 169
Pitch contour
  analysis 299
  shape [see Shape: pitch contour]
Planetary Biodiversity Inventory (PBI) projects 19
Planktonic foraminifera [see Foraminifera: planktonic]
Plants 278
Plastic self-organizing map (PSOM) [see Neural network(s): plastic self-organizing map (PSOM)]
Points of interest (PoIs) 124,125,126
Pollen 83, 92, 110
Pollution
  chemical 192
  thermal 192
  tolerance 192
Polymorphism xi, 241
Polyphasic identification [Identification: polyphasic]
Pop-out 28, 29, 31
Praat 301, 308
Prelingual infants (see Infants: prelingual)
Principal
  component analysis (PCA) 102, 120, 301, 157-161, 248, 250, 262
  coordinate analysis (PCoA) 281
  warp(s) [see Warp(s): principal]
Procrustes
  distance(s) [see Distance(s): Procrustes]
    metric [see Metric Procrustes distance]
  shape coordinates [see Shape Coordinates: Procrustes]
Proteins 154
PubMed 283

R

Radial basis function(s) [see Functions: radial basis]
Radial cells [see Cell(s): radial]
Ragozin, David 73
Raven, Peter 9
rDNA 277
Recognition accuracy (Rtest) 216-217
Recognition
  object-class 203
Recognizable taxonomic unit (RTU) 87,88
Reeves' muntjac 266
Reference vector 57
Region detector(s)
  Harris-affine 198
  Kadir 198
Riemannian shape space [see Shape space: Riemannian]
Relationships
  ancestor-descendant 241
Remote sensing 190
Research
  psychological 299
Revisions
  taxonomic 14
Reyment, Richard 70
Rieppel, Olivier 18

Roe deer 91
Royal Botanic Gardens, Kew 210, 212

S

Saccades 27
Scalability 139
Scale-invariant feature transform (SIFT)
 transform [see Transform(s): scale-invariant feature]
 vectors [see Vector(s): Scale-invariant feature transform (SIFT)
Scanner(s) 290
Scanning electron microscope (SEM) 83, 87, 139
Science fiction 1
Sedimentation 192
Segmentation: automated 195
Selective feeding [see Feeding: selective]
self-organizing map (SOM) [see Neural network(s): self-organizing map (SOM)]
Sensor(s)
 active 87
 passive 87
Servlets
 Java 142
Shape
 coefficients
  Fourier 196
  Wing [see Wing shape]
 coordinates 167
  Procrustes 181
 indices 89
 space
  Riemannian 70
 pitch contour 300, 301
 theory [see Theory: shape]
Sharp, David 11
Signal-to-noise ratio (SNR) 87
Similarity
 Bray-Curtis 36
 Matrix [see Matrix: similarity]
Simple modelling by class analogy (SIMCA) 160, 161, 167, 169
Single-sided identification [see Identification: single-sided]
SIPPER 37, 41
Sneath, Peter ix
Soil mesofauna 190-191, 194, 203
Sokal, Robert ix
Solitary bees [see Bees: solitary]
Sound waveforms 300
Squashing function(s) [see Function(s): squashing]
Species Identification, Automated (SPIDA) 5, 102, 165, 183, 133-150
Sphingid moths [see Moths: sphingid]
SPIDA-Web 133-150, 165
Spider(s) 92, 133-150
 Australasian ground 133, 134
 lycosid 133
Spin images [see Image(s): spin]
Springtails
 entomobryid 195
Stag beetle(s) 91
Standard eigenshape analysis [see Eigenshape analysis: standard]
Static neural network [see Neural network(s): static]
Statistics
 bootstrap 158
 jacknife 110, 158
Stoneflies 190-204
Stonefly larvae [see Larvae: stonefly]
Stoneham, Graham 104
Stopping criteria 56
Stuttgart Neural Network Simulator 141
Superposition
 Generalized Least Squares (GLS) 171, 174, 176
Supervised learning [see Learning: supervised]
Support vector machine (SVM) 122, 124, 138
Surface(s) 69,70
Survivorship 240
Synapomorphy 11
Systematics 3, 5, 6, 25,153, 154, 155, 190, 220
 Association 4
 community 84
System(s)
 automated x
 biometric 69
 computer-aided identification 132
 expert 93, 132
 vision 25
Systematist(s) 1, 4, 6, 132, 154, 280

T

Tapir 267
Tarsometatarsus 225
Taxa
 hyperdiverse 20
Taxonomic
 impediment 3, 83
 keys [see Key(s): taxonomic]
 revisions 12
Taxonomist(s) ix, 9, 11, 15, 22, 83, 116
Taxonomy ix, 1, 10, 14, 15, 16, 17, 19, 21, 22, 69, 101, 117, 240
 applied 13, 20, 22
 automated x
 computer-aided 84
 conodont 256
 descriptive 12, 14, 17, 19
 DNA-based 11
 integrated 17
 numerical ix, x, 69, 84
 user 14
Template(s)
 genus 120, 121
 wing 117, 120
Tentorial pit 170, 172, 173
Texture 89
The Natural History Museum
 (London) 4, 210, 301
 (Tring) 227

Theorem
 Bayes' 280
Theoretical morphospace [see Morphospace: geometric]
Theory
 shape 155
Thermal pollution [see Pollution: thermal]
Thin-plate spline(s) 69, 70, 73, 75
Time-delay pulse 89
Timofeeff-Ressovsky, N.W. 12
Traditional neural network(s) [see Neural network(s): traditional]
Training set 50
Transform(s): scale-invariant feature (SIFT) 124, 125, 197
Transformation(s)
 affine 122
 Fourier 137
 Fast Fourier (FFT) 88
 wavelet 71, 88, 89, 137, 142
Tsukuba University, Japan xi
Turing, Alan 161
Turk, Matthew 102

U

Uddenberg, Nils ix
Ungulate(s) 265
 suiform 275
University of Costa Rica 103
Unsurpervised
 learning [see Learning: unsupervised]
 artificial neural network(s) (uANN) [see Neural network(s): unsupervised, artificial (uANN)]

V

Validation 60, 66
Validation set 50, 209
Vector(s)
 descriptor 197
 input 212
 output 141
 Scale-invariant feature transform SIFT 203
Video
 camera(s) [see Camera[s): video]
 Plankton Recorder (VPR) 36, 41
Vision systems [see Systems: vision]
von Neumann
 John 161
 machines 161

W

Wägele, 18
Warp(s)
 intrinsic (IW) 70, 75-77
 partial 70, 76, 77
 principal 70, 75
Warthog 273
Wasp(s) 123, 165-180
 braconid 92
 ichnumonid 92
 parasitic 92, 103
Water
 chevrotain 266
 quality 191
  monitoring [see Monitoring: water quality]
WAV format 301
Web: World Wide 4, 11, 12, 13, 15, 39
WEKA machine learning system 201
Williams, D. 18
Wilson, Edward O. 16, 17
Wing
 shape 110
 template [see Template: wing]
World Wide Web [see Web: World Wide]
Worms 14

X

Xerophytes 209

Y

Yeast(s) 133, 277

Z

Zooplankton 31
 marine 92
ZooSCAN 36, 41

Taxon Index

A

Adhemarius gannascus 107
Aedes aegypti 101
Aegeseucoela 177
Ammodorcas clarkei 270, 271, 272
Anastrepha 101
Andrena 123
Andricus gallaeurnaeformis 166
Anoplophora glabripennis 91
Apis mellifera carnica 293
Apis mellifera ligustica 293
Apocrita 165
Aptenodytes 227-236
Aptenodytes fosteri 227-236
Aptenodytes patagonicus 227-236
Aricia
 agestis 107
 artaxerxe 107
Artiodactyla 266
Ascomycetes 277
Aspicera scutellata 173

B

Biorhiza pallida 173
Bombus 109
 cryptarum 123,127
 lapidarius 109
 lucorum 109
 lucorum 123,127
 magnus 109, 123,127
 terrestris 109, 123,127
Bos taurus 91
Bufo marinus 91

C

Calinueria 202, 203
 californica 194, 201, 204
Callirhytis erythrocephala 180
Capreolus capreolus 91
Catagonus wagneri 272, 274
Cephalophus
 dorsalis 267
 monticola 266
Ceratium
 arcticum 34-35
 longipes 34-35
Chrestosema erythropum 166, 169
Colletes 123
Cymatocylis 42
Cynipidae 165, 166, 169
Cynipoidea 165
Cynomys gunnisoni 91

D

Dama dama 91
Desognaphosa 135, 136
 bartle 135, 136
 finnigan 135, 136
 halycon 136
 karnak 135, 136
 kuranda 134
 massey 134
 millaa 134
 windsor 136
 yabbra 134
Dinophysis
 acuminata 39, 40
 caudata 39, 40
 fortii 39, 38
 rotundata 39, 39
 sacculus 39, 41
 tripos 41
Diplodocus 48-67
Diplolepis triforma 166, 169
Diptera 92
Doroneuria 202, 203
 baumanni 194, 201, 204

E

Enicospilus 103
Euceroptres montanus 170, 171
Eudyptes 227-236
 chrysocome 227-236
 pachyrhynchus 227-236
 robustus 227-236
 schlegeli 227-236
 sclateri 227-236
Eudyptula 227-236
 minor 227-236
 minor albosignata 227-236

F

Figitidae 165, 166, 168, 169

G

Gonaspis potentillae 173

H

Hemicloena julatten 134
Hesperoperla 202, 203
  pacifica 194, 201
Hyemoschus aquaticus 266
Hyles 111
Hylochoerus meinertzhageni 267
Hylotrupes bajulus 91
Hymenoptera 83, 92
Hyracoidea 266

L

Lepidoptera 92
Lingula anatina xi
Lithops 209
Longrita
  insidiosa 134
  millewa 136
  yuinmery 136
Loxondonta africana 85
Lucanus cervus 91

M

Megadyptes antipodes 227-236
Melanips 170
  opacus 168, 173
Morebilus
  diversus 134
Morebilus
  fumosus 134
  plagusius 134
Muntiacus
  muntjak 266
  reevesi 266
Mylohyus sp. 274

N

Nerium oleander 102
Neuroterus numismalis 173

O

Orthoptera 91,93
Osmia 123
Ozarkodina excavata 240-256

P

Perissodactyle 266
Perlidae 202, 203
Phacochoerus aethiopicus 273
Phaenoglyphis villosa 170
Phanacis centaureae 173
Plagiotrochus quercusilicis 177
Platorish
  churchillae 136
  flavitarsus 136
  nebo 136
Plectopera 191, 196
Precis octavia 109
Prosthennops xiphodonticus 262, 265, 271, 273
Pseudorca crassidens 91
Pygoscelis 227-236
  adeliae 227-236
  antarctica 227-236
  papua 227-236

R

Rebilus
  bilpin 136
  brooklana 136
  bulburin 134,136
  credition 136
  lugubris 136
Rhizopertha dominica 91

S

Saccharomyces 278
Sphaeroidinella dehiscens 109
Sphenisciformes 227-236
Spheniscus 227-236
  demersus 227-236
  humboldti 227-236
  magellanicus 227-236
  mendiculus 227-236
Stegosaurus 48-67
Synergus crassicornis 173

T

Tilla 209, 210-223
  americana 209-223
  amurensis 210-223
  caroliniana 210-223
  chinensis 210-223
  cordata 210-223
  dasystyla 210-223
  henryana 210-223
  heterophylla 210-223
  insularis 210-223
  japonica 210-223
  kiusiana 210-223
  mandshurica 210-223

maximowicziana 210-223
miqueliana 210-223
mongolica 210-223
neglecta 209-223
oliveri 210-223
platyphyllos 210-223
tomentosa 210-223
tuan 210-223
Trachyrema
castaneum 136
allyn 134
garnet 134,136
sculptilis 134
Tragulus
javanicus 266
memmina 266
napu 266
Trochanteriidae 133, 134, 138, 142, 144
Tyrannosaurus rex 48-67

X

Xylophanes
cryptolibya 110
libya 109, 110
loelia 109
titania 107

Y

Yoraperla 194, 201

Z

Zaeucoila 173

Systematics Association Publications

1. Bibliography of Key Works for the Identification of the British Fauna and Flora, 3rd edition (1967)†
   Edited by G.J. Kerrich, R.D. Meikie and N. Tebble
2. Function and Taxonomic Importance (1959)†
   Edited by A.J. Cain
3. The Species Concept in Palaeontology (1956)†
   Edited by P.C. Sylvester-Bradley
4. Taxonomy and Geography (1962)†
   Edited by D. Nichols
5. Speciation in the Sea (1963)†
   Edited by J.P. Harding and N. Tebble
6. Phenetic and Phylogenetic Classification (1964)†
   Edited by V.H. Heywood and J. McNeill
7. Aspects of Tethyan Biogeography (1967)†
   Edited by C.G. Adams and D.V. Ager
8. The Soil Ecosystem (1969)†
   Edited by H. Sheals
9. Organisms and Continents through Time (1973)†
   Edited by N.F. Hughes
10. Cladistics: A Practical Course in Systematics (1992)*
    P.L. Forey, C.J. Humphries, I.J. Kitching, R.W. Scotland, D.J. Siebert and D.M. Williams
11. Cladistics: The Theory and Practice of Parsimony Analysis (2nd edition) (1998)*
    I.J. Kitching, P.L. Forey, C.J. Humphries and D.M. Williams

* Published by Oxford University Press for the Systematics Association

† Published by the Association (out of print)

SYSTEMATICS ASSOCIATION SPECIAL VOLUMES

1. The New Systematics (1940)
   Edited by J.S. Huxley (reprinted 1971)
2. Chemotaxonomy and Serotaxonomy (1968)*
   Edited by J.C. Hawkes
3. Data Processing in Biology and Geology (1971)*
   Edited by J.L. Cutbill

4. Scanning Electron Microscopy (1971)*
   Edited by V.H. Heywood
5. Taxonomy and Ecology (1973)*
   Edited by V.H. Heywood
6. The Changing Flora and Fauna of Britain (1974)*
   Edited by D.L. Hawksworth
7. Biological Identification with Computers (1975)*
   Edited by R.J. Pankhurst
8. Lichenology: Progress and Problems (1976)*
   Edited by D.H. Brown, D.L. Hawksworth and R.H. Bailey
9. Key Works to the Fauna and Flora of the British Isles and Northwestern Europe, 4th edition (1978)*
   Edited by G.J. Kerrich, D.L. Hawksworth and R.W. Sims
10. Modern Approaches to the Taxonomy of Red and Brown Algae (1978)
    Edited by D.E.G. Irvine and J.H. Price
11. Biology and Systematics of Colonial Organisms (1979)*
    Edited by C. Larwood and B.R. Rosen
12. The Origin of Major Invertebrate Groups (1979)*
    Edited by M.R. House
13. Advances in Bryozoology (1979)*
    Edited by G.P. Larwood and M.B. Abbott
14. Bryophyte Systematics (1979)*
    Edited by G.C.S. Clarke and J.G. Duckett
15. The Terrestrial Environment and the Origin of Land Vertebrates (1980)
    Edited by A.L. Pachen
16 Chemosystematics: Principles and Practice (1980)*
    Edited by F.A. Bisby, J.G. Vaughan and C.A. Wright
17. The Shore Environment: Methods and Ecosystems (2 volumes) (1980)*
    Edited by J.H. Price, D.E.C. Irvine and W.F. Farnham
18. The Ammonoidea (1981)*
    Edited by M.R. House and J.R. Senior
19. Biosystematics of Social Insects (1981)*
    Edited by P.E. House and J.-L. Clement
20. Genome Evolution (1982)*
    Edited by G.A. Dover and R.B. Flavell
21. Problems of Phylogenetic Reconstruction (1982)
    Edited by K.A. Joysey and A.E. Friday
22. Concepts in Nematode Systematics (1983)*
    Edited by A.R. Stone, H.M. Platt and L.F. Khalil
23. Evolution, Time and Space: The Emergence of the Biosphere (1983)*
    Edited by R.W. Sims, J.H. Price and P.E.S. Whalley
24. Protein Polymorphism: Adaptive and Taxonomic Significance (1983)*
    Edited by G.S. Oxford and D. Rollinson
25. Current Concepts in Plant Taxonomy (1983)*
    Edited by V.H. Heywood and D.M. Moore

26. Databases in Systematics (1984)*
Edited by R. Allkin and F.A. Bisby

27. Systematics of the Green Algae (1984)*
Edited by D.E.G. Irvine and D.M. John

28. The Origins and Relationships of Lower Invertebrates (1985)‡
Edited by S. Conway Morris, J.D. George, R. Gibson and H.M. Platt

29. Infraspecific Classification of Wild and Cultivated Plants (1986)‡
Edited by B.T. Styles

30. Biomineralization in Lower Plants and Animals (1986)‡
Edited by B.S.C. Leadbeater and R. Riding

31. Systematic and Taxonomic Approaches in Palaeobotany (1986)‡
Edited by R.A. Spicer and B.A. Thomas

32. Coevolution and Systematics (1986)‡
Edited by A.R. Stone and D.L. Hawksworth

33. Key Works to the Fauna and Flora of the British Isles and Northwestern Europe, 5th edition (1988)‡
Edited by R.W. Sims, P. Freeman and D.L. Hawksworth

34. Extinction and Survival in the Fossil Record (1988)‡
Edited by G.P. Larwood

35. The Phylogeny and Classification of the Tetrapods (2 volumes) (1988)‡
Edited by M.J. Benton

36. Prospects in Systematics (1988)‡
Edited by J.L. Hawksworth

37. Biosystematics of Haematophagous Insects (1988)‡
Edited by M.W. Service

38. The Chromophyte Algae: Problems and Perspective (1989)‡
Edited by J.C. Green, B.S.C. Leadbeater and W.L. Diver

39. Electrophoretic Studies on Agricultural Pests (1989)‡
Edited by H.D. Loxdale and J. den Hollander

40. Evolution, Systematics, and Fossil History of the Hamamelidae (2 volumes) (1989)‡
Edited by P.R. Crane and S. Blackmore

41. Scanning Electron Microscopy in Taxonomy and Functional Morphology (1990)‡
Edited by D. Claugher

42. Major Evolutionary Radiations (1990)‡
Edited by P.D. Taylor and G.P. Larwood

43. Tropical Lichens: Their Systematics, Conservation and Ecology (1991)‡
Edited by G.J. Galloway

44. Pollen and Spores: Patterns and Diversification (1991)‡
Edited by S. Blackmore and S.H. Barnes

45. The Biology of Free-Living Heterotrophic Flagellates (1991)‡
Edited by D.J. Patterson and J. Larsen

46. Plant–Animal Interactions in the Marine Benthos (1992)‡
Edited by D.M. John, S.J. Hawkins and J.H. Price

47. The Ammonoidea: Environment, Ecology and Evolutionary Change (1993)‡
Edited by M.R. House

48. Designs for a Global Plant Species Information System (1993)‡
Edited by F.A. Bisby, G.F. Russell and R.J. Pankhurst

49. Plant Galls: Organisms, Interactions, Populations (1994)‡
Edited by M.A.J. Williams

50. Systematics and Conservation Evaluation (1994)‡
Edited by P.L. Forey, C.J. Humphries and R.I. Vane-Wright

51. The Haptophyte Algae (1994)‡
Edited by J.C. Green and B.S.C. Leadbeater

52. Models in Phylogeny Reconstruction (1994)‡
Edited by R. Scotland, D.I. Siebert and D.M. Williams

53. The Ecology of Agricultural Pests: Biochemical Approaches (1996)**
Edited by W.O.C. Symondson and J.E. Liddell

54. Species: the Units of Diversity (1997)**
Edited by M.F. Claridge, H.A. Dawah and M.R. Wilson

55. Arthropod Relationships (1998)**
Edited by R.A. Fortey and R.H. Thomas

56. Evolutionary Relationships among Protozoa (1998)**
Edited by G.H. Coombs, K. Vickerman, M.A. Sleigh and A. Warren

57. Molecular Systematics and Plant Evolution (1999)
Edited by P.M. Hollingsworth, R.M. Bateman and R.J. Gornall

58. Homology and Systematics (2000)
Edited by R. Scotland and R.T. Pennington

59. The Flagellates: Unity, Diversity and Evolution (2000)
Edited by B.S.C. Leadbeater and J.C. Green

60. Interrelationships of the Platyhelminthes (2001)
Edited by D.T.J. Littlewood and R.A. Bray

61. Major Events in Early Vertebrate Evolution (2001)
Edited by P.E. Ahlberg

62. The Changing Wildlife of Great Britain and Ireland (2001)
Edited by D.L. Hawksworth

63. Brachiopods Past and Present (2001)
Edited by H. Brunton, L.R.M. Cocks and S.L. Long

64. Morphology, Shape and Phylogeny (2002)
Edited by N. MacLeod and P.L. Forey

65. Developmental Genetics and Plant Evolution (2002)
Edited by Q.C.B. Cronk, R.M. Bateman and J.A. Hawkins

66. Telling the Evolutionary Time: Molecular Clocks and the Fossil Record (2003)
Edited by P.C.J. Donoghue and M.P. Smith

67. Milestones in Systematics (2004)
Edited by D.M. Williams and P.L. Forey

68. Organelles, Genomes and Eukaryote Phylogeny (2004)
Edited by R.P. Hirt and D.S. Horner

69. Neotropical Savannas and Seasonally Dry Forests: Plant Diversity, Biogeography and Conservation (2006)
Edited by R.T. Pennington, G.P. Lewis and J.A. Rattan

70. Biogeography in a Changing World (2006)
Edited by M.C. Ebach and R.S. Tangney

71. Pleurocarpous Mosses: Systematics & Evolution (2006)
Edited by A.E. Newton and R.S. Tangney

72. Reconstructing the Tree of Life: Taxonomy and Systematics of Species Rich Taxa (2006)
Edited by T.R. Hodkinson and J.A.N. Parnell

73. Biodiversity Databases: Techniques, Politics, and Applications (2007)
Edited by G.B. Curry and C.J. Humphries

74. Automated Taxon Identification in Systematics: Theory, Approaches and Applications (2007)
Edited by N. MacLeod

* Published by Academic Press for the Systematics Association
† Published by the Palaeontological Association in conjunction with Systematics Association
‡ Published by the Oxford University Press for the Systematics Association
** Published by Chapman & Hall for the Systematics Association